Defect

Detect

Windows API for Software Diagnostics Accelerated

With Category Theory in View

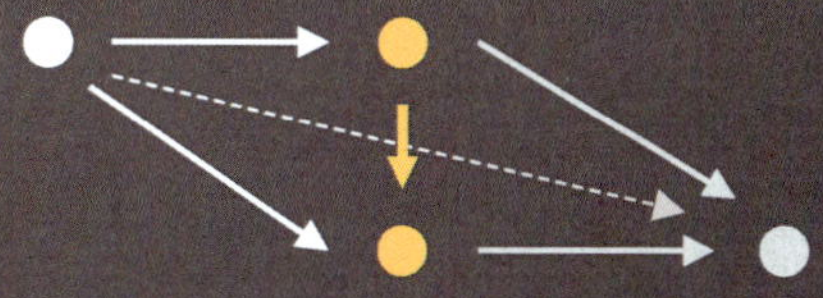

Second Edition

Dmitry Vostokov
Software Diagnostics Services

Accelerated Windows API for Software Diagnostics: With Category Theory in View, Second Edition

Published by OpenTask, Republic of Ireland

OpenTask books and magazines are available through booksellers and distributors worldwide. For further information or comments, send requests to press@opentask.com.

A CIP catalog record for this book is available from the British Library.

ISBN-13: 978-1-912636-88-4 (Paperback)

Revision 2.01 (December 2024)

Contents

About the Author

Dmitry Vostokov is an internationally recognized expert, speaker, educator, scientist, inventor, and author. He founded the pattern-oriented software diagnostics, forensics, and prognostics discipline (Systematic Software Diagnostics) and Software Diagnostics Institute (DA+TA: DumpAnalysis.org + TraceAnalysis.org). Vostokov has also authored over 50 books on software diagnostics, anomaly detection and analysis, software and memory forensics, root cause analysis and problem solving, memory dump analysis, debugging, software trace and log analysis, reverse engineering, and malware analysis. He has over 30 years of experience in software architecture, design, development, and maintenance in various industries, including leadership, technical, and people management roles. Dmitry founded OpenTask Iterative and Incremental Publishing (OpenTask.com) and Software Diagnostics Technology and Services (former Memory Dump Analysis Services) PatternDiagnostics.com. In his spare time, he explores Software Narratology and Quantum Software Diagnostics. His interest areas are theoretical software diagnostics and its mathematical and computer science foundations, application of formal logic, semiotics, artificial intelligence, machine learning, and data mining to diagnostics and anomaly detection, software diagnostics engineering and diagnostics-driven development, diagnostics workflow and interaction. Recent interest areas also include functional programming, cloud native computing, monitoring, observability, visualization, security, automation, applications of category theory to software diagnostics, development and big data, and diagnostics of artificial intelligence.

Introduction

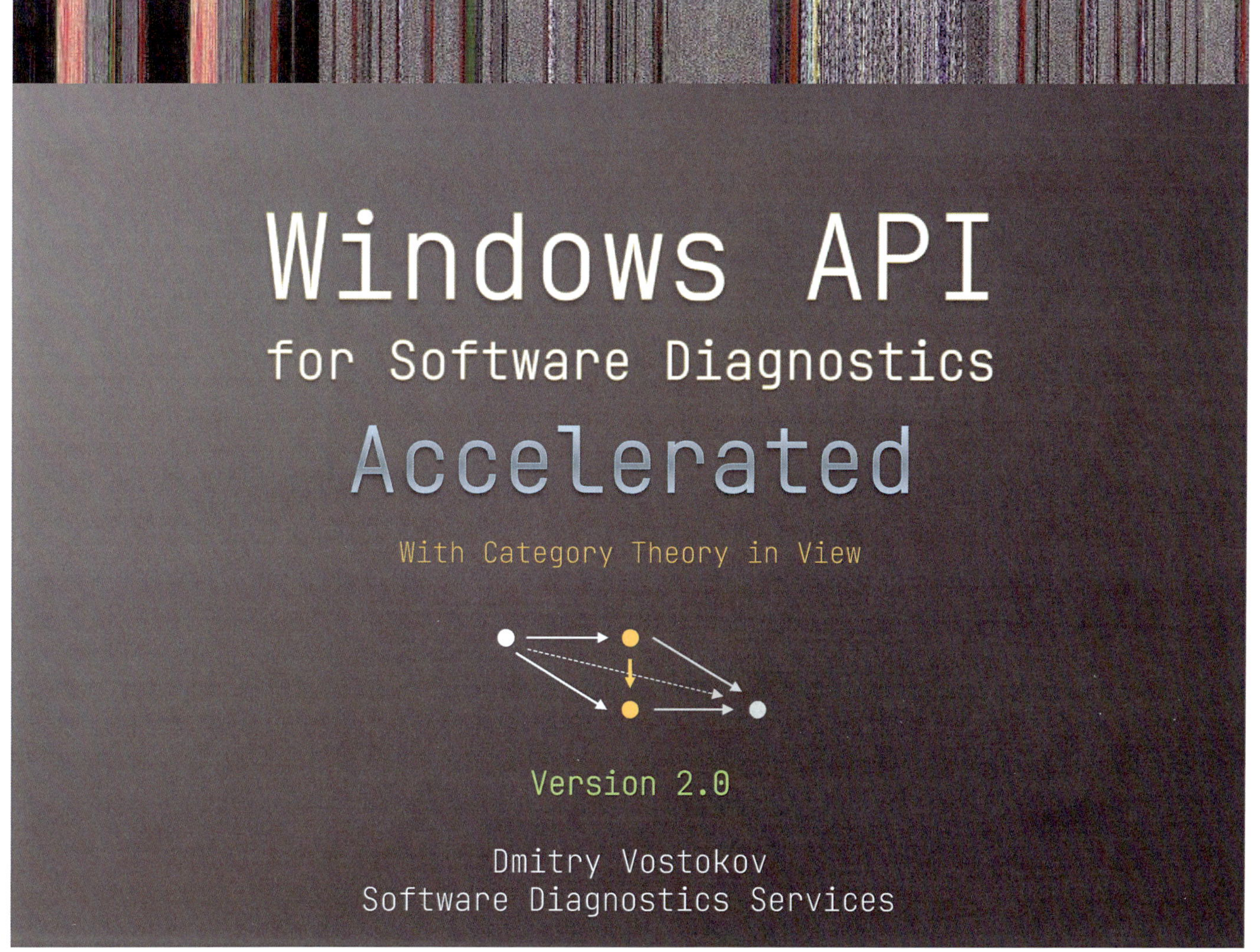
Windows API
for Software Diagnostics
Accelerated
With Category Theory in View
Version 2.0
Dmitry Vostokov
Software Diagnostics Services

Hello everyone, my name is Dmitry Vostokov, and I teach this training course.

Prerequisites

- Development experience

and (optional)

- Basic memory dump analysis

To get most of this training, you are expected to have basic development experience and optional basic memory dump analysis experience. I assume you know what types, functions, and their parameters are. If you don't have a memory dump analysis experience, then you also learn some basics too, because we use the Microsoft debugger, WinDbg, for some exercises. If you haven't got the *Practical Foundations of Windows Debugging, Disassembling, and Reversing* book, which also uses WinDbg, or haven't had a chance to read it, I explain some concepts when necessary during the course. The second edition also contains the necessary review of x64 disassembly.

Training Goals

- Review fundamentals of Windows API
- Learn diagnostic analysis techniques
- See how Windows API knowledge is used during diagnostics and debugging

Our primary goal is to learn Windows API in an accelerated fashion. So, first, we review Windows API fundamentals necessary for software diagnostics. Then, we learn various analysis techniques for Windows API exploration. And finally, we see examples of how the knowledge of Windows API helps in diagnostics and debugging.

Training Principles

- Talk only about what I can show
- Lots of pictures
- Lots of examples
- Original content and examples

There were many training formats to consider, and I decided that the best way is to concentrate on slides and hands-on demonstrations you can repeat yourself as homework. Some of them are available as step-by-step exercises.

Schedule

- Review of relevant x64 disassembly
- General Windows API aspects
- Windows API formalization
- Windows API and languages
- Windows API classes
- Practical exercises

The rough coverage or schedule includes general API aspects that can also be applicable to other operating systems. We also take a radical detour and introduce category theory in the API context. Our coverage is not only theoretical. We also take a tour through different API subsets and classes. An integral part of this training is practical exercises.

Training Idea

- Cybersecurity
- Memory dump analysis
- Reading Windows-based Code training

I started thinking about this training when security professionals mentioned the need for Windows API knowledge. Later, some attendees of my memory dump analysis training courses asked questions, and I realized they would have benefitted if they had this training. In addition, this training may also fill some gaps between different training courses. And finally, I recalled that I developed Reading Windows-based Code training in the past (see links to it in the References part of this training) for software technical support and escalation engineers and that the new training would benefit from some aspects of it.

General Windows API Aspects

- Header-technology view
- Naming convention
- Basic type system
- Hungarian notation
- Call types
- Export/import functions
- IAT
- Virtual process address space
- Calling convention
- API sequences
- API layers
- Documented/undocumented API
- Exports and imports
- API and system calls
- API source code
- API and functional programming
- API and versioning
- Modules
- API usage
- API internals
- Delay-loaded API
- API sets
- API name patterns
- API namespaces
- API syntagms/paradigms
- Marked API
- ADDR patterns
- DebugWare patterns
- Memory analysis patterns
- Trace and log analysis patterns
- WOW64
- API and errors
- API and security
- API and Unicode

The general API aspects we plan to discuss are listed on this slide. In this and a few further slides, I highlighted the new topics for this edition.

Windows API Formalization

- API compositionality
- Category theory language
- A view of category theory
- Category theory square
- API category
- API functor
- API diagram
- API natural transformation
- Cross-platform API
- API adjunction
- Informal n-API
- API and trace categories
- API I/O
- Monoidal API category

This training also includes an API formalization via Category Theory. We start with a brief overview of categories, functors, and other aspects using Windows API examples.

Windows API and Languages

- C#
- API metadata
- Scala Native
- Golang
- Rust
- Python

We also cover the basics of how Windows API is used in languages other than C and C++ with template code examples.

Windows API Classes

- GUI
- Windowing
- GDI
- GDI+
- Module/Library
- Process/Thread
- Services
- Security
- Process Heap
- Error Handling and Debugging
- Virtual Memory
- IPC
- RPC
- Synchronization
- I/O
- Runtime
- COM
- Networking
- Console
- Event Tracing
- Strings

A part of this training includes a tour through different API subsets and classes.

Links

- Memory Dumps

Included in Exercise W0

- Exercise Transcripts

Included in this book

Exercise W0

- **Goal:** Install WinDbg or Debugging Tools for Windows, or pull Docker image, and check that symbols are set up correctly
- **Memory Analysis Patterns:** Stack Trace; Incorrect Stack Trace
- \AWAPI-Dumps\Exercise-W0.pdf

Exercise W0: Download, set up, and verify your WinDbg or Debugging Tools for Windows installation, or Docker Debugging Tools for Windows image

Goal: Install WinDbg or Debugging Tools for Windows, or pull Docker image, and check that symbols are set up correctly.

Memory Analysis Patterns: Stack Trace; Incorrect Stack Trace.

1. Download memory dump files if you haven't done that already and unpack the archive:

https://www.patterndiagnostics.com/Training/AWAPI/AWAPI-Dumps.zip

2. Install WinDbg (or upgrade existing WinDbg Preview) from https://learn.microsoft.com/en-gb/windows-hardware/drivers/debugger. Run WinDbg.

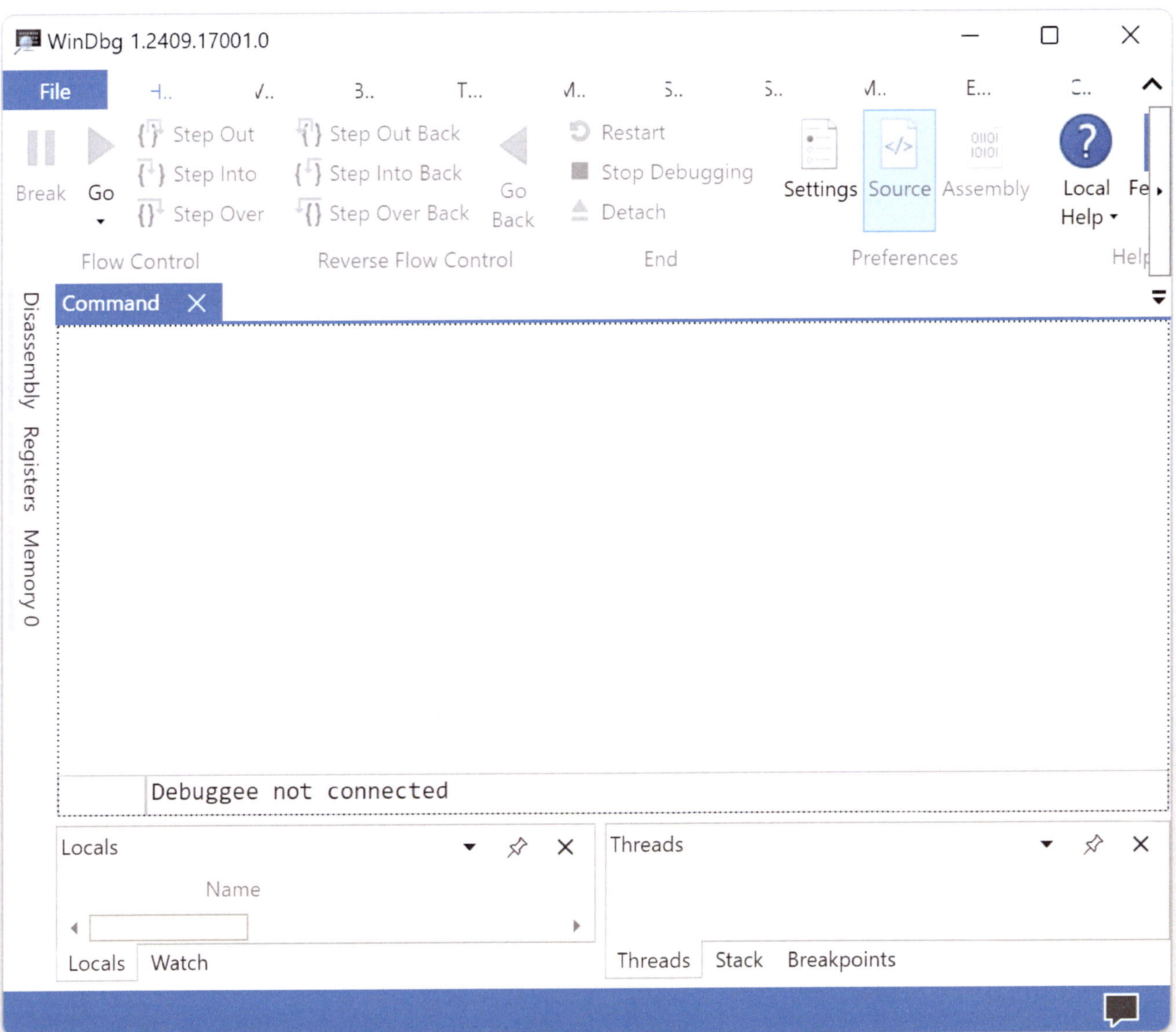

3. Open \AWAPI-Dumps\Process\wordpad.DMP:

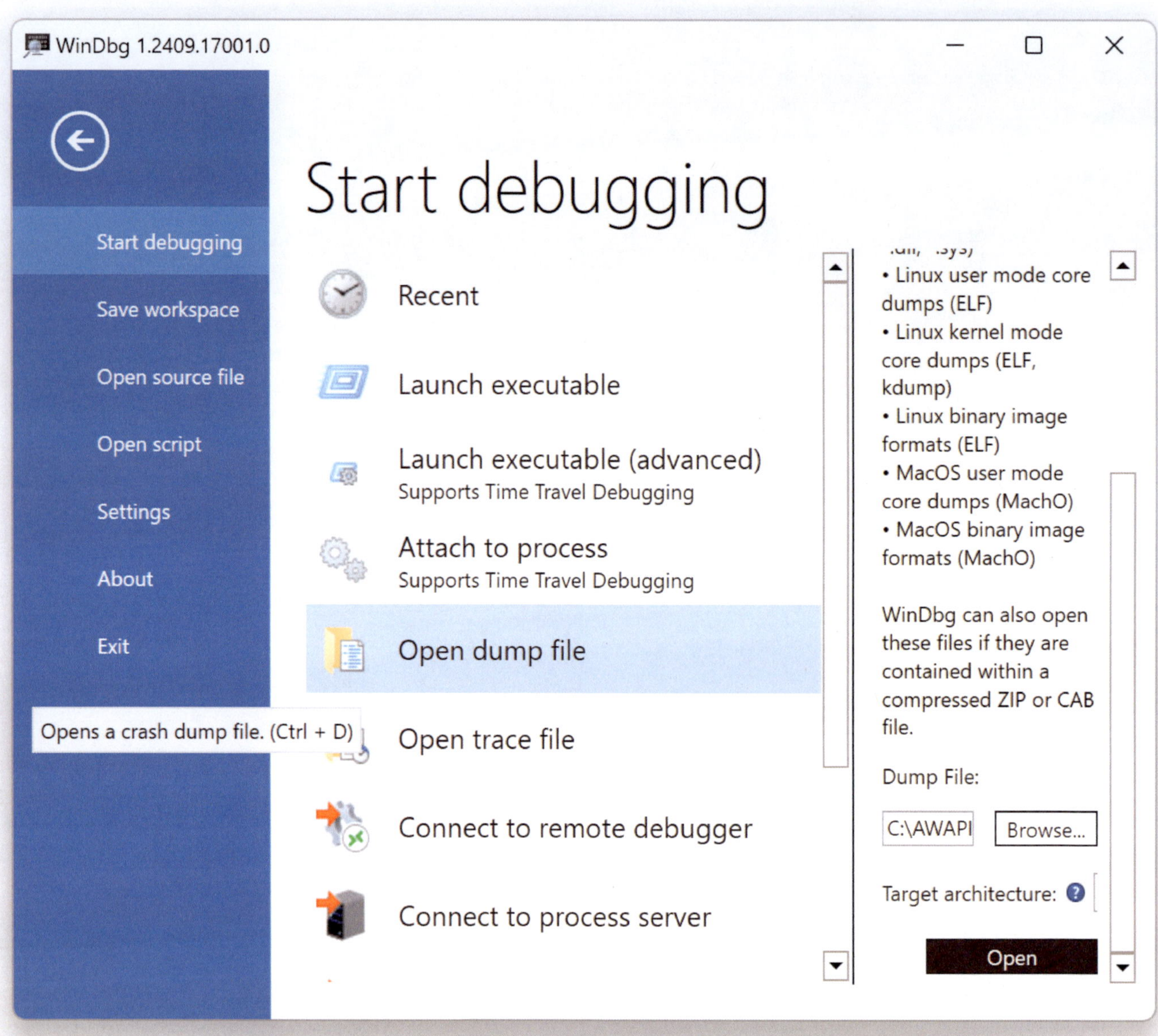
WinDbg 1.2409.17001.0
Start debugging
Start debugging
Save workspace
Open source file
Open script
Settings
About
Exit
Recent
Launch executable
Launch executable (advanced)
Supports Time Travel Debugging
Attach to process
Supports Time Travel Debugging
Open dump file
Opens a crash dump file. (Ctrl + D)
Open trace file
Connect to remote debugger
Connect to process server
• Linux user mode core dumps (ELF)
• Linux kernel mode core dumps (ELF, kdump)
• Linux binary image formats (ELF)
• MacOS user mode core dumps (MachO)
• MacOS binary image formats (MachO)
WinDbg can also open these files if they are contained within a compressed ZIP or CAB file.
Dump File:
C:\AWAPI
Browse...
Target architecture:
Open

4. We get the dump file loaded:

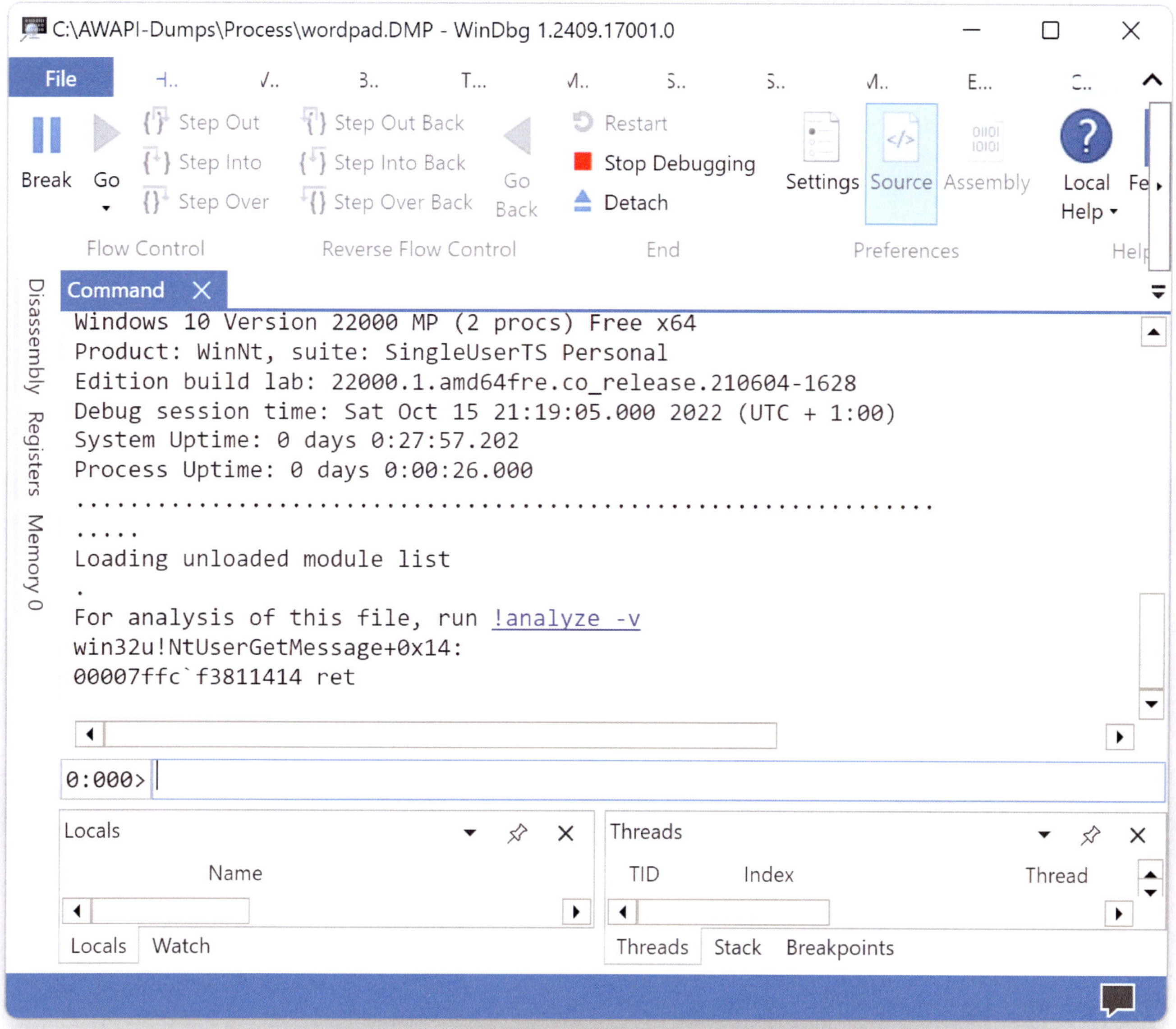
C:\AWAPI-Dumps\Process\wordpad.DMP - WinDbg 1.2409.17001.0
File
Break
Go
Step Out
Step Into
Step Over
Step Out Back
Step Into Back
Step Over Back
Go Back
Restart
Stop Debugging
Detach
Settings
Source
Assembly
Local Help
Flow Control
Reverse Flow Control
End
Preferences
Command
Disassembly
Registers
Memory 0
Windows 10 Version 22000 MP (2 procs) Free x64
Product: WinNt, suite: SingleUserTS Personal
Edition build lab: 22000.1.amd64fre.co_release.210604-1628
Debug session time: Sat Oct 15 21:19:05.000 2022 (UTC + 1:00)
System Uptime: 0 days 0:27:57.202
Process Uptime: 0 days 0:00:26.000
Loading unloaded module list
For analysis of this file, run !analyze -v
win32u!NtUserGetMessage+0x14:
00007ffc`f3811414 ret
0:000>
Locals
Name
Locals
Watch
Threads
TID
Index
Thread
Threads
Stack
Breakpoints

5. We can execute the **k** command to get the stack trace:

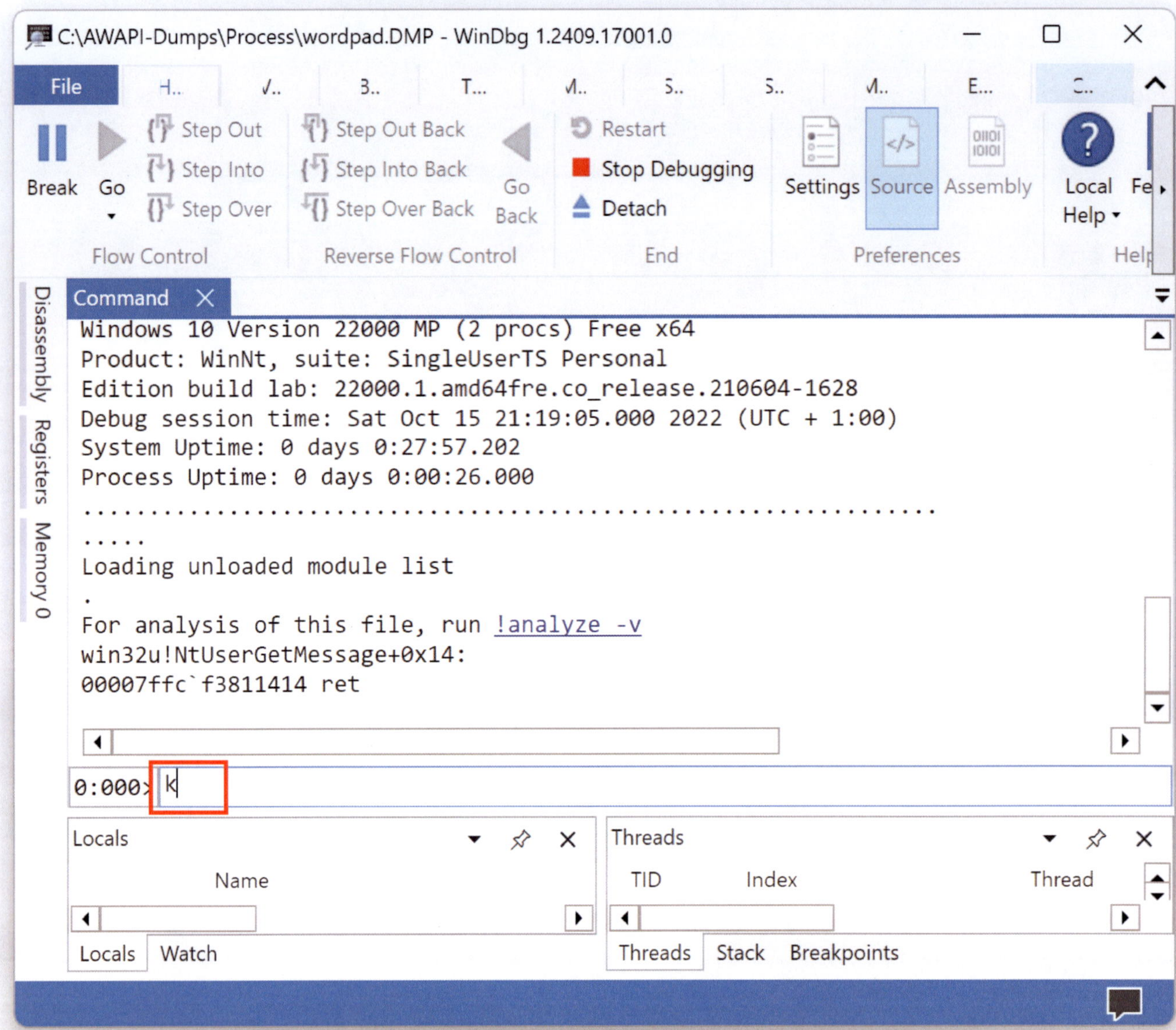

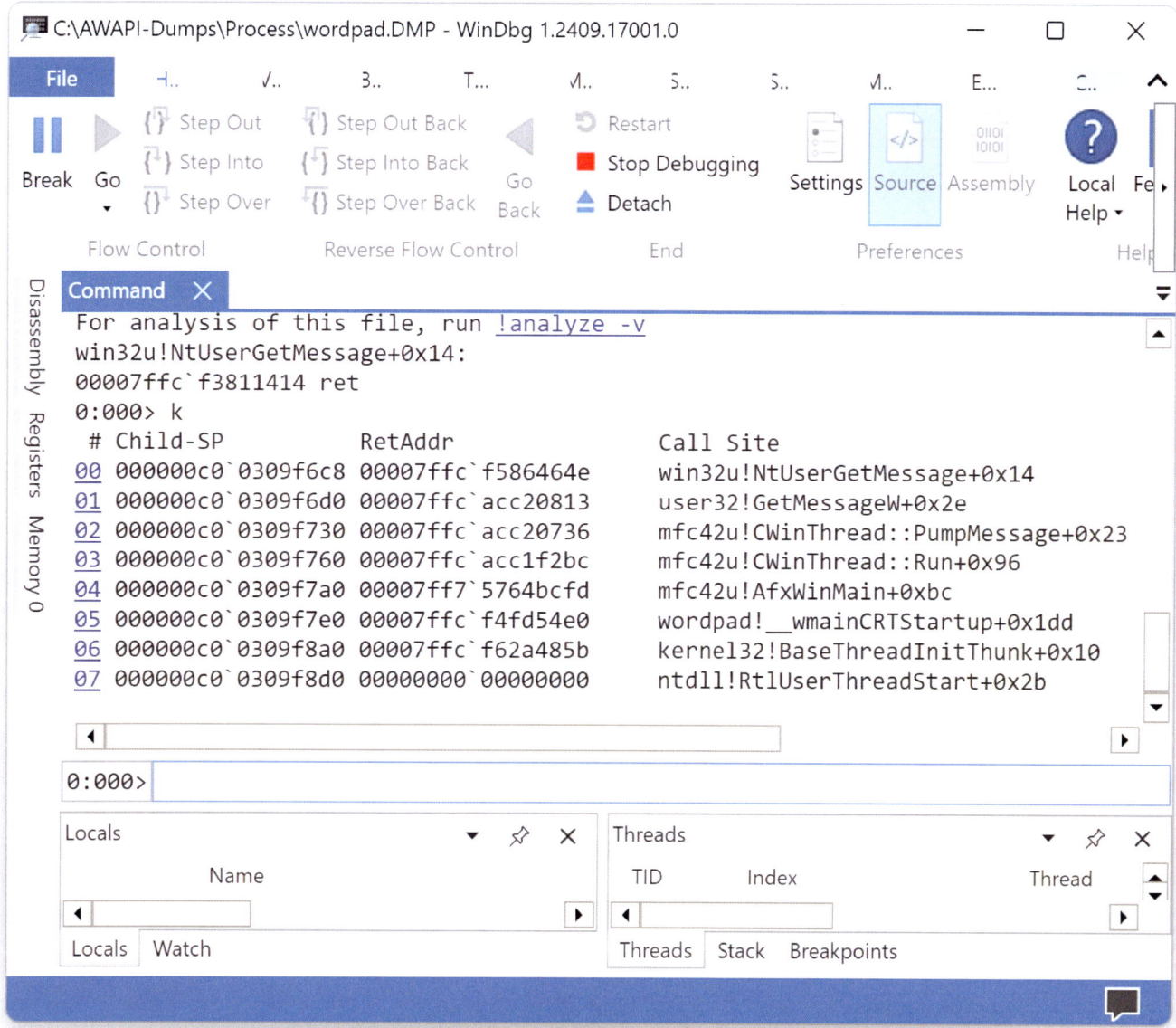

6. The output of the **k** command should be this:

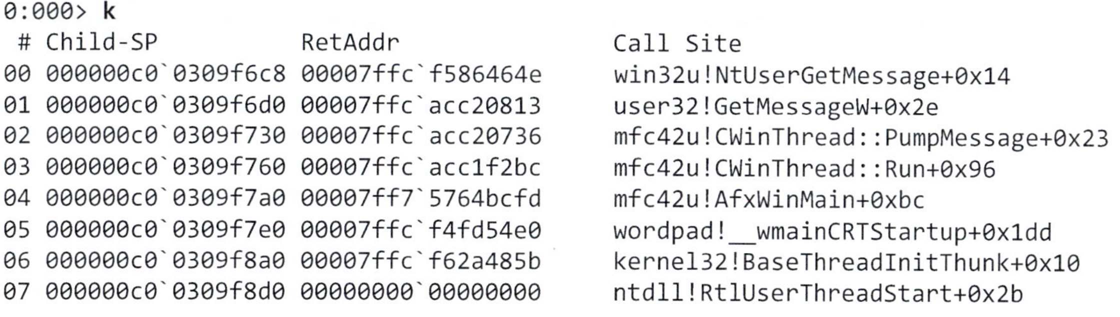

```
0:000> k
 # Child-SP          RetAddr               Call Site
00 000000c0`0309f6c8 00007ffc`f586464e     win32u!NtUserGetMessage+0x14
01 000000c0`0309f6d0 00007ffc`acc20813     user32!GetMessageW+0x2e
02 000000c0`0309f730 00007ffc`acc20736     mfc42u!CWinThread::PumpMessage+0x23
03 000000c0`0309f760 00007ffc`acc1f2bc     mfc42u!CWinThread::Run+0x96
04 000000c0`0309f7a0 00007ff7`5764bcfd     mfc42u!AfxWinMain+0xbc
05 000000c0`0309f7e0 00007ffc`f4fd54e0     wordpad!__wmainCRTStartup+0x1dd
06 000000c0`0309f8a0 00007ffc`f62a485b     kernel32!BaseThreadInitThunk+0x10
07 000000c0`0309f8d0 00000000`00000000     ntdll!RtlUserThreadStart+0x2b
```

If it has this form below with a large offset, then your symbol files were not set up correctly – **Incorrect Stack Trace** pattern:

```
0:000> k
 # Child-SP          RetAddr               Call Site
00 000000c0`0309f6c8 00007ffc`f586464e     win32u!NtUserGetMessage+0x14
01 000000c0`0309f6d0 00007ffc`acc20813     user32!GetMessageW+0x2e
02 000000c0`0309f730 00007ffc`acc20736     mfc42u+0x10813
03 000000c0`0309f760 00007ffc`acc1f2bc     mfc42u+0x10736
04 000000c0`0309f7a0 00007ff7`5764bcfd     mfc42u+0xf2bc
05 000000c0`0309f7e0 00007ffc`f4fd54e0     wordpad+0xbcfd
06 000000c0`0309f8a0 00007ffc`f62a485b     kernel32!BaseThreadInitThunk+0x10
07 000000c0`0309f8d0 00000000`00000000     ntdll!RtlUserThreadStart+0x2b
```

7. [Optional] Download and install the recommended version of Debugging Tools for Windows (See windbg.org for quick links, WinDbg Quick Links \ Download Debugging Tools for Windows). For this part, we use WinDbg 10.0.26100.1 from Windows SDK 10.0.26100 for Windows 11, version 24H2.

8. Launch WinDbg from Windows Kits \ WinDbg (X64).

9. Open \AWAPI-Dumps\Process\wordpad.DMP:

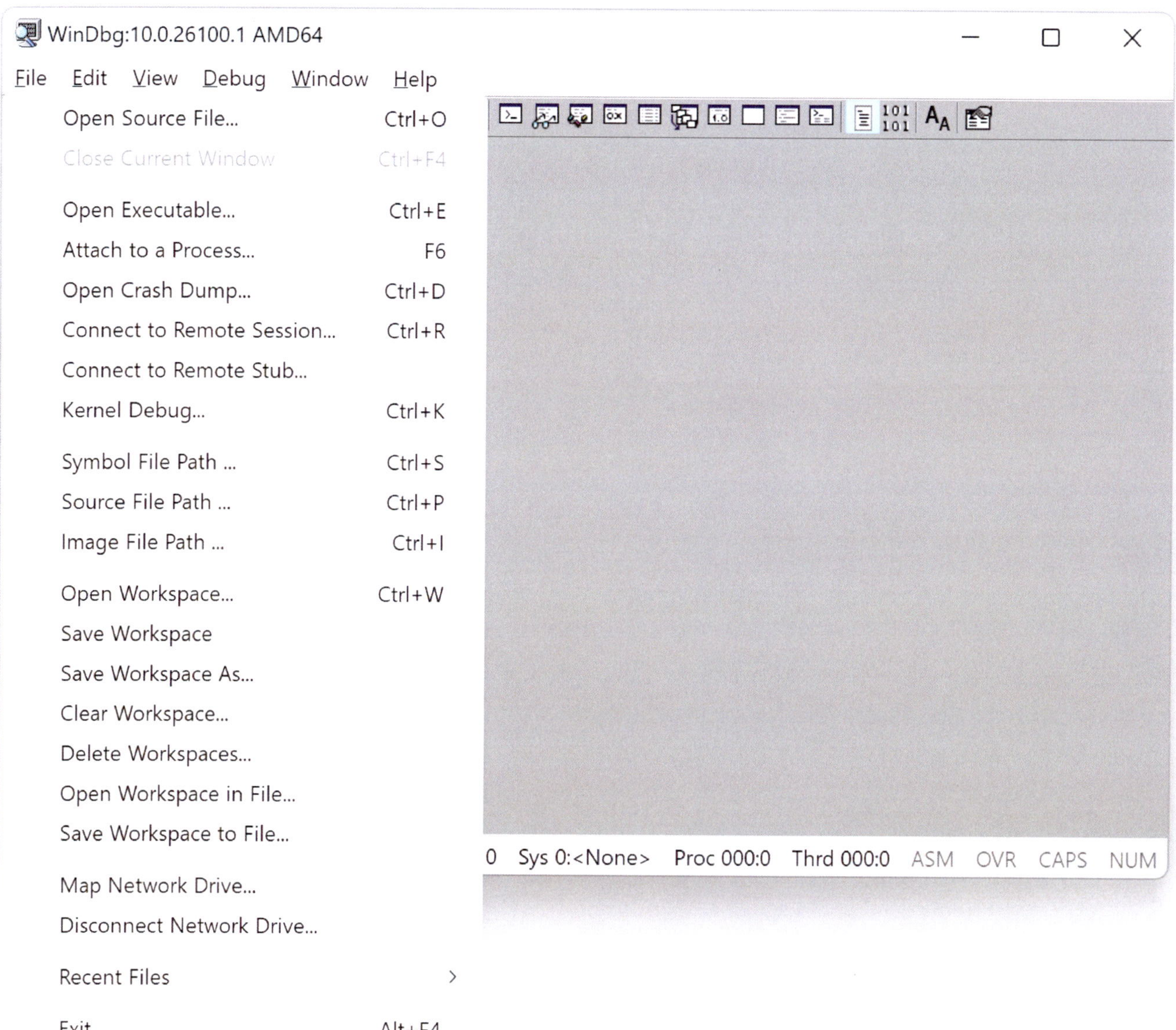

10. We get the dump file loaded:

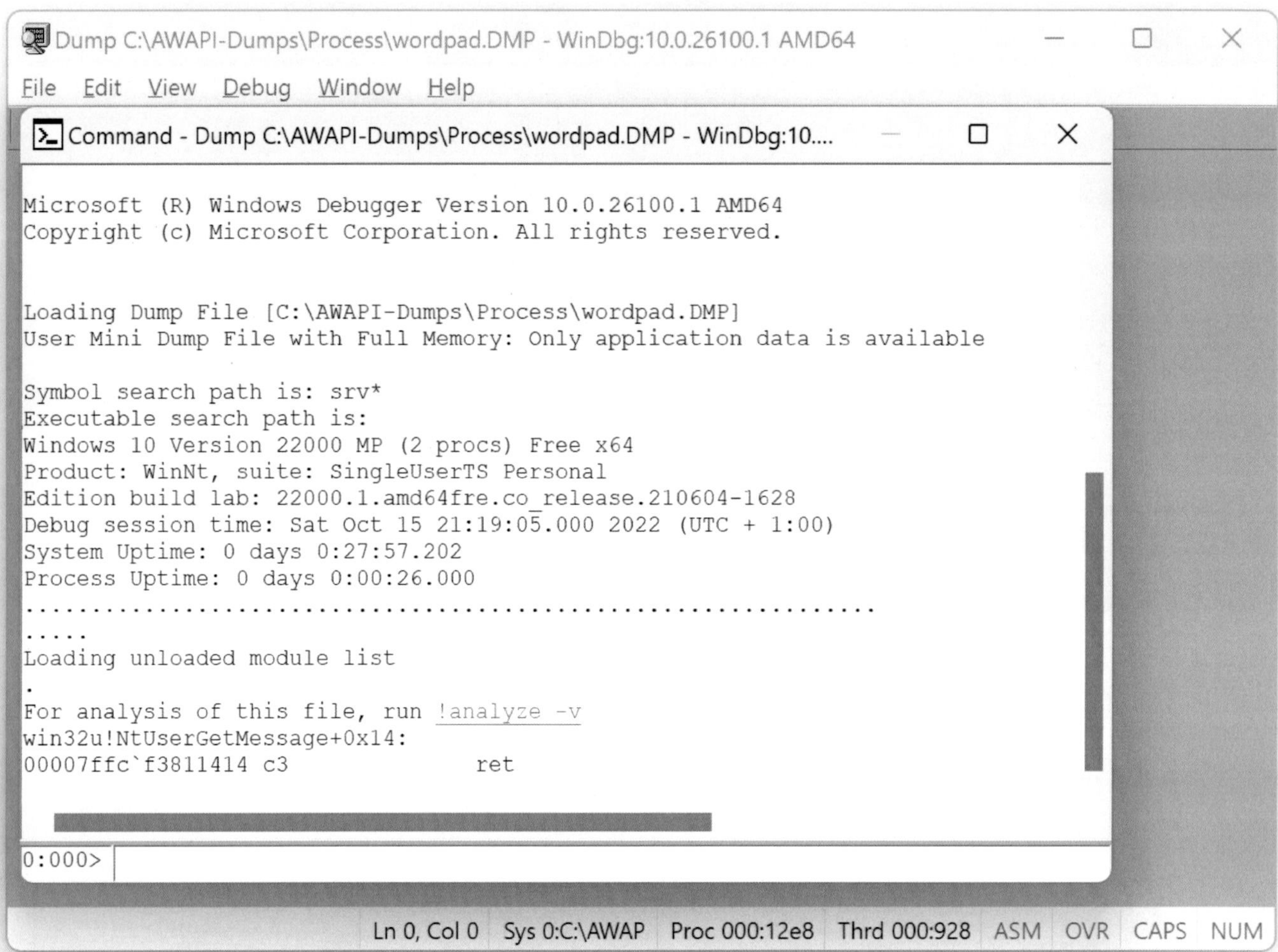
Dump C:\AWAPI-Dumps\Process\wordpad.DMP - WinDbg:10.0.26100.1 AMD64
File Edit View Debug Window Help
Command - Dump C:\AWAPI-Dumps\Process\wordpad.DMP - WinDbg:10....
Microsoft (R) Windows Debugger Version 10.0.26100.1 AMD64
Copyright (c) Microsoft Corporation. All rights reserved.
Loading Dump File [C:\AWAPI-Dumps\Process\wordpad.DMP]
User Mini Dump File with Full Memory: Only application data is available
Symbol search path is: srv*
Executable search path is:
Windows 10 Version 22000 MP (2 procs) Free x64
Product: WinNt, suite: SingleUserTS Personal
Edition build lab: 22000.1.amd64fre.co_release.210604-1628
Debug session time: Sat Oct 15 21:19:05.000 2022 (UTC + 1:00)
System Uptime: 0 days 0:27:57.202
Process Uptime: 0 days 0:00:26.000
....
.....
Loading unloaded module list
.
For analysis of this file, run !analyze -v
win32u!NtUserGetMessage+0x14:
00007ffc`f3811414 c3 ret
0:000>
Ln 0, Col 0 Sys 0:C:\AWAP Proc 000:12e8 Thrd 000:928 ASM OVR CAPS NUM

11. Type the **k** command to verify the correctness of the stack trace:

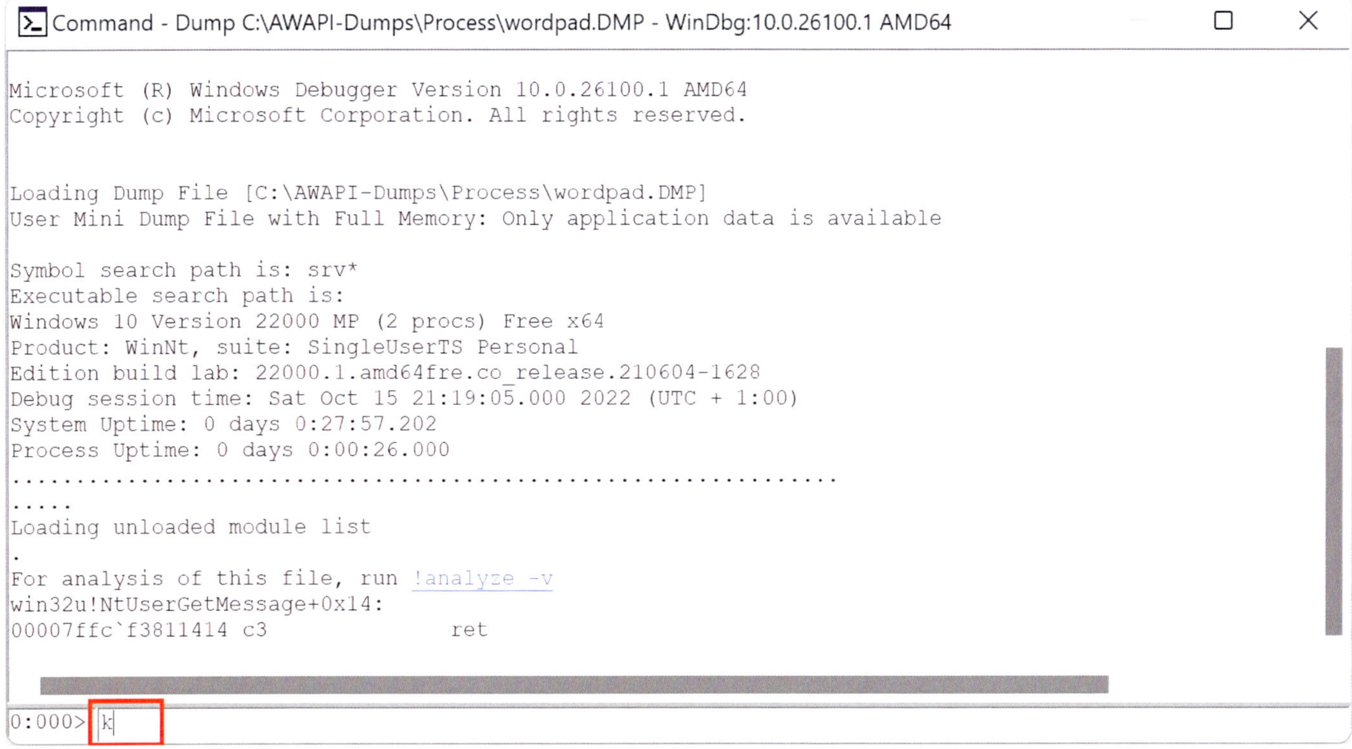

```
Microsoft (R) Windows Debugger Version 10.0.26100.1 AMD64
Copyright (c) Microsoft Corporation. All rights reserved.


Loading Dump File [C:\AWAPI-Dumps\Process\wordpad.DMP]
User Mini Dump File with Full Memory: Only application data is available

Symbol search path is: srv*
Executable search path is:
Windows 10 Version 22000 MP (2 procs) Free x64
Product: WinNt, suite: SingleUserTS Personal
Edition build lab: 22000.1.amd64fre.co_release.210604-1628
Debug session time: Sat Oct 15 21:19:05.000 2022 (UTC + 1:00)
System Uptime: 0 days 0:27:57.202
Process Uptime: 0 days 0:00:26.000
................................................................
.....
Loading unloaded module list
.
For analysis of this file, run !analyze -v
win32u!NtUserGetMessage+0x14:
00007ffc`f3811414 c3              ret

0:000> k
```

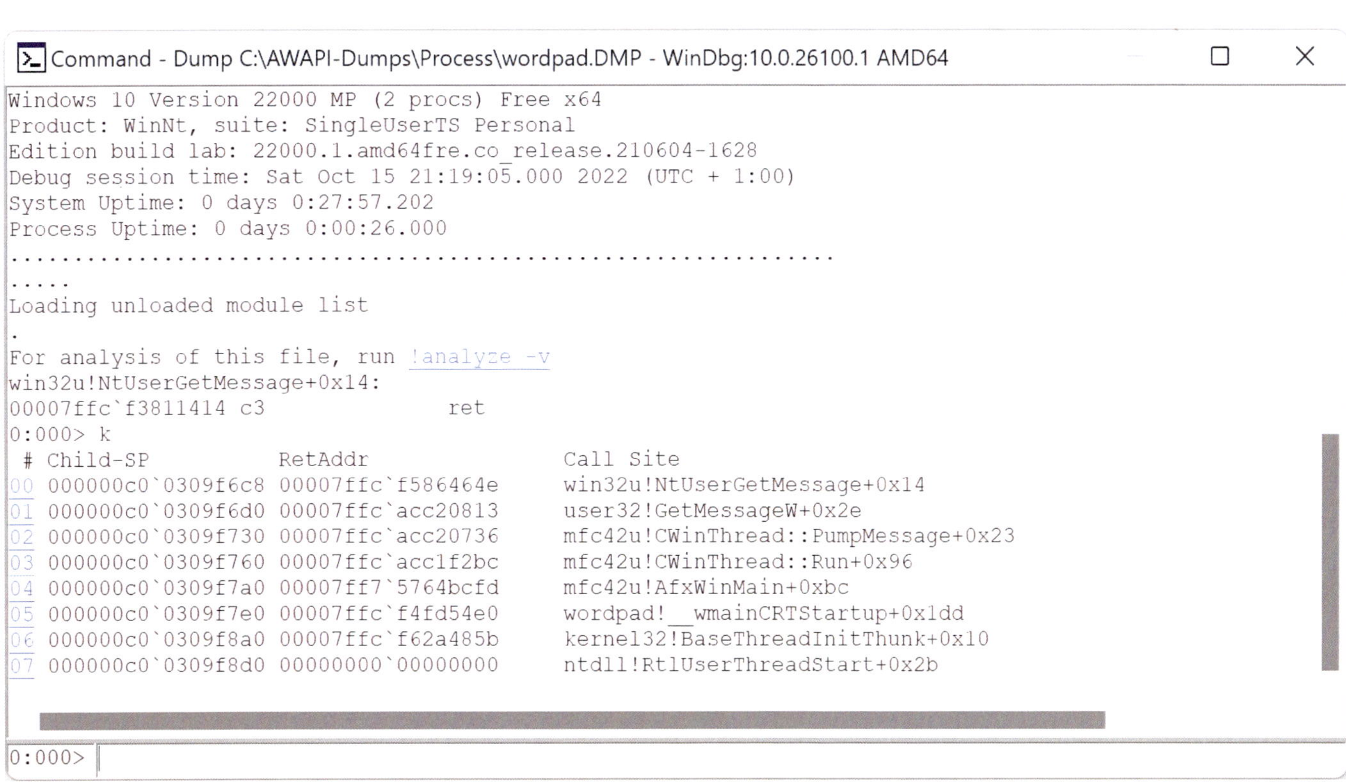

```
Windows 10 Version 22000 MP (2 procs) Free x64
Product: WinNt, suite: SingleUserTS Personal
Edition build lab: 22000.1.amd64fre.co_release.210604-1628
Debug session time: Sat Oct 15 21:19:05.000 2022 (UTC + 1:00)
System Uptime: 0 days 0:27:57.202
Process Uptime: 0 days 0:00:26.000
................................................................
.....
Loading unloaded module list
.
For analysis of this file, run !analyze -v
win32u!NtUserGetMessage+0x14:
00007ffc`f3811414 c3              ret
0:000> k
 # Child-SP          RetAddr               Call Site
00 000000c0`0309f6c8 00007ffc`f586464e     win32u!NtUserGetMessage+0x14
01 000000c0`0309f6d0 00007ffc`acc20813     user32!GetMessageW+0x2e
02 000000c0`0309f730 00007ffc`acc20736     mfc42u!CWinThread::PumpMessage+0x23
03 000000c0`0309f760 00007ffc`acc1f2bc     mfc42u!CWinThread::Run+0x96
04 000000c0`0309f7a0 00007ff7`5764bcfd     mfc42u!AfxWinMain+0xbc
05 000000c0`0309f7e0 00007ffc`f4fd54e0     wordpad!__wmainCRTStartup+0x1dd
06 000000c0`0309f8a0 00007ffc`f62a485b     kernel32!BaseThreadInitThunk+0x10
07 000000c0`0309f8d0 00000000`00000000     ntdll!RtlUserThreadStart+0x2b

0:000>
```

12. [Optional] Another approach is to use a Docker container image that contains preinstalled WinDbg x64 with the required symbol files for this course's memory dump files:

```
C:\AWAPI-Dumps>docker pull patterndiagnostics/windbg:10.0.26100.1-awapi

C:\AWAPI-Dumps>docker run -it -v C:\AWAPI-Dumps:C:\AWAPI-Dumps
patterndiagnostics/windbg:10.0.26100.1-awapi

Microsoft Windows [Version 10.0.20348.2655]
(c) Microsoft Corporation. All rights reserved.

C:\WinDbg>windbg C:\AWAPI-Dumps\Process\wordpad.DMP

Microsoft (R) Windows Debugger Version 10.0.26100.1 AMD64
Copyright (c) Microsoft Corporation. All rights reserved.


Loading Dump File [C:\AWAPI-Dumps\Process\wordpad.DMP]
User Mini Dump File with Full Memory: Only application data is available


************* Path validation summary **************
Response                         Time (ms)     Location
OK                                             C:\WinDbg\mss
Symbol search path is: C:\WinDbg\mss
Executable search path is:
Windows 10 Version 22000 MP (2 procs) Free x64
Product: WinNt, suite: SingleUserTS Personal
Edition build lab: 22000.1.amd64fre.co_release.210604-1628
Debug session time: Sat Oct 15 21:19:05.000 2022 (UTC + 1:00)
System Uptime: 0 days 0:27:57.202
Process Uptime: 0 days 0:00:26.000
................................................................
.....
Loading unloaded module list
.
For analysis of this file, run !analyze -v
win32u!NtUserGetMessage+0x14:
00007ffc`f3811414 c3              ret

0:000> k
Child-SP          RetAddr               Call Site
000000c0`0309f6c8 00007ffc`f586464e     win32u!NtUserGetMessage+0x14
000000c0`0309f6d0 00007ffc`acc20813     user32!GetMessageW+0x2e
000000c0`0309f730 00007ffc`acc20736     mfc42u!CWinThread::PumpMessage+0x23
000000c0`0309f760 00007ffc`acc1f2bc     mfc42u!CWinThread::Run+0x96
000000c0`0309f7a0 00007ff7`5764bcfd     mfc42u!AfxWinMain+0xbc
000000c0`0309f7e0 00007ffc`f4fd54e0     wordpad!__wmainCRTStartup+0x1dd
000000c0`0309f8a0 00007ffc`f62a485b     kernel32!BaseThreadInitThunk+0x10
000000c0`0309f8d0 00000000`00000000     ntdll!RtlUserThreadStart+0x2b

0:000> q
quit:
NatVis script unloaded from 'C:\Program Files\Windows
Kits\10\Debuggers\x64\Visualizers\atlmfc.natvis'
NatVis script unloaded from 'C:\Program Files\Windows
Kits\10\Debuggers\x64\Visualizers\ObjectiveC.natvis'
```

```
NatVis script unloaded from 'C:\Program Files\Windows
Kits\10\Debuggers\x64\Visualizers\concurrency.natvis'
NatVis script unloaded from 'C:\Program Files\Windows
Kits\10\Debuggers\x64\Visualizers\cpp_rest.natvis'
NatVis script unloaded from 'C:\Program Files\Windows
Kits\10\Debuggers\x64\Visualizers\stl.natvis'
NatVis script unloaded from 'C:\Program Files\Windows
Kits\10\Debuggers\x64\Visualizers\Windows.Data.Json.natvis'
NatVis script unloaded from 'C:\Program Files\Windows
Kits\10\Debuggers\x64\Visualizers\Windows.Devices.Geolocation.natvis'
NatVis script unloaded from 'C:\Program Files\Windows
Kits\10\Debuggers\x64\Visualizers\Windows.Devices.Sensors.natvis'
NatVis script unloaded from 'C:\Program Files\Windows
Kits\10\Debuggers\x64\Visualizers\Windows.Media.natvis'
NatVis script unloaded from 'C:\Program Files\Windows
Kits\10\Debuggers\x64\Visualizers\windows.natvis'
NatVis script unloaded from 'C:\Program Files\Windows
Kits\10\Debuggers\x64\Visualizers\winrt.natvis'

C:\WinDbg>exit

C:\AWAPI-Dumps>
```

If you find any symbol problems, please use the Contact form on www.patterndiagnostics.com to report them.

We recommend exiting WinDbg after each exercise.

Why Windows API?

- Development
- Malware analysis
- Vulnerability analysis and exploitation
- Reversing
- Diagnostics
- Debugging
- Memory forensics
- Crash and hang analysis
- Secure coding
- Static code analysis
- Trace and log analysis

First, why did we create this course? The knowledge of Windows API is necessary for many software construction and post-construction activities listed on this slide. In this training, we look at Windows API from a software diagnostics perspective. This perspective includes memory dump analysis and, partially, trace and log analysis. The knowledge of Windows API is tacitly assumed in many of my courses. Sometimes, course attendees ask why I chose a particular function from a stack trace for analysis or what this or that function is doing. Of course, there is an intersection of what we learn with other areas as well. During this course, we also do a bit of live debugging.

My History of Windows API

- I started using Windows API in 1990 (Old CV)
- Windows SDK since 1990
- Win16 1990 – 1999
- Win32 since 1995
- Win64 since 2006 (WindowHistory64, earlier than that)
- Windows NT since 1996
- Windows NT/2000/XP DDK and WDK since 2003
- Daily programming using Windows API 1990 – 2003
- DebugWare (DiagWare) tools and patterns 2004 – 2017
- Daily programming using Windows API 2017 – 2020
- Daily reading of Windows API for dump analysis since 2003
- Return to system programming using Windows API in 2024

It is hard for me to recall when I started using Windows API. It is such a distant memory. I believe it was 1990 and Windows 3.0 (although, as a user, I worked with Windows 2.0 before). Fortunately, I recalled that I meticulously documented my work history before 2003 in my old CV, which I no longer maintain. So, from the record, I see I started using Windows API in 1990, and I recall that it was March when I got a copy of Programming Windows by Charles Petzold. Although I started with 16-bit Windows API (and had to use it up to 1999), I moved to 32-bit Windows API in 1995, initially for native Windows 95 development and in 1996 for native Windows NT development, which was Windows NT 4. It was still SDK for user space. In 2003, I started using NT/2000/XP DDK first as a part of some security-related projects (such as encryption), and later, it was necessary for kernel memory dump analysis onwards, and I also wrote my first kernel driver circa 2006 and used Windows API for communication. DDK (Driver Development Kit) is now called WDK (Windows Driver Kit). Software Diagnostics Library has posts that started only in August 2006 and already included material from x64 Windows. From 1990, I used Windows API daily until 2003, when I started doing full-time memory dump analysis until 2017. But even during that time, I used Windows API occasionally to write various troubleshooting, diagnostics, and debugging tools, which I collectively called DebugWare and later DiagWare. In 2017 – 2020 I again used Windows API, including Windows Services and Security, while working on Windows monitoring tools and later native Windows desktop UI interfaces. And although in 2020, I switched to full-time Linux development, I still use Windows API almost daily for my own projects and books, teaching Windows diagnostics and debugging, and doing Windows memory dump analysis for my own needs or for customers who need my services. Also, in 2024, I returned to system programming using Windows API.

Old CV
https://opentask.com/Vostokov/CV.htm

WindowHistory64
https://support.citrix.com/article/CTX109235/windowhistory-tool

Perspectives of Windows API

- Memory analysis: dumps / live debugging
- Disassembly, reconstruction, reversing
- Trace and log analysis (Procmon)
- Category theory

The perspective we take is from software diagnostics, particularly memory analysis (this may include memory forensics), which includes both dump analysis and live debugging, disassembly, reconstruction, reversing (the so-called ADDR patterns), and trace and log analysis (this may include event monitoring) using tools such as Procmon. We do not look at Windows API from a development perspective, such as writing applications and services, although there is some necessary intersection, especially when doing diagnostics and vulnerability analysis. We also take a view of Windows API from a category theory perspective.

Procmon
https://learn.microsoft.com/en-us/sysinternals/downloads/procmon

What Windows API?

- Source code perspective (SDK and/or WDK)
- ABI (Application Binary Interface) perspective

Let's determine what we mean by Windows API. There are two general perspectives on this question. The first is the source code (or software construction) perspective. What is included in SDK and WDK is Windows API. In this training, we take a different software post-construction perspective: what is seen from already compiled and linked modules is what we call Windows API. This perspective also includes what is not included in SDK and WDK descriptions: undocumented and internal API.

API and Language Levels

- Conceptual cross-platform level
- High-level and assembly language
- Machine language

Conceptually, there are many similarities between different APIs, such as Windows and Linux. This is understandable since a large part of API centers on interfacing with OS, which must provide common resource abstractions. The next level is the high-level and assembly languages level, which differs across platforms. And finally, the machine language level erases all differences in the previous level.

x64 Disassembly Review

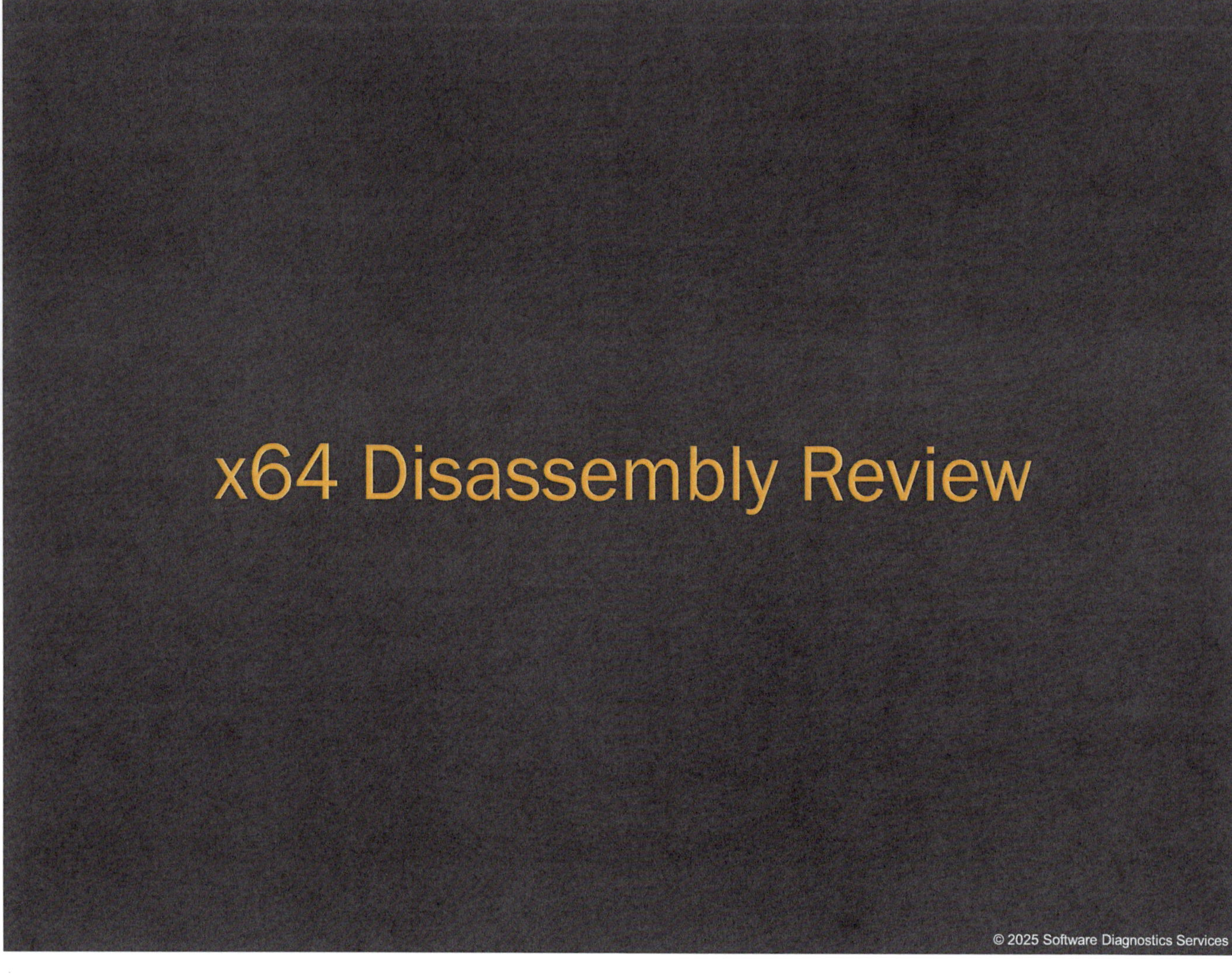
x64 Disassembly Review
© 2025 Software Diagnostics Services

x64 CPU Registers

- RAX ⊃ EAX ⊃ AX ⊇ {AH, AL}
- ALU: RAX, RDX
- Counter: RCX
- Memory copy: RSI (src), RDI (dst)
- Stack: RSP
- Next instruction: RIP
- New: R8 – R15, Rx(D|W|B)

RAX 64-bit	EAX 32-bit

There are familiar 32-bit CPU register names, such as **EAX,** that are extended to 64-bit names, such as **RAX**. Most of them are traditionally specialized, such as ALU, counter, and memory copy registers. Although, now they all can be used as general-purpose registers. There is, of course, a stack pointer, **RSP**, and it also takes the role of a frame pointer, which is also used to address local variables and saved parameters. It can be used for stack reconstruction. In Microsoft compiler code generation implementations, **RBP** is also used as a general-purpose register. An instruction pointer **RIP** is saved in the stack memory region with every function call, then restored on return from the called function. In addition, the x64 platform features another eight general-purpose registers, from **R8** to **R15**.

Instructions and Registers

- Opcode DST, SRC
- Examples:

```
mov   rax, 10h              ; RAX ← 0x10
mov   r13, rdx              ; R13 ← RDX
add   r10, 10h              ; R10 ← R10 + 0x10
imul  edx, ecx              ; EDX ← EDX * ECX
call  rdx                   ; RDX already contains
                            ;    the address of func (&func)
                            ; PUSH RIP; RIP ← &func
sub   rsp, 30h              ; RSP ← RSP-0x30
                            ; make room for local variables
```


This slide shows a few examples of CPU instructions involving operations with registers, such as moving a value and doing arithmetic. The direction of operands is opposite to the AT&T x64 disassembly flavor if you are accustomed to default GDB disassembly on Linux.

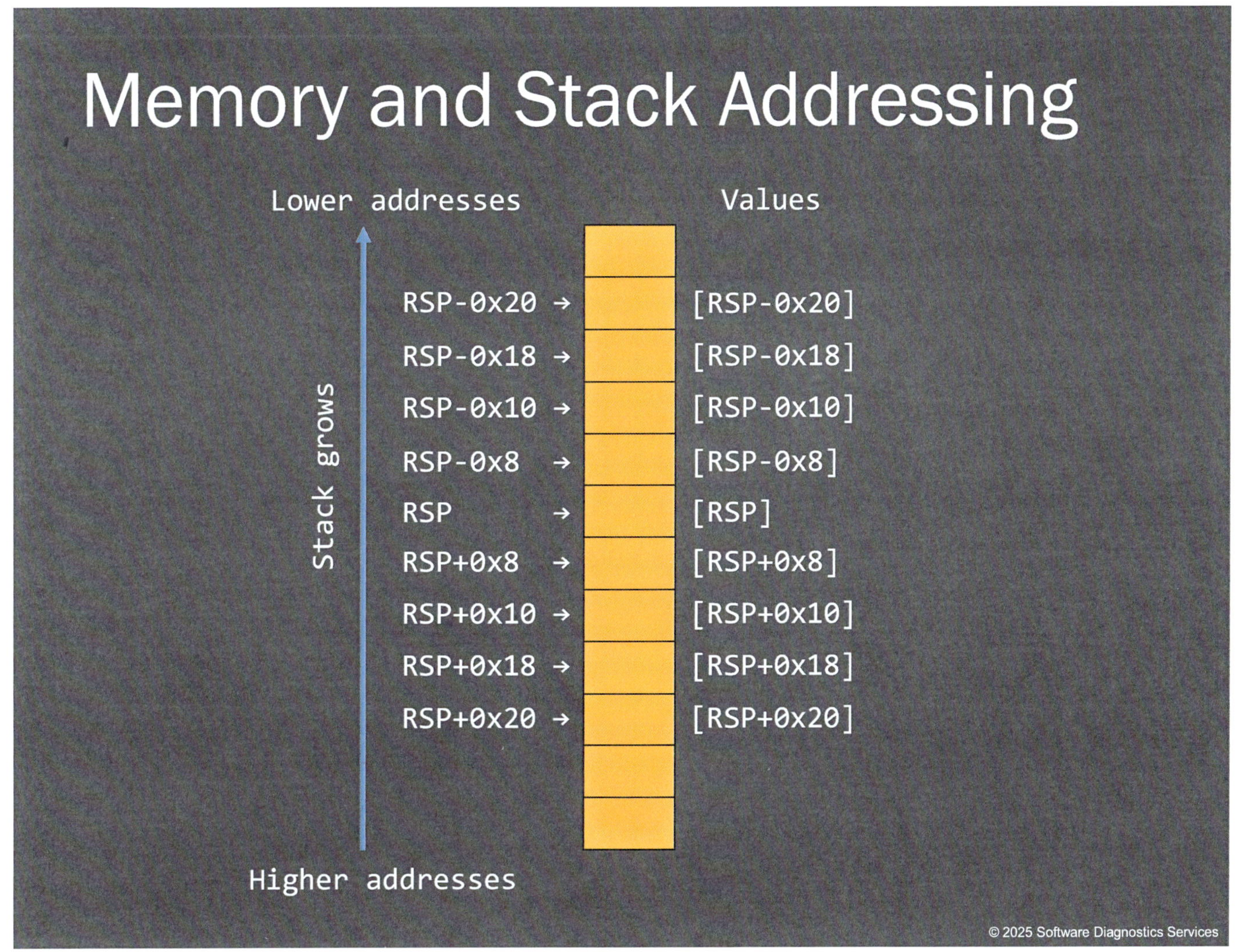

Before we look at operations with memory, let's look at a graphical representation of memory addressing where, for simplicity, I use 64-bit (or 8-byte) memory cells. A thread stack is just any other memory region, so instead of **RSP,** any other register can be used. Please note that the stack grows towards lower addresses, so to access the previously pushed values, you need to use positive offsets from **RSP**.

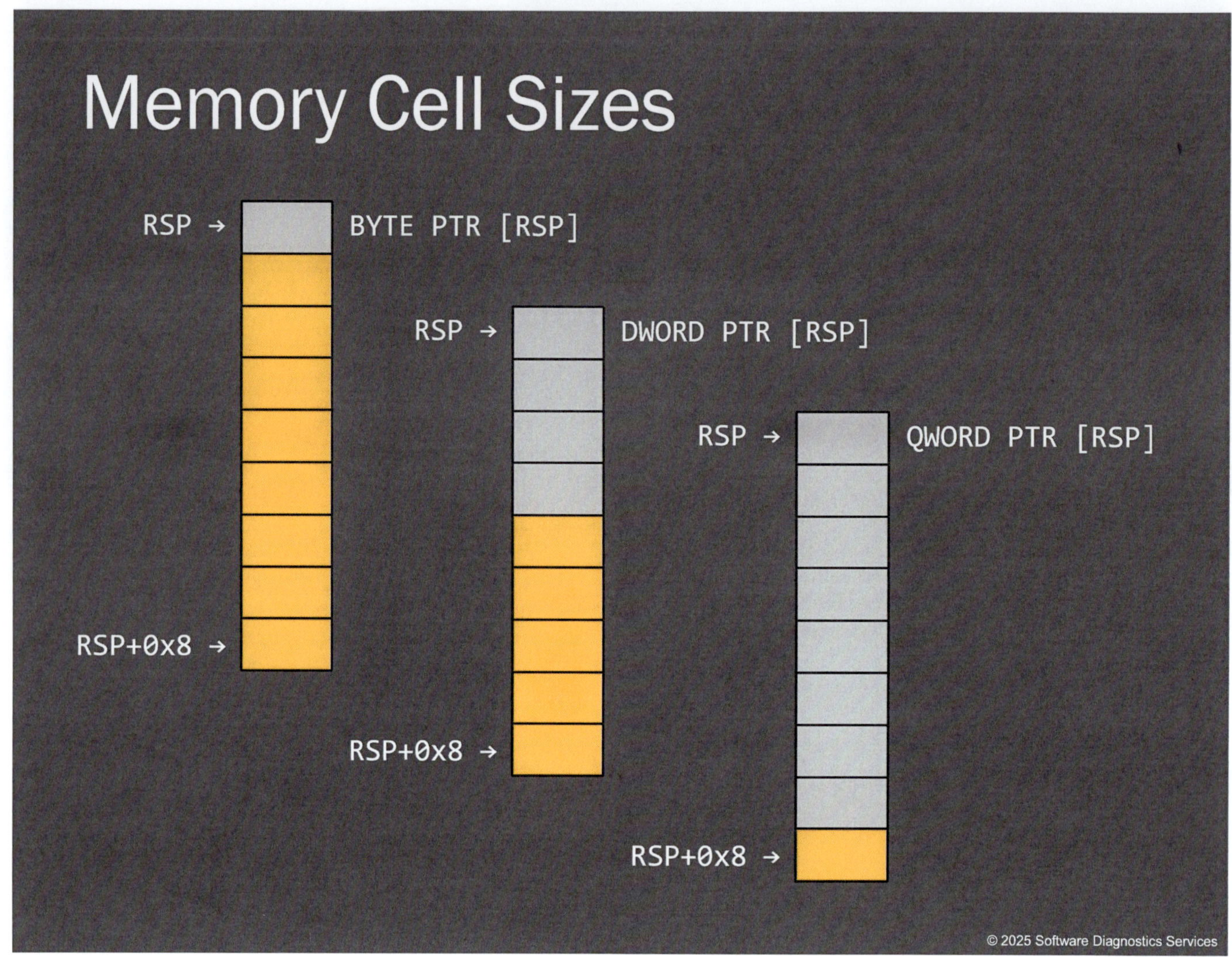

Here, each memory cell is 8-bit (or one byte). When we have a register pointing to memory, and we want to work with the value at that address, we need to specify the size of memory cells to work with, for example, **BYTE PTR** if we want to work with a byte, **DWORD PTR** if we want to work with 32-bit double words, and **QWORD PTR** if we want to work with 64-bit quad words. There's also **WORD PTR** for 16-bit values. This notation is different from Linux GDB, where we have bytes, half-words, words, and double words.

Memory Load Instructions

- Opcode DST, PTR [SRC+Offset]
- Opcode DST
- Examples:

```
mov   rax, qword ptr [rsp+10h] ; RAX ←
                               ; 64-bit value at address RSP+0x10
mov   ecx, dword ptr [20]      ; ECX ←
                               ; 32-bit value at address 0x20
pop   rdi                      ; RDI ← value at address RSP
                               ; RSP ← RSP + 8
lea   r8, [rsp+20h]            ; R8 ← address RSP+0x20
```


Constants are encoded in instructions, but if we need arbitrary values, we must get them from memory. Square brackets show memory access relative to an address stored in a register.

Memory Store Instructions

- Opcode PTR [DST+Offset], SRC
- Opcode DST|SRC
- Examples:

```
mov   qword ptr [rbp-20h], rcx ; 64-bit value at address RBP-0x20
                               ;   ← RCX
mov   byte ptr [0], 1          ; 8-bit value at address 0 ← 1
push  rsi                      ; RSP ← RSP - 8
                               ; value at address RSP ← RSI
inc   dword ptr [rcx]          ; 32-bit value at address RCX ←
                               ;   1 + 32-bit value at address RCX
```


Storing is similar to loading.

Flow Instructions

- Opcode DST
- Opcode PTR [DST]
- Examples:

```
jmp   00007ff6`9ef2f008    ; RIP ← 0x7ff69ef2f008
                           ; (goto 0x7ff69ef2f008)
jmp   qword ptr [rax+10h]  ; RIP ← value at address RAX+0x10
call  00007ff6`9ef21400    ; RSP ← RSP - 8
00007ff6`9ef21057:         ; value at address RSP ← 0x7ff69ef21057
                           ; RIP ← 0x7ff69ef21400
                           ; (goto 0x7ff69ef21400)
```


Goto (an unconditional jump) is implemented via the **JMP** instruction. Function calls are implemented via **CALL** instruction. For conditional branches, please look at the official Intel documentation. We don't use these instructions in our exercises.

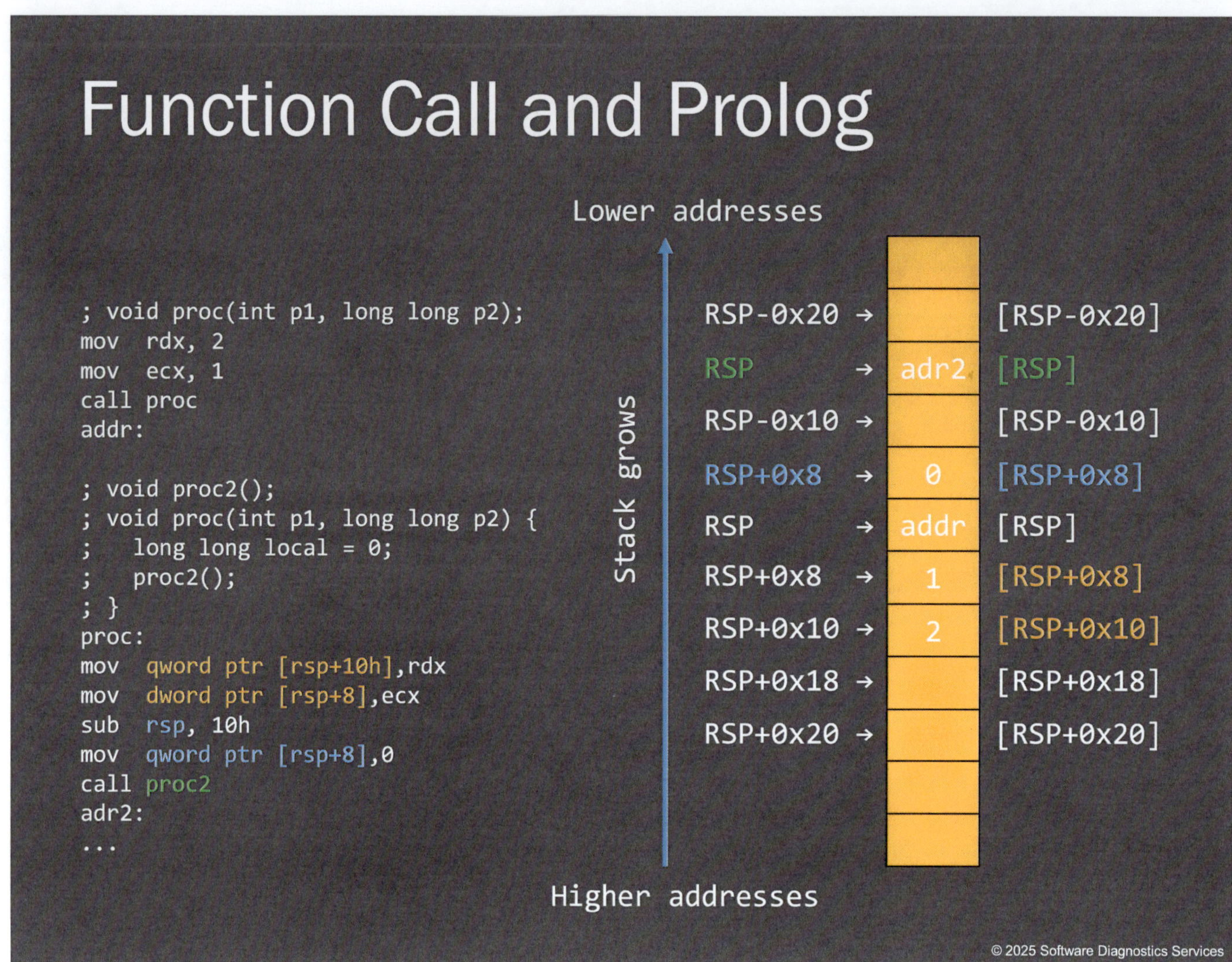

When a function is called from the caller, a callee needs to do certain operations to make room for local variables on the thread stack. There are different ways to do that, and the assembly language code on the left is one of them. I use different colors in the diagram on the right to highlight the updated **RSP** values from the start of the *proc* function up to the moment when the *proc2* function is called. For simplicity of illustration, I only use 64-bit values.

General Windows API Aspects

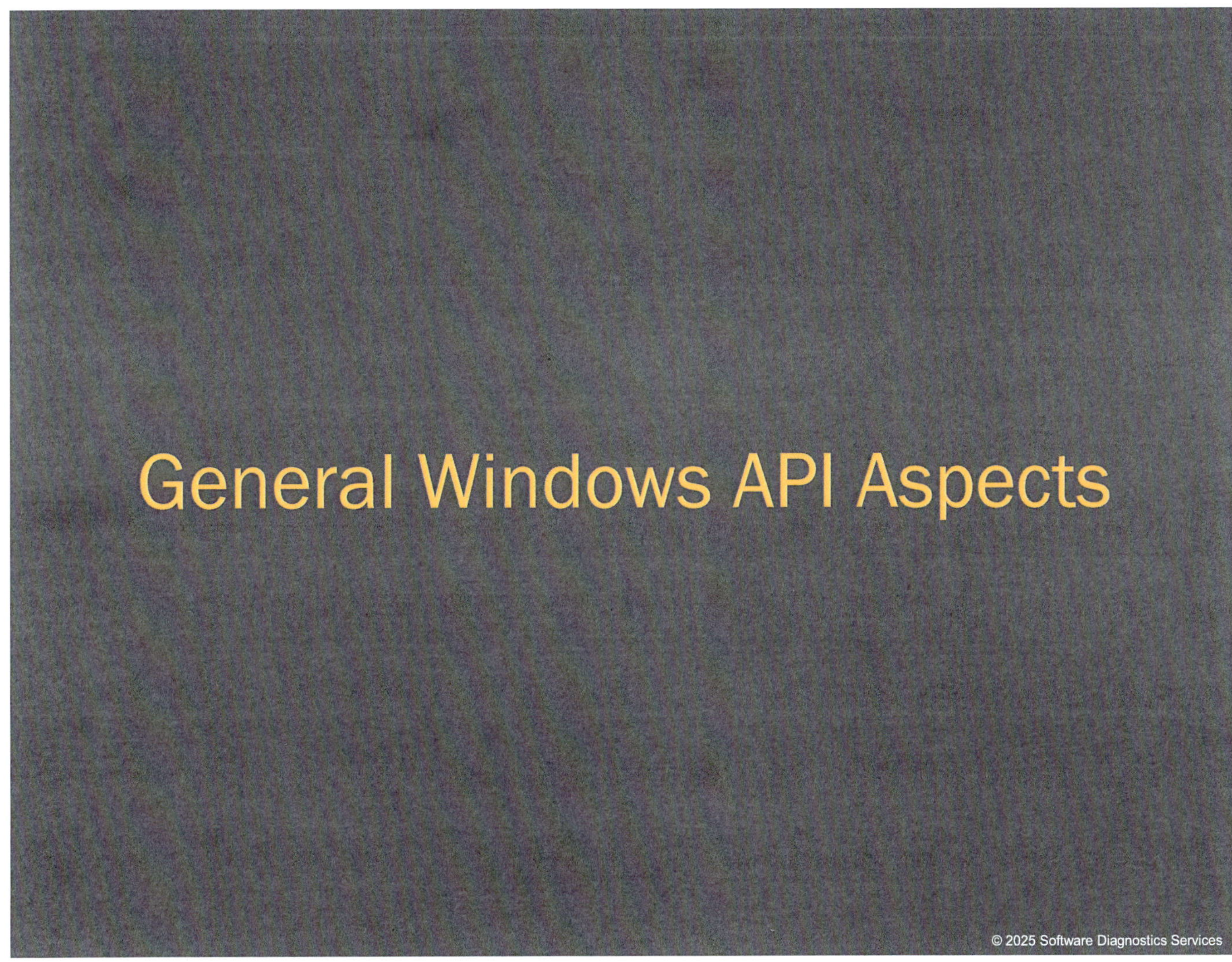
General Windows API Aspects
© 2025 Software Diagnostics Services

Header-Technology View

- Programming reference for the Win32 API

Header 1..* — 1..* Technology

- CreateThread is included in:
  - processthreadsapi.h header is used by:
    - Remote Desktop Services
    - Security and Identity
    - Processes and threads

If you are unfamiliar with C or C++, a header is a textual file referenced in source code and inserted during compilation. It can contain anything but most likely some programming language source code such as function declarations, in our context, for example, declarations of Windows API functions and types. Initially, there were very few headers, but now there are many more, and Microsoft reorganized Windows API help information along the many-to-many relationship between headers and the so-called technologies. So, for example, if we are interested in creating threads, the *CreateThread* API function is declared in the *processthreadsapi.h* header file, which is used by several technologies, including Security and Identity. But if we look at Security and Identity, we see that it uses many other header files.

Programming reference for the Win32 API
https://learn.microsoft.com/en-us/windows/win32/api/

CreateThread
https://learn.microsoft.com/en-us/windows/win32/api/processthreadsapi/nf-processthreadsapi-createthread

processthreadsapi.h header
https://learn.microsoft.com/en-us/windows/win32/api/processthreadsapi/

Processes and threads
https://learn.microsoft.com/en-us/windows/win32/api/_processthreadsapi/

Security and Identity
https://learn.microsoft.com/en-us/windows/win32/api/_security/

Remote Desktop Services
https://learn.microsoft.com/en-us/windows/win32/api/_termserv/

Naming Convention

- Naming conventions
- Functions, parameters, fields: PascalCase, UpperCamelCase
  - GetCurrentThread
  - CreateWindowExA / CreateWindowExW
- Types: SCREAMING SNAKE CASE
  - SECURITY ATTRIBUTES

There are several naming conventions. The Windows API convention differs from the Linux API convention; also, there are differences between different programming languages; please see the first link on the slide. Windows API naming convention also differs from the Standard C++ library naming convention. Windows API uses **UpperCamelCase** for functions, parameters, and structure fields. It is also called **PascalCase**, and we highlighted this name because we would see the connection with the **Pascal** calling convention later on. If a function takes strings as parameters, there are usually two variants, having **A** and **W** suffixes, differentiating between ASCII and UNICODE string parameters. Types use a different **snake case** convention called **screaming**, according to a Wikipedia naming conventions link.

Naming conventions
https://en.wikipedia.org/wiki/Naming_convention_(programming)

GetCurrentThread
https://learn.microsoft.com/en-us/windows/win32/api/processthreadsapi/nf-processthreadsapi-getcurrentthread

CreateWindowExA
https://learn.microsoft.com/en-us/windows/win32/api/winuser/nf-winuser-createwindowexa

CreateWindowExW
https://learn.microsoft.com/en-us/windows/win32/api/winuser/nf-winuser-createwindowexw

Snake Case
https://en.wikipedia.org/wiki/Snake_case

SECURITY_ATTRIBUTES
https://learn.microsoft.com/en-us/previous-versions/windows/desktop/legacy/aa379560(v=vs.85)

Basic Type System

- With a few exceptions (`int`), basic types are `typedef-ed`
- `minwindef.h`
  - `BOOL, DWORD, LPDWORD, UINT, WPARAM, LPARAM`
- `winnt.h`
  - `CHAR, PSTR, PCSTR, HANDLE`
- `basetsd.h`
  - `LONG_PTR, UINT64`
- `windows.h`

WinDbg Commands

```
0:000> dt WinTypes!DWORD
Uint4B

0:000> dt WinTypes!WPARAM
Uint8B

0:000> dt WinTypes!LONG
Int4B
```


Unlike some APIs, which may use a C language type system, Windows API (with a few exceptions such as int) uses its own custom types, usually *typedef*-ed from standard C types. Basic type definitions are scattered across a few files. Examples are shown on this slide. Usually, the catch-all header file, *windows.h* includes all such necessary header files. Windows OS also includes *WinTypes.dll* with a corresponding symbol file with byte size information of basic types. So, during memory dump analysis, you don't need to rush for a chain of header files.

Hungarian Notation

- Wikipedia reference
- Microsoft reference
- CreateWindowExW / WNDCLASSW

```
HWND CreateWindowExW(
  [in]           DWORD     dwExStyle,
  [in, optional] LPCWSTR   lpClassName,
  [in, optional] LPCWSTR   lpWindowName,
  [in]           DWORD     dwStyle,
  [in]           int       X,
  [in]           int       Y,
  [in]           int       nWidth,
  [in]           int       nHeight,
  [in, optional] HWND      hWndParent,
  [in, optional] HMENU     hMenu,
  [in, optional] HINSTANCE hInstance,
  [in, optional] LPVOID    lpParam
);
```

```
typedef struct tagWNDCLASSW {
  UINT      style;
  WNDPROC   lpfnWndProc;
  int       cbClsExtra;
  int       cbWndExtra;
  HINSTANCE hInstance;
  HICON     hIcon;
  HCURSOR   hCursor;
  HBRUSH    hbrBackground;
  LPCWSTR   lpszMenuName;
  LPCWSTR   lpszClassName;
} WNDCLASSW, *PWNDCLASSW,
*NPWNDCLASSW, *LPWNDCLASSW;
```


If you come from the Linux world, you find the **Hungarian** notation weird. The main idea is to prefix names such as parameters, fields, and type names themselves with type information. But these prefixes are excellent hints when you look at memory because they show low-level implementation details like whether a parameter is a pointer to a null-terminated string, a pointer to a 32-bit value, or just a 32-bit value itself. Sometimes, different prefixes refer to the same implementation, for example, **LP** (Long Pointer), **NP** (Near Pointer), and **P** (Pointer). Such different prefix names are kept for compilation of legacy source code when distinction among them existed.

Wikipedia reference
https://en.wikipedia.org/wiki/Hungarian_notation

Microsoft reference
https://learn.microsoft.com/en-us/windows/win32/stg/coding-style-conventions

CreateWindowExW
https://learn.microsoft.com/en-us/windows/win32/api/winuser/nf-winuser-createwindowexw

WNDCLASSW
https://learn.microsoft.com/en-us/windows/win32/api/winuser/ns-winuser-wndclassw

Call Types

- Direct (the same module, non-exported functions)

```
USER32!CreateWindowExW:
…
0007ffa`5e0cf20d e812000000 call USER32!CreateWindowInternal (00007ffa`5e0cf224)
```

- Indirect
  - Pointer (memory or register)
    - IAT (Import Address Table) inter-module call

```
App!wWinMain:
…
00007ff6`77741101 ff15c1e10000 call qword ptr [App!_imp_CreateWindowExW (00007ff6`7774f2c8)]
```


Now, we have come from source code to implementation details. Windows API calls are ultimately implemented as call instructions on Intel and branch and link instructions on ARM platforms. In our training, we are only concerned with x64. ARM64 is left as a homework exercise if you have access to the platform. The compiler chooses different variants of call instructions based on whether the destination is in the same module or not. Direct calls may be emitted when a callee in the same module, a relative offset to the current instruction pointer is known, or an address is absolute. An internal function called from an exported function is one example illustrated in this slide.

API as Interface

- Provided by (exported from) some DLL module (may have different file extensions)
- Used by (imported by) EXE or DLL
- Can be functional or object-oriented

EXE/DLL → ● — DLL

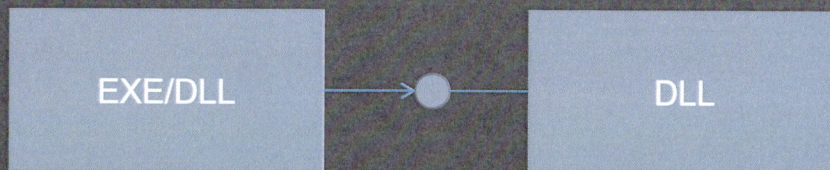

Most of the time, Windows API means a set of interfaces provided by or exported from some DLL, Dynamic Link Library. These interfaces are used by or imported by some other DLL or executable. Interfaces can be functional or object-oriented conceptually. Let's look at how these API interfaces are implemented.

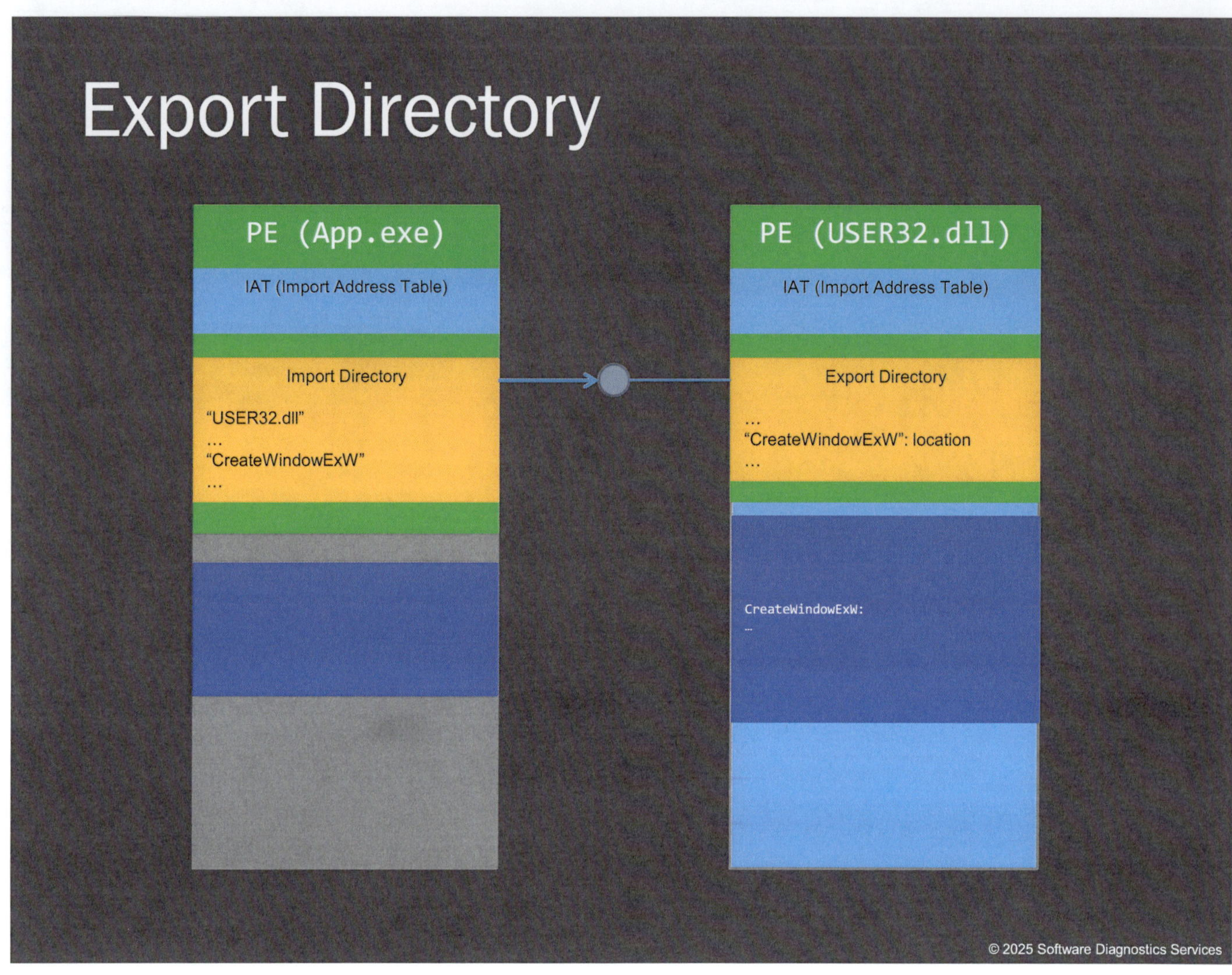

A module or DLL that wants to provide an interface has the so-called **Export Directory** of names and corresponding locations in code. It is located in the module header, **PE**, Portable Executable header. On the other side, if some module, DLL, or executable file wants to use some names from that interface, it has the so-called **Import Directory** in the **PE** header with module names and names from modules' interfaces, as illustrated on this slide where *App.exe* wants to use *CreateWindowExW* from *USER32.dll*.

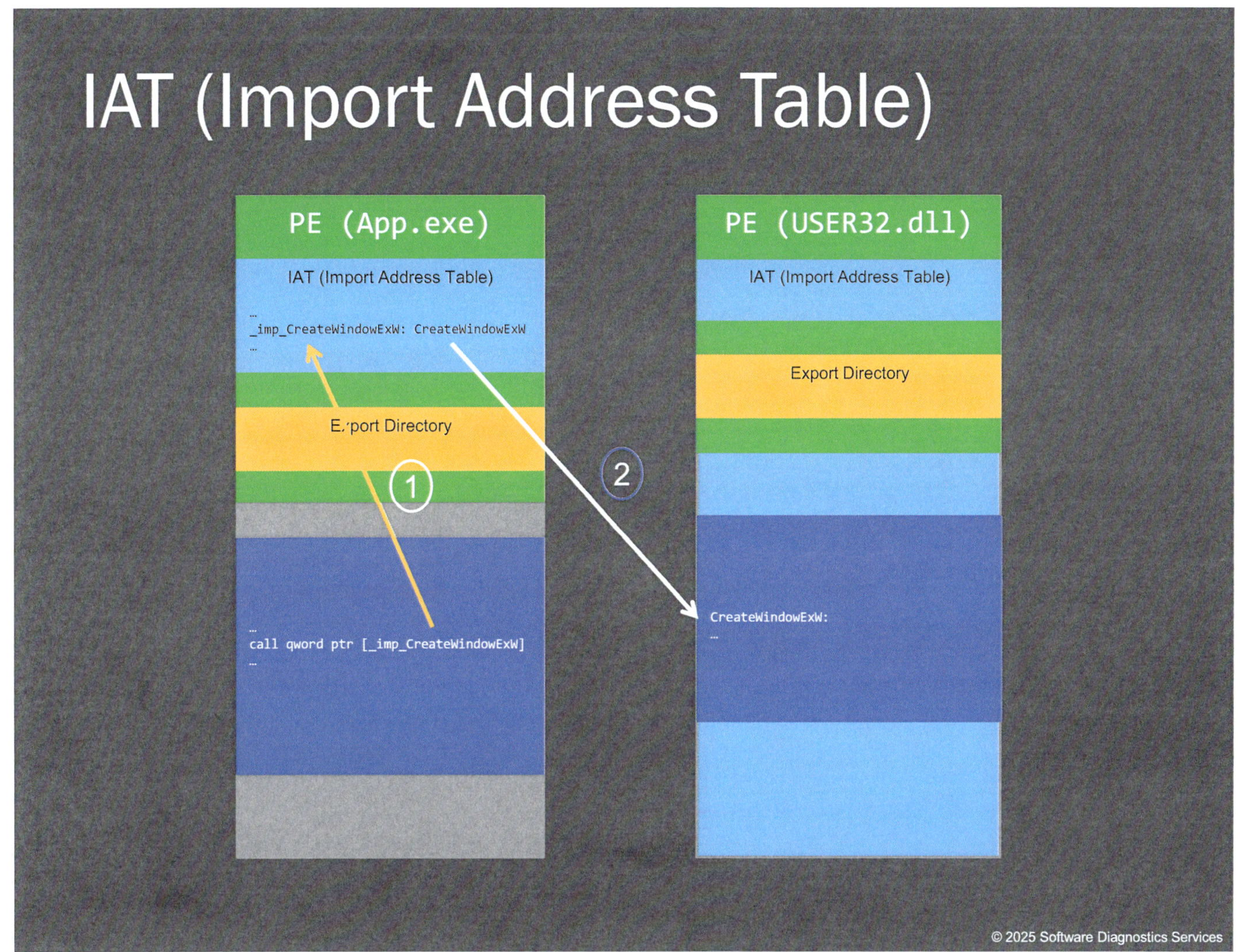

When a module that wants to use interfaces from other modules is loaded, the dynamic linker modifies the so-called **Import Address Table** directory in the PE header by populating each entry there with the real addresses from other already loaded DLL modules that export the required interface entries. For example, if *App.exe* wanted to use the *CreateWindowExW* function from *USER32.dll*, the corresponding entry in *App.exe Import Address Table* will be filled with the real address of that function inside the *USER32.dll* code. If we want to call an interface entry, indirect addressing is used to transfer execution to the interface module code. It uses an address from the **Import Address Table** that points to code in another module that implements the interface entry. This indirection allows calling interfaces independently from their implementation code location in memory since the relative address of the **Import Address Table** is the same. This mechanism also allows interface modules to be loaded at different addresses during each program run, as illustrated in the next slide.

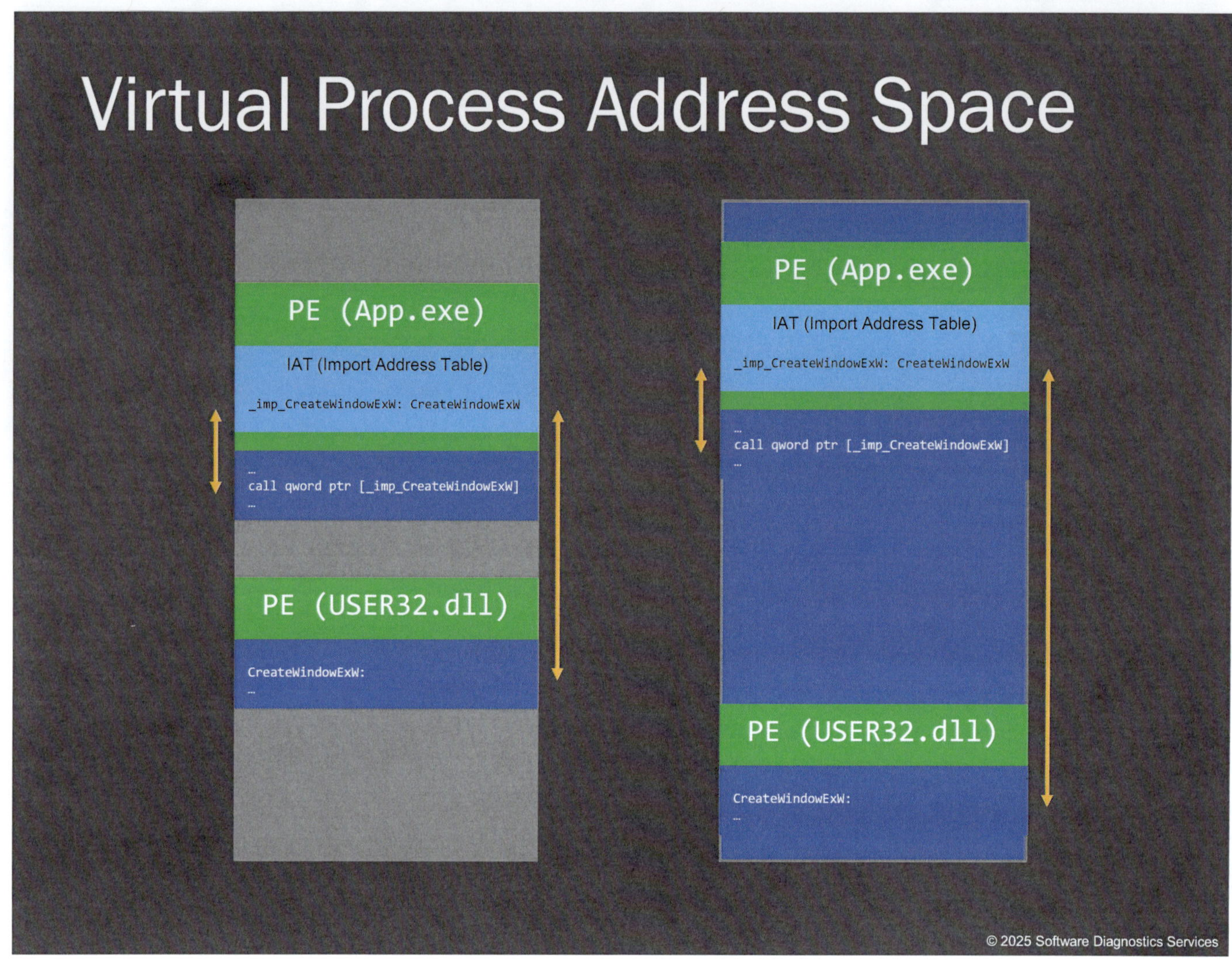

This slide shows different program executions where, each time, both modules are loaded at different addresses. We see that indirect addressing allows the *App.exe* code to call the *USER32.dll* code loaded at different addresses in virtual memory.

Calling Convention

- GetMessageW (from documentation)
- Actual declaration (WinUser.h)

```
WINUSERAPI
BOOL
WINAPI
GetMessageW(
    _Out_ LPMSG lpMsg,
    _In_opt_ HWND hWnd,
    _In_ UINT wMsgFilterMin,
    _In_ UINT wMsgFilterMax);
```

- WINAPI is defined as __stdcall (minwindef.h) vs. default __cdecl (C/C++)
- Argument passing order
  - x86: pushed to stack right-to-left, the callee cleans the stack (in __cdecl – the caller)
  - x64 (__stdcall and __cdecl): left-to-right via RCX, RDX, R8, R9, [RSP+20], [RSP+28], ...

Except for some API calls that do not need arguments, most API calls require them. How such a source code is implemented on a particular platform is a subject of the so-called calling convention. On a 32-bit Intel platform, argument values (call parameters) are passed using stack, and, in callee, to see them linearly from left to right starting from the lower stack address, you need to push them to the stack from right to left because stack grows down in memory. It is illustrated in the next slide. This convention is the same for C/C++ code (**_cdecl**) and WINAPI (**_stdcall**, a variation of the **Pascal** calling convention used in 16-bit Windows). The difference between 32-bit **_cdecl** and **_stdcall** is who clears the stack. Windows API uses **_stdcall**, and the callee clears the stack, for example, by adding the required value to the stack pointer. In **_cdecl**, the caller clears the stack. The latter allows passing the variable number of arguments. On the 64-bit Intel platform, **_stdcall** and **_cdecl** are ignored, and the Microsoft x64 calling convention is used: the first four parameters are passed via registers from left to right, and the rest – via direct stack addressing. It is also illustrated on the slide after the next one, and both platforms are illustrated in Exercise W1. ARM platforms also use registers for parameter passing.

GetMessageW
https://learn.microsoft.com/en-us/windows/win32/api/winuser/nf-winuser-getmessagew

__stdcall
https://learn.microsoft.com/en-us/cpp/cpp/stdcall?view=msvc-170

__cdecl
https://learn.microsoft.com/en-us/cpp/cpp/cdecl?view=msvc-170

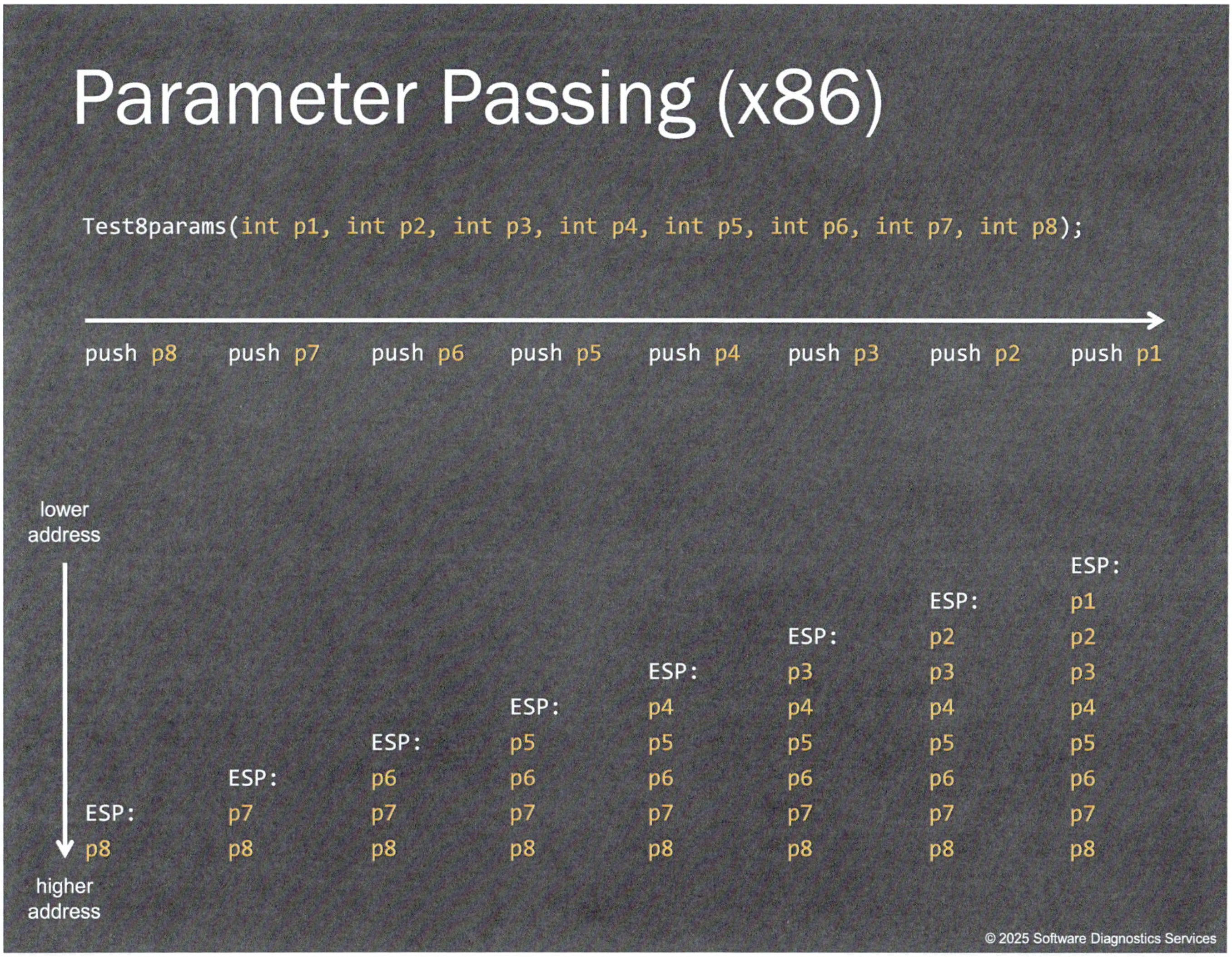

Parameter Passing (x86)

```
Test8params(int p1, int p2, int p3, int p4, int p5, int p6, int p7, int p8);
```

push p8	push p7	push p6	push p5	push p4	push p3	push p2	push p1
							ESP:
						ESP:	p1
					ESP:	p2	p2
				ESP:	p3	p3	p3
			ESP:	p4	p4	p4	p4
		ESP:	p5	p5	p5	p5	p5
	ESP:	p6	p6	p6	p6	p6	p6
ESP:	p7	p7	p7	p7	p7	p7	p7
p8	p8	p8	p8	p8	p8	p8	p8

lower address

higher address

Parameter Passing (x64)

```
Test8params(int p1, int p2, int p3, int p4, int p5, int p6, int p7, int p8);
```

Caller

```
ECX (p1)
EDX (p2)
R8D (p3)
R9D (p4)

RSP:     0`0
RSP+8:   0`0
RSP+10:  0`0
RSP+18:  0`0
RSP+20:  0`p5
RSP+28:  0`p6
RSP+30:  0`p7
RSP+38:  0`p8
```

call →

Callee

```
ECX (p1)
EDX (p2)
R8D (p3)
R9D (p4)

RSP:     return address
RSP+8:   0`0
RSP+10:  0`0
RSP+18:  0`0
RSP+20:  0`0
RSP+28:  0`p5
RSP+30:  0`p6
RSP+38:  0`p7
RSP+40:  0`p8
```

Exercise W1

- **Goal:** Compare calling conventions on x86 and x64 platforms
- **ADDR Patterns:** Call Prologue; Call Parameter; Function Prologue
- \AWAPI-Dumps\Exercise-W1.pdf

Exercise W1

Goal: Compare calling conventions on x86 and x64 platforms.

ADDR Patterns: Call Prologue; Call Parameter; Function Prologue.

1. Let's look at a 64-bit process. Launch WinDbg.

2. Open \AWAPI-Dumps\Process\wordpad.DMP

3. We get the dump file loaded:

```
Microsoft (R) Windows Debugger Version 10.0.27704.1001 AMD64
Copyright (c) Microsoft Corporation. All rights reserved.


Loading Dump File [C:\AWAPI-Dumps\Process\wordpad.DMP]
User Mini Dump File with Full Memory: Only application data is available


************* Path validation summary **************
Response                         Time (ms)     Location
Deferred                                       srv*
Symbol search path is: srv*
Executable search path is:
Windows 10 Version 22000 MP (2 procs) Free x64
Product: WinNt, suite: SingleUserTS Personal
Edition build lab: 22000.1.amd64fre.co_release.210604-1628
Debug session time: Sat Oct 15 21:19:05.000 2022 (UTC + 1:00)
System Uptime: 0 days 0:27:57.202
Process Uptime: 0 days 0:00:26.000
................................................................
.....
Loading unloaded module list
.
For analysis of this file, run !analyze -v
win32u!NtUserGetMessage+0x14:
00007ffc`f3811414 ret
```

4. Open a log file using **.logopen**:

```
0:000> .logopen C:\AWAPI-Dumps\W1.log
Opened log file 'C:\AWAPI-Dumps\W1.log'
```

5. There are several places in Wordpad code where windows are created. We look at two of them. First, disassemble the *InitMainThreadWnd* function from the *combase* module and find the *CreateWindowExW* API call:

```
0:000> uf combase!InitMainThreadWnd
combase!InitMainThreadWnd [onecore\com\combase\objact\mainthrd.cxx @ 114]:
  114 00007ffc`f5e14770 4c8bdc          mov     r11,rsp
  114 00007ffc`f5e14773 53              push    rbx
  114 00007ffc`f5e14774 4883ec60        sub     rsp,60h
  148 00007ffc`f5e14778 498363f000      and     qword ptr [r11-10h],0
  148 00007ffc`f5e1477d 4c8d05b45e1d00  lea     r8,[combase!`string' (00007ffc`f5fea638)]
  148 00007ffc`f5e14784 488b059d1a2400  mov     rax,qword ptr [combase!g_hinst (00007ffc`f6056228)]
```

```
  148 00007ffc`f5e1478b 41b900000088     mov     r9d,88000000h
  148 00007ffc`f5e14791 488b15c0232400   mov     rdx,qword ptr [combase!gOleWindowClass (00007ffc`f6056b58)]
  148 00007ffc`f5e14798 33c9             xor     ecx,ecx
  148 00007ffc`f5e1479a 498943e8         mov     qword ptr [r11-18h],rax
  148 00007ffc`f5e1479e b800000080       mov     eax,80000000h
  148 00007ffc`f5e147a3 498363e000       and     qword ptr [r11-20h],0
  148 00007ffc`f5e147a8 49c743d8fdffffff mov     qword ptr [r11-28h],0FFFFFFFFFFFFFFFDh
  148 00007ffc`f5e147b0 89442438         mov     dword ptr [rsp+38h],eax
  148 00007ffc`f5e147b4 89442430         mov     dword ptr [rsp+30h],eax
  148 00007ffc`f5e147b8 89442428         mov     dword ptr [rsp+28h],eax
  148 00007ffc`f5e147bc 89442420         mov     dword ptr [rsp+20h],eax
  148 00007ffc`f5e147c0 48ff15411e2700   call    qword ptr [combase!_imp_CreateWindowExW (00007ffc`f6086608)]
  148 00007ffc`f5e147c7 0f1f440000       nop     dword ptr [rax+rax]
  148 00007ffc`f5e147cc 4889057d232400   mov     qword ptr [combase!ghwndOleMainThread (00007ffc`f6056b50)],rax
  163 00007ffc`f5e147d3 4885c0           test    rax,rax
  163 00007ffc`f5e147d6 0f84e8950b00     je      combase!InitMainThreadWnd+0xb9654 (00007ffc`f5ecddc4)  Branch

combase!InitMainThreadWnd+0x6c [onecore\com\combase\object\mainthrd.cxx @ 180]:
  180 00007ffc`f5e147dc 48ff15fd081c00   call    qword ptr [combase!_imp_GetCurrentThreadId (00007ffc`f5fd50e0)]
  180 00007ffc`f5e147e3 0f1f440000       nop     dword ptr [rax+rax]
  180 00007ffc`f5e147e8 890572232400     mov     dword ptr [combase!gdwMainThreadId (00007ffc`f6056b60)],eax
  181 00007ffc`f5e147ee 33db             xor     ebx,ebx

combase!InitMainThreadWnd+0x80 [onecore\com\combase\object\mainthrd.cxx @ 186]:
  186 00007ffc`f5e147f0 8bc3             mov     eax,ebx
  187 00007ffc`f5e147f2 4883c460         add     rsp,60h
  187 00007ffc`f5e147f6 5b               pop     rbx
  187 00007ffc`f5e147f7 c3               ret

combase!InitMainThreadWnd+0xb9654 [onecore\com\combase\object\mainthrd.cxx @ 168]:
  168 00007ffc`f5ecddc4 48ff1525701000   call    qword ptr [combase!_imp_GetLastError (00007ffc`f5fd4df0)]
  168 00007ffc`f5ecddcb 0f1f440000       nop     dword ptr [rax+rax]
  168 00007ffc`f5ecddd0 8bd8             mov     ebx,eax
  168 00007ffc`f5ecddd2 85c0             test    eax,eax
  168 00007ffc`f5ecddd4 7e09             jle     combase!InitMainThreadWnd+0xb966f (00007ffc`f5ecdddf)  Branch

combase!InitMainThreadWnd+0xb9666 [onecore\com\combase\object\mainthrd.cxx @ 168]:
  168 00007ffc`f5ecddd6 0fb7d8           movzx   ebx,ax
  168 00007ffc`f5ecddd9 81cb00000780     or      ebx,80070000h

combase!InitMainThreadWnd+0xb966f [onecore\com\combase\object\mainthrd.cxx @ 169]:
  169 00007ffc`f5ecdddf 833d0265180000   cmp     dword ptr [combase!_Tlgg_hCombaseTraceLoggingProviderProv
(00007ffc`f60542e8)],0
  169 00007ffc`f5ecdde6 7714             ja      combase!InitMainThreadWnd+0xb968c (00007ffc`f5ecddfc)  Branch

combase!InitMainThreadWnd+0xb9678 [onecore\com\combase\object\mainthrd.cxx @ 169]:
  169 00007ffc`f5ecdde8 833dbd85180000   cmp     dword ptr [combase!gfEnableTracing (00007ffc`f60563ac)],0
  169 00007ffc`f5ecddef 743b             je      combase!InitMainThreadWnd+0xb96bc (00007ffc`f5ecde2c)  Branch

combase!InitMainThreadWnd+0xb9681 [onecore\com\combase\object\mainthrd.cxx @ 169]:
  169 00007ffc`f5ecddf1 33c9             xor     ecx,ecx
  169 00007ffc`f5ecddf3 e8b4ecedff       call    combase!IsWppLevelEnabled (00007ffc`f5dacaac)
  169 00007ffc`f5ecddf8 84c0             test    al,al
  169 00007ffc`f5ecddfa 7430             je      combase!InitMainThreadWnd+0xb96bc (00007ffc`f5ecde2c)  Branch

combase!InitMainThreadWnd+0xb968c [onecore\com\combase\object\mainthrd.cxx @ 169]:
  169 00007ffc`f5ecddfc 488d05c5b91100   lea     rax,[combase!`string' (00007ffc`f5fe97c8)]
  169 00007ffc`f5ecde03 895c2430         mov     dword ptr [rsp+30h],ebx
  169 00007ffc`f5ecde07 4889442428       mov     qword ptr [rsp+28h],rax
  169 00007ffc`f5ecde0c 4c8d055d7b1400   lea     r8,[combase!`string' (00007ffc`f6015970)]
  169 00007ffc`f5ecde13 8364242000       and     dword ptr [rsp+20h],0
  169 00007ffc`f5ecde18 488d15297b1400   lea     rdx,[combase!`string' (00007ffc`f6015948)]
  169 00007ffc`f5ecde1f 41b9a9000000     mov     r9d,0A9h
  169 00007ffc`f5ecde25 8bcb             mov     ecx,ebx
  169 00007ffc`f5ecde27 e8ec9aeaff       call    combase!ComTraceMessage (00007ffc`f5d77918)

combase!InitMainThreadWnd+0xb96bc [onecore\com\combase\object\mainthrd.cxx @ 170]:
  170 00007ffc`f5ecde2c 85db             test    ebx,ebx
  170 00007ffc`f5ecde2e 0f88bc69f4ff     js      combase!InitMainThreadWnd+0x80 (00007ffc`f5e147f0)  Branch

combase!InitMainThreadWnd+0xb96c4 [onecore\com\combase\object\mainthrd.cxx @ 172]:
  172 00007ffc`f5ecde34 bb0e000780       mov     ebx,8007000Eh
  176 00007ffc`f5ecde39 e9b269f4ff       jmp     combase!InitMainThreadWnd+0x80 (00007ffc`f5e147f0)  Branch
```

The API function definition is:

```
HWND CreateWindowExW(
  [in]           DWORD     dwExStyle,
  [in, optional] LPCWSTR   lpClassName,
  [in, optional] LPCWSTR   lpWindowName,
  [in]           DWORD     dwStyle,
  [in]           int       X,
  [in]           int       Y,
  [in]           int       nWidth,
  [in]           int       nHeight,
  [in, optional] HWND      hWndParent,
  [in, optional] HMENU     hMenu,
  [in, optional] HINSTANCE hInstance,
  [in, optional] LPVOID    lpParam
);
```

We see the first 4 parameters are passed via ECX, RDX, R8, and R9D according to their size, 32-bit or 64-bit:

```
148 00007ffc`f5e1477d 4c8d05b45e1d00  lea     r8,[combase!`string' (00007ffc`f5fea638)]
148 00007ffc`f5e1478b 41b900000088    mov     r9d,88000000h
148 00007ffc`f5e14791 488b15c0232400  mov     rdx,qword ptr [combase!gOleWindowClass (00007ffc`f6056b58)]
148 00007ffc`f5e14798 33c9            xor     ecx,ecx
```

R8 contains the 64-bit pointer to a window name (UNICODE):

```
0:000> du 00007ffc`f5fea638
00007ffc`f5fea638  "OleMainThreadWndName"
```

The next 4 parameters are passed via direct stack storage (a default rectangle value):

```
148 00007ffc`f5e147b0 89442438        mov     dword ptr [rsp+38h],eax
148 00007ffc`f5e147b4 89442430        mov     dword ptr [rsp+30h],eax
148 00007ffc`f5e147b8 89442428        mov     dword ptr [rsp+28h],eax
148 00007ffc`f5e147bc 89442420        mov     dword ptr [rsp+20h],eax
```

It is interesting to see that the last four parameters are passed via direct stack storage as well but using R11 as saved RSP value during function prologue (here, both prologues overlap in purpose):

```
114 00007ffc`f5e14770 4c8bdc          mov     r11,rsp
114 00007ffc`f5e14773 53              push    rbx
114 00007ffc`f5e14774 4883ec60        sub     rsp,60h
148 00007ffc`f5e14778 498363f000      and     qword ptr [r11-10h],0
148 00007ffc`f5e1479a 498943e8        mov     qword ptr [r11-18h],rax
148 00007ffc`f5e147a3 498363e000      and     qword ptr [r11-20h],0
148 00007ffc`f5e147a8 49c743d8fdffffff mov     qword ptr [r11-28h],0FFFFFFFFFFFFFFFDh
```

Before the *InitMainThreadWnd* function prologue is executed, RSP points to the save return address. So, R11 points to that address as well. RSP is then decremented by 8 via the next push instruction. Then 60h bytes (0x60) are reserved, and RSP now points down 0x68 bytes down its original value:

```
R11:               return address
R11-8h  (RSP+60h): saved RBX
R11-10h (RSP+58h):
R11-18h (RSP+50h):
R11-20h (RSP+48h):
R11-28h (RSP+40h):
RSP+38h:
RSP+30h:
RSP+28h:
```

RSP+20h:

So, by this illustration above, we see the continuity of stack-based parameters despite different CPU registers and offsets used. We also see that all parameters are passed left-to-right.

6. Second, disassemble the *CWnd::CreateEx* function from the *MFC42u* module and find the *CreateWindowExW* API call:

```
0:000> uf MFC42u!CWnd::CreateEx
mfc42u!CWnd::CreateEx:
00007ffc`acc1ca50 488bc4          mov     rax,rsp
00007ffc`acc1ca53 48895808        mov     qword ptr [rax+8],rbx
00007ffc`acc1ca57 48897010        mov     qword ptr [rax+10h],rsi
00007ffc`acc1ca5b 48897818        mov     qword ptr [rax+18h],rdi
00007ffc`acc1ca5f 55              push    rbp
00007ffc`acc1ca60 488d68e1        lea     rbp,[rax-1Fh]
00007ffc`acc1ca64 4881ecb0000000  sub     rsp,0B0h
00007ffc`acc1ca6b 8b4547          mov     eax,dword ptr [rbp+47h]
00007ffc`acc1ca6e 488bd9          mov     rbx,rcx
00007ffc`acc1ca71 8945f7          mov     dword ptr [rbp-9],eax
00007ffc`acc1ca74 8b454f          mov     eax,dword ptr [rbp+4Fh]
00007ffc`acc1ca77 8945f3          mov     dword ptr [rbp-0Dh],eax
00007ffc`acc1ca7a 8b4557          mov     eax,dword ptr [rbp+57h]
00007ffc`acc1ca7d 8945ef          mov     dword ptr [rbp-11h],eax
00007ffc`acc1ca80 8b455f          mov     eax,dword ptr [rbp+5Fh]
00007ffc`acc1ca83 8945eb          mov     dword ptr [rbp-15h],eax
00007ffc`acc1ca86 8b4567          mov     eax,dword ptr [rbp+67h]
00007ffc`acc1ca89 8945e7          mov     dword ptr [rbp-19h],eax
00007ffc`acc1ca8c 488b456f        mov     rax,qword ptr [rbp+6Fh]
00007ffc`acc1ca90 488945df        mov     qword ptr [rbp-21h],rax
00007ffc`acc1ca94 488b4577        mov     rax,qword ptr [rbp+77h]
00007ffc`acc1ca98 488945d7        mov     qword ptr [rbp-29h],rax
00007ffc`acc1ca9c 89550f          mov     dword ptr [rbp+0Fh],edx
00007ffc`acc1ca9f 4c894507        mov     qword ptr [rbp+7],r8
00007ffc`acc1caa3 4c894dff        mov     qword ptr [rbp-1],r9
00007ffc`acc1caa7 e8c4ebffff      call    mfc42u!AfxGetModuleState (00007ffc`acc1b670)
00007ffc`acc1caac 488b4810        mov     rcx,qword ptr [rax+10h]
00007ffc`acc1cab0 488b457f        mov     rax,qword ptr [rbp+7Fh]
00007ffc`acc1cab4 488945c7        mov     qword ptr [rbp-39h],rax
00007ffc`acc1cab8 488b03          mov     rax,qword ptr [rbx]
00007ffc`acc1cabb 48894dcf        mov     qword ptr [rbp-31h],rcx
00007ffc`acc1cabf 49ba709bdb303627698b mov r10,8B69273630DB9B70h
00007ffc`acc1cac9 488b80c8000000  mov     rax,qword ptr [rax+0C8h]
00007ffc`acc1cad0 488d55c7        lea     rdx,[rbp-39h]
00007ffc`acc1cad4 488bcb          mov     rcx,rbx
00007ffc`acc1cad7 ff159b071000    call    qword ptr [mfc42u!_guard_xfg_dispatch_icall_fptr (00007ffc`acd1d278)]
00007ffc`acc1cadd 33ff            xor     edi,edi
00007ffc`acc1cadf 85c0            test    eax,eax
00007ffc`acc1cae1 0f843bc10100    je      mfc42u!CWnd::CreateEx+0x1c1d2 (00007ffc`acc38c22)  Branch

mfc42u!CWnd::CreateEx+0x97:
00007ffc`acc1cae7 488bcb          mov     rcx,rbx
00007ffc`acc1caea e891a5ffff      call    mfc42u!AfxHookWindowCreate (00007ffc`acc17080)
00007ffc`acc1caef 488b45c7        mov     rax,qword ptr [rbp-39h]
00007ffc`acc1caf3 448b4df7        mov     r9d,dword ptr [rbp-9]
00007ffc`acc1caf7 4c8b45ff        mov     r8,qword ptr [rbp-1]
00007ffc`acc1cafb 488b5507        mov     rdx,qword ptr [rbp+7]
00007ffc`acc1caff 8b4d0f          mov     ecx,dword ptr [rbp+0Fh]
00007ffc`acc1cb02 4889442458      mov     qword ptr [rsp+58h],rax
00007ffc`acc1cb07 488b45cf        mov     rax,qword ptr [rbp-31h]
00007ffc`acc1cb0b 4889442450      mov     qword ptr [rsp+50h],rax
00007ffc`acc1cb10 488b45d7        mov     rax,qword ptr [rbp-29h]
00007ffc`acc1cb14 4889442448      mov     qword ptr [rsp+48h],rax
00007ffc`acc1cb19 488b45df        mov     rax,qword ptr [rbp-21h]
00007ffc`acc1cb1d 4889442440      mov     qword ptr [rsp+40h],rax
00007ffc`acc1cb22 8b45e7          mov     eax,dword ptr [rbp-19h]
00007ffc`acc1cb25 89442438        mov     dword ptr [rsp+38h],eax
00007ffc`acc1cb29 8b45eb          mov     eax,dword ptr [rbp-15h]
00007ffc`acc1cb2c 89442430        mov     dword ptr [rsp+30h],eax
00007ffc`acc1cb30 8b45ef          mov     eax,dword ptr [rbp-11h]
00007ffc`acc1cb33 89442428        mov     dword ptr [rsp+28h],eax
```

```
00007ffc`acc1cb37 8b45f3          mov     eax,dword ptr [rbp-0Dh]
00007ffc`acc1cb3a 89442420        mov     dword ptr [rsp+20h],eax
00007ffc`acc1cb3e 48ff158bfa0f00  call    qword ptr [mfc42u!_imp_CreateWindowExW (00007ffc`acd1c5d0)]
00007ffc`acc1cb45 0f1f440000      nop     dword ptr [rax+rax]
00007ffc`acc1cb4a 488bf0          mov     rsi,rax
00007ffc`acc1cb4d e8eea5ffff      call    mfc42u!AfxUnhookWindowCreate (00007ffc`acc17140)
00007ffc`acc1cb52 85c0            test    eax,eax
00007ffc`acc1cb54 0f84ecc00100    je      mfc42u!CWnd::CreateEx+0x1c1f6 (00007ffc`acc38c46)  Branch

mfc42u!CWnd::CreateEx+0x10a:
00007ffc`acc1cb5a 4885f6          test    rsi,rsi
00007ffc`acc1cb5d 400f95c7        setne   dil
00007ffc`acc1cb61 8bc7            mov     eax,edi

mfc42u!CWnd::CreateEx+0x113:
00007ffc`acc1cb63 4c8d9c24b0000000 lea     r11,[rsp+0B0h]
00007ffc`acc1cb6b 498b5b10        mov     rbx,qword ptr [r11+10h]
00007ffc`acc1cb6f 498b7318        mov     rsi,qword ptr [r11+18h]
00007ffc`acc1cb73 498b7b20        mov     rdi,qword ptr [r11+20h]
00007ffc`acc1cb77 498be3          mov     rsp,r11
00007ffc`acc1cb7a 5d              pop     rbp
00007ffc`acc1cb7b c3              ret

mfc42u!CWnd::CreateEx+0x1c1d2:
00007ffc`acc38c22 488b03          mov     rax,qword ptr [rbx]
00007ffc`acc38c25 49ba7011d03a0b36b89a mov r10,9AB8360B3AD01170h
00007ffc`acc38c2f 488b8058010000  mov     rax,qword ptr [rax+158h]
00007ffc`acc38c36 488bcb          mov     rcx,rbx
00007ffc`acc38c39 ff1539460e00    call    qword ptr [mfc42u!_guard_xfg_dispatch_icall_fptr (00007ffc`acd1d278)]
00007ffc`acc38c3f 33c0            xor     eax,eax
00007ffc`acc38c41 e91d3ffeff      jmp     mfc42u!CWnd::CreateEx+0x113 (00007ffc`acc1cb63)  Branch

mfc42u!CWnd::CreateEx+0x1c1f6:
00007ffc`acc38c46 488b0b          mov     rcx,qword ptr [rbx]
00007ffc`acc38c49 488b8158010000  mov     rax,qword ptr [rcx+158h]
00007ffc`acc38c50 49ba7011d03a0b36b89a mov r10,9AB8360B3AD01170h
00007ffc`acc38c5a 488bcb          mov     rcx,rbx
00007ffc`acc38c5d ff1515460e00    call    qword ptr [mfc42u!_guard_xfg_dispatch_icall_fptr (00007ffc`acd1d278)]
00007ffc`acc38c63 90              nop
00007ffc`acc38c64 e9f13efeff      jmp     mfc42u!CWnd::CreateEx+0x10a (00007ffc`acc1cb5a)  Branch
```

We see that 8 parameters are passed using the RSP pointer and consecutive offsets.

7. Let's look at an indirect call through IAT:

```
0:000> dps 00007ffc`acd1c5d0 L1
00007ffc`acd1c5d0  00007ffc`f5848030 user32!CreateWindowExW

0:000> !dh mfc42u

File Type: DLL
FILE HEADER VALUES
    8664 machine (X64)
       7 number of sections
F91A937D time date stamp Fri Jun  9 04:54:37 2102

       0 file pointer to symbol table
       0 number of symbols
      F0 size of optional header
    2022 characteristics
            Executable
            App can handle >2gb addresses
            DLL

OPTIONAL HEADER VALUES
     20B magic #
```

```
   14.28 linker version
   F2000 size of code
   81000 size of initialized data
       0 size of uninitialized data
   21730 address of entry point
    1000 base of code
         ----- new -----
00007ffcacc10000 image base
    1000 section alignment
    1000 file alignment
       3 subsystem (Windows CUI)
   10.00 operating system version
   10.00 image version
   10.00 subsystem version
  174000 size of image
    1000 size of headers
  17BE15 checksum
0000000000040000 size of stack reserve
0000000000001000 size of stack commit
0000000000100000 size of heap reserve
0000000000001000 size of heap commit
    4160  DLL characteristics
            High entropy VA supported
            Dynamic base
            NX compatible
            Guard
  139980 [    6CC0] address [size] of Export Directory
  140640 [     21C] address [size] of Import Directory
  163000 [    A4B8] address [size] of Resource Directory
  151000 [   100A4] address [size] of Exception Directory
       0 [       0] address [size] of Security Directory
  16E000 [    5734] address [size] of Base Relocation Directory
  116B60 [      70] address [size] of Debug Directory
       0 [       0] address [size] of Description Directory
       0 [       0] address [size] of Special Directory
   FD4A0 [      28] address [size] of Thread Storage Directory
   FBC00 [     138] address [size] of Load Configuration Directory
       0 [       0] address [size] of Bound Import Directory
  10BD18 [    1548] address [size] of Import Address Table Directory
  13838C [     180] address [size] of Delay Import Directory
       0 [       0] address [size] of COR20 Header Directory
       0 [       0] address [size] of Reserved Directory


SECTION HEADER #1
   .text name
   F124E virtual size
    1000 virtual address
   F2000 size of raw data
    1000 file pointer to raw data
       0 file pointer to relocation table
       0 file pointer to line numbers
       0 number of relocations
       0 number of line numbers
60000020 flags
         Code
         (no align specified)
         Execute Read
```

```
SECTION HEADER #2
  .rdata name
   51894 virtual size
   F3000 virtual address
   52000 size of raw data
   F3000 file pointer to raw data
       0 file pointer to relocation table
       0 file pointer to line numbers
       0 number of relocations
       0 number of line numbers
40000040 flags
         Initialized Data
         (no align specified)
         Read Only


Debug Directories(4)
      Type       Size      Address  Pointer
      cv           23       126810   126810    Format: RSDS, guid, 1, mfc42u.pdb
      (    13)    5e4       126834   126834
      (    16)     24       126e18   126e18
      dllchar       4       126e3c   126e3c

00000001 extended DLL characteristics
          CET compatible


SECTION HEADER #3
   .data name
    B19C virtual size
  145000 virtual address
    5000 size of raw data
  145000 file pointer to raw data
       0 file pointer to relocation table
       0 file pointer to line numbers
       0 number of relocations
       0 number of line numbers
C0000040 flags
         Initialized Data
         (no align specified)
         Read Write

SECTION HEADER #4
  .pdata name
   100A4 virtual size
  151000 virtual address
   11000 size of raw data
  14A000 file pointer to raw data
       0 file pointer to relocation table
       0 file pointer to line numbers
       0 number of relocations
       0 number of line numbers
40000040 flags
         Initialized Data
         (no align specified)
         Read Only

SECTION HEADER #5
  .didat name
```

```
     548 virtual size
  162000 virtual address
    1000 size of raw data
  15B000 file pointer to raw data
       0 file pointer to relocation table
       0 file pointer to line numbers
       0 number of relocations
       0 number of line numbers
C0000040 flags
         Initialized Data
         (no align specified)
         Read Write

SECTION HEADER #6
   .rsrc name
    A4B8 virtual size
  163000 virtual address
    B000 size of raw data
  15C000 file pointer to raw data
       0 file pointer to relocation table
       0 file pointer to line numbers
       0 number of relocations
       0 number of line numbers
40000040 flags
         Initialized Data
         (no align specified)
         Read Only

SECTION HEADER #7
  .reloc name
    5734 virtual size
  16E000 virtual address
    6000 size of raw data
  167000 file pointer to raw data
       0 file pointer to relocation table
       0 file pointer to line numbers
       0 number of relocations
       0 number of line numbers
42000040 flags
         Initialized Data
         Discardable
         (no align specified)
         Read Only
```

We see that the *mfc42u!_imp_CreateWindowExW* address is inside the IAT memory region of length 0x1548 bytes:

```
0:000> ? 00007ffcacc10000 + 10BD18
Evaluate expression: 140723207912728 = 00007ffc`acd1bd18

0:000> * mfc42u!_imp_CreateWindowExW (00007ffc`acd1c5d0)

0:000> ? 00007ffcacc10000 + 10BD18 + 1548
Evaluate expression: 140723207918176 = 00007ffc`acd1d260
```

8. We close logging before exiting WinDbg:

```
0:000> .logclose
Closing open log file C:\AWAPI-Dumps\W1.log
```

Note: To avoid possible confusion and glitches, we recommend exiting WinDbg after each exercise.

9. Let's now look at a 32-bit process. Launch the new instance of WinDbg.

10. Open \AWAPI-Dumps\Process\x86\wordpad.DMP

11. We get the dump file loaded:

```
Microsoft (R) Windows Debugger Version 10.0.27704.1001 X86
Copyright (c) Microsoft Corporation. All rights reserved.


Loading Dump File [C:\AWAPI-Dumps\Process\x86\wordpad.DMP]
User Mini Dump File with Full Memory: Only application data is available


************* Path validation summary **************
Response                         Time (ms)     Location
Deferred                                       srv*
Symbol search path is: srv*
Executable search path is:
Windows 10 Version 22000 MP (2 procs) Free x86 compatible
Product: WinNt, suite: SingleUserTS Personal
Edition build lab: 22000.1.amd64fre.co_release.210604-1628
Debug session time: Fri Nov  4 22:05:05.000 2022 (UTC + 1:00)
System Uptime: 0 days 0:39:06.687
Process Uptime: 20 days 0:46:35.000
................................................................
.....
Loading unloaded module list
.
For analysis of this file, run !analyze -v
eax=00000000 ebx=010593c8 ecx=00000000 edx=00000000 esi=010593fc edi=010593fc
eip=758c10cc esp=0013f870 ebp=0013f8a8 iopl=0         nv up ei pl nz ac po nc
cs=0023  ss=002b  ds=002b  es=002b  fs=0053  gs=002b             efl=00000212
win32u!NtUserGetMessage+0xc:
758c10cc ret     10h
```

12. Open the same log file using **.logappend**:

```
0:000> .logappend C:\AWAPI-Dumps\W1.log
Opened log file 'C:\AWAPI-Dumps\W1.log'
```

13. Let's look at the *InitMainThreadWnd* function from the *combase* module and find the *CreateWindowExW* API call:

```
0:000> uf combase!InitMainThreadWnd
0:000> uf combase!InitMainThreadWnd
combase!InitMainThreadWnd [onecore\com\combase\object\mainthrd.cxx @ 114]:
  114 75c032e3 8bff            mov     edi,edi
  114 75c032e5 51              push    ecx
  114 75c032e6 56              push    esi
  115 75c032e7 33f6            xor     esi,esi
  115 75c032e9 57              push    edi
  115 75c032ea 46              inc     esi
  125 75c032eb e8235ef8ff      call    combase!IsWOWThread (75b89113)
```

```
  125 75c032f0 85c0            test    eax,eax
  125 75c032f2 7546            jne     combase!InitMainThreadWnd+0x57 (75c0333a)  Branch

combase!InitMainThreadWnd+0x11 [onecore\com\combase\objact\mainthrd.cxx @ 148]:
  148 75c032f4 33ff            xor     edi,edi
  148 75c032f6 b800000080      mov     eax,80000000h
  148 75c032fb 57              push    edi
  148 75c032fc ff35e8ccd275    push    dword ptr [combase!g_hinst (75d2cce8)]
  148 75c03302 57              push    edi
  148 75c03303 6afd            push    0FFFFFFFDh
  148 75c03305 50              push    eax
  148 75c03306 50              push    eax
  148 75c03307 50              push    eax
  148 75c03308 50              push    eax
  148 75c03309 6800000088      push    88000000h
  148 75c0330e 6870f0b075      push    offset combase!`string' (75b0f070)
  148 75c03313 ff353cd3d275    push    dword ptr [combase!gOleWindowClass (75d2d33c)]
  148 75c03319 57              push    edi
  148 75c0331a ff15e842d375    call    dword ptr [combase!_imp__CreateWindowExW (75d342e8)]
  148 75c03320 a340d3d275      mov     dword ptr [combase!ghwndOleMainThread (75d2d340)],eax
  163 75c03325 85c0            test    eax,eax
  163 75c03327 0f848aa30800    je      combase!InitMainThreadWnd+0x8a3d4 (75c8d6b7)  Branch

combase!InitMainThreadWnd+0x4a [onecore\com\combase\objact\mainthrd.cxx @ 180]:
  180 75c0332d ff15ecf3d275    call    dword ptr [combase!_imp__GetCurrentThreadId (75d2f3ec)]
  180 75c03333 a334d3d275      mov     dword ptr [combase!gdwMainThreadId (75d2d334)],eax
  181 75c03338 8bf7            mov     esi,edi

combase!InitMainThreadWnd+0x57 [onecore\com\combase\objact\mainthrd.cxx @ 187]:
  187 75c0333a 5f              pop     edi
  187 75c0333b 8bc6            mov     eax,esi
  187 75c0333d 5e              pop     esi
  187 75c0333e 59              pop     ecx
  187 75c0333f c3              ret

combase!InitMainThreadWnd+0x8a3d4 [onecore\com\combase\objact\mainthrd.cxx @ 168]:
  168 75c8d6b7 ff1538f2d275    call    dword ptr [combase!_imp__GetLastError (75d2f238)]
  168 75c8d6bd 8bf0            mov     esi,eax
  168 75c8d6bf 85f6            test    esi,esi
  168 75c8d6c1 7e09            jle     combase!InitMainThreadWnd+0x8a3e9 (75c8d6cc)  Branch

combase!InitMainThreadWnd+0x8a3e0 [onecore\com\combase\objact\mainthrd.cxx @ 168]:
  168 75c8d6c3 0fb7f6          movzx   esi,si
  168 75c8d6c6 81ce00000780    or      esi,80070000h

combase!InitMainThreadWnd+0x8a3e9 [onecore\com\combase\objact\mainthrd.cxx @ 169]:
  169 75c8d6cc 393d70c2d275    cmp     dword ptr [combase!_Tlgg_hCombaseTraceLoggingProviderProv (75d2c270)],edi
  169 75c8d6d2 7713            ja      combase!InitMainThreadWnd+0x8a404 (75c8d6e7)  Branch

combase!InitMainThreadWnd+0x8a3f1 [onecore\com\combase\objact\mainthrd.cxx @ 169]:
  169 75c8d6d4 393d48ced275    cmp     dword ptr [combase!gfEnableTracing (75d2ce48)],edi
  169 75c8d6da 742a            je      combase!InitMainThreadWnd+0x8a423 (75c8d706)  Branch

combase!InitMainThreadWnd+0x8a3f9 [onecore\com\combase\objact\mainthrd.cxx @ 169]:
  169 75c8d6dc 33c9            xor     ecx,ecx
  169 75c8d6de e86a4df2ff      call    combase!IsWppLevelEnabled (75bb244d)
  169 75c8d6e3 84c0            test    al,al
  169 75c8d6e5 741f            je      combase!InitMainThreadWnd+0x8a423 (75c8d706)  Branch

combase!InitMainThreadWnd+0x8a404 [onecore\com\combase\objact\mainthrd.cxx @ 169]:
  169 75c8d6e7 56              push    esi
  169 75c8d6e8 682cabb075      push    offset combase!`string' (75b0ab2c)
  169 75c8d6ed 57              push    edi
  169 75c8d6ee 68a9000000      push    0A9h
  169 75c8d6f3 6854eeb375      push    offset combase!`string' (75b3ee54)
  169 75c8d6f8 687ce6b175      push    offset combase!`string' (75b1e67c)
  169 75c8d6fd 56              push    esi
  169 75c8d6fe e8662dedff      call    combase!ComTraceMessage (75b60469)
  169 75c8d703 83c41c          add     esp,1Ch

combase!InitMainThreadWnd+0x8a423 [onecore\com\combase\objact\mainthrd.cxx @ 170]:
  170 75c8d706 85f6            test    esi,esi
  170 75c8d708 0f882c5cf7ff    js      combase!InitMainThreadWnd+0x57 (75c0333a)  Branch
```

```
combase!InitMainThreadWnd+0x8a42b [onecore\com\combase\objact\mainthrd.cxx @ 172]:
  172 75c8d70e be0e000780      mov     esi,8007000Eh
  176 75c8d713 e9225cf7ff      jmp     combase!InitMainThreadWnd+0x57 (75c0333a)  Branch
```

To remind you, the API function definition is:

```
HWND CreateWindowExW(
  [in]           DWORD     dwExStyle,
  [in, optional] LPCWSTR   lpClassName,
  [in, optional] LPCWSTR   lpWindowName,
  [in]           DWORD     dwStyle,
  [in]           int       X,
  [in]           int       Y,
  [in]           int       nWidth,
  [in]           int       nHeight,
  [in, optional] HWND      hWndParent,
  [in, optional] HMENU     hMenu,
  [in, optional] HINSTANCE hInstance,
  [in, optional] LPVOID    lpParam
);
```

We see that all parameters are pushed right-to-left with the third parameter from the left containing the 64-bit pointer to a window name (UNICODE):

```
0:000> du 75b0f070
75b0f070  "OleMainThreadWndName"
```

14. We close logging before exiting WinDbg:

```
0:000> .logclose
Closing open log file C:\AWAPI-Dumps\W1.log
```

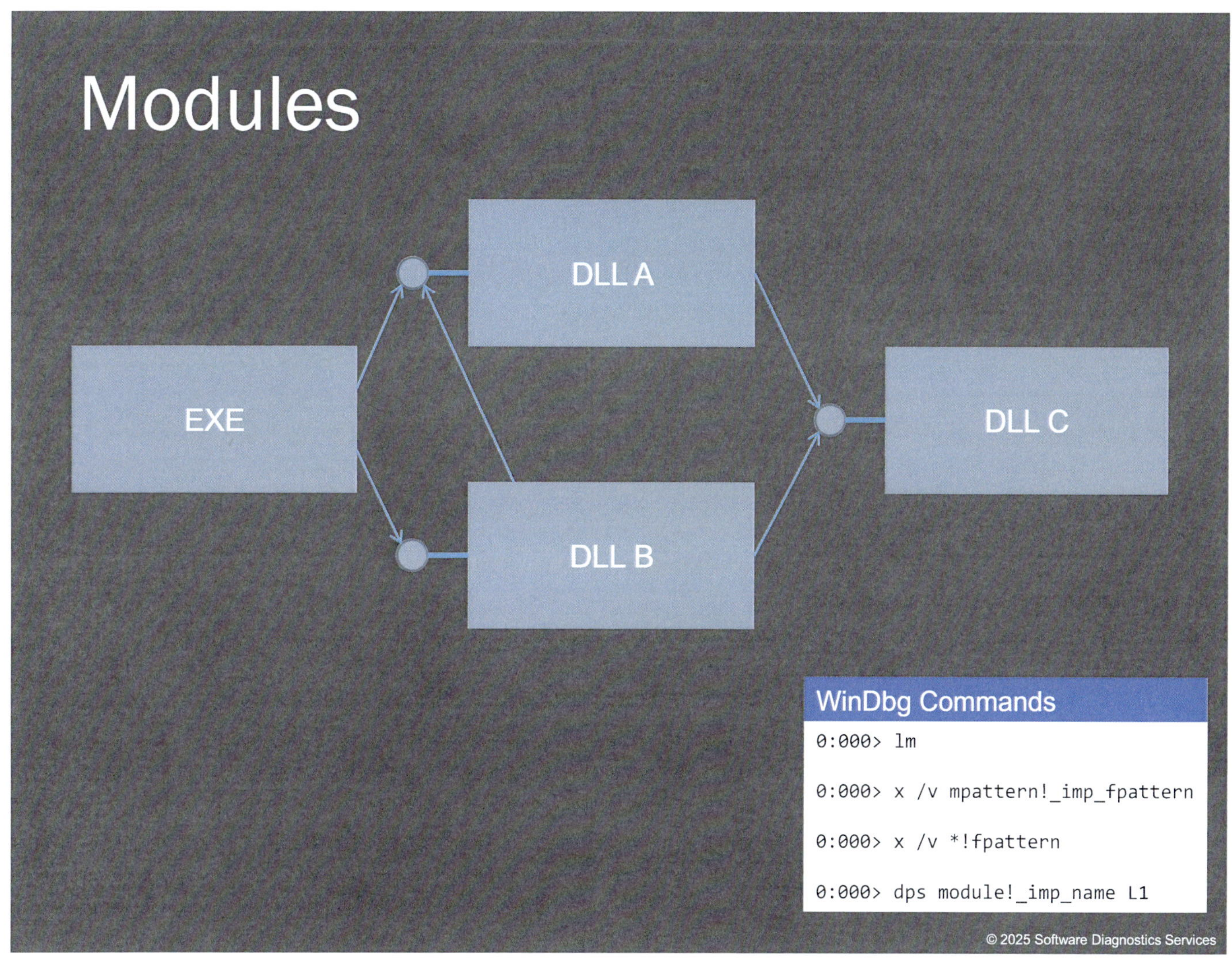
Modules
DLL A
EXE
DLL C
DLL B
WinDbg Commands
0:000> lm
0:000> x /v mpattern!_imp_fpattern
0:000> x /v *!fpattern
0:000> dps module!_imp_name L1
© 2025 Software Diagnostics Services

Modules can import functions from other modules. But function names are not unique. Several modules may export the same function names. Such functions may or may not have the same parameters. So, function syntax, semantics, and pragmatics are subject to API design. Also, not all functions from modules are imported – only those that are actually used in code.

Modules and Analysis Patterns

- Module memory analysis patterns
  - Module Collection
  - Coupled Modules
  - Duplicated Module
- Namespace malware analysis pattern

There are many memory analysis patterns related to modules. On this slide, I list a few of them related to module dependencies and imported functions. **Module Collection** lists loaded modules. **Coupled Modules** is about module dependencies. Sometimes, the same module name may be loaded twice, for example, from different locations, the so-called **Duplicated Module**. Imported functions constitute the so-called **Namespace** malware analysis pattern which can give hints at overall module functionality and purpose. For example, an innocuous module name may have functions imported from GDI that reveal potential screen-grabbing functionality or network connectivity.

Module memory analysis patterns
https://www.dumpanalysis.org/blog/index.php/2012/07/15/module-patterns/

Module Collection
https://www.dumpanalysis.org/blog/index.php/2012/12/24/crash-dump-analysis-patterns-part-190/

Coupled Modules
https://www.dumpanalysis.org/blog/index.php/2011/06/20/crash-dump-analysis-patterns-part-140/

Duplicated Module
https://www.dumpanalysis.org/blog/index.php/2008/06/19/crash-dump-analysis-patterns-part-64/

Namespace malware analysis pattern
https://www.dumpanalysis.org/blog/index.php/2013/02/05/malware-analysis-patterns-part-20/

Note: all referenced patterns from dumpanalysis.org (and hundreds of others) are also available in **Memory Dump Analysis Anthology** volumes or **Encyclopedia of Crash Dump Analysis Patterns** (see References slides).

Exercise W2

- **Goal:** Explore modules and their dependencies
- **Memory Analysis Patterns:** Module Collection; Coupled Modules
- **Malware Analysis Patterns:** Namespace
- \AWAPI-Dumps\Exercise-W2.pdf

Exercise W2

Goal: Explore modules and their dependencies.

Memory Analysis Patterns: Module Collection; Coupled Modules.

Malware Analysis Patterns: Namespace.

1. Launch WinDbg.

2. Open \AWAPI-Dumps\Process\wordpad.DMP

3. We get the dump file loaded:

```
Microsoft (R) Windows Debugger Version 10.0.27725.1000 AMD64
Copyright (c) Microsoft Corporation. All rights reserved.


Loading Dump File [C:\AWAPI-Dumps\Process\wordpad.DMP]
User Mini Dump File with Full Memory: Only application data is available


************* Path validation summary **************
Response                         Time (ms)     Location
Deferred                                       srv*
Symbol search path is: srv*
Executable search path is:
Windows 10 Version 22000 MP (2 procs) Free x64
Product: WinNt, suite: SingleUserTS Personal
Edition build lab: 22000.1.amd64fre.co_release.210604-1628
Debug session time: Sat Oct 15 20:19:05.000 2022 (UTC + 0:00)
System Uptime: 0 days 0:27:57.202
Process Uptime: 0 days 0:00:26.000
................................................................
.....
Loading unloaded module list
.
For analysis of this file, run !analyze -v
win32u!NtUserGetMessage+0x14:
00007ffc`f3811414 ret
```

4. Open a log file using **.logopen**:

```
0:000> .logopen C:\AWAPI-Dumps\W2.log
Opened log file 'C:\AWAPI-Dumps\W2.log'
```

5. We can get the list of loaded modules and their addresses in memory using the **lm** command:

```
0:000> lm
start             end                 module name
00007ff7`57640000 00007ff7`5792d000   wordpad    (deferred)
00007ffc`abd60000 00007ffc`ac19c000   UIRibbon   (deferred)
00007ffc`acc10000 00007ffc`acd84000   mfc42u     (deferred)
00007ffc`bcad0000 00007ffc`bcd3e000   msxml3     (deferred)
```

```
00007ffc`cf460000 00007ffc`cf46d000   atlthunk   (deferred)
00007ffc`cf470000 00007ffc`cf623000   GdiPlus    (deferred)
00007ffc`d0010000 00007ffc`d0071000   AcGenral   (deferred)
00007ffc`d3500000 00007ffc`d3526000   globinputhost   (deferred)
00007ffc`d35d0000 00007ffc`d3936000   msftedit   (deferred)
00007ffc`d85a0000 00007ffc`d863b000   winspool   (deferred)
00007ffc`d8d90000 00007ffc`d9012000   msxml6     (deferred)
00007ffc`da390000 00007ffc`da3ed000   dataexchange   (deferred)
00007ffc`da670000 00007ffc`da6d9000   oleacc     (deferred)
00007ffc`db8f0000 00007ffc`db95a000   ninput     (deferred)
00007ffc`e1880000 00007ffc`e1b25000   comctl32   (deferred)
00007ffc`e22d0000 00007ffc`e22ed000   mpr        (deferred)
00007ffc`e74c0000 00007ffc`e756e000   TextShaping   (deferred)
00007ffc`e7570000 00007ffc`e7582000   npmproxy   (deferred)
00007ffc`e7800000 00007ffc`e792d000   textinputframework   (deferred)
00007ffc`e85b0000 00007ffc`e85e3000   winmm      (deferred)
00007ffc`e8690000 00007ffc`e8852000   Windows_Globalization   (deferred)
00007ffc`ea8a0000 00007ffc`eaa4e000   windowscodecs   (deferred)
00007ffc`eaac0000 00007ffc`eac48000   Windows_UI   (deferred)
00007ffc`eb930000 00007ffc`eb995000   Bcp47Langs   (deferred)
00007ffc`ec290000 00007ffc`ec4f6000   twinapi_appcore   (deferred)
00007ffc`ec700000 00007ffc`ec728000   srvcli     (deferred)
00007ffc`ec7d0000 00007ffc`eca82000   iertutil   (deferred)
00007ffc`eca90000 00007ffc`ecc7e000   urlmon     (deferred)
00007ffc`ee0e0000 00007ffc`ee151000   netprofm   (deferred)
00007ffc`ee420000 00007ffc`ee4b2000   msvcp110_win   (deferred)
00007ffc`ee5f0000 00007ffc`ee95d000   CoreUIComponents   (deferred)
00007ffc`ef420000 00007ffc`ef457000   xmllite    (deferred)
00007ffc`f0360000 00007ffc`f0492000   CoreMessaging   (deferred)
00007ffc`f0990000 00007ffc`f0a21000   apphelp    (deferred)
00007ffc`f0a50000 00007ffc`f0afc000   uxtheme    (deferred)
00007ffc`f0cf0000 00007ffc`f0d1f000   dwmapi     (deferred)
00007ffc`f14f0000 00007ffc`f15e7000   propsys    (deferred)
00007ffc`f1790000 00007ffc`f18f6000   WinTypes   (deferred)
00007ffc`f1900000 00007ffc`f2167000   windows_storage   (deferred)
00007ffc`f2290000 00007ffc`f229c000   netutils   (deferred)
00007ffc`f2890000 00007ffc`f28a8000   kernel_appcore   (deferred)
00007ffc`f2ab0000 00007ffc`f2af2000   sspicli    (deferred)
00007ffc`f2db0000 00007ffc`f2dd9000   userenv    (deferred)
00007ffc`f2f20000 00007ffc`f2f2c000   CRYPTBASE   (deferred)
00007ffc`f3100000 00007ffc`f3127000   bcrypt     (deferred)
00007ffc`f3790000 00007ffc`f380f000   bcryptPrimitives   (deferred)
00007ffc`f3810000 00007ffc`f3836000   win32u     (pdb symbols)
C:\WinDbg.Docker.AWAPI\mss\win32u.pdb\045A07FC5CC3A90DFCCC8B4C1918F7421\win32u.pdb
00007ffc`f38b0000 00007ffc`f394d000   msvcp_win   (deferred)
00007ffc`f3950000 00007ffc`f3a61000   ucrtbase   (deferred)
00007ffc`f3b30000 00007ffc`f3ea4000   KERNELBASE   (deferred)
00007ffc`f3eb0000 00007ffc`f3fc2000   gdi32full   (deferred)
00007ffc`f4140000 00007ffc`f41e3000   msvcrt     (deferred)
00007ffc`f41f0000 00007ffc`f4219000   gdi32      (deferred)
00007ffc`f4230000 00007ffc`f434e000   msctf      (deferred)
00007ffc`f4350000 00007ffc`f4470000   rpcrt4     (deferred)
00007ffc`f4640000 00007ffc`f46ef000   clbcatq    (deferred)
00007ffc`f46f0000 00007ffc`f4e9e000   shell32    (deferred)
00007ffc`f4fc0000 00007ffc`f507d000   kernel32   (deferred)
00007ffc`f5080000 00007ffc`f516a000   SHCore     (deferred)
00007ffc`f5240000 00007ffc`f52ee000   advapi32   (deferred)
00007ffc`f52f0000 00007ffc`f538e000   sechost    (deferred)
```

```
00007ffc`f5840000 00007ffc`f59ec000   user32     (pdb symbols)
C:\WinDbg.Docker.AWAPI\mss\user32.pdb\9479B9C8D8218B8972152084F4D6840C1\user32.pdb
00007ffc`f59f0000 00007ffc`f5adc000   comdlg32   (deferred)
00007ffc`f5b40000 00007ffc`f5c16000   oleaut32   (deferred)
00007ffc`f5c20000 00007ffc`f5c7d000   shlwapi    (deferred)
00007ffc`f5c80000 00007ffc`f5cb1000   imm32      (deferred)
00007ffc`f5d30000 00007ffc`f60a8000   combase    (deferred)
00007ffc`f60c0000 00007ffc`f625a000   ole32      (deferred)
00007ffc`f62a0000 00007ffc`f64a9000   ntdll      (deferred)
```

6. We can list all functions imported by a particular module from all other modules using the **x** command and *_imp_** pattern. We use the **/v** parameter to find out **pub global** functions. These functions are the imported functions:

```
0:000> x /v wordpad!_imp_*
pub global 00007ff7`576f6688    0 wordpad!_imp_?DeleteCWinThreadUEAAXXZ = <no type information>
pub global 00007ff7`576f62d0    0 wordpad!_imp_?GetRuntimeClassCPenUEBAPEAUCRuntimeClassXZ = <no type information>
pub global 00007ff7`576f6210    0 wordpad!_imp_?GetPositionCFileUEBAKXZ = <no type information>
pub global 00007ff7`576f4dd8    0 wordpad!_imp_EventSetInformation = <no type information>
pub global 00007ff7`576f6508    0 wordpad!_imp_??1CCommandLineInfoUEAAXZ = <no type information>
pub global 00007ff7`576f58a8    0 wordpad!_imp_??0CFindReplaceDialogQEAAXZ = <no type information>
pub global 00007ff7`576f4f80    0 wordpad!_imp_TextOutW = <no type information>
pub global 00007ff7`576f6970    0 wordpad!_imp_?IsSelectedCViewUEBAHPEBVCObjectZ = <no type information>
...
pub global 00007ff7`576f6fe0    0 wordpad!_imp_free = <no type information>
pub global 00007ff7`576f6a78    0 wordpad!_imp_ShellExecuteW = <no type information>
pub global 00007ff7`576f66a0    0 wordpad!_imp_?SaveAllModifiedCWinAppUEAAHXZ = <no type information>
pub global 00007ff7`576f5e70    0 wordpad!_imp_?GetRuntimeClassCListBoxUEBAPEAUCRuntimeClassXZ = <no type information>
pub global 00007ff7`576f70f0    0 wordpad!_imp_memcpy_s = <no type information>
pub global 00007ff7`576f4fb0    0 wordpad!_imp_GetDeviceCaps = <no type information>
pub global 00007ff7`576f63d0    0 wordpad!_imp_?DDV_MinMaxIntYAXPEAVCDataExchangeHHHZ = <no type information>
pub global 00007ff7`576f6150    0 wordpad!_imp_?GetPrinterDeviceDefaultsCWinAppQEAAHPEAUtagPDWZ = <no type
information>
pub global 00007ff7`576f6948    0 wordpad!_imp_?SetScaleRatioCPreviewDCQEAAXHHZ = <no type information>
pub global 00007ff7`576f64f0    0 wordpad!_imp_?OnSetDataCOleServerItemUEAAHPEAUtagFORMATETCPEAUtagSTGMEDIUMHZ = <no
type information>
pub global 00007ff7`576f5088    0 wordpad!_imp_AddAtomW = <no type information>
pub global 00007ff7`576f5a50    0 wordpad!_imp_?OnNotifyCWndMEAAH_K_JPEA_JZ = <no type information>
pub global 00007ff7`576f6038    0 wordpad!_imp_?SetTextColorCDCUEAAKKZ = <no type information>
pub global 00007ff7`576f6c88    0 wordpad!_imp_GetPropW = <no type information>
pub global 00007ff7`576f6000    0 wordpad!_imp_?ScaleWindowExtCDCUEAA?AVCSizeHHHHZ = <no type information>
pub global 00007ff7`57745040    0 wordpad!_imp_GdipGetImageGraphicsContext = <no type information>
pub global 00007ff7`576f60d0    0 wordpad!_imp_?ActivateFrameCFrameWndUEAAXHZ = <no type information>
pub func   00007ff7`5764c8a0    0 wordpad!_imp_load_GdipCloneImage (__imp_load_GdipCloneImage)
pub global 00007ff7`576f57c0    0 wordpad!_imp_??0COleDataSourceQEAAXZ = <no type information>
pub global 00007ff7`576f5fc0    0 wordpad!_imp_?messageMapCFrameWnd = <no type information>
pub global 00007ff7`576f6f68    0 wordpad!_imp_?gbump?$basic_streambufGU?$char_traitsGstdstdIEAAXHZ = <no type
information>
pub global 00007ff7`576f4ff0    0 wordpad!_imp_ReadFile = <no type information>
pub global 00007ff7`576f5c20    0 wordpad!_imp_?OnShowDocumentCOleServerDocMEAAXHZ = <no type information>
pub global 00007ff7`576f6ab0    0 wordpad!_imp_PathFindFileNameW = <no type information>
pub global 00007ff7`576f6778    0 wordpad!_imp_?SubclassWindowCWndQEAAHPEAUHWND__Z = <no type information>
```

7. For any such a function, we can find modules that export it using the same **x** command and the * module name pattern, and the function name without the *_imp_* prefix. We are interested in **pub** entries only:

```
0:000> x /v *!TextOutW
prv func   00007ffc`f3c8a460   1d KERNELBASE!TextOutW (void)
prv func   00007ffc`f3eb5c00   66 gdi32full!TextOutW (void)
pub func   00007ffc`f41f7610    0 gdi32!TextOutW (TextOutW)

0:000> x /v *!ReadFile
prv func   00007ffc`f3b64420  19b KERNELBASE!ReadFile (void)
pub func   00007ffc`f47fb8a4    0 shell32!ReadFile (ReadFile)
pub func   00007ffc`f4fe3270    0 kernel32!ReadFile (ReadFile)
```

8. To see from which module the function is imported, we can dereference the indirect *_imp_* address:

```
0:000> dps wordpad!_imp_ReadFile L1
00007ff7`576f4ff0  00007ffc`f4fe3270 kernel32!ReadFile
```

Note: We also see functions imported from C language runtime. These functions are not considered Windows API. They also have a different naming convention:

```
0:000> dps wordpad!_imp_free L1
00007ff7`576f6fe0  00007ffc`f415c690 msvcrt!free
```

```
0:000> dps wordpad!_imp_memcpy_s L1
00007ff7`576f70f0  00007ffc`f41a05e0 msvcrt!memcpy_s
```

9. Let's see which modules use *ReadFile* and *CreateWindowExW* API functions:

```
0:000> x /v *!_imp_ReadFile
pub global 00007ff7`576f4ff0    0 wordpad!_imp_ReadFile = <no type information>
pub global 00007ffc`ac007560    0 UIRibbon!_imp_ReadFile = <no type information>
pub global 00007ffc`acd1cbc8    0 mfc42u!_imp_ReadFile = <no type information>
pub global 00007ffc`bcc34980    0 msxml3!_imp_ReadFile = <no type information>
pub global 00007ffc`cf5cde50    0 GdiPlus!_imp_ReadFile = <no type information>
pub global 00007ffc`d38add20    0 msftedit!_imp_ReadFile = <no type information>
pub global 00007ffc`d8603528    0 winspool!_imp_ReadFile = <no type information>
pub global 00007ffc`d8f72470    0 msxml6!_imp_ReadFile = <no type information>
pub global 00007ffc`ea9fe5f8    0 windowscodecs!_imp_ReadFile = <no type information>
pub global 00007ffc`ecbd6ce8    0 urlmon!_imp_ReadFile = <no type information>
pub global 00007ffc`f0441320    0 CoreMessaging!_imp_ReadFile = <no type information>
pub global 00007ffc`f0ac0a08    0 uxtheme!_imp_ReadFile = <no type information>
pub global 00007ffc`f1fc5380    0 windows_storage!_imp_ReadFile = <no type information>
pub global 00007ffc`f3a186f0    0 ucrtbase!_imp_ReadFile = <no type information>
pub global 00007ffc`f3f5dc80    0 gdi32full!_imp_ReadFile = <no type information>
pub global 00007ffc`f41bb200    0 msvcrt!_imp_ReadFile = <no type information>
pub global 00007ffc`f46c1030    0 clbcatq!_imp_ReadFile = <no type information>
pub global 00007ffc`f4d3ad80    0 shell32!_imp_ReadFile = <no type information>
pub global 00007ffc`f5040310    0 kernel32!_imp_ReadFile = <no type information>
pub global 00007ffc`f51307f8    0 SHCore!_imp_ReadFile = <no type information>
pub global 00007ffc`f5363f58    0 sechost!_imp_ReadFile = <no type information>
pub global 00007ffc`f58d6178    0 user32!_imp_ReadFile = <no type information>
pub global 00007ffc`f61a4bc0    0 ole32!_imp_ReadFile = <no type information>
```

```
0:000> x /v *!_imp_CreateWindowExW
pub global 00007ffc`ac007790    0 UIRibbon!_imp_CreateWindowExW = <no type information>
pub global 00007ffc`acd1c5d0    0 mfc42u!_imp_CreateWindowExW = <no type information>
pub global 00007ffc`bcd1e288    0 msxml3!_imp_CreateWindowExW = <no type information>
pub global 00007ffc`cf5cdce0    0 GdiPlus!_imp_CreateWindowExW = <no type information>
pub global 00007ffc`d39147b8    0 msftedit!_imp_CreateWindowExW = <no type information>
pub global 00007ffc`da6d0318    0 oleacc!_imp_CreateWindowExW = <no type information>
pub global 00007ffc`db954178    0 ninput!_imp_CreateWindowExW = <no type information>
pub global 00007ffc`e1a8d560    0 comctl32!_imp_CreateWindowExW = <no type information>
pub global 00007ffc`e7928130    0 textinputframework!_imp_CreateWindowExW = <no type information>
pub global 00007ffc`eac39448    0 Windows_UI!_imp_CreateWindowExW = <no type information>
pub global 00007ffc`ecc27640    0 urlmon!_imp_CreateWindowExW = <no type information>
pub global 00007ffc`ee90e250    0 CoreUIComponents!_imp_CreateWindowExW = <no type information>
pub global 00007ffc`f04842a8    0 CoreMessaging!_imp_CreateWindowExW = <no type information>
pub global 00007ffc`f2141a28    0 windows_storage!_imp_CreateWindowExW = <no type information>
pub global 00007ffc`f434a428    0 msctf!_imp_CreateWindowExW = <no type information>
pub global 00007ffc`f4d3a490    0 shell32!_imp_CreateWindowExW = <no type information>
pub global 00007ffc`f51641d0    0 SHCore!_imp_CreateWindowExW = <no type information>
```

```
pub global 00007ffc`f5aac638    0 comdlg32!_imp_CreateWindowExW = <no type information>
pub global 00007ffc`f5caa220    0 imm32!_imp_CreateWindowExW = <no type information>
pub global 00007ffc`f6086608    0 combase!_imp_CreateWindowExW = <no type information>
pub global 00007ffc`f61a47a0    0 ole32!_imp_CreateWindowExW = <no type information>

0:000> x /v *!CreateWindowExW
prv func   00007ffc`f3c999b0   1d KERNELBASE!CreateWindowExW (void)
pub func   00007ffc`f5848030    0 user32!CreateWindowExW (CreateWindowExW)
```

Note: We see that the *wordpad* module doesn't create windows itself, only through other modules.

10. We can see all module-qualified imported functions by dumping IAT for a specific module:

```
0:000> !dh UIRibbon

File Type: DLL
FILE HEADER VALUES
    8664 machine (X64)
       8 number of sections
9A0B9171 time date stamp Fri Nov 24 14:36:33 2051

       0 file pointer to symbol table
       0 number of symbols
      F0 size of optional header
    2022 characteristics
            Executable
            App can handle >2gb addresses
            DLL

OPTIONAL HEADER VALUES
     20B magic #
   14.28 linker version
  264000 size of code
  1D7000 size of initialized data
       0 size of uninitialized data
    8560 address of entry point
    1000 base of code
         ----- new -----
00007ffcabd60000 image base
    1000 section alignment
    1000 file alignment
       3 subsystem (Windows CUI)
   10.00 operating system version
   10.00 image version
   10.00 subsystem version
  43C000 size of image
    1000 size of headers
  43B025 checksum
0000000000040000 size of stack reserve
0000000000001000 size of stack commit
0000000000100000 size of heap reserve
0000000000001000 size of heap commit
    4160  DLL characteristics
            High entropy VA supported
            Dynamic base
            NX compatible
            Guard
  3AA2F0 [      80] address [size] of Export Directory
```

```
  3AA370 [       F0] address [size] of Import Directory
  3EB000 [    3F020] address [size] of Resource Directory
  3D3000 [    15F0C] address [size] of Exception Directory
       0 [        0] address [size] of Security Directory
  42B000 [    10E58] address [size] of Base Relocation Directory
  375B60 [       70] address [size] of Debug Directory
       0 [        0] address [size] of Description Directory
       0 [        0] address [size] of Special Directory
  2A6DA8 [       28] address [size] of Thread Storage Directory
  2A6C70 [      138] address [size] of Load Configuration Directory
       0 [        0] address [size] of Bound Import Directory
  2A6DD0 [     1350] address [size] of Import Address Table Directory
  3A999C [       E0] address [size] of Delay Import Directory
       0 [        0] address [size] of COR20 Header Directory
       0 [        0] address [size] of Reserved Directory
```

```
SECTION HEADER #1
   .text name
  263A5B virtual size
    1000 virtual address
  264000 size of raw data
    1000 file pointer to raw data
       0 file pointer to relocation table
       0 file pointer to line numbers
       0 number of relocations
       0 number of line numbers
60000020 flags
         Code
         (no align specified)
         Execute Read

SECTION HEADER #2
  .rdata name
  148FA8 virtual size
  265000 virtual address
  149000 size of raw data
  265000 file pointer to raw data
       0 file pointer to relocation table
       0 file pointer to line numbers
       0 number of relocations
       0 number of line numbers
40000040 flags
         Initialized Data
         (no align specified)
         Read Only


Debug Directories(4)
      Type       Size     Address  Pointer
      cv           25      39fe1c   39fe1c    Format: RSDS, guid, 1, UIRibbon.pdb
      (   13)     3fc      39fe44   39fe44
      (   16)      24      3a0240   3a0240
      dllchar       4      3a0264   3a0264

00000001 extended DLL characteristics
          CET compatible
```

```
SECTION HEADER #3
   .data name
   24CC0 virtual size
  3AE000 virtual address
   22000 size of raw data
  3AE000 file pointer to raw data
       0 file pointer to relocation table
       0 file pointer to line numbers
       0 number of relocations
       0 number of line numbers
C0000040 flags
         Initialized Data
         (no align specified)
         Read Write

SECTION HEADER #4
  .pdata name
   15F0C virtual size
  3D3000 virtual address
   16000 size of raw data
  3D0000 file pointer to raw data
       0 file pointer to relocation table
       0 file pointer to line numbers
       0 number of relocations
       0 number of line numbers
40000040 flags
         Initialized Data
         (no align specified)
         Read Only

SECTION HEADER #5
  .didat name
     1D0 virtual size
  3E9000 virtual address
    1000 size of raw data
  3E6000 file pointer to raw data
       0 file pointer to relocation table
       0 file pointer to line numbers
       0 number of relocations
       0 number of line numbers
C0000040 flags
         Initialized Data
         (no align specified)
         Read Write

SECTION HEADER #6
.bootdat name
     E40 virtual size
  3EA000 virtual address
    1000 size of raw data
  3E7000 file pointer to raw data
       0 file pointer to relocation table
       0 file pointer to line numbers
       0 number of relocations
       0 number of line numbers
C0000040 flags
         Initialized Data
         (no align specified)
         Read Write
```

```
SECTION HEADER #7
   .rsrc name
   3F020 virtual size
  3EB000 virtual address
   40000 size of raw data
  3E8000 file pointer to raw data
       0 file pointer to relocation table
       0 file pointer to line numbers
       0 number of relocations
       0 number of line numbers
40000040 flags
         Initialized Data
         (no align specified)
         Read Only

SECTION HEADER #8
  .reloc name
   10E58 virtual size
  42B000 virtual address
   11000 size of raw data
  428000 file pointer to raw data
       0 file pointer to relocation table
       0 file pointer to line numbers
       0 number of relocations
       0 number of line numbers
42000040 flags
         Initialized Data
         Discardable
         (no align specified)
         Read Only
```

```
0:000> dps 00007ffcabd60000 + 2A6DD0 L1350/8
00007ffc`ac006dd0  00007ffc`f5246b40 advapi32!IsTextUnicode
00007ffc`ac006dd8  00007ffc`f525f970 advapi32!RegSetKeyValueWStub
00007ffc`ac006de0  00007ffc`f5248d20 advapi32!RegGetValueWStub
00007ffc`ac006de8  00007ffc`f5246b20 advapi32!RegCloseKeyStub
00007ffc`ac006df0  00007ffc`f5247310 advapi32!RegQueryValueExAStub
00007ffc`ac006df8  00007ffc`f52472d0 advapi32!RegOpenKeyExAStub
00007ffc`ac006e00  00007ffc`f62a4f40 ntdll!EtwEventWriteTransfer
00007ffc`ac006e08  00007ffc`f62b5520 ntdll!EtwEventSetInformation
00007ffc`ac006e10  00007ffc`f62b59f0 ntdll!EtwEventRegister
00007ffc`ac006e18  00007ffc`f62a65e0 ntdll!EtwEventUnregister
00007ffc`ac006e20  00007ffc`f62a6590 ntdll!EtwUnregisterTraceGuids
00007ffc`ac006e28  00007ffc`f632e130 ntdll!EtwLogTraceEvent
00007ffc`ac006e30  00007ffc`f62b5360 ntdll!EtwRegisterTraceGuidsW
00007ffc`ac006e38  00007ffc`f6327240 ntdll!EtwGetTraceEnableFlags
00007ffc`ac006e40  00007ffc`f6327200 ntdll!EtwGetTraceEnableLevel
00007ffc`ac006e48  00007ffc`f63271c0 ntdll!EtwGetTraceLoggerHandle
00007ffc`ac006e50  00000000`00000000
00007ffc`ac006e58  00007ffc`f41f4e90 gdi32!SetBitmapBitsStub
00007ffc`ac006e60  00007ffc`f41f5880 gdi32!RealizePaletteStub
00007ffc`ac006e68  00007ffc`f41f16e0 gdi32!SetMapModeStub
00007ffc`ac006e70  00007ffc`f41fbc60 gdi32!DeleteEnhMetaFileStub
00007ffc`ac006e78  00007ffc`f41f1740 gdi32!OffsetRgn
00007ffc`ac006e80  00007ffc`f3ec3a50 gdi32full!ScriptBreak
00007ffc`ac006e88  00007ffc`f41f3fb0 gdi32!GetTextMetricsWStub
00007ffc`ac006e90  00007ffc`f41f6b50 gdi32!CreateDIBPatternBrushPt
00007ffc`ac006e98  00007ffc`f41f45c0 gdi32!LPtoDPStub
00007ffc`ac006ea0  00007ffc`f41f41b0 gdi32!PatBltStub
00007ffc`ac006ea8  00007ffc`f41fbcb0 gdi32!EllipseStub
00007ffc`ac006eb0  00007ffc`f41fb890 gdi32!CreatePolygonRgn
00007ffc`ac006eb8  00007ffc`f41f43e0 gdi32!SetViewportOrgExStub
00007ffc`ac006ec0  00007ffc`f41f5060 gdi32!GetBkColorStub
00007ffc`ac006ec8  00007ffc`f41f5010 gdi32!GetTextColorStub
```

```
00007ffc`ac006ed0  00007ffc`f41f13a0 gdi32!CreateDCW
00007ffc`ac006ed8  00007ffc`f41f6f10 gdi32!SetLayout
00007ffc`ac006ee0  00007ffc`f41f6780 gdi32!GetDCOrgEx
00007ffc`ac006ee8  00007ffc`f41f4570 gdi32!RectVisibleStub
00007ffc`ac006ef0  00007ffc`f41f5490 gdi32!GetSystemPaletteEntriesStub
00007ffc`ac006ef8  00007ffc`f41f50d0 gdi32!SelectPaletteStub
00007ffc`ac006f00  00007ffc`f3ee1810 gdi32full!ScriptItemize
00007ffc`ac006f08  00007ffc`f41f1c70 gdi32!DeleteObject
00007ffc`ac006f10  00007ffc`f41f4020 gdi32!GetObjectTypeStub
00007ffc`ac006f18  00007ffc`f41f4310 gdi32!GetObjectWStub
00007ffc`ac006f20  00007ffc`f41f3eb0 gdi32!CreateCompatibleDCStub
00007ffc`ac006f28  00007ffc`f41f2a40 gdi32!CreateDIBSectionStub
00007ffc`ac006f30  00007ffc`f41f1350 gdi32!CreateFontIndirectW
00007ffc`ac006f38  00007ffc`f41f3a90 gdi32!SelectObject
00007ffc`ac006f40  00007ffc`f41f40a0 gdi32!SetBkColorStub
00007ffc`ac006f48  00007ffc`f41f3e60 gdi32!SetTextColor
00007ffc`ac006f50  00007ffc`f41f4c50 gdi32!CreateSolidBrushStub
00007ffc`ac006f58  00007ffc`f41f2ef0 gdi32!DeleteDC
00007ffc`ac006f60  00007ffc`f41f2b40 gdi32!CreateRectRgn
00007ffc`ac006f68  00007ffc`f41f22b0 gdi32!CombineRgn
00007ffc`ac006f70  00007ffc`f41fd9f0 gdi32!FillRgnStub
00007ffc`ac006f78  00007ffc`f41f46b0 gdi32!CreateBitmapStub
00007ffc`ac006f80  00007ffc`f41f3c80 gdi32!BitBltStub
00007ffc`ac006f88  00007ffc`f41f53d0 gdi32!CreatePatternBrushStub
00007ffc`ac006f90  00007ffc`f41f3ba0 gdi32!GetStockObjectStub
00007ffc`ac006f98  00007ffc`f41f4e20 gdi32!CreateDCA
00007ffc`ac006fa0  00007ffc`f41f33d0 gdi32!GetDeviceCaps
00007ffc`ac006fa8  00007ffc`f41fbe20 gdi32!ExtTextOutA
00007ffc`ac006fb0  00007ffc`f41f5270 gdi32!GetPixelStub
00007ffc`ac006fb8  00007ffc`f41f4940 gdi32!GetClipBox
00007ffc`ac006fc0  00007ffc`f41f4b00 gdi32!CreateCompatibleBitmapStub
00007ffc`ac006fc8  00007ffc`f41f4780 gdi32!GetDIBitsStub
00007ffc`ac006fd0  00007ffc`f41f5650 gdi32!SetDIBitsToDeviceStub
00007ffc`ac006fd8  00007ffc`f41f48f0 gdi32!GetCurrentObjectStub
00007ffc`ac006fe0  00007ffc`f41f4810 gdi32!GetObjectAStub
00007ffc`ac006fe8  00007ffc`f41f6db0 gdi32!GetDIBColorTable
00007ffc`ac006ff0  00007ffc`f41f6ed0 gdi32!SetDIBColorTable
00007ffc`ac006ff8  00007ffc`f41f4230 gdi32!StretchDIBits
00007ffc`ac007000  00007ffc`f41f3d60 gdi32!SetBkMode
00007ffc`ac007008  00007ffc`f41fbb60 gdi32!CreateFontIndirectAStub
00007ffc`ac007010  00007ffc`f41f1920 gdi32!SetDIBitsStub
00007ffc`ac007018  00007ffc`f41f4870 gdi32!CreateDIBitmapStub
00007ffc`ac007020  00007ffc`f41f44f0 gdi32!SetStretchBltMode
00007ffc`ac007028  00007ffc`f41f4d60 gdi32!StretchBltStub
00007ffc`ac007030  00007ffc`f41f4730 gdi32!GetViewportOrgExStub
00007ffc`ac007038  00007ffc`f41f6e10 gdi32!GetWindowOrgEx
00007ffc`ac007040  00007ffc`f41f5360 gdi32!SetPixelStub
00007ffc`ac007048  00007ffc`f41f6bf0 gdi32!GdiAlphaBlend
00007ffc`ac007050  00007ffc`f41f6d50 gdi32!GdiTransparentBlt
00007ffc`ac007058  00007ffc`f41f6d30 gdi32!GdiGradientFill
00007ffc`ac007060  00007ffc`f41f4160 gdi32!GetClipRgnStub
00007ffc`ac007068  00007ffc`f41f3e00 gdi32!ExtSelectClipRgn
00007ffc`ac007070  00007ffc`f41f40f0 gdi32!SelectClipRgn
00007ffc`ac007078  00007ffc`f41f6820 gdi32!GetLayout
00007ffc`ac007080  00007ffc`f41f5130 gdi32!CreateRoundRectRgnStub
00007ffc`ac007088  00007ffc`f41f4620 gdi32!SetWindowOrgExStub
00007ffc`ac007090  00007ffc`f41f4540 gdi32!IntersectClipRect
00007ffc`ac007098  00007ffc`f41f3db0 gdi32!SetTextAlign
00007ffc`ac0070a0  00007ffc`f41f5160 gdi32!GetTextAlignStub
00007ffc`ac0070a8  00007ffc`f41f8140 gdi32!ExcludeClipRect
00007ffc`ac0070b0  00007ffc`f41f3b00 gdi32!ExtTextOutW
00007ffc`ac0070b8  00007ffc`f41f6840 gdi32!SetDCBrushColor
00007ffc`ac0070c0  00007ffc`f41f6d90 gdi32!GetBrushOrgEx
00007ffc`ac0070c8  00007ffc`f41f6eb0 gdi32!SetBrushOrgEx
00007ffc`ac0070d0  00007ffc`f41fde30 gdi32!PlayEnhMetaFileStub
00007ffc`ac0070d8  00007ffc`f41f44a0 gdi32!SaveDC
00007ffc`ac0070e0  00007ffc`f41f4390 gdi32!RestoreDC
00007ffc`ac0070e8  00007ffc`f41f12d0 gdi32!GetTextExtentPoint32WStub
00007ffc`ac0070f0  00007ffc`f41f79f0 gdi32!GetEnhMetaFileHeader
00007ffc`ac0070f8  00007ffc`f41f5210 gdi32!CreatePenStub
00007ffc`ac007100  00007ffc`f41f3c10 gdi32!MoveToEx
00007ffc`ac007108  00007ffc`f41f4d00 gdi32!LineTo
00007ffc`ac007110  00007ffc`f41f1010 gdi32!EqualRgn
00007ffc`ac007118  00007ffc`f41f1c10 gdi32!SetRectRgn
```

```
00007ffc`ac007120  00000000`00000000
00007ffc`ac007128  00007ffc`f4fe2d60 kernel32!InitializeCriticalSectionEx
00007ffc`ac007130  00007ffc`f62bef70 ntdll!TpWaitForTimer
00007ffc`ac007138  00007ffc`f6310180 ntdll!TpReleaseTimer
00007ffc`ac007140  00007ffc`f62bc940 ntdll!TpSetTimer
00007ffc`ac007148  00007ffc`f4fda720 kernel32!CreateThreadpoolTimerStub
00007ffc`ac007150  00007ffc`f4fde6f0 kernel32!GetModuleHandleAStub
00007ffc`ac007158  00007ffc`f4fdb7d0 kernel32!IsProcessorFeaturePresentStub
00007ffc`ac007160  00007ffc`f4fdd680 kernel32!LoadLibraryExAStub
00007ffc`ac007168  00007ffc`f631a2b0 ntdll!RtlEncodePointer
00007ffc`ac007170  00007ffc`f63139d0 ntdll!RtlDecodePointer
00007ffc`ac007178  00007ffc`f4fd9390 kernel32!FlushInstructionCacheStub
00007ffc`ac007180  00007ffc`f4fdfca0 kernel32!GetLocaleInfoAStub
00007ffc`ac007188  00007ffc`f4fe0620 kernel32!IsDBCSLeadByteStub
00007ffc`ac007190  00007ffc`f3b94530 KERNELBASE!IsWow64Process2
00007ffc`ac007198  00007ffc`f4ffa590 kernel32!Wow64DisableWow64FsRedirectionStub
00007ffc`ac0071a0  00007ffc`f4ffa5d0 kernel32!Wow64RevertWow64FsRedirectionStub
00007ffc`ac0071a8  00007ffc`f4fd9eb0 kernel32!lstrcmpWStub
00007ffc`ac0071b0  00007ffc`f4fe3160 kernel32!GetFullPathNameW
00007ffc`ac0071b8  00007ffc`f62da4e0 ntdll!RtlEnterCriticalSection
00007ffc`ac0071c0  00007ffc`f62db4d0 ntdll!RtlLeaveCriticalSection
00007ffc`ac0071c8  00007ffc`f6308fe0 ntdll!RtlInitializeCriticalSection
00007ffc`ac0071d0  00007ffc`f4fe00c0 kernel32!HeapDestroyStub
00007ffc`ac0071d8  00007ffc`f62be080 ntdll!RtlDeleteCriticalSection
00007ffc`ac0071e0  00007ffc`f4fd5ef0 kernel32!HeapFreeStub
00007ffc`ac0071e8  00007ffc`f4fd6340 kernel32!GetProcessHeapStub
00007ffc`ac0071f0  00007ffc`f4fe2c50 kernel32!CloseHandle
00007ffc`ac0071f8  00007ffc`f4fd6360 kernel32!SetLastErrorStub
00007ffc`ac007200  00007ffc`f4fd62e0 kernel32!GetLastErrorStub
00007ffc`ac007208  00007ffc`f4fd93b0 kernel32!GetProcAddressStub
00007ffc`ac007210  00007ffc`f4fde090 kernel32!GetModuleHandleExWStub
00007ffc`ac007218  00007ffc`f4fde070 kernel32!GetVersionExWStub
00007ffc`ac007220  00007ffc`f4fdb790 kernel32!GetModuleHandleWStub
00007ffc`ac007228  00007ffc`f4fdbf80 kernel32!FormatMessageWStub
00007ffc`ac007230  00007ffc`f4fc6170 kernel32!GetCurrentThreadId
00007ffc`ac007238  00007ffc`f62c8ac0 ntdll!RtlAllocateHeap
00007ffc`ac007240  00007ffc`f4fdd600 kernel32!GetModuleFileNameAStub
00007ffc`ac007248  00007ffc`f4ff8630 kernel32!DebugBreakStub
00007ffc`ac007250  00007ffc`f4fde730 kernel32!IsDebuggerPresentStub
00007ffc`ac007258  00007ffc`f4fdaf10 kernel32!OutputDebugStringWStub
00007ffc`ac007260  00007ffc`f4fe2dd0 kernel32!ReleaseSemaphore
00007ffc`ac007268  00007ffc`f4fe2dc0 kernel32!ReleaseMutex
00007ffc`ac007270  00007ffc`f4fe2e50 kernel32!WaitForSingleObjectEx
00007ffc`ac007278  00007ffc`f4fe2e40 kernel32!WaitForSingleObject
00007ffc`ac007280  00007ffc`f4fe2da0 kernel32!OpenSemaphoreW
00007ffc`ac007288  00007ffc`f4fe2be0 kernel32!GetCurrentProcessId
00007ffc`ac007290  00007ffc`f4fe2d00 kernel32!CreateMutexExW
00007ffc`ac007298  00007ffc`f4fe2d20 kernel32!CreateSemaphoreExW
00007ffc`ac0072a0  00007ffc`f4fc8520 kernel32!FindAtomW
00007ffc`ac0072a8  00007ffc`f4fda390 kernel32!VirtualProtectStub
00007ffc`ac0072b0  00007ffc`f4fd82f0 kernel32!VirtualAllocStub
00007ffc`ac0072b8  00007ffc`f4fda5d0 kernel32!VirtualQueryStub
00007ffc`ac0072c0  00007ffc`f4fdbf60 kernel32!GetSystemInfoStub
00007ffc`ac0072c8  00007ffc`f4fc6160 kernel32!TlsGetValueStub
00007ffc`ac0072d0  00007ffc`f4fdb880 kernel32!TlsAllocStub
00007ffc`ac0072d8  00007ffc`f4fdc0b0 kernel32!TlsFreeStub
00007ffc`ac0072e0  00007ffc`f4fd6300 kernel32!TlsSetValueStub
00007ffc`ac0072e8  00007ffc`f4fd7b20 kernel32!CompareStringOrdinalStub
00007ffc`ac0072f0  00007ffc`f4fdedf0 kernel32!GetLocaleInfoWStub
00007ffc`ac0072f8  00007ffc`f4fe58e0 kernel32!GetNumberFormatWStub
00007ffc`ac007300  00007ffc`f4fdf2e0 kernel32!GetUserDefaultLCIDStub
00007ffc`ac007308  00007ffc`f4fdb390 kernel32!CompareStringWStub
00007ffc`ac007310  00007ffc`f62daa90 ntdll!RtlReleaseSRWLockShared
00007ffc`ac007318  00007ffc`f62da8d0 ntdll!RtlAcquireSRWLockShared
00007ffc`ac007320  00007ffc`f4fdbe40 kernel32!RaiseExceptionStub
00007ffc`ac007328  00007ffc`f4fe2cd0 kernel32!CreateEventW
00007ffc`ac007330  00007ffc`f4fd6580 kernel32!CompareStringExStub
00007ffc`ac007338  00007ffc`f4fe2df0 kernel32!SetEvent
00007ffc`ac007340  00007ffc`f4fe2de0 kernel32!ResetEvent
00007ffc`ac007348  00007ffc`f4fd9e40 kernel32!CreateThreadStub
00007ffc`ac007350  00007ffc`f4fd9e90 kernel32!SetThreadPriorityStub
00007ffc`ac007358  00007ffc`f4fd62a0 kernel32!GetCurrentThread
00007ffc`ac007360  00007ffc`f4fdfc80 kernel32!FreeLibraryAndExitThreadStub
00007ffc`ac007368  00007ffc`f4fda080 kernel32!GetThreadPriorityStub
```

```
00007ffc`ac007370  00007ffc`f62bb270 ntdll!RtlReleaseSRWLockExclusive
00007ffc`ac007378  00007ffc`f62b9860 ntdll!RtlAcquireSRWLockExclusive
00007ffc`ac007380  00007ffc`f4fda650 kernel32!FreeLibraryStub
00007ffc`ac007388  00007ffc`f4fd95d0 kernel32!GetSystemDirectoryWStub
00007ffc`ac007390  00007ffc`f4fdee70 kernel32!DeactivateActCtxStub
00007ffc`ac007398  00007ffc`f4fde880 kernel32!LoadLibraryWStub
00007ffc`ac0073a0  00007ffc`f4fdee50 kernel32!ActivateActCtxStub
00007ffc`ac0073a8  00007ffc`f4fdffe0 kernel32!FindActCtxSectionStringWStub
00007ffc`ac0073b0  00007ffc`f4fdfe20 kernel32!CreateActCtxWStub
00007ffc`ac0073b8  00007ffc`f4fdbfa0 kernel32!GetModuleFileNameWStub
00007ffc`ac0073c0  00007ffc`f4fdbfe0 kernel32!QueryActCtxWStub
00007ffc`ac0073c8  00007ffc`f4fe1c50 kernel32!OutputDebugStringAStub
00007ffc`ac0073d0  00007ffc`f4fdedb0 kernel32!GetVersionExAStub
00007ffc`ac0073d8  00007ffc`f4fd9310 kernel32!VirtualFreeStub
00007ffc`ac0073e0  00007ffc`f4fdf500 kernel32!LoadLibraryAStub
00007ffc`ac0073e8  00007ffc`f4ff8ef0 kernel32!GetUserDefaultLangIDStub
00007ffc`ac0073f0  00007ffc`f3b450c0 KERNELBASE!InitOnceBeginInitialize
00007ffc`ac0073f8  00007ffc`f3b99220 KERNELBASE!InitOnceComplete
00007ffc`ac007400  00007ffc`f4fd9f30 kernel32!SleepStub
00007ffc`ac007408  00007ffc`f4fd5fe0 kernel32!GetTickCountKernel32
00007ffc`ac007410  00007ffc`f4fd9370 kernel32!GlobalAllocStub
00007ffc`ac007418  00007ffc`f4fd6b30 kernel32!GlobalFreeStub
00007ffc`ac007420  00007ffc`f4fd6100 kernel32!GlobalLock
00007ffc`ac007428  00007ffc`f4fd6030 kernel32!GlobalUnlock
00007ffc`ac007430  00007ffc`f4fc7580 kernel32!FindResourceA
00007ffc`ac007438  00007ffc`f4fd9570 kernel32!LoadResourceStub
00007ffc`ac007440  00007ffc`f4fd9a80 kernel32!LockResourceStub
00007ffc`ac007448  00007ffc`f4fd9aa0 kernel32!SizeofResourceStub
00007ffc`ac007450  00007ffc`f4fe33a0 kernel32!MulDiv
00007ffc`ac007458  00007ffc`f4fde050 kernel32!GetACPStub
00007ffc`ac007460  00007ffc`f4fd6010 kernel32!WideCharToMultiByteStub
00007ffc`ac007468  00007ffc`f4fe2ed0 kernel32!CreateFileW
00007ffc`ac007470  00007ffc`f4fe3120 kernel32!GetFileType
00007ffc`ac007478  00007ffc`f4fdbac0 kernel32!GlobalMemoryStatusExStub
00007ffc`ac007480  00007ffc`f4fdc090 kernel32!FindResourceWStub
00007ffc`ac007488  00007ffc`f3b80ab0 KERNELBASE!InitOnceExecuteOnce
00007ffc`ac007490  00007ffc`f6311d40 ntdll!RtlInitializeSRWLock
00007ffc`ac007498  00007ffc`f4fd5fc0 kernel32!MultiByteToWideCharStub
00007ffc`ac0074a0  00007ffc`f4fc8110 kernel32!GlobalAddAtomW
00007ffc`ac0074a8  00007ffc`f4fdffc0 kernel32!GetModuleHandleExAStub
00007ffc`ac0074b0  00007ffc`f4fe2e20 kernel32!WaitForMultipleObjects
00007ffc`ac0074b8  00007ffc`f62d5eb0 ntdll!LdrResolveDelayLoadedAPI
00007ffc`ac0074c0  00007ffc`f4ff8650 kernel32!DelayLoadFailureHookStub
00007ffc`ac0074c8  00007ffc`f62a45a0 ntdll!RtlWakeAllConditionVariable
00007ffc`ac0074d0  00007ffc`f3b8d4f0 KERNELBASE!SleepConditionVariableSRW
00007ffc`ac0074d8  00007ffc`f4fe2a00 kernel32!RtlCaptureContext
00007ffc`ac0074e0  00007ffc`f4fe0b20 kernel32!RtlLookupFunctionEntryStub
00007ffc`ac0074e8  00007ffc`f4fe5ab0 kernel32!RtlVirtualUnwindStub
00007ffc`ac0074f0  00007ffc`f4ffa370 kernel32!UnhandledExceptionFilterStub
00007ffc`ac0074f8  00007ffc`f4fde6d0 kernel32!SetUnhandledExceptionFilterStub
00007ffc`ac007500  00007ffc`f4fe2bd0 kernel32!GetCurrentProcess
00007ffc`ac007508  00007ffc`f4fdf800 kernel32!TerminateProcessStub
00007ffc`ac007510  00007ffc`f4fd6670 kernel32!QueryPerformanceCounterStub
00007ffc`ac007518  00007ffc`f4fd7a90 kernel32!GetSystemTimeAsFileTimeStub
00007ffc`ac007520  00007ffc`f4fda0e0 kernel32!ExpandEnvironmentStringsWStub
00007ffc`ac007528  00007ffc`f4fe2320 kernel32!GetStringTypeExWStub
00007ffc`ac007530  00007ffc`f4fc7c50 kernel32!GetAtomNameW
00007ffc`ac007538  00007ffc`f4fc80f0 kernel32!AddAtomW
00007ffc`ac007540  00007ffc`f4fc8060 kernel32!DeleteAtom
00007ffc`ac007548  00007ffc`f4fe3360 kernel32!WriteFile
00007ffc`ac007550  00007ffc`f4fe2d50 kernel32!InitializeCriticalSectionAndSpinCount
00007ffc`ac007558  00007ffc`f4fe30f0 kernel32!GetFileSize
00007ffc`ac007560  00007ffc`f4fe3270 kernel32!ReadFile
00007ffc`ac007568  00007ffc`f4fde840 kernel32!HeapCreateStub
00007ffc`ac007570  00007ffc`f62c4480 ntdll!RtlReAllocateHeap
00007ffc`ac007578  00007ffc`f4fe30d0 kernel32!GetFileAttributesW
00007ffc`ac007580  00007ffc`f63155b0 ntdll!RtlInitializeSListHead
00007ffc`ac007588  00007ffc`f6316880 ntdll!RtlInterlockedFlushSList
00007ffc`ac007590  00007ffc`f62a6c60 ntdll!RtlInterlockedPushEntrySList
00007ffc`ac007598  00007ffc`f63473f0 ntdll!ExpInterlockedPopEntrySList
00007ffc`ac0075a0  00007ffc`f63138a0 ntdll!RtlQueryDepthSList
00007ffc`ac0075a8  00007ffc`f4fe30a0 kernel32!GetFileAttributesA
00007ffc`ac0075b0  00007ffc`f4fe2ef0 kernel32!DeleteFileA
00007ffc`ac0075b8  00007ffc`f4fe2c40 kernel32!SetProcessWorkingSetSize
```

```
00007ffc`ac0075c0  00000000`00000000
00007ffc`ac0075c8  00007ffc`f5b9b5d0 oleaut32!VarDecAdd
00007ffc`ac0075d0  00007ffc`f5b65230 oleaut32!VarDecInt
00007ffc`ac0075d8  00007ffc`f5b41700 oleaut32!VarDecRound
00007ffc`ac0075e0  00007ffc`f5b54e20 oleaut32!VarI4FromStr
00007ffc`ac0075e8  00007ffc`f5b46780 oleaut32!VarBstrFromI4
00007ffc`ac0075f0  00007ffc`f5b9bc90 oleaut32!VarDecSub
00007ffc`ac0075f8  00007ffc`f5b65270 oleaut32!VarDecMul
00007ffc`ac007600  00007ffc`f5b9b5e0 oleaut32!VarDecDiv
00007ffc`ac007608  00007ffc`f5b575e0 oleaut32!VarCmp
00007ffc`ac007610  00007ffc`f5b499b0 oleaut32!VariantChangeTypeEx
00007ffc`ac007618  00007ffc`f5b5a340 oleaut32!VarDecFromR8
00007ffc`ac007620  00007ffc`f5bdcd70 oleaut32!VariantInit
00007ffc`ac007628  00007ffc`f5b49980 oleaut32!VariantChangeType
00007ffc`ac007630  00007ffc`f5bdc910 oleaut32!VariantCopy
00007ffc`ac007638  00007ffc`f5b42670 oleaut32!OleCreateFontIndirect
00007ffc`ac007640  00007ffc`f5bdcb60 oleaut32!VariantClear
00007ffc`ac007648  00007ffc`f5b629e0 oleaut32!SetErrorInfo
00007ffc`ac007650  00007ffc`f5b4a6c0 oleaut32!SysAllocStringLen
00007ffc`ac007658  00007ffc`f5b57300 oleaut32!SysStringLen
00007ffc`ac007660  00007ffc`f5b495e0 oleaut32!SafeArrayPutElement
00007ffc`ac007668  00007ffc`f5b57220 oleaut32!SafeArrayGetUBound
00007ffc`ac007670  00007ffc`f5b57450 oleaut32!SafeArrayGetLBound
00007ffc`ac007678  00007ffc`f5b575c0 oleaut32!SafeArrayGetDim
00007ffc`ac007680  00007ffc`f5b497b0 oleaut32!SafeArrayGetElement
00007ffc`ac007688  00007ffc`f5b4b070 oleaut32!SafeArrayCopy
00007ffc`ac007690  00007ffc`f5b57490 oleaut32!SafeArrayGetVartype
00007ffc`ac007698  00007ffc`f5bac4f0 oleaut32!VarBstrFromDec
00007ffc`ac0076a0  00007ffc`f5bacdc0 oleaut32!VarDecFromStr
00007ffc`ac0076a8  00007ffc`f5b5ad70 oleaut32!VarUI4FromDec
00007ffc`ac0076b0  00007ffc`f5bacd40 oleaut32!VarDecFromI4
00007ffc`ac0076b8  00007ffc`f5b59000 oleaut32!VarDecCmp
00007ffc`ac0076c0  00007ffc`f5b4a7a0 oleaut32!SysAllocString
00007ffc`ac0076c8  00007ffc`f5b4a940 oleaut32!SysFreeString
00007ffc`ac0076d0  00007ffc`f5b49290 oleaut32!SafeArrayDestroy
00007ffc`ac0076d8  00007ffc`f5b4dc80 oleaut32!SafeArrayCreateVector
00007ffc`ac0076e0  00000000`00000000
00007ffc`ac0076e8  00007ffc`f48f51b0 shell32!ShellExecuteW
00007ffc`ac0076f0  00000000`00000000
00007ffc`ac0076f8  00007ffc`f5857070 user32!PostMessageW
00007ffc`ac007700  00007ffc`f5868be0 user32!SetWindowsHookExW
00007ffc`ac007708  00007ffc`f586b8d0 user32!UnhookWindowsHookEx
00007ffc`ac007710  00007ffc`f58563e0 user32!TranslateMessage
00007ffc`ac007718  00007ffc`f5850bf0 user32!DispatchMessageW
00007ffc`ac007720  00007ffc`f585a9a0 user32!GetPropW
00007ffc`ac007728  00007ffc`f5850600 user32!SendMessageW
00007ffc`ac007730  00007ffc`f585da80 user32!CallNextHookEx
00007ffc`ac007738  00007ffc`f5864990 user32!IsIconic
00007ffc`ac007740  00007ffc`f5856b90 user32!GetClientRect
00007ffc`ac007748  00007ffc`f5855120 user32!MonitorFromRect
00007ffc`ac007750  00007ffc`f5865860 user32!SetTimer
00007ffc`ac007758  00007ffc`f58728d0 user32!NtUserKillTimer
00007ffc`ac007760  00007ffc`f5869010 user32!LoadStringW
00007ffc`ac007768  00007ffc`f5849080 user32!LoadImageW
00007ffc`ac007770  00007ffc`f5866bf0 user32!GetSysColor
00007ffc`ac007778  00007ffc`f5865b20 user32!FillRect
00007ffc`ac007780  00007ffc`f585bbd0 user32!DrawTextW
00007ffc`ac007788  00007ffc`f5868f10 user32!GetAsyncKeyState
00007ffc`ac007790  00007ffc`f5848030 user32!CreateWindowExW
00007ffc`ac007798  00007ffc`f5855d00 user32!GetWindowThreadProcessId
00007ffc`ac0077a0  00007ffc`f5849760 user32!LoadIconW
00007ffc`ac0077a8  00007ffc`f584b110 user32!LoadCursorW
00007ffc`ac0077b0  00007ffc`f5847bf0 user32!RegisterClassW
00007ffc`ac0077b8  00007ffc`f585d6f0 user32!ClientToScreen
00007ffc`ac0077c0  00007ffc`f5871dd0 user32!NtUserBeginPaint
00007ffc`ac0077c8  00007ffc`f5872180 user32!NtUserEndPaint
00007ffc`ac0077d0  00007ffc`f5872810 user32!NtUserInvalidateRect
00007ffc`ac0077d8  00007ffc`f58731d0 user32!NtUserTrackMouseEvent
00007ffc`ac0077e0  00007ffc`f5872db0 user32!NtUserSetCapture
00007ffc`ac0077e8  00007ffc`f5842860 user32!ReleaseCapture
00007ffc`ac0077f0  00007ffc`f5864820 user32!GetSystemMetrics
00007ffc`ac0077f8  00007ffc`f58424e0 user32!LoadCursorA
00007ffc`ac007800  00007ffc`f58667a0 user32!RegisterWindowMessageA
00007ffc`ac007808  00007ffc`f58cb840 user32!GetProcessDefaultLayout
```

```
00007ffc`ac007810  00007ffc`f5854a20 user32!SystemParametersInfoA
00007ffc`ac007818  00007ffc`f586d240 user32!GetMonitorInfoA
00007ffc`ac007820  00007ffc`f5872190 user32!NtUserEnumDisplayMonitors
00007ffc`ac007828  00007ffc`f5868410 user32!GetSysColorBrush
00007ffc`ac007830  00007ffc`f5853c10 user32!IntersectRect
00007ffc`ac007838  00007ffc`f584fd40 user32!GetSystemMetricsForDpi
00007ffc`ac007840  00007ffc`f5872940 user32!NtUserLogicalToPerMonitorDPIPhysicalPoint
00007ffc`ac007848  00007ffc`f5855160 user32!SetThreadDpiAwarenessContext
00007ffc`ac007850  00007ffc`f5855390 user32!MonitorFromPoint
00007ffc`ac007858  00007ffc`f58479b0 user32!GetClassInfoExW
00007ffc`ac007860  00007ffc`f5847bb0 user32!RegisterClassExW
00007ffc`ac007868  00007ffc`f586b1a0 user32!EnableWindow
00007ffc`ac007870  00007ffc`f5859a20 user32!GetParent
00007ffc`ac007878  00007ffc`f585ace0 user32!GetKeyState
00007ffc`ac007880  00007ffc`f5876c80 user32!GetClassLongA
00007ffc`ac007888  00007ffc`f586fe50 user32!UnregisterClassA
00007ffc`ac007890  00007ffc`f5859480 user32!GetFocus
00007ffc`ac007898  00007ffc`f586b6c0 user32!SendMessageA
00007ffc`ac0078a0  00007ffc`f5871fe0 user32!NtUserDestroyWindow
00007ffc`ac0078a8  00007ffc`f586b7b0 user32!PostQuitMessage
00007ffc`ac0078b0  00007ffc`f58732c0 user32!NtUserWaitMessage
00007ffc`ac0078b8  00007ffc`f585ab40 user32!IsWindowVisible
00007ffc`ac0078c0  00007ffc`f5855320 user32!AdjustWindowRectEx
00007ffc`ac0078c8  00007ffc`f58533d0 user32!GetWindowLongA
00007ffc`ac0078d0  00007ffc`f58503b0 user32!GetWindowRect
00007ffc`ac0078d8  00007ffc`f58730d0 user32!NtUserSetWindowPos
00007ffc`ac0078e0  00007ffc`f584d440 user32!SetWindowLongW
00007ffc`ac0078e8  00007ffc`f5865160 user32!SetRect
00007ffc`ac0078f0  00007ffc`f5864b80 user32!IsRectEmpty
00007ffc`ac0078f8  00007ffc`f584c710 user32!IsChild
00007ffc`ac007900  00007ffc`f58725c0 user32!NtUserGetSystemMenu
00007ffc`ac007908  00007ffc`f58726a0 user32!NtUserGetWindowPlacement
00007ffc`ac007910  00007ffc`f58690c0 user32!EnableMenuItem
00007ffc`ac007918  00007ffc`f584cac0 user32!OffsetRect
00007ffc`ac007920  00007ffc`f5855be0 user32!SetWindowTextW
00007ffc`ac007928  00007ffc`f5864bb0 user32!UnregisterClassW
00007ffc`ac007930  00007ffc`f58cbbf0 user32!MessageBeep
00007ffc`ac007938  00007ffc`f58650c0 user32!EqualRect
00007ffc`ac007940  00007ffc`f5872f60 user32!NtUserSetLayeredWindowAttributes
00007ffc`ac007948  00007ffc`f584ff10 user32!MapWindowPoints
00007ffc`ac007950  00007ffc`f5872980 user32!NtUserMoveWindow
00007ffc`ac007958  00007ffc`f5855670 user32!PtInRect
00007ffc`ac007960  00007ffc`f586af40 user32!PostMessageA
00007ffc`ac007968  00007ffc`f5866a60 user32!InflateRect
00007ffc`ac007970  00007ffc`f584cde0 user32!IsWindowUnicode
00007ffc`ac007978  00007ffc`f5865520 user32!SendMessageTimeoutW
00007ffc`ac007980  00007ffc`f58cfc80 user32!SendMessageTimeoutA
00007ffc`ac007988  00007ffc`f58cfd80 user32!SetWindowTextA
00007ffc`ac007990  00007ffc`f585aea0 user32!AppendMenuW
00007ffc`ac007998  00007ffc`f58562f0 user32!EnumThreadWindows
00007ffc`ac0079a0  00007ffc`f5855a00 user32!EnumWindows
00007ffc`ac0079a8  00007ffc`f58ced40 user32!GetClassNameA
00007ffc`ac0079b0  00007ffc`f58531e0 user32!GetWindow
00007ffc`ac0079b8  00007ffc`f58646a0 user32!MonitorFromWindow
00007ffc`ac0079c0  00007ffc`f5851810 user32!CallWindowProcW
00007ffc`ac0079c8  00007ffc`f5866d60 user32!GetDC
00007ffc`ac0079d0  00007ffc`f586c350 user32!SetWindowLongPtrA
00007ffc`ac0079d8  00007ffc`f586a2b0 user32!GetActiveWindow
00007ffc`ac0079e0  00007ffc`f58732f0 user32!NtUserWindowFromPoint
00007ffc`ac0079e8  00007ffc`f5868520 user32!GetCursorPos
00007ffc`ac0079f0  00007ffc`f584c780 user32!ScreenToClient
00007ffc`ac0079f8  00007ffc`f586b3d0 user32!SetParentStub
00007ffc`ac007a00  00007ffc`f5857fd0 user32!NotifyWinEvent
00007ffc`ac007a08  00007ffc`f5872ec0 user32!NtUserSetFocus
00007ffc`ac007a10  00007ffc`f5865820 user32!GetClassNameW
00007ffc`ac007a18  00007ffc`f58683b0 user32!DeferWindowPos
00007ffc`ac007a20  00007ffc`f58c4cc0 user32!AnimateWindow
00007ffc`ac007a28  00007ffc`f586a010 user32!SetCursorStub
00007ffc`ac007a30  00007ffc`f6343410 ntdll!NtdllDefWindowProc_A
00007ffc`ac007a38  00007ffc`f584dfb0 user32!GetWindowTextW
00007ffc`ac007a40  00007ffc`f5859bc0 user32!PeekMessageA
00007ffc`ac007a48  00007ffc`f5859d40 user32!PeekMessageW
00007ffc`ac007a50  00007ffc`f5864620 user32!GetMessageW
00007ffc`ac007a58  00007ffc`f585d090 user32!MsgWaitForMultipleObjects
```

```
00007ffc`ac007a60  00007ffc`f586bd30 user32!GetMessagePos
00007ffc`ac007a68  00007ffc`f5872310 user32!NtUserGetForegroundWindow
00007ffc`ac007a70  00007ffc`f5864ee0 user32!SystemParametersInfoW
00007ffc`ac007a78  00007ffc`f58677b0 user32!GetKeyboardLayout
00007ffc`ac007a80  00007ffc`f588ff40 user32!LoadImageA
00007ffc`ac007a88  00007ffc`f5869940 user32!UpdateWindow
00007ffc`ac007a90  00007ffc`f58722f0 user32!NtUserGetDoubleClickTime
00007ffc`ac007a98  00007ffc`f58684b0 user32!IsWindowEnabled
00007ffc`ac007aa0  00007ffc`f586bc80 user32!SetForegroundWindow
00007ffc`ac007aa8  00007ffc`f586cca0 user32!GetPropA
00007ffc`ac007ab0  00007ffc`f58656a0 user32!UnionRect
00007ffc`ac007ab8  00007ffc`f5869ec0 user32!BeginDeferWindowPos
00007ffc`ac007ac0  00007ffc`f586a130 user32!GetCapture
00007ffc`ac007ac8  00007ffc`f588da60 user32!SetWindowsHookExA
00007ffc`ac007ad0  00007ffc`f5872be0 user32!NtUserRedrawWindow
00007ffc`ac007ad8  00007ffc`f586a1b0 user32!EndDeferWindowPos
00007ffc`ac007ae0  00007ffc`f5854ac0 user32!SetWindowRgn
00007ffc`ac007ae8  00007ffc`f58657f0 user32!CopyRect
00007ffc`ac007af0  00007ffc`f58cc6a0 user32!UpdateLayeredWindow
00007ffc`ac007af8  00007ffc`f58cf670 user32!GetClassLongPtrA
00007ffc`ac007b00  00007ffc`f5853090 user32!IsMenu
00007ffc`ac007b08  00007ffc`f58661d0 user32!GetClassInfoA
00007ffc`ac007b10  00007ffc`f5872660 user32!NtUserGetWindowDC
00007ffc`ac007b18  00007ffc`f5873290 user32!NtUserValidateRect
00007ffc`ac007b20  00007ffc`f5872900 user32!NtUserLockWindowUpdate
00007ffc`ac007b28  00007ffc`f58731e0 user32!NtUserTrackPopupMenuEx
00007ffc`ac007b30  00007ffc`f5866110 user32!SetPropW
00007ffc`ac007b38  00007ffc`f5866050 user32!RemovePropW
00007ffc`ac007b40  00007ffc`f5876250 user32!FrameRect
00007ffc`ac007b48  00007ffc`f5842780 user32!VkKeyScanExW
00007ffc`ac007b50  00007ffc`f5867130 user32!RegisterWindowMessageW
00007ffc`ac007b58  00007ffc`f58650f0 user32!IsProcessDPIAware
00007ffc`ac007b60  00007ffc`f5868bc0 user32!DestroyCursor
00007ffc`ac007b68  00007ffc`f58725e0 user32!NtUserGetTitleBarInfo
00007ffc`ac007b70  00007ffc`f5865e90 user32!SetRectEmpty
00007ffc`ac007b78  00007ffc`f584f720 user32!GetWindowInfo
00007ffc`ac007b80  00007ffc`f5852210 user32!IsZoomed
00007ffc`ac007b88  00007ffc`f5867730 user32!InternalGetWindowText
00007ffc`ac007b90  00007ffc`f5866d90 user32!DrawIconEx
00007ffc`ac007b98  00007ffc`f58d2da0 user32!DrawFrameControl
00007ffc`ac007ba0  00007ffc`f585bba0 user32!DrawTextExW
00007ffc`ac007ba8  00007ffc`f586ce90 user32!DrawEdge
00007ffc`ac007bb0  00007ffc`f58670c0 user32!GetIconInfo
00007ffc`ac007bb8  00007ffc`f584c160 user32!CreateIconIndirect
00007ffc`ac007bc0  00007ffc`f5852ca0 user32!GetClassLongW
00007ffc`ac007bc8  00007ffc`f586b790 user32!GetMessageTime
00007ffc`ac007bd0  00007ffc`f5842320 user32!DrawFocusRect
00007ffc`ac007bd8  00007ffc`f5853d70 user32!GetWindowTextLengthW
00007ffc`ac007be0  00007ffc`f586c4e0 user32!CreatePopupMenu
00007ffc`ac007be8  00007ffc`f5871fc0 user32!NtUserDestroyMenu
00007ffc`ac007bf0  00007ffc`f5866cb0 user32!InvertRect
00007ffc`ac007bf8  00007ffc`f585d210 user32!GetMonitorInfoW
00007ffc`ac007c00  00007ffc`f5865900 user32!IsWinEventHookInstalled
00007ffc`ac007c08  00007ffc`f5871e70 user32!NtUserChildWindowFromPointEx
00007ffc`ac007c10  00007ffc`f58723c0 user32!NtUserGetKeyboardState
00007ffc`ac007c18  00007ffc`f58522f0 user32!GetWindowLongPtrW
00007ffc`ac007c20  00007ffc`f58501e0 user32!SetWindowLongPtrW
00007ffc`ac007c28  00007ffc`f586c0d0 user32!SetMenu
00007ffc`ac007c30  00007ffc`f5855b40 user32!GetWindowDPI
00007ffc`ac007c38  00007ffc`f5872890 user32!NtUserIsTopLevelWindow
00007ffc`ac007c40  00007ffc`f58728b0 user32!NtUserIsWindowBroadcastingDpiToChildren
00007ffc`ac007c48  00007ffc`f5872830 user32!NtUserIsChildWindowDpiMessageEnabled
00007ffc`ac007c50  00007ffc`f58720b0 user32!NtUserEnableChildWindowDpiMessage
00007ffc`ac007c58  00007ffc`f5872430 user32!NtUserGetOwnerTransformedMonitorRect
00007ffc`ac007c60  00007ffc`f58cbd20 user32!RegisterSystemThread
00007ffc`ac007c68  00007ffc`f5853040 user32!IsWindowInDestroy
00007ffc`ac007c70  00007ffc`f5873150 user32!NtUserShowWindow
00007ffc`ac007c78  00007ffc`f5852280 user32!GetWindowLongPtrA
00007ffc`ac007c80  00007ffc`f58525f0 user32!GetWindowLongW
00007ffc`ac007c88  00007ffc`f584cf10 user32!GetCurrentThreadDesktopHwnd
00007ffc`ac007c90  00007ffc`f58721f0 user32!NtUserGetAncestor
00007ffc`ac007c98  00007ffc`f58569b0 user32!IsWindow
00007ffc`ac007ca0  00007ffc`f5849850 user32!CopyImage
00007ffc`ac007ca8  00007ffc`f584af60 user32!ReleaseDC
```

```
00007ffc`ac007cb0  00007ffc`f6343420 ntdll!NtdllDefWindowProc_W
00007ffc`ac007cb8  00007ffc`f588feb0 user32!LoadBitmapA
00007ffc`ac007cc0  00000000`00000000
00007ffc`ac007cc8  00007ffc`ef42bfe0 xmllite!CreateXmlReader
00007ffc`ac007cd0  00007ffc`ef42cc20 xmllite!CreateXmlWriter
00007ffc`ac007cd8  00000000`00000000
00007ffc`ac007ce0  00007ffc`f3b9f120 KERNELBASE!PathCchAppend
00007ffc`ac007ce8  00007ffc`f3b7d840 KERNELBASE!PathCchRemoveFileSpec
00007ffc`ac007cf0  00000000`00000000
00007ffc`ac007cf8  00007ffc`cf4c9e00 GdiPlus!GdipGetImagePixelFormat
00007ffc`ac007d00  00007ffc`cf486ef0 GdiPlus!GdipCreateBitmapFromScan0
00007ffc`ac007d08  00007ffc`cf485ad0 GdiPlus!GdipGetImageGraphicsContext
00007ffc`ac007d10  00007ffc`cf4c9f40 GdiPlus!GdipBitmapLockBits
00007ffc`ac007d18  00007ffc`cf4ca040 GdiPlus!GdipBitmapUnlockBits
00007ffc`ac007d20  00007ffc`cf4abc70 GdiPlus!GdipSetClipRect
00007ffc`ac007d28  00007ffc`cf4b48c0 GdiPlus!GdipDrawImageRectRect
00007ffc`ac007d30  00007ffc`cf477050 GdiPlus!GdiplusShutdown
00007ffc`ac007d38  00007ffc`cf4a9310 GdiPlus!GdipFillRectangle
00007ffc`ac007d40  00007ffc`cf4eebf0 GdiPlus!GdipCreateRegion
00007ffc`ac007d48  00007ffc`cf4f42a0 GdiPlus!GdipGetClip
00007ffc`ac007d50  00007ffc`cf4fab60 GdiPlus!GdipIsClipEmpty
00007ffc`ac007d58  00007ffc`cf4ff8a0 GdiPlus!GdipSetClipRegion
00007ffc`ac007d60  00007ffc`cf4c16e0 GdiPlus!GdipDeleteRegion
00007ffc`ac007d68  00007ffc`cf4a7020 GdiPlus!GdipDeletePen
00007ffc`ac007d70  00007ffc`cf4bda80 GdiPlus!GdipCreateMatrix
00007ffc`ac007d78  00007ffc`cf4bdff0 GdiPlus!GdipDeleteMatrix
00007ffc`ac007d80  00007ffc`cf4b5000 GdiPlus!GdipSetCompositingMode
00007ffc`ac007d88  00007ffc`cf4c0f50 GdiPlus!GdipGetCompositingMode
00007ffc`ac007d90  00007ffc`cf4ff9a0 GdiPlus!GdipSetCompositingQuality
00007ffc`ac007d98  00007ffc`cf4b7620 GdiPlus!GdipCreateBitmapFromFile
00007ffc`ac007da0  00007ffc`cf4b4f20 GdiPlus!GdipSetInterpolationMode
00007ffc`ac007da8  00007ffc`cf4f5e70 GdiPlus!GdipGetInterpolationMode
00007ffc`ac007db0  00007ffc`cf4be790 GdiPlus!GdipSetSmoothingMode
00007ffc`ac007db8  00007ffc`cf4bfa80 GdiPlus!GdipGetSmoothingMode
00007ffc`ac007dc0  00007ffc`cf4b51b0 GdiPlus!GdipSetPixelOffsetMode
00007ffc`ac007dc8  00007ffc`cf4f91d0 GdiPlus!GdipGetPixelOffsetMode
00007ffc`ac007dd0  00007ffc`cf5028f0 GdiPlus!GdipSetTextContrast
00007ffc`ac007dd8  00007ffc`cf4fa430 GdiPlus!GdipGetTextContrast
00007ffc`ac007de0  00007ffc`cf502970 GdiPlus!GdipSetTextRenderingHint
00007ffc`ac007de8  00007ffc`cf4fa4c0 GdiPlus!GdipGetTextRenderingHint
00007ffc`ac007df0  00007ffc`cf4c2800 GdiPlus!GdipGetWorldTransform
00007ffc`ac007df8  00007ffc`cf5035c0 GdiPlus!GdipTranslateRegionI
00007ffc`ac007e00  00007ffc`cf4c3cf0 GdiPlus!GdipSetWorldTransform
00007ffc`ac007e08  00007ffc`cf4ae910 GdiPlus!GdipGetDC
00007ffc`ac007e10  00007ffc`cf4b5d00 GdiPlus!GdipReleaseDC
00007ffc`ac007e18  00007ffc`cf4fe7f0 GdiPlus!GdipResetWorldTransform
00007ffc`ac007e20  00007ffc`cf4bd8b0 GdiPlus!GdipCreateMatrix2
00007ffc`ac007e28  00007ffc`cf4fee80 GdiPlus!GdipSaveGraphics
00007ffc`ac007e30  00007ffc`cf4fe870 GdiPlus!GdipRestoreGraphics
00007ffc`ac007e38  00007ffc`cf473950 GdiPlus!GdipCreateBitmapFromHBITMAP
00007ffc`ac007e40  00007ffc`cf4c7a90 GdiPlus!GdipCreateBitmapFromStream
00007ffc`ac007e48  00007ffc`cf4b4ae0 GdiPlus!GdipSetImageAttributesColorKeys
00007ffc`ac007e50  00007ffc`cf4b45a0 GdiPlus!GdipDisposeImageAttributes
00007ffc`ac007e58  00007ffc`cf4b4be0 GdiPlus!GdipCreateImageAttributes
00007ffc`ac007e60  00007ffc`cf4b47a0 GdiPlus!GdipDrawImageRectRectI
00007ffc`ac007e68  00007ffc`cf4b2bb0 GdiPlus!GdipDrawImagePointRectI
00007ffc`ac007e70  00007ffc`cf486bf0 GdiPlus!GdipDeleteGraphics
00007ffc`ac007e78  00007ffc`cf4bf070 GdiPlus!GdipCreateFromHDC
00007ffc`ac007e80  00007ffc`cf47fb60 GdiPlus!GdipImageRotateFlip
00007ffc`ac007e88  00007ffc`cf4eb280 GdiPlus!GdipCloneBitmapAreaI
00007ffc`ac007e90  00007ffc`cf4bf910 GdiPlus!GdipCloneImage
00007ffc`ac007e98  00007ffc`cf473870 GdiPlus!GdipAlloc
00007ffc`ac007ea0  00007ffc`cf476820 GdiPlus!GdipDisposeImage
00007ffc`ac007ea8  00007ffc`cf473830 GdiPlus!GdipFree
00007ffc`ac007eb0  00007ffc`cf4eb020 GdiPlus!GdipBitmapGetPixel
00007ffc`ac007eb8  00007ffc`cf4ca240 GdiPlus!GdipGetImageHeight
00007ffc`ac007ec0  00007ffc`cf4f43e0 GdiPlus!GdipGetCompositingQuality
00007ffc`ac007ec8  00007ffc`cf4ca100 GdiPlus!GdipGetImageWidth
00007ffc`ac007ed0  00000000`00000000
00007ffc`ac007ed8  00007ffc`f41b2460 msvcrt!log
00007ffc`ac007ee0  00007ffc`f41b2740 msvcrt!log10
00007ffc`ac007ee8  00007ffc`f41b2c80 msvcrt!logf
00007ffc`ac007ef0  00007ffc`f41a0500 msvcrt!memcmp
00007ffc`ac007ef8  00007ffc`f41b7d40 msvcrt!memcpy
```

```
00007ffc`ac007f00  00007ffc`f41b7d40 msvcrt!memcpy
00007ffc`ac007f08  00007ffc`f41b8000 msvcrt!memset
00007ffc`ac007f10  00007ffc`f41b6920 msvcrt!pow
00007ffc`ac007f18  00007ffc`f41af800 msvcrt!sin
00007ffc`ac007f20  00007ffc`f41b0320 msvcrt!sinf
00007ffc`ac007f28  00007ffc`f41b3930 msvcrt!sqrt
00007ffc`ac007f30  00007ffc`f416d750 msvcrt!onexit
00007ffc`ac007f38  00007ffc`f416d670 msvcrt!_dllonexit
00007ffc`ac007f40  00007ffc`f417e2e0 msvcrt!unlock
00007ffc`ac007f48  00007ffc`f417e0a0 msvcrt!lock
00007ffc`ac007f50  00007ffc`f4156a50 msvcrt!type_info::~type_info
00007ffc`ac007f58  00007ffc`f417d520 msvcrt!initterm
00007ffc`ac007f60  00007ffc`f417d190 msvcrt!amsg_exit
00007ffc`ac007f68  00007ffc`f416aac0 msvcrt!XcptFilter
00007ffc`ac007f70  00007ffc`f41a1960 msvcrt!wcsncmp
00007ffc`ac007f78  00007ffc`f41a1b60 msvcrt!wcsrchr
00007ffc`ac007f80  00007ffc`f4172850 msvcrt!qsort
00007ffc`ac007f88  00007ffc`f41b19c0 msvcrt!fmod
00007ffc`ac007f90  00007ffc`f4179a70 msvcrt!swprintf_s
00007ffc`ac007f98  00007ffc`f4190080 msvcrt!fclose
00007ffc`ac007fa0  00007ffc`f418acb0 msvcrt!fopen
00007ffc`ac007fa8  00007ffc`f418ae00 msvcrt!fwprintf
00007ffc`ac007fb0  00007ffc`f41457e0 msvcrt!wcstoul
00007ffc`ac007fb8  00007ffc`f415c500 msvcrt!resetstkoflw
00007ffc`ac007fc0  00007ffc`f41a1240 msvcrt!strstr
00007ffc`ac007fc8  00007ffc`f41a01e0 msvcrt!wcsupr
00007ffc`ac007fd0  00007ffc`f4190310 msvcrt!fgets
00007ffc`ac007fd8  00007ffc`f4143600 msvcrt!iswascii
00007ffc`ac007fe0  00007ffc`f41436a0 msvcrt!iswprint
00007ffc`ac007fe8  00007ffc`f41a1c60 msvcrt!wcstok
00007ffc`ac007ff0  00007ffc`f41a0680 msvcrt!memmove_s
00007ffc`ac007ff8  00007ffc`f41b17e0 msvcrt!floor
00007ffc`ac008000  00007ffc`f41b1430 msvcrt!expf
00007ffc`ac008008  00007ffc`f41aff40 msvcrt!cosf
00007ffc`ac008010  00007ffc`f41af3e0 msvcrt!cos
00007ffc`ac008018  00007ffc`f41af300 msvcrt!ceilf
00007ffc`ac008020  00007ffc`f41af1e0 msvcrt!ceil
00007ffc`ac008028  00007ffc`f414db90 msvcrt!_RTDynamicCast
00007ffc`ac008030  00007ffc`f415c690 msvcrt!free
00007ffc`ac008038  00007ffc`f41ac080 msvcrt!isnan
00007ffc`ac008040  00007ffc`f41abf10 msvcrt!finite
00007ffc`ac008048  00007ffc`f4146030 msvcrt!wtoi
00007ffc`ac008050  00007ffc`f41a05e0 msvcrt!memcpy_s
00007ffc`ac008058  00007ffc`f418e200 msvcrt!vsnwprintf
00007ffc`ac008060  00007ffc`f415c7c0 msvcrt!realloc
00007ffc`ac008068  00007ffc`f416f0b0 msvcrt!purecall
00007ffc`ac008070  00007ffc`f416acc0 msvcrt!_C_specific_handler
00007ffc`ac008078  00007ffc`f415bc60 msvcrt!callnewh
00007ffc`ac008080  00007ffc`f415c6e0 msvcrt!malloc
00007ffc`ac008088  00007ffc`f41b18f0 msvcrt!floorf
00007ffc`ac008090  00007ffc`f41b3a40 msvcrt!sqrtf
00007ffc`ac008098  00000000`00000000
00007ffc`ac0080a0  00007ffc`f60d8490 ole32!CoInitialize [com\ole32\com\class\coapi.cxx @ 45]
00007ffc`ac0080a8  00007ffc`f5d65b30 combase!CLSIDFromString [onecore\com\combase\class\compapi.cxx @ 381]
00007ffc`ac0080b0  00007ffc`f5ef4350 combase!CoLockObjectExternal [onecore\com\combase\dcomrem\coapi.cxx @ 2285]
00007ffc`ac0080b8  00007ffc`f60eddc0 ole32!OleDraw [com\ole32\ole232\base\api.cpp @ 1910]
00007ffc`ac0080c0  00007ffc`f60cac50 ole32!OleUninitialize [com\ole32\ole232\base\ole2.cpp @ 514]
00007ffc`ac0080c8  00007ffc`f5da15f0 combase!CoUninitialize [onecore\com\combase\class\compobj.cxx @ 3793]
00007ffc`ac0080d0  00007ffc`f5da0f00 combase!CoInitializeEx [onecore\com\combase\class\compobj.cxx @ 3734]
00007ffc`ac0080d8  00007ffc`f5d509e0 combase!CoDisconnectObject [onecore\com\combase\dcomrem\coapi.cxx @ 2171]
00007ffc`ac0080e0  00007ffc`f5df8320 combase!CreateStreamOnHGlobal [onecore\com\combase\util\memstm.cpp @ 1229]
00007ffc`ac0080e8  00007ffc`f5de54d0 combase!CoTaskMemFree [onecore\com\combase\class\memapi.cxx @ 453]
00007ffc`ac0080f0  00007ffc`f5da3f70 combase!CoCreateInstance [onecore\com\combase\objact\actapi.cxx @ 252]
00007ffc`ac0080f8  00007ffc`f5de6640 combase!CoTaskMemAlloc [onecore\com\combase\class\memapi.cxx @ 437]
00007ffc`ac008100  00007ffc`f5de4d40 combase!PropVariantCopy [onecore\com\combase\util\propvar.cxx @ 544]
00007ffc`ac008108  00007ffc`f5de4340 combase!PropVariantClear [onecore\com\combase\util\propvar.cxx @ 278]
00007ffc`ac008110  00007ffc`f60cc4f0 ole32!RevokeDragDrop [com\ole32\ole232\drag\drag.cpp @ 3000]
00007ffc`ac008118  00000000`00000000
```

```
0:000> !dh xmllite

File Type: DLL
FILE HEADER VALUES
    8664 machine (X64)
       6 number of sections
CED9EC48 time date stamp Thu Dec 21 12:44:56 2079

       0 file pointer to symbol table
       0 number of symbols
      F0 size of optional header
    2C22 characteristics
            Executable
            App can handle >2gb addresses
            Run from swap file if image is on removable media
            Run from swap file if image is on net
            DLL

OPTIONAL HEADER VALUES
     20B magic #
   14.28 linker version
   28000 size of code
    E000 size of initialized data
       0 size of uninitialized data
    FD40 address of entry point
    1000 base of code
         ----- new -----
00007ffcef420000 image base
    1000 section alignment
    1000 file alignment
       2 subsystem (Windows GUI)
   10.00 operating system version
   10.00 image version
   10.00 subsystem version
   37000 size of image
    1000 size of headers
   3B6B2 checksum
0000000000040000 size of stack reserve
0000000000001000 size of stack commit
0000000000100000 size of heap reserve
0000000000001000 size of heap commit
    4160  DLL characteristics
            High entropy VA supported
            Dynamic base
            NX compatible
            Guard
   2FD20 [     130] address [size] of Export Directory
   2FE50 [     118] address [size] of Import Directory
   35000 [     420] address [size] of Resource Directory
   32000 [    2394] address [size] of Exception Directory
   37000 [    2958] address [size] of Security Directory
   36000 [     5F4] address [size] of Base Relocation Directory
   2B958 [      70] address [size] of Debug Directory
       0 [       0] address [size] of Description Directory
       0 [       0] address [size] of Special Directory
       0 [       0] address [size] of Thread Storage Directory
   29B00 [     138] address [size] of Load Configuration Directory
       0 [       0] address [size] of Bound Import Directory
    2A8F0 [     188] address [size] of Import Address Table Directory
```

```
       0 [       0] address [size] of Delay Import Directory
       0 [       0] address [size] of COR20 Header Directory
       0 [       0] address [size] of Reserved Directory


SECTION HEADER #1
   .text name
   27A2F virtual size
    1000 virtual address
   28000 size of raw data
    1000 file pointer to raw data
       0 file pointer to relocation table
       0 file pointer to line numbers
       0 number of relocations
       0 number of line numbers
60000020 flags
         Code
         (no align specified)
         Execute Read

SECTION HEADER #2
  .rdata name
    7536 virtual size
   29000 virtual address
    8000 size of raw data
   29000 file pointer to raw data
       0 file pointer to relocation table
       0 file pointer to line numbers
       0 number of relocations
       0 number of line numbers
40000040 flags
         Initialized Data
         (no align specified)
         Read Only


Debug Directories(4)
      Type       Size      Address  Pointer
      cv           24        2da3c    2da3c    Format: RSDS, guid, 1, XmlLite.pdb
      (   13)     3c8        2da60    2da60
      (   16)      24        2de28    2de28
      dllchar       4        2de4c    2de4c

00000001 extended DLL characteristics
          CET compatible


SECTION HEADER #3
   .data name
     F14 virtual size
   31000 virtual address
    1000 size of raw data
   31000 file pointer to raw data
       0 file pointer to relocation table
       0 file pointer to line numbers
       0 number of relocations
       0 number of line numbers
C0000040 flags
         Initialized Data
```

```
         (no align specified)
         Read Write

SECTION HEADER #4
  .pdata name
    2394 virtual size
   32000 virtual address
    3000 size of raw data
   32000 file pointer to raw data
       0 file pointer to relocation table
       0 file pointer to line numbers
       0 number of relocations
       0 number of line numbers
40000040 flags
         Initialized Data
         (no align specified)
         Read Only

SECTION HEADER #5
   .rsrc name
     420 virtual size
   35000 virtual address
    1000 size of raw data
   35000 file pointer to raw data
       0 file pointer to relocation table
       0 file pointer to line numbers
       0 number of relocations
       0 number of line numbers
40000040 flags
         Initialized Data
         (no align specified)
         Read Only

SECTION HEADER #6
  .reloc name
     5F4 virtual size
   36000 virtual address
    1000 size of raw data
   36000 file pointer to raw data
       0 file pointer to relocation table
       0 file pointer to line numbers
       0 number of relocations
       0 number of line numbers
42000040 flags
         Initialized Data
         Discardable
         (no align specified)
         Read Only

0:000> dps 00007ffcef420000 + 2A8F0 L188/8
00007ffc`ef44a8f0  00007ffc`f3b8cf40 KERNELBASE!SetUnhandledExceptionFilter
00007ffc`ef44a8f8  00007ffc`f3b74700 KERNELBASE!GetLastError
00007ffc`ef44a900  00007ffc`f3c727a0 KERNELBASE!UnhandledExceptionFilter
00007ffc`ef44a908  00000000`00000000
00007ffc`ef44a910  00007ffc`f62c8ac0 ntdll!RtlAllocateHeap
00007ffc`ef44a918  00007ffc`f3b7f1d0 KERNELBASE!GetProcessHeap
00007ffc`ef44a920  00007ffc`f62c75e0 ntdll!RtlFreeHeap
00007ffc`ef44a928  00000000`00000000
00007ffc`ef44a930  00007ffc`f3b915b0 KERNELBASE!DisableThreadLibraryCalls
```

```
00007ffc`ef44a938  00000000`00000000
00007ffc`ef44a940  00007ffc`f3b815a0 KERNELBASE!GetCPInfo
00007ffc`ef44a948  00000000`00000000
00007ffc`ef44a950  00007ffc`f3b8a2b0 KERNELBASE!VirtualQuery
00007ffc`ef44a958  00000000`00000000
00007ffc`ef44a960  00007ffc`f4fc6170 kernel32!GetCurrentThreadId
00007ffc`ef44a968  00007ffc`f4fdf800 kernel32!TerminateProcessStub
00007ffc`ef44a970  00007ffc`f4fe2bd0 kernel32!GetCurrentProcess
00007ffc`ef44a978  00007ffc`f4fe2be0 kernel32!GetCurrentProcessId
00007ffc`ef44a980  00000000`00000000
00007ffc`ef44a988  00007ffc`f4fdb7d0 kernel32!IsProcessorFeaturePresentStub
00007ffc`ef44a990  00000000`00000000
00007ffc`ef44a998  00007ffc`f62b28f0 ntdll!RtlQueryPerformanceCounter
00007ffc`ef44a9a0  00000000`00000000
00007ffc`ef44a9a8  00007ffc`f6347970 ntdll!RtlCaptureContext
00007ffc`ef44a9b0  00007ffc`f62d9ca0 ntdll!RtlLookupFunctionEntry
00007ffc`ef44a9b8  00007ffc`f62d8f10 ntdll!RtlVirtualUnwind
00007ffc`ef44a9c0  00000000`00000000
00007ffc`ef44a9c8  00007ffc`f3b7b6c0 KERNELBASE!WideCharToMultiByte
00007ffc`ef44a9d0  00007ffc`f3b6d750 KERNELBASE!MultiByteToWideChar
00007ffc`ef44a9d8  00000000`00000000
00007ffc`ef44a9e0  00007ffc`f3b7b320 KERNELBASE!Sleep
00007ffc`ef44a9e8  00000000`00000000
00007ffc`ef44a9f0  00007ffc`f3bb4ce0 KERNELBASE!GetTickCount
00007ffc`ef44a9f8  00007ffc`f3b86f00 KERNELBASE!GetSystemTimeAsFileTime
00007ffc`ef44aa00  00000000`00000000
00007ffc`ef44aa08  00007ffc`f415bc60 msvcrt!callnewh
00007ffc`ef44aa10  00007ffc`f416aac0 msvcrt!XcptFilter
00007ffc`ef44aa18  00007ffc`f416acc0 msvcrt!_C_specific_handler
00007ffc`ef44aa20  00007ffc`f417d520 msvcrt!initterm
00007ffc`ef44aa28  00007ffc`f415c6e0 msvcrt!malloc
00007ffc`ef44aa30  00007ffc`f41b7d40 msvcrt!memcpy
00007ffc`ef44aa38  00007ffc`f416f0b0 msvcrt!purecall
00007ffc`ef44aa40  00007ffc`f415c690 msvcrt!free
00007ffc`ef44aa48  00007ffc`f417d190 msvcrt!amsg_exit
00007ffc`ef44aa50  00007ffc`f418e200 msvcrt!vsnwprintf
00007ffc`ef44aa58  00007ffc`f41a0500 msvcrt!memcmp
00007ffc`ef44aa60  00007ffc`f41b7d40 msvcrt!memcpy
00007ffc`ef44aa68  00007ffc`f41b8000 msvcrt!memset
00007ffc`ef44aa70  00000000`00000000
```

11. We close logging before exiting WinDbg:

```
0:000> .logclose
Closing open log file C:\AWAPI-Dumps\W2.log
```

API Usage

- Module usage (static analysis)
  - Hidden Module
- Function usage (dynamic analysis)

WinDbg Commands

```
0:000> .imgscan

0:000> bm mpattern!fpattern
```


To find out whether a particular module is potentially used, we can look at **Module Collection** or scan in memory for **Hidden Modules** using the **.imgscan** WinDbg command. To find out whether a particular module uses this or that function, we can just look at its Import Address Table. It can be done statically, looking at a memory dump, for example. However, if we want to find out where in code a particular function is used, we can't search for its address or its import table address because usually relative addressing is used. Here we can find the location in code dynamically, for example, using a breakpoint.

Hidden Module
https://www.dumpanalysis.org/blog/index.php/2008/08/07/crash-dump-analysis-patterns-part-75/

Warning

Because of live debugging, due to differences in actual systems and ASLR (Address Space Layout Randomization), when you launch applications, actual addresses and even the number and order of threads in WinDbg command output may differ from those shown in exercise transcripts.

Exercise W3

- **Goal:** Find usage of specific Windows API functions
- **Debugging Implementation Patterns:** Code Breakpoint; Breakpoint Action
- \AWAPI-Dumps\Exercise-W3.pdf

Exercise W3

Goal: Find usage of specific Windows API functions.

Debugging Implementation Patterns: Code Breakpoint; Breakpoint Action.

1. Launch WinDbg.

2. Choose Launch Executable and \Program Files\Windows NT\Accessories\wordpad.exe (you can also choose any other process and API functions).

3. We get the process launched with an initial breakpoint:

```
Microsoft (R) Windows Debugger Version 10.0.27725.1000 AMD64
Copyright (c) Microsoft Corporation. All rights reserved.

CommandLine: C:\Program Files\Windows NT\Accessories\wordpad.exe

************* Path validation summary **************
Response                         Time (ms)     Location
Deferred                                       srv*
Symbol search path is: srv*
Executable search path is:
ModLoad: 00007ff7`84f60000 00007ff7`8522e000   wordpad.exe
ModLoad: 00007ffd`50350000 00007ffd`50567000   ntdll.dll
ModLoad: 00007ffd`4e900000 00007ffd`4e9c4000   C:\WINDOWS\System32\KERNEL32.DLL
ModLoad: 00007ffd`4dbe0000 00007ffd`4df9a000   C:\WINDOWS\System32\KERNELBASE.dll
ModLoad: 00007ffd`46c70000 00007ffd`46d07000   C:\WINDOWS\SYSTEM32\apphelp.dll
ModLoad: 00007ffd`23640000 00007ffd`236b0000   C:\WINDOWS\SYSTEM32\AcGenral.dll
ModLoad: 00007ffd`4e7a0000 00007ffd`4e847000   C:\WINDOWS\System32\msvcrt.dll
ModLoad: 00007ffd`4f140000 00007ffd`4f1e7000   C:\WINDOWS\System32\sechost.dll
ModLoad: 00007ffd`4dfa0000 00007ffd`4dfc8000   C:\WINDOWS\System32\bcrypt.dll
ModLoad: 00007ffd`4ee00000 00007ffd`4ee5e000   C:\WINDOWS\System32\SHLWAPI.dll
ModLoad: 00007ffd`4e090000 00007ffd`4e23e000   C:\WINDOWS\System32\USER32.dll
ModLoad: 00007ffd`4d740000 00007ffd`4d766000   C:\WINDOWS\System32\win32u.dll
ModLoad: 00007ffd`4f110000 00007ffd`4f139000   C:\WINDOWS\System32\GDI32.dll
ModLoad: 00007ffd`4d770000 00007ffd`4d88b000   C:\WINDOWS\System32\gdi32full.dll
ModLoad: 00007ffd`4d620000 00007ffd`4d6ba000   C:\WINDOWS\System32\msvcp_win.dll
ModLoad: 00007ffd`4da00000 00007ffd`4db11000   C:\WINDOWS\System32\ucrtbase.dll
ModLoad: 00007ffd`4e560000 00007ffd`4e705000   C:\WINDOWS\System32\ole32.dll
ModLoad: 00007ffd`4fb00000 00007ffd`4fe8f000   C:\WINDOWS\System32\combase.dll
ModLoad: 00007ffd`4ee60000 00007ffd`4ef74000   C:\WINDOWS\System32\RPCRT4.dll
ModLoad: 00007ffd`4e9d0000 00007ffd`4ea82000   C:\WINDOWS\System32\advapi32.dll
ModLoad: 00007ffd`4e3d0000 00007ffd`4e4ca000   C:\WINDOWS\System32\shcore.dll
ModLoad: 00007ffd`4f1f0000 00007ffd`4fa68000   C:\WINDOWS\System32\SHELL32.dll
ModLoad: 00007ffd`4cb30000 00007ffd`4cb58000   C:\WINDOWS\SYSTEM32\USERENV.dll
ModLoad: 00007ffd`3e3d0000 00007ffd`3e3ee000   C:\WINDOWS\SYSTEM32\MPR.dll
ModLoad: 00007ffd`4c880000 00007ffd`4c8c3000   C:\WINDOWS\SYSTEM32\SspiCli.dll
ModLoad: 00007ffd`4e050000 00007ffd`4e081000   C:\WINDOWS\System32\IMM32.DLL
ModLoad: 00007ffd`4e2d0000 00007ffd`4e3d0000   C:\WINDOWS\System32\COMDLG32.dll
ModLoad: 00007ffd`0e3c0000 00007ffd`0e535000   C:\WINDOWS\SYSTEM32\MFC42u.dll
ModLoad: 00007ffd`4eb40000 00007ffd`4ec17000   C:\WINDOWS\System32\OLEAUT32.dll
ModLoad: 00007ffd`339a0000 00007ffd`33c32000
C:\WINDOWS\WinSxS\amd64_microsoft.windows.common-
controls_6595b64144ccf1df_6.0.22621.4541_none_2710d1c57384c085\COMCTL32.dll
```

```
ModLoad: 00000295`27ab0000 00000295`27d42000
C:\WINDOWS\WinSxS\amd64_microsoft.windows.common-
controls_6595b64144ccf1df_6.0.22621.4541_none_2710d1c57384c085\COMCTL32.dll
ModLoad: 00007ffd`4a0f0000 00007ffd`4a1f1000   C:\WINDOWS\SYSTEM32\PROPSYS.dll
ModLoad: 00007ffd`42aa0000 00007ffd`42ad4000   C:\WINDOWS\SYSTEM32\WINMM.dll
ModLoad: 00007ffd`43650000 00007ffd`43687000   C:\WINDOWS\SYSTEM32\XmlLite.dll
ModLoad: 00007ffd`36c00000 00007ffd`36df0000   C:\WINDOWS\SYSTEM32\urlmon.dll
ModLoad: 00007ffd`36940000 00007ffd`36bfe000   C:\WINDOWS\SYSTEM32\iertutil.dll
ModLoad: 00007ffd`4c070000 00007ffd`4c07c000   C:\WINDOWS\SYSTEM32\netutils.dll
ModLoad: 00007ffd`3e370000 00007ffd`3e398000   C:\WINDOWS\SYSTEM32\srvcli.dll
(7020.166ec): Break instruction exception - code 80000003 (first chance)
ntdll!LdrpDoDebuggerBreak+0x30:
00007ffd`5042c134 int     3
```

4. Open a log file using **.logopen**:

```
0:000> .logopen C:\AWAPI-Dumps\W3.log
Opened log file 'C:\AWAPI-Dumps\W3.log'
```

5. Let's see why we can't search for the *module!_imp_ CreateWindowExW* function address to see its usages (calls) in a particular *module*. We disassemble again the *combase!InitMainThreadWnd* function we looked at in Exercise W1:

```
0:000> uf combase!InitMainThreadWnd
combase!InitMainThreadWnd [onecore\com\combase\objact\mainthrd.cxx @ 114]:
  114 00007ff8`09f9aca4 4c8bdc          mov     r11,rsp
  114 00007ff8`09f9aca7 53              push    rbx
  114 00007ff8`09f9aca8 4883ec60        sub     rsp,60h
  148 00007ff8`09f9acac 498363f000      and     qword ptr [r11-10h],0
  148 00007ff8`09f9acb1 4c8d0580842500  lea     r8,[combase!`string' (00007ff8`0a1f3138)]
  148 00007ff8`09f9acb8 488b0529c62c00  mov     rax,qword ptr [combase!g_hinst (00007ff8`0a2672e8)]
  148 00007ff8`09f9acbf 41b900000088    mov     r9d,88000000h
  148 00007ff8`09f9acc5 488b159ccf2c00  mov     rdx,qword ptr [combase!gOleWindowClass (00007ff8`0a267c68)]
  148 00007ff8`09f9accc 33c9            xor     ecx,ecx
  148 00007ff8`09f9acce 498943e8        mov     qword ptr [r11-18h],rax
  148 00007ff8`09f9acd2 b800000080      mov     eax,80000000h
  148 00007ff8`09f9acd7 498363e000      and     qword ptr [r11-20h],0
  148 00007ff8`09f9acdc 49c743d8fdffffff mov     qword ptr [r11-28h],0FFFFFFFFFFFFFFFDh
  148 00007ff8`09f9ace4 89442438        mov     dword ptr [rsp+38h],eax
  148 00007ff8`09f9ace8 89442430        mov     dword ptr [rsp+30h],eax
  148 00007ff8`09f9acec 89442428        mov     dword ptr [rsp+28h],eax
  148 00007ff8`09f9acf0 89442420        mov     dword ptr [rsp+20h],eax
  148 00007ff8`09f9acf4 48ff1505c92f00  call    qword ptr [combase!_imp_CreateWindowExW (00007ff8`0a297600)]
  148 00007ff8`09f9acfb 0f1f440000      nop     dword ptr [rax+rax]
  148 00007ff8`09f9ad00 48890559cf2c00  mov     qword ptr [combase!ghwndOleMainThread (00007ff8`0a267c60)],rax
  163 00007ff8`09f9ad07 4885c0          test    rax,rax
  163 00007ff8`09f9ad0a 0f847ec81000    je      combase!InitMainThreadWnd+0x10c8ea (00007ff8`0a0a758e)  Branch

combase!InitMainThreadWnd+0x6c [onecore\com\combase\objact\mainthrd.cxx @ 180]:
  180 00007ff8`09f9ad10 48ff1529652400  call    qword ptr [combase!_imp_GetCurrentThreadId (00007ff8`0a1e1240)]
  180 00007ff8`09f9ad17 0f1f440000      nop     dword ptr [rax+rax]
  180 00007ff8`09f9ad1c 890536cf2c00    mov     dword ptr [combase!gdwMainThreadId (00007ff8`0a267c58)],eax
  181 00007ff8`09f9ad22 33db            xor     ebx,ebx

combase!InitMainThreadWnd+0x80 [onecore\com\combase\objact\mainthrd.cxx @ 186]:
  186 00007ff8`09f9ad24 8bc3            mov     eax,ebx
  187 00007ff8`09f9ad26 4883c460        add     rsp,60h
  187 00007ff8`09f9ad2a 5b              pop     rbx
  187 00007ff8`09f9ad2b c3              ret

combase!InitMainThreadWnd+0x10c8ea [onecore\com\combase\objact\mainthrd.cxx @ 168]:
  168 00007ff8`0a0a758e 48ff15bb991300  call    qword ptr [combase!_imp_GetLastError (00007ff8`0a1e0f50)]
  168 00007ff8`0a0a7595 0f1f440000      nop     dword ptr [rax+rax]
  168 00007ff8`0a0a759a 8bd8            mov     ebx,eax
  168 00007ff8`0a0a759c 85c0            test    eax,eax
  168 00007ff8`0a0a759e 7e09            jle     combase!InitMainThreadWnd+0x10c905 (00007ff8`0a0a75a9)  Branch
```

```
combase!InitMainThreadWnd+0x10c8fc [onecore\com\combase\object\mainthrd.cxx @ 168]:
  168 00007ff8`0a0a75a0 0fb7d8          movzx   ebx,ax
  168 00007ff8`0a0a75a3 81cb00000780    or      ebx,80070000h

combase!InitMainThreadWnd+0x10c905 [onecore\com\combase\object\mainthrd.cxx @ 169]:
  169 00007ff8`0a0a75a9 833d28dd1b0000  cmp     dword ptr [combase!_Tlgg_hCombaseTraceLoggingProviderProv
(00007ff8`0a2652d8)],0
  169 00007ff8`0a0a75b0 7714            ja      combase!InitMainThreadWnd+0x10c922 (00007ff8`0a0a75c6)  Branch

combase!InitMainThreadWnd+0x10c90e [onecore\com\combase\object\mainthrd.cxx @ 169]:
  169 00007ff8`0a0a75b2 833db3fe1b0000  cmp     dword ptr [combase!gfEnableTracing (00007ff8`0a26746c)],0
  169 00007ff8`0a0a75b9 743b            je      combase!InitMainThreadWnd+0x10c952 (00007ff8`0a0a75f6)  Branch

combase!InitMainThreadWnd+0x10c917 [onecore\com\combase\object\mainthrd.cxx @ 169]:
  169 00007ff8`0a0a75bb 33c9            xor     ecx,ecx
  169 00007ff8`0a0a75bd e892f3f2ff      call    combase!IsWppLevelEnabled (00007ff8`09fd6954)
  169 00007ff8`0a0a75c2 84c0            test    al,al
  169 00007ff8`0a0a75c4 7430            je      combase!InitMainThreadWnd+0x10c952 (00007ff8`0a0a75f6)  Branch

combase!InitMainThreadWnd+0x10c922 [onecore\com\combase\object\mainthrd.cxx @ 169]:
  169 00007ff8`0a0a75c6 488d05239a1400  lea     rax,[combase!`string' (00007ff8`0a1f0ff0)]
  169 00007ff8`0a0a75cd 895c2430        mov     dword ptr [rsp+30h],ebx
  169 00007ff8`0a0a75d1 4889442428      mov     qword ptr [rsp+28h],rax
  169 00007ff8`0a0a75d6 4c8d05bbe61700  lea     r8,[combase!`string' (00007ff8`0a225c98)]
  169 00007ff8`0a0a75dd 8364242000      and     dword ptr [rsp+20h],0
  169 00007ff8`0a0a75e2 488d1587e61700  lea     rdx,[combase!`string' (00007ff8`0a225c70)]
  169 00007ff8`0a0a75e9 41b9a9000000    mov     r9d,0A9h
  169 00007ff8`0a0a75ef 8bcb            mov     ecx,ebx
  169 00007ff8`0a0a75f1 e8aa78f5ff      call    combase!ComTraceMessage (00007ff8`09ffeea0)

combase!InitMainThreadWnd+0x10c952 [onecore\com\combase\object\mainthrd.cxx @ 170]:
  170 00007ff8`0a0a75f6 85db            test    ebx,ebx
  170 00007ff8`0a0a75f8 0f882637efff    js      combase!InitMainThreadWnd+0x80 (00007ff8`09f9ad24)  Branch

combase!InitMainThreadWnd+0x10c95a [onecore\com\combase\object\mainthrd.cxx @ 172]:
  172 00007ff8`0a0a75fe bb0e000780      mov     ebx,8007000Eh
  176 00007ff8`0a0a7603 e91c37efff      jmp     combase!InitMainThreadWnd+0x80 (00007ff8`09f9ad24)  Branch
```

Note: We don't see the address encoded in code instruction bytes, even accounting for little-endian Intel architecture (the least significant byte is at the lowest address). Only a relative offset from the current instruction pointer address is encoded, which differs in different caller functions.

```
0:000> ? 00007ff8`0a297600 - 00007ff8`09f9acfb
Evaluate expression: 3131653 = 00000000`002fc905
```

Compare the result with (note the reverse order of bytes):

```
  148 00007ff8`09f9acf4 48ff1505c92f00  call    qword ptr [combase!_imp_CreateWindowExW (00007ff8`0a297600)]
```

Note: If you don't see code bytes in the output of the **uf** command, check your asm options:

```
0:000> .asm
Assembly options: no_code_bytes

0:000> .asm- no_code_bytes
Assembly options: <default>
```

6. But we can find usages by creating pattern-matching breakpoints:

```
0:000> bm *!CreateWindowEx*
  0: 00007ff8`09d0d7d0 @!"KERNELBASE!CreateWindowExA"
  1: 00007ff8`09d0d800 @!"KERNELBASE!CreateWindowExW"
```

```
  2: 00007ff8`0be4af90 @!"USER32!CreateWindowExA"
  3: 00007ff8`0be2f190 @!"USER32!CreateWindowExW"

0:000> bl
     0 e Disable Clear  00007ff8`09d0d7d0     0001 (0001)  0:**** KERNELBASE!CreateWindowExA
     1 e Disable Clear  00007ff8`09d0d800     0001 (0001)  0:**** KERNELBASE!CreateWindowExW
     2 e Disable Clear  00007ff8`0be4af90     0001 (0001)  0:**** USER32!CreateWindowExA
     3 e Disable Clear  00007ff8`0be2f190     0001 (0001)  0:**** USER32!CreateWindowExW
```

7. Now we can resume execution until a breakpoint is reached and print stack trace (**L** omits source code references if any):

```
0:000> g
ModLoad: 00007fff`d63b0000 00007fff`d641b000   C:\WINDOWS\SYSTEM32\ninput.dll
ModLoad: 00007ff8`07690000 00007ff8`076a8000   C:\WINDOWS\SYSTEM32\kernel.appcore.dll
ModLoad: 00007ff8`09560000 00007ff8`095db000   C:\WINDOWS\System32\bcryptPrimitives.dll
Breakpoint 3 hit
USER32!CreateWindowExW:
00007ff8`0be2f190 4c8bdc          mov     r11,rsp

0:000> kL
 # Child-SP          RetAddr               Call Site
00 00000040`3607fa28 00007ff8`09f9acfb     USER32!CreateWindowExW
01 00000040`3607fa30 00007ff8`09fc0a69     combase!InitMainThreadWnd+0x57
02 00000040`3607faa0 00007ff8`09fc25b5     combase!ThreadFirstInitialize+0x181
03 00000040`3607faf0 00007ff8`09fc2287     combase!_CoInitializeEx+0x1bd
04 00000040`3607fbe0 00007ff7`f383d22b     combase!CoInitializeEx+0x37
05 00000040`3607fc30 00007ff7`f37f9b19     wordpad!CWordPadApp::CWordPadApp+0x263
06 00000040`3607fc60 00007ff8`0a57dbbd     wordpad!`dynamic initializer for 'theApp''+0x9
07 00000040`3607fc90 00007ff7`f37fb3d7     msvcrt!initterm+0x2d
08 00000040`3607fcc0 00007ff8`0b03244d     wordpad!__wmainCRTStartup+0x117
09 00000040`3607fd80 00007ff8`0c18df78     KERNEL32!BaseThreadInitThunk+0x1d
0a 00000040`3607fdb0 00000000`00000000     ntdll!RtlUserThreadStart+0x28

0:000> g
ModLoad: 00007ff8`05b40000 00007ff8`05beb000   C:\WINDOWS\system32\uxtheme.dll
ModLoad: 00007ff8`0ba10000 00007ff8`0bac7000   C:\WINDOWS\System32\clbcatq.dll
ModLoad: 00007fff`84ff0000 00007fff`8525c000   C:\WINDOWS\System32\msxml3.dll
ModLoad: 00007ff8`08cd0000 00007ff8`08cf8000   C:\WINDOWS\System32\bcrypt.dll
ModLoad: 00007fff`ccf50000 00007fff`cd2bd000   C:\WINDOWS\SYSTEM32\MSFTEDIT.DLL
ModLoad: 00007fff`e0300000 00007fff`e039a000   C:\WINDOWS\SYSTEM32\WINSPOOL.DRV
ModLoad: 00007fff`5ce90000 00007fff`5d2af000   C:\WINDOWS\system32\UIRibbon.dll
ModLoad: 00007fff`e1140000 00007fff`e12f9000
C:\WINDOWS\WinSxS\amd64_microsoft.windows.gdiplus_6595b64144ccf1df_1.1.22621.674_none_da2b3d691
ba1c6c6\gdiplus.dll
ModLoad: 00007ff8`02320000 00007ff8`02397000   C:\WINDOWS\System32\netprofm.dll
Breakpoint 3 hit
USER32!CreateWindowExW:
00007ff8`0be2f190 4c8bdc          mov     r11,rsp

0:000> kL
 # Child-SP          RetAddr               Call Site
00 00000040`3607e458 00007fff`7c7ec835     USER32!CreateWindowExW
01 00000040`3607e460 00007fff`7c7ef6c5     MFC42u!CWnd::CreateEx+0xf5
02 00000040`3607e520 00007fff`7c7ef7b3     MFC42u!CFrameWnd::Create+0xe5
03 00000040`3607e590 00007fff`7c7ef0ea     MFC42u!CFrameWnd::LoadFrame+0xb3
04 00000040`3607e610 00007fff`7c7fa1ec     MFC42u!CDocTemplate::CreateNewFrame+0x7a
05 00000040`3607e680 00007ff7`f383e331     MFC42u!CSingleDocTemplate::OpenDocumentFile+0x8c
```

```
06 00000040`3607e6e0 00007fff`7c7efa57     wordpad!CWordPadApp::InitInstance+0x521
07 00000040`3607fc80 00007ff7`f37fb49e     MFC42u!AfxWinMain+0x97
08 00000040`3607fcc0 00007ff8`0b03244d     wordpad!__wmainCRTStartup+0x1de
09 00000040`3607fd80 00007ff8`0c18df78     KERNEL32!BaseThreadInitThunk+0x1d
0a 00000040`3607fdb0 00000000`00000000     ntdll!RtlUserThreadStart+0x28
```

```
0:000> g
ModLoad: 00007ff8`0bfd0000 00007ff8`0c0ee000   C:\WINDOWS\System32\MSCTF.dll
Breakpoint 3 hit
USER32!CreateWindowExW:
00007ff8`0be2f190 4c8bdc          mov     r11,rsp
```

```
0:000> kL
 # Child-SP          RetAddr               Call Site
00 00000040`3607d588 00007fff`7c7ec835     USER32!CreateWindowExW
01 00000040`3607d590 00007fff`7c7ec919     MFC42u!CWnd::CreateEx+0xf5
02 00000040`3607d650 00007fff`7c7ef1b0     MFC42u!CWnd::Create+0x99
03 00000040`3607d6c0 00007fff`7c7ef263     MFC42u!CFrameWnd::CreateView+0x80
04 00000040`3607d750 00007fff`7c7f2a25     MFC42u!CFrameWnd::OnCreateClient+0x23
05 00000040`3607d780 00007ff7`f38351f9     MFC42u!CFrameWnd::OnCreateHelper+0x45
06 00000040`3607d7b0 00007fff`7c7e8f27     wordpad!CMainFrame::OnCreate+0x19
07 00000040`3607d800 00007ff7`f3836f5c     MFC42u!CWnd::OnWndMsg+0x387
08 00000040`3607d930 00007fff`7c7f0bff     wordpad!CMainFrame::OnWndMsg+0x14c
09 00000040`3607d9c0 00007fff`7c7eb42a     MFC42u!CWnd::WindowProc+0x4f
0a 00000040`3607da00 00007fff`7c7ea76b     MFC42u!AfxCallWndProc+0x14a
0b 00000040`3607db00 00007ff8`0be38161     MFC42u!AfxWndProcBase+0x15b
0c 00000040`3607db70 00007ff8`0be37e1c     USER32!UserCallWinProcCheckWow+0x2d1
0d 00000040`3607dcd0 00007ff8`0be44edc     USER32!DispatchClientMessage+0x9c
0e 00000040`3607dd30 00007ff8`0c1d2db4     USER32!__fnINLPCREATESTRUCT+0xac
0f 00000040`3607dd90 00007ff8`097f2294     ntdll!KiUserCallbackDispatcherContinue
10 00000040`3607ded8 00007ff8`0be2f6b0     win32u!NtUserCreateWindowEx+0x14
11 00000040`3607dee0 00007ff8`0be2f3cc     USER32!VerNtUserCreateWindowEx+0x210
12 00000040`3607e270 00007ff8`0be2f212     USER32!CreateWindowInternal+0x1a8
13 00000040`3607e3d0 00007fff`7c7ec835     USER32!CreateWindowExW+0x82
14 00000040`3607e460 00007fff`7c7ef6c5     MFC42u!CWnd::CreateEx+0xf5
15 00000040`3607e520 00007fff`7c7ef7b3     MFC42u!CFrameWnd::Create+0xe5
16 00000040`3607e590 00007fff`7c7ef0ea     MFC42u!CFrameWnd::LoadFrame+0xb3
17 00000040`3607e610 00007fff`7c7fa1ec     MFC42u!CDocTemplate::CreateNewFrame+0x7a
18 00000040`3607e680 00007ff7`f383e331     MFC42u!CSingleDocTemplate::OpenDocumentFile+0x8c
19 00000040`3607e6e0 00007fff`7c7efa57     wordpad!CWordPadApp::InitInstance+0x521
1a 00000040`3607fc80 00007ff7`f37fb49e     MFC42u!AfxWinMain+0x97
1b 00000040`3607fcc0 00007ff8`0b03244d     wordpad!__wmainCRTStartup+0x1de
1c 00000040`3607fd80 00007ff8`0c18df78     KERNEL32!BaseThreadInitThunk+0x1d
1d 00000040`3607fdb0 00000000`00000000     ntdll!RtlUserThreadStart+0x28
```

Note: We see that stack trace may have several usages of the same API function.

8. We can add an action to breakpoints to automate printing stack traces:

```
0:000> bm *!CreateWindowEx* "kL; g"
breakpoint 0 redefined
  0: 00007ff8`09d0d7d0 @!"KERNELBASE!CreateWindowExA"
breakpoint 1 redefined
  1: 00007ff8`09d0d800 @!"KERNELBASE!CreateWindowExW"
breakpoint 2 redefined
  2: 00007ff8`0be4af90 @!"USER32!CreateWindowExA"
breakpoint 3 redefined
  3: 00007ff8`0be2f190 @!"USER32!CreateWindowExW"
```

```
0:000> bl
     0 e Disable Clear  00007ff8`09d0d7d0     0001 (0001)  0:**** KERNELBASE!CreateWindowExA "kL; g"
     1 e Disable Clear  00007ff8`09d0d800     0001 (0001)  0:**** KERNELBASE!CreateWindowExW "kL; g"
     2 e Disable Clear  00007ff8`0be4af90     0001 (0001)  0:**** USER32!CreateWindowExA "kL; g"
     3 e Disable Clear  00007ff8`0be2f190     0001 (0001)  0:**** USER32!CreateWindowExW "kL; g"

0:000> g
 # Child-SP          RetAddr               Call Site
00 00000040`3607b258 00007ff8`0bfe92a0     USER32!CreateWindowExW
01 00000040`3607b260 00007ff8`0bfe91de     MSCTF!SYSTHREAD::LockThreadMessageWindow+0xa8
02 00000040`3607b340 00007ff8`0bfe90c1     MSCTF!CCtfClientPort::CCtfClientPort+0x52
03 00000040`3607b370 00007ff8`0c01eb06     MSCTF!CCtfClientPort::CreateInstance+0x4d
04 00000040`3607b3a0 00007ff8`0bff5d14     MSCTF!CThreadInputMgr::OnActivationChange+0x116
05 00000040`3607b420 00007ff8`0c011771     MSCTF!CThreadInputMgr::ActivateEx_P+0x814
06 00000040`3607bd40 00007fff`ccf94522     MSCTF!CThreadInputMgr::Activate+0x21
07 00000040`3607bd70 00007fff`ccf94376     MSFTEDIT!CreateUIM+0x5a
08 00000040`3607bdb0 00007fff`ccf5f270     MSFTEDIT!CTextMsgFilter::StartUIM+0xe
09 00000040`3607bde0 00007fff`ccfcba65     MSFTEDIT!CTextMsgFilter::HandleMessage+0x1020
0a 00000040`3607bec0 00007fff`ccfcbd0e     MSFTEDIT!CTxtEdit::FilterMessageIfPossible+0x65
0b 00000040`3607bf20 00007fff`cd00f63c     MSFTEDIT!CTxtEdit::TxSendMessage+0x9e
0c 00000040`3607c220 00007ff8`0be38161     MSFTEDIT!RichEditWndProc+0x38c
0d 00000040`3607c330 00007ff8`0be379ab     USER32!UserCallWinProcCheckWow+0x2d1
0e 00000040`3607c490 00007fff`7c7f0c8e     USER32!CallWindowProcW+0x8b
0f 00000040`3607c4e0 00007fff`7c7f0c28     MFC42u!CWnd::DefWindowProcW+0x3e
10 00000040`3607c520 00007fff`7c7eb42a     MFC42u!CWnd::WindowProc+0x78
11 00000040`3607c560 00007fff`7c7ea76b     MFC42u!AfxCallWndProc+0x14a
12 00000040`3607c660 00007ff8`0be38161     MFC42u!AfxWndProcBase+0x15b
13 00000040`3607c6d0 00007ff8`0be3768d     USER32!UserCallWinProcCheckWow+0x2d1
14 00000040`3607c830 00007ff8`0be37447     USER32!SendMessageWorker+0x1dd
15 00000040`3607c8e0 00007ff7`f37f44c7     USER32!SendMessageW+0x137
16 00000040`3607c940 00007ff7`f3851812     wordpad!CRichEdit2View::OnCreate+0x177
17 00000040`3607c9b0 00007fff`7c7e8f27     wordpad!CWordPadView::OnCreate+0x12
18 00000040`3607c9e0 00007fff`7c7f0bff     MFC42u!CWnd::OnWndMsg+0x387
19 00000040`3607cb10 00007fff`7c7eb42a     MFC42u!CWnd::WindowProc+0x4f
1a 00000040`3607cb50 00007fff`7c7ea76b     MFC42u!AfxCallWndProc+0x14a
1b 00000040`3607cc50 00007ff8`0be38161     MFC42u!AfxWndProcBase+0x15b
1c 00000040`3607ccc0 00007ff8`0be37e1c     USER32!UserCallWinProcCheckWow+0x2d1
1d 00000040`3607ce20 00007ff8`0be44edc     USER32!DispatchClientMessage+0x9c
1e 00000040`3607ce80 00007ff8`0c1d2db4     USER32!__fnINLPCREATESTRUCT+0xac
1f 00000040`3607cee0 00007ff8`097f2294     ntdll!KiUserCallbackDispatcherContinue
20 00000040`3607d008 00007ff8`0be2f6b0     win32u!NtUserCreateWindowEx+0x14
21 00000040`3607d010 00007ff8`0be2f3cc     USER32!VerNtUserCreateWindowEx+0x210
22 00000040`3607d3a0 00007ff8`0be2f212     USER32!CreateWindowInternal+0x1a8
23 00000040`3607d500 00007fff`7c7ec835     USER32!CreateWindowExW+0x82
24 00000040`3607d590 00007fff`7c7ec919     MFC42u!CWnd::CreateEx+0xf5
25 00000040`3607d650 00007fff`7c7ef1b0     MFC42u!CWnd::Create+0x99
26 00000040`3607d6c0 00007fff`7c7ef263     MFC42u!CFrameWnd::CreateView+0x80
27 00000040`3607d750 00007fff`7c7f2a25     MFC42u!CFrameWnd::OnCreateClient+0x23
28 00000040`3607d780 00007ff7`f38351f9     MFC42u!CFrameWnd::OnCreateHelper+0x45
29 00000040`3607d7b0 00007fff`7c7e8f27     wordpad!CMainFrame::OnCreate+0x19
2a 00000040`3607d800 00007ff7`f3836f5c     MFC42u!CWnd::OnWndMsg+0x387
2b 00000040`3607d930 00007fff`7c7f0bff     wordpad!CMainFrame::OnWndMsg+0x14c
2c 00000040`3607d9c0 00007fff`7c7eb42a     MFC42u!CWnd::WindowProc+0x4f
2d 00000040`3607da00 00007fff`7c7ea76b     MFC42u!AfxCallWndProc+0x14a
2e 00000040`3607db00 00007ff8`0be38161     MFC42u!AfxWndProcBase+0x15b
2f 00000040`3607db70 00007ff8`0be37e1c     USER32!UserCallWinProcCheckWow+0x2d1
30 00000040`3607dcd0 00007ff8`0be44edc     USER32!DispatchClientMessage+0x9c
31 00000040`3607dd30 00007ff8`0c1d2db4     USER32!__fnINLPCREATESTRUCT+0xac
32 00000040`3607dd90 00007ff8`097f2294     ntdll!KiUserCallbackDispatcherContinue
33 00000040`3607ded8 00007ff8`0be2f6b0     win32u!NtUserCreateWindowEx+0x14
34 00000040`3607dee0 00007ff8`0be2f3cc     USER32!VerNtUserCreateWindowEx+0x210
35 00000040`3607e270 00007ff8`0be2f212     USER32!CreateWindowInternal+0x1a8
36 00000040`3607e3d0 00007fff`7c7ec835     USER32!CreateWindowExW+0x82
37 00000040`3607e460 00007fff`7c7ef6c5     MFC42u!CWnd::CreateEx+0xf5
38 00000040`3607e520 00007fff`7c7ef7b3     MFC42u!CFrameWnd::Create+0xe5
39 00000040`3607e590 00007fff`7c7ef0ea     MFC42u!CFrameWnd::LoadFrame+0xb3
3a 00000040`3607e610 00007fff`7c7fa1ec     MFC42u!CDocTemplate::CreateNewFrame+0x7a
3b 00000040`3607e680 00007ff7`f383e331     MFC42u!CSingleDocTemplate::OpenDocumentFile+0x8c
3c 00000040`3607e6e0 00007fff`7c7efa57     wordpad!CWordPadApp::InitInstance+0x521
3d 00000040`3607fc80 00007ff7`f37fb49e     MFC42u!AfxWinMain+0x97
3e 00000040`3607fcc0 00007ff8`0b03244d     wordpad!__wmainCRTStartup+0x1de
3f 00000040`3607fd80 00007ff8`0c18df78     KERNEL32!BaseThreadInitThunk+0x1d
40 00000040`3607fdb0 00000000`00000000     ntdll!RtlUserThreadStart+0x28

 # Child-SP          RetAddr               Call Site
00 00000040`3627de98 00007fff`f3cce048     USER32!CreateWindowExW
01 00000040`3627dea0 00007fff`f3ccdf1f     urlmon!NotificationWindow::CreateHandle+0x68
```

```
02 00000040`3627df10 00007fff`f3c7cc9b     urlmon!CTransaction::GetNotificationWnd+0x13f
03 00000040`3627df50 00007fff`f3cb98ee     urlmon!GetTransactionObjects+0x2fb
04 00000040`3627dfd0 00007fff`f3cba81c     urlmon!CBinding::StartBinding+0x53e
05 00000040`3627e150 00007fff`f3cb9276     urlmon!CUrlMon::StartBinding+0x1b0
06 00000040`3627e220 00007fff`f1c05529     urlmon!CUrlMon::BindToStorage+0x96
07 00000040`3627e270 00007fff`f1c04acb     msxml6!URLMONStream::deferedOpen+0x159
08 00000040`3627e2d0 00007fff`f1c048df     msxml6!XMLParser::PushURL+0x1af
09 00000040`3627e350 00007fff`f1bbf1c6     msxml6!XMLParser::SetURL+0x7f
0a 00000040`3627e3a0 00007fff`f1bbed48     msxml6!Document::_load+0x162
0b 00000040`3627e420 00007fff`f1b87712     msxml6!Document::load+0x198
0c 00000040`3627e490 00007fff`7c46afae     msxml6!DOMDocumentWrapper::load+0x182
0d 00000040`3627e560 00007fff`7c414968     PrintConfig!PrintConfig::PrivateDevmodeMapFile::Parse+0x14e
0e 00000040`3627e6d0 00007fff`7c558d7b     PrintConfig!PrintConfigPlugIn::DevMode+0x458
0f 00000040`3627e7f0 00007fff`7c512b53     PrintConfig!BCalcTotalOEMDMSize+0xa3
10 00000040`3627e890 00007fff`7c5120ec     PrintConfig!UniDrvUI::LSimpleDocumentProperties+0x10f
11 00000040`3627e900 00007fff`7c40f54b     PrintConfig!UniDrvUI::DrvDocumentPropertySheets+0x60
12 00000040`3627edb0 00007fff`7c4249ad     PrintConfig!PrintConfig::DrvDocumentPropertySheets+0x123
13 00000040`3627ee20 00007fff`7c425543     PrintConfig!ExceptionBoundary<<lambda_ce41999e58d4ed069b2790d632f6e3b5> >+0x3d
14 00000040`3627ee70 00007fff`e032e9ba     PrintConfig!DrvDocumentPropertySheets+0x53
15 00000040`3627eee0 00007fff`e031593f     WINSPOOL!DocumentPropertySheets+0x3fa
16 00000040`3627ef30 00007fff`e0324aff     WINSPOOL!DocumentPropertiesWNative+0x193
17 00000040`3627efc0 00007ff8`0a340f67     WINSPOOL!DocumentPropertiesW+0xcf
18 00000040`3627f020 00007ff8`0a344d59     COMDLG32!PrintGetDevMode+0x2b
19 00000040`3627f060 00007ff8`0a34058b     COMDLG32!PrintReturnDefault+0x91
1a 00000040`3627f090 00007ff8`0a340407     COMDLG32!PrintDlgX+0x163
1b 00000040`3627f530 00007fff`7c7ff697     COMDLG32!PrintDlgW+0x47
1c 00000040`3627fa40 00007fff`7c7ff591     MFC42u!CWinApp::UpdatePrinterSelection+0x57
1d 00000040`3627fbe0 00007ff7`f383d742     MFC42u!CWinApp::GetPrinterDeviceDefaults+0x21
1e 00000040`3627fc10 00007ff7`f383d938     wordpad!CWordPadApp::CreateDevNames+0xb2
1f 00000040`3627fcf0 00007fff`7c7ffc1d     wordpad!CWordPadApp::DoDeferredInitialization+0x18
20 00000040`3627fd20 00007ff8`0a57e634     MFC42u!_AfxThreadEntry+0xdd
21 00000040`3627fde0 00007ff8`0a57e70c     msvcrt!_callthreadstartex+0x28
22 00000040`3627fe10 00007ff8`0b03244d     msvcrt!_threadstartex+0x7c
23 00000040`3627fe40 00007ff8`0c18df78     KERNEL32!BaseThreadInitThunk+0x1d
24 00000040`3627fe70 00000000`00000000     ntdll!RtlUserThreadStart+0x28
 # Child-SP          RetAddr               Call Site
00 00000040`364ff948 00007ff8`09f91ca5     USER32!CreateWindowExW
01 00000040`364ff950 00007ff8`09f91bb5     combase!GetOrCreateSTAWindow+0x91
02 00000040`364ff9c0 00007ff8`0a023ab4     combase!OXIDEntry::StartServer+0x3d
03 (Inline Function) --------`--------     combase!CComApartment::StartServer+0x9d
04 00000040`364ff9f0 00007ff8`09ff76ad     combase!InitChannelIfNecessary+0x114
05 00000040`364ffa20 00007ff8`09f88477     combase!MarshalInternalObjRef+0x29
06 00000040`364ffaa0 00007ff8`09f8828b     combase!CDllHost::Marshal+0x2b
07 00000040`364ffae0 00007ff8`09ff80cb     combase!CDllHost::WorkerThread+0x9f
08 00000040`364ffb20 00007ff8`09ff8049     combase!CRpcThread::WorkerLoop+0x57
09 00000040`364ffba0 00007ff8`0b03244d     combase!CRpcThreadCache::RpcWorkerThreadEntry+0x29
0a 00000040`364ffbd0 00007ff8`0c18df78     KERNEL32!BaseThreadInitThunk+0x1d
0b 00000040`364ffc00 00000000`00000000     ntdll!RtlUserThreadStart+0x28
ModLoad: 00007fff`e1400000 00007fff`e1469000   C:\Windows\System32\oleacc.dll
ModLoad: 00007ff8`034f0000 00007ff8`03503000   C:\Program Files (x86)\Microsoft Office\root\VFS\ProgramFilesCommonX64\Microsoft Shared\Office16\MSOXMLMF.DLL
ModLoad: 00007fff`f1ef0000 00007fff`f1f0b000   C:\Program Files (x86)\Microsoft Office\root\VFS\ProgramFilesCommonX64\Microsoft
Shared\Office16\VCRUNTIME140.dll
ModLoad: 00007ff8`03270000 00007ff8`0327c000   C:\Program Files (x86)\Microsoft Office\root\VFS\ProgramFilesCommonX64\Microsoft
Shared\Office16\VCRUNTIME140_1.dll
 # Child-SP          RetAddr               Call Site
00 00000040`3607d608 00007fff`7c7ec835     USER32!CreateWindowExW
01 00000040`3607d610 00007fff`7c7ec919     MFC42u!CWnd::CreateEx+0xf5
02 00000040`3607d6d0 00007ff7`f3866bdc     MFC42u!CWnd::Create+0x99
03 00000040`3607d740 00007ff7`f383522b     wordpad!CRibbonBar::Create+0x98
04 00000040`3607d7b0 00007fff`7c7e8f27     wordpad!CMainFrame::OnCreate+0x4b
05 00000040`3607d800 00007ff7`f3836f5c     MFC42u!CWnd::OnWndMsg+0x387
06 00000040`3607d930 00007fff`7c7f0bff     wordpad!CMainFrame::OnWndMsg+0x14c
07 00000040`3607d9c0 00007fff`7c7eb42a     MFC42u!CWnd::WindowProc+0x4f
08 00000040`3607da00 00007fff`7c7ea76b     MFC42u!AfxCallWndProc+0x14a
09 00000040`3607db00 00007ff8`0be38161     MFC42u!AfxWndProcBase+0x15b
0a 00000040`3607db70 00007ff8`0be37e1c     USER32!UserCallWinProcCheckWow+0x2d1
0b 00000040`3607dcd0 00007ff8`0be44edc     USER32!DispatchClientMessage+0x9c
0c 00000040`3607dd30 00007ff8`0c1d2db4     USER32!__fnINLPCREATESTRUCT+0xac
0d 00000040`3607dd90 00007ff8`097f2294     ntdll!KiUserCallbackDispatcherContinue
0e 00000040`3607ded8 00007ff8`0be2f6b0     win32u!NtUserCreateWindowEx+0x14
0f 00000040`3607dee0 00007ff8`0be2f3cc     USER32!VerNtUserCreateWindowEx+0x210
10 00000040`3607e270 00007ff8`0be2f212     USER32!CreateWindowInternal+0x1a8
11 00000040`3607e3d0 00007fff`7c7ec835     USER32!CreateWindowExW+0x82
12 00000040`3607e460 00007fff`7c7ef6c5     MFC42u!CWnd::CreateEx+0xf5
13 00000040`3607e520 00007fff`7c7ef7b3     MFC42u!CFrameWnd::Create+0xe5
14 00000040`3607e590 00007fff`7c7ef0ea     MFC42u!CFrameWnd::LoadFrame+0xb3
15 00000040`3607e610 00007fff`7c7fa1ec     MFC42u!CDocTemplate::CreateNewFrame+0x7a
16 00000040`3607e680 00007ff7`f383e331     MFC42u!CSingleDocTemplate::OpenDocumentFile+0x8c
17 00000040`3607e6e0 00007fff`7c7efa57     wordpad!CWordPadApp::InitInstance+0x521
18 00000040`3607fc80 00007ff7`f37fb49e     MFC42u!AfxWinMain+0x97
19 00000040`3607fcc0 00007ff8`0b03244d     wordpad!__wmainCRTStartup+0x1de
1a 00000040`3607fd80 00007ff8`0c18df78     KERNEL32!BaseThreadInitThunk+0x1d
1b 00000040`3607fdb0 00000000`00000000     ntdll!RtlUserThreadStart+0x28
(1a28.44f0): Windows Runtime Originate Error - code 40080201 (first chance)
 # Child-SP          RetAddr               Call Site
00 00000040`3607d478 00007fff`7c7ec835     USER32!CreateWindowExW
01 00000040`3607d480 00007fff`7c7ec919     MFC42u!CWnd::CreateEx+0xf5
02 00000040`3607d540 00007fff`7c7ff9c4     MFC42u!CWnd::Create+0x99
03 00000040`3607d5b0 00007fff`7c7ff8f4     MFC42u!CStatusBar::CreateEx+0xb4
04 00000040`3607d630 00007ff7`f386ac6c     MFC42u!CStatusBar::Create+0x14
05 00000040`3607d670 00007ff7`f385763a     wordpad!CAppletStatusBar::Create+0x20
06 00000040`3607d6d0 00007ff7`f383525c     wordpad!CWpadStatusBar::Create+0x2a
07 00000040`3607d7b0 00007fff`7c7e8f27     wordpad!CMainFrame::OnCreate+0x7c
08 00000040`3607d800 00007ff7`f3836f5c     MFC42u!CWnd::OnWndMsg+0x387
09 00000040`3607d930 00007fff`7c7f0bff     wordpad!CMainFrame::OnWndMsg+0x14c
0a 00000040`3607d9c0 00007fff`7c7eb42a     MFC42u!CWnd::WindowProc+0x4f
0b 00000040`3607da00 00007fff`7c7ea76b     MFC42u!AfxCallWndProc+0x14a
0c 00000040`3607db00 00007ff8`0be38161     MFC42u!AfxWndProcBase+0x15b
```

```
0d 00000040`3607db70 00007ff8`0be37e1c     USER32!UserCallWinProcCheckWow+0x2d1
0e 00000040`3607dcd0 00007ff8`0be44edc     USER32!DispatchClientMessage+0x9c
0f 00000040`3607dd30 00007ff8`0c1d2db4     USER32!__fnINLPCREATESTRUCT+0xac
10 00000040`3607dd90 00007ff8`097f2294     ntdll!KiUserCallbackDispatcherContinue
11 00000040`3607ded8 00007ff8`0be2f6b0     win32u!NtUserCreateWindowEx+0x14
12 00000040`3607dee0 00007ff8`0be2f3cc     USER32!VerNtUserCreateWindowEx+0x210
13 00000040`3607e270 00007ff8`0be2f212     USER32!CreateWindowInternal+0x1a8
14 00000040`3607e3d0 00007fff`7c7ec835     USER32!CreateWindowExW+0x82
15 00000040`3607e460 00007fff`7c7ef6c5     MFC42u!CWnd::CreateEx+0xf5
16 00000040`3607e520 00007fff`7c7ef7b3     MFC42u!CFrameWnd::Create+0xe5
17 00000040`3607e590 00007fff`7c7ef0ea     MFC42u!CFrameWnd::LoadFrame+0xb3
18 00000040`3607e610 00007fff`7c7fa1ec     MFC42u!CDocTemplate::CreateNewFrame+0x7a
19 00000040`3607e680 00007ff7`f383e331     MFC42u!CSingleDocTemplate::OpenDocumentFile+0x8c
1a 00000040`3607e6e0 00007fff`7c7efa57     wordpad!CWordPadApp::InitInstance+0x521
1b 00000040`3607fc80 00007ff7`f37fb49e     MFC42u!AfxWinMain+0x97
1c 00000040`3607fcc0 00007ff8`0b03244d     wordpad!__wmainCRTStartup+0x1de
1d 00000040`3607fd80 00007ff8`0c18df78     KERNEL32!BaseThreadInitThunk+0x1d
1e 00000040`3607fdb0 00000000`00000000     ntdll!RtlUserThreadStart+0x28
onecore\com\combase\dcomrem\marshal.cxx(1283)\combase.dll!00007FF80A07C6A0: (caller: 00007FF809FC1626) ReturnHr(1) tid(4440) 80040155 Interface not registered
    Msg:[Failed to marshal with IID={79EAC9E4-BAF9-11CE-8C82-00AA004BA90B}]
onecore\com\combase\dcomrem\marshal.cxx(1178)\combase.dll!00007FF809FC176E: (caller: 00007FF80A020F2D) LogHr(1) tid(4440) 80040155 Interface not registered
onecore\com\combase\dcomrem\marshal.cxx(1118)\combase.dll!00007FF80A0D2EB8: (caller: 00007FF80A027F79) ReturnHr(2) tid(4440) 80040155 Interface not registered
(1a28.4440): Unknown exception - code 80040155 (first chance)
onecore\com\combase\dcomrem\servercall.cpp(373)\combase.dll!00007FF80A0D3EE4: (caller: 00007FF809FBA530) ReturnHr(3) tid(4440) 80040155 Interface not
registered
(1a28.623c): Unknown exception - code 80040155 (first chance)
 # Child-SP          RetAddr               Call Site
00 00000040`3607d438 00007fff`7c7ec835     USER32!CreateWindowExW
01 00000040`3607d440 00007fff`7c7ec919     MFC42u!CWnd::CreateEx+0xf5
02 00000040`3607d500 00007ff7`f386bb4a     MFC42u!CWnd::Create+0x99
03 00000040`3607d570 00007ff7`f3857700     wordpad!CAppletZoomControl::Create+0x9a
04 00000040`3607d6d0 00007ff7`f383525c     wordpad!CWpadStatusBar::Create+0xf0
05 00000040`3607d7b0 00007fff`7c7e8f27     wordpad!CMainFrame::OnCreate+0x7c
06 00000040`3607d800 00007ff7`f3836f5c     MFC42u!CWnd::OnWndMsg+0x387
07 00000040`3607d930 00007fff`7c7f0bff     wordpad!CMainFrame::OnWndMsg+0x14c
08 00000040`3607d9c0 00007fff`7c7eb42a     MFC42u!CWnd::WindowProc+0x4f
09 00000040`3607da00 00007fff`7c7ea76b     MFC42u!AfxCallWndProc+0x14a
0a 00000040`3607db00 00007ff8`0be38161     MFC42u!AfxWndProcBase+0x15b
0b 00000040`3607db70 00007ff8`0be37e1c     USER32!UserCallWinProcCheckWow+0x2d1
0c 00000040`3607dcd0 00007ff8`0be44edc     USER32!DispatchClientMessage+0x9c
0d 00000040`3607dd30 00007ff8`0c1d2db4     USER32!__fnINLPCREATESTRUCT+0xac
0e 00000040`3607dd90 00007ff8`097f2294     ntdll!KiUserCallbackDispatcherContinue
0f 00000040`3607ded8 00007ff8`0be2f6b0     win32u!NtUserCreateWindowEx+0x14
10 00000040`3607dee0 00007ff8`0be2f3cc     USER32!VerNtUserCreateWindowEx+0x210
11 00000040`3607e270 00007ff8`0be2f212     USER32!CreateWindowInternal+0x1a8
12 00000040`3607e3d0 00007fff`7c7ec835     USER32!CreateWindowExW+0x82
13 00000040`3607e460 00007fff`7c7ef6c5     MFC42u!CWnd::CreateEx+0xf5
14 00000040`3607e520 00007fff`7c7ef7b3     MFC42u!CFrameWnd::Create+0xe5
15 00000040`3607e590 00007fff`7c7ef0ea     MFC42u!CFrameWnd::LoadFrame+0xb3
16 00000040`3607e610 00007fff`7c7fa1ec     MFC42u!CDocTemplate::CreateNewFrame+0x7a
17 00000040`3607e680 00007ff7`f383e331     MFC42u!CSingleDocTemplate::OpenDocumentFile+0x8c
18 00000040`3607e6e0 00007fff`7c7efa57     wordpad!CWordPadApp::InitInstance+0x521
19 00000040`3607fc80 00007ff7`f37fb49e     MFC42u!AfxWinMain+0x97
1a 00000040`3607fcc0 00007ff8`0b03244d     wordpad!__wmainCRTStartup+0x1de
1b 00000040`3607fd80 00007ff8`0c18df78     KERNEL32!BaseThreadInitThunk+0x1d
1c 00000040`3607fdb0 00000000`00000000     ntdll!RtlUserThreadStart+0x28
 # Child-SP          RetAddr               Call Site
00 00000040`3607d3d8 00007fff`7c7ec835     USER32!CreateWindowExW
01 00000040`3607d3e0 00007fff`7c7ec919     MFC42u!CWnd::CreateEx+0xf5
02 00000040`3607d4a0 00007fff`7c7f2e30     MFC42u!CWnd::Create+0x99
03 00000040`3607d510 00007ff7`f386bca3     MFC42u!CStatic::Create+0x50
04 00000040`3607d570 00007ff7`f3857700     wordpad!CAppletZoomControl::Create+0x1f3
05 00000040`3607d6d0 00007ff7`f383525c     wordpad!CWpadStatusBar::Create+0xf0
06 00000040`3607d7b0 00007fff`7c7e8f27     wordpad!CMainFrame::OnCreate+0x7c
07 00000040`3607d800 00007ff7`f3836f5c     MFC42u!CWnd::OnWndMsg+0x387
08 00000040`3607d930 00007fff`7c7f0bff     wordpad!CMainFrame::OnWndMsg+0x14c
09 00000040`3607d9c0 00007fff`7c7eb42a     MFC42u!CWnd::WindowProc+0x4f
0a 00000040`3607da00 00007fff`7c7ea76b     MFC42u!AfxCallWndProc+0x14a
0b 00000040`3607db00 00007ff8`0be38161     MFC42u!AfxWndProcBase+0x15b
0c 00000040`3607db70 00007ff8`0be37e1c     USER32!UserCallWinProcCheckWow+0x2d1
0d 00000040`3607dcd0 00007ff8`0be44edc     USER32!DispatchClientMessage+0x9c
0e 00000040`3607dd30 00007ff8`0c1d2db4     USER32!__fnINLPCREATESTRUCT+0xac
0f 00000040`3607dd90 00007ff8`097f2294     ntdll!KiUserCallbackDispatcherContinue
10 00000040`3607ded8 00007ff8`0be2f6b0     win32u!NtUserCreateWindowEx+0x14
11 00000040`3607dee0 00007ff8`0be2f3cc     USER32!VerNtUserCreateWindowEx+0x210
12 00000040`3607e270 00007ff8`0be2f212     USER32!CreateWindowInternal+0x1a8
13 00000040`3607e3d0 00007fff`7c7ec835     USER32!CreateWindowExW+0x82
14 00000040`3607e460 00007fff`7c7ef6c5     MFC42u!CWnd::CreateEx+0xf5
15 00000040`3607e520 00007fff`7c7ef7b3     MFC42u!CFrameWnd::Create+0xe5
16 00000040`3607e590 00007fff`7c7ef0ea     MFC42u!CFrameWnd::LoadFrame+0xb3
17 00000040`3607e610 00007fff`7c7fa1ec     MFC42u!CDocTemplate::CreateNewFrame+0x7a
18 00000040`3607e680 00007ff7`f383e331     MFC42u!CSingleDocTemplate::OpenDocumentFile+0x8c
19 00000040`3607e6e0 00007fff`7c7efa57     wordpad!CWordPadApp::InitInstance+0x521
1a 00000040`3607fc80 00007ff7`f37fb49e     MFC42u!AfxWinMain+0x97
1b 00000040`3607fcc0 00007ff8`0b03244d     wordpad!__wmainCRTStartup+0x1de
1c 00000040`3607fd80 00007ff8`0c18df78     KERNEL32!BaseThreadInitThunk+0x1d
1d 00000040`3607fdb0 00000000`00000000     ntdll!RtlUserThreadStart+0x28
 # Child-SP          RetAddr               Call Site
00 00000040`3607d3d8 00007fff`7c7ec835     USER32!CreateWindowExW
01 00000040`3607d3e0 00007fff`7c7ec919     MFC42u!CWnd::CreateEx+0xf5
02 00000040`3607d4a0 00007fff`7c7f2e30     MFC42u!CWnd::Create+0x99
03 00000040`3607d510 00007ff7`f386bd06     MFC42u!CStatic::Create+0x50
04 00000040`3607d570 00007ff7`f3857700     wordpad!CAppletZoomControl::Create+0x256
05 00000040`3607d6d0 00007ff7`f383525c     wordpad!CWpadStatusBar::Create+0xf0
06 00000040`3607d7b0 00007fff`7c7e8f27     wordpad!CMainFrame::OnCreate+0x7c
07 00000040`3607d800 00007ff7`f3836f5c     MFC42u!CWnd::OnWndMsg+0x387
08 00000040`3607d930 00007fff`7c7f0bff     wordpad!CMainFrame::OnWndMsg+0x14c
09 00000040`3607d9c0 00007fff`7c7eb42a     MFC42u!CWnd::WindowProc+0x4f
```

```
0a 00000040`3607da00 00007fff`7c7ea76b     MFC42u!AfxCallWndProc+0x14a
0b 00000040`3607db00 00007ff8`0be38161     MFC42u!AfxWndProcBase+0x15b
0c 00000040`3607db70 00007ff8`0be37e1c     USER32!UserCallWinProcCheckWow+0x2d1
0d 00000040`3607dcd0 00007ff8`0be44edc     USER32!DispatchClientMessage+0x9c
0e 00000040`3607dd30 00007ff8`0c1d2db4     USER32!__fnINLPCREATESTRUCT+0xac
0f 00000040`3607dd90 00007ff8`097f2294     ntdll!KiUserCallbackDispatcherContinue
10 00000040`3607ded8 00007ff8`0be2f6b0     win32u!NtUserCreateWindowEx+0x14
11 00000040`3607dee0 00007ff8`0be2f3cc     USER32!VerNtUserCreateWindowEx+0x210
12 00000040`3607e270 00007ff8`0be2f212     USER32!CreateWindowInternal+0x1a8
13 00000040`3607e3d0 00007fff`7c7ec835     USER32!CreateWindowExW+0x82
14 00000040`3607e460 00007fff`7c7ef6c5     MFC42u!CWnd::CreateEx+0xf5
15 00000040`3607e520 00007fff`7c7ef7b3     MFC42u!CFrameWnd::Create+0xe5
16 00000040`3607e590 00007fff`7c7ef0ea     MFC42u!CFrameWnd::LoadFrame+0xb3
17 00000040`3607e610 00007fff`7c7fa1ec     MFC42u!CDocTemplate::CreateNewFrame+0x7a
18 00000040`3607e680 00007ff7`f383e331     MFC42u!CSingleDocTemplate::OpenDocumentFile+0x8c
19 00000040`3607e6e0 00007fff`7c7efa57     wordpad!CWordPadApp::InitInstance+0x521
1a 00000040`3607fc80 00007ff7`f37fb49e     MFC42u!AfxWinMain+0x97
1b 00000040`3607fcc0 00007ff8`0b03244d     wordpad!__wmainCRTStartup+0x1de
1c 00000040`3607fd80 00007ff8`0c18df78     KERNEL32!BaseThreadInitThunk+0x1d
1d 00000040`3607fdb0 00000000`00000000     ntdll!RtlUserThreadStart+0x28
 # Child-SP          RetAddr               Call Site
00 00000040`3607d358 00007fff`7c7ec835     USER32!CreateWindowExW
01 00000040`3607d360 00007fff`7c7ec919     MFC42u!CWnd::CreateEx+0xf5
02 00000040`3607d420 00007fff`7c800150     MFC42u!CWnd::Create+0x99
03 00000040`3607d490 00007ff7`f386d32a     MFC42u!CButton::Create+0x50
04 00000040`3607d4f0 00007ff7`f386be27     wordpad!CAppletPopoutButton::Create+0x3a
05 00000040`3607d570 00007ff7`f3857700     wordpad!CAppletZoomControl::Create+0x377
06 00000040`3607d6d0 00007ff7`f383525c     wordpad!CWpadStatusBar::Create+0xf0
07 00000040`3607d7b0 00007fff`7c7e8f27     wordpad!CMainFrame::OnCreate+0x7c
08 00000040`3607d800 00007ff7`f3836f5c     MFC42u!CWnd::OnWndMsg+0x387
09 00000040`3607d930 00007fff`7c7f0bff     wordpad!CMainFrame::OnWndMsg+0x14c
0a 00000040`3607d9c0 00007fff`7c7eb42a     MFC42u!CWnd::WindowProc+0x4f
0b 00000040`3607da00 00007fff`7c7ea76b     MFC42u!AfxCallWndProc+0x14a
0c 00000040`3607db00 00007ff8`0be38161     MFC42u!AfxWndProcBase+0x15b
0d 00000040`3607db70 00007ff8`0be37e1c     USER32!UserCallWinProcCheckWow+0x2d1
0e 00000040`3607dcd0 00007ff8`0be44edc     USER32!DispatchClientMessage+0x9c
0f 00000040`3607dd30 00007ff8`0c1d2db4     USER32!__fnINLPCREATESTRUCT+0xac
10 00000040`3607dd90 00007ff8`097f2294     ntdll!KiUserCallbackDispatcherContinue
11 00000040`3607ded8 00007ff8`0be2f6b0     win32u!NtUserCreateWindowEx+0x14
12 00000040`3607dee0 00007ff8`0be2f3cc     USER32!VerNtUserCreateWindowEx+0x210
13 00000040`3607e270 00007ff8`0be2f212     USER32!CreateWindowInternal+0x1a8
14 00000040`3607e3d0 00007fff`7c7ec835     USER32!CreateWindowExW+0x82
15 00000040`3607e460 00007fff`7c7ef6c5     MFC42u!CWnd::CreateEx+0xf5
16 00000040`3607e520 00007fff`7c7ef7b3     MFC42u!CFrameWnd::Create+0xe5
17 00000040`3607e590 00007fff`7c7ef0ea     MFC42u!CFrameWnd::LoadFrame+0xb3
18 00000040`3607e610 00007fff`7c7fa1ec     MFC42u!CDocTemplate::CreateNewFrame+0x7a
19 00000040`3607e680 00007ff7`f383e331     MFC42u!CSingleDocTemplate::OpenDocumentFile+0x8c
1a 00000040`3607e6e0 00007fff`7c7efa57     wordpad!CWordPadApp::InitInstance+0x521
1b 00000040`3607fc80 00007ff7`f37fb49e     MFC42u!AfxWinMain+0x97
1c 00000040`3607fcc0 00007ff8`0b03244d     wordpad!__wmainCRTStartup+0x1de
1d 00000040`3607fd80 00007ff8`0c18df78     KERNEL32!BaseThreadInitThunk+0x1d
1e 00000040`3607fdb0 00000000`00000000     ntdll!RtlUserThreadStart+0x28
 # Child-SP          RetAddr               Call Site
00 00000040`3607d3d8 00007fff`7c7ec835     USER32!CreateWindowExW
01 00000040`3607d3e0 00007fff`7c7ec919     MFC42u!CWnd::CreateEx+0xf5
02 00000040`3607d4a0 00007fff`7c800551     MFC42u!CWnd::Create+0x99
03 00000040`3607d510 00007ff7`f386bea1     MFC42u!CSliderCtrl::Create+0x71
04 00000040`3607d570 00007ff7`f3857700     wordpad!CAppletZoomControl::Create+0x3f1
05 00000040`3607d6d0 00007ff7`f383525c     wordpad!CWpadStatusBar::Create+0xf0
06 00000040`3607d7b0 00007fff`7c7e8f27     wordpad!CMainFrame::OnCreate+0x7c
07 00000040`3607d800 00007ff7`f3836f5c     MFC42u!CWnd::OnWndMsg+0x387
08 00000040`3607d930 00007fff`7c7f0bff     wordpad!CMainFrame::OnWndMsg+0x14c
09 00000040`3607d9c0 00007fff`7c7eb42a     MFC42u!CWnd::WindowProc+0x4f
0a 00000040`3607da00 00007fff`7c7ea76b     MFC42u!AfxCallWndProc+0x14a
0b 00000040`3607db00 00007ff8`0be38161     MFC42u!AfxWndProcBase+0x15b
0c 00000040`3607db70 00007ff8`0be37e1c     USER32!UserCallWinProcCheckWow+0x2d1
0d 00000040`3607dcd0 00007ff8`0be44edc     USER32!DispatchClientMessage+0x9c
0e 00000040`3607dd30 00007ff8`0c1d2db4     USER32!__fnINLPCREATESTRUCT+0xac
0f 00000040`3607dd90 00007ff8`097f2294     ntdll!KiUserCallbackDispatcherContinue
10 00000040`3607ded8 00007ff8`0be2f6b0     win32u!NtUserCreateWindowEx+0x14
11 00000040`3607dee0 00007ff8`0be2f3cc     USER32!VerNtUserCreateWindowEx+0x210
12 00000040`3607e270 00007ff8`0be2f212     USER32!CreateWindowInternal+0x1a8
13 00000040`3607e3d0 00007fff`7c7ec835     USER32!CreateWindowExW+0x82
14 00000040`3607e460 00007fff`7c7ef6c5     MFC42u!CWnd::CreateEx+0xf5
15 00000040`3607e520 00007fff`7c7ef7b3     MFC42u!CFrameWnd::Create+0xe5
16 00000040`3607e590 00007fff`7c7ef0ea     MFC42u!CFrameWnd::LoadFrame+0xb3
17 00000040`3607e610 00007fff`7c7fa1ec     MFC42u!CDocTemplate::CreateNewFrame+0x7a
18 00000040`3607e680 00007ff7`f383e331     MFC42u!CSingleDocTemplate::OpenDocumentFile+0x8c
19 00000040`3607e6e0 00007fff`7c7efa57     wordpad!CWordPadApp::InitInstance+0x521
1a 00000040`3607fc80 00007ff7`f37fb49e     MFC42u!AfxWinMain+0x97
1b 00000040`3607fcc0 00007ff8`0b03244d     wordpad!__wmainCRTStartup+0x1de
1c 00000040`3607fd80 00007ff8`0c18df78     KERNEL32!BaseThreadInitThunk+0x1d
1d 00000040`3607fdb0 00000000`00000000     ntdll!RtlUserThreadStart+0x28
 # Child-SP          RetAddr               Call Site
00 00000040`3607d358 00007fff`7c7ec835     USER32!CreateWindowExW
01 00000040`3607d360 00007fff`7c7ec919     MFC42u!CWnd::CreateEx+0xf5
02 00000040`3607d420 00007fff`7c800150     MFC42u!CWnd::Create+0x99
03 00000040`3607d490 00007ff7`f386d32a     MFC42u!CButton::Create+0x50
04 00000040`3607d4f0 00007ff7`f386c042     wordpad!CAppletPopoutButton::Create+0x3a
05 00000040`3607d570 00007ff7`f3857700     wordpad!CAppletZoomControl::Create+0x592
06 00000040`3607d6d0 00007ff7`f383525c     wordpad!CWpadStatusBar::Create+0xf0
07 00000040`3607d7b0 00007fff`7c7e8f27     wordpad!CMainFrame::OnCreate+0x7c
08 00000040`3607d800 00007ff7`f3836f5c     MFC42u!CWnd::OnWndMsg+0x387
09 00000040`3607d930 00007fff`7c7f0bff     wordpad!CMainFrame::OnWndMsg+0x14c
0a 00000040`3607d9c0 00007fff`7c7eb42a     MFC42u!CWnd::WindowProc+0x4f
0b 00000040`3607da00 00007fff`7c7ea76b     MFC42u!AfxCallWndProc+0x14a
0c 00000040`3607db00 00007ff8`0be38161     MFC42u!AfxWndProcBase+0x15b
0d 00000040`3607db70 00007ff8`0be37e1c     USER32!UserCallWinProcCheckWow+0x2d1
```

```
0e 00000040`3607dcd0 00007ff8`0be44edc     USER32!DispatchClientMessage+0x9c
0f 00000040`3607dd30 00007ff8`0c1d2db4     USER32!__fnINLPCREATESTRUCT+0xac
10 00000040`3607dd90 00007ff8`097f2294     ntdll!KiUserCallbackDispatcherContinue
11 00000040`3607ded8 00007ff8`0be2f6b0     win32u!NtUserCreateWindowEx+0x14
12 00000040`3607dee0 00007ff8`0be2f3cc     USER32!VerNtUserCreateWindowEx+0x210
13 00000040`3607e270 00007ff8`0be2f212     USER32!CreateWindowInternal+0x1a8
14 00000040`3607e3d0 00007fff`7c7ec835     USER32!CreateWindowExW+0x82
15 00000040`3607e460 00007fff`7c7ef6c5     MFC42u!CWnd::CreateEx+0xf5
16 00000040`3607e520 00007fff`7c7ef7b3     MFC42u!CFrameWnd::Create+0xe5
17 00000040`3607e590 00007fff`7c7ef0ea     MFC42u!CFrameWnd::LoadFrame+0xb3
18 00000040`3607e610 00007fff`7c7fa1ec     MFC42u!CDocTemplate::CreateNewFrame+0x7a
19 00000040`3607e680 00007ff7`f383e331     MFC42u!CSingleDocTemplate::OpenDocumentFile+0x8c
1a 00000040`3607e6e0 00007fff`7c7efa57     wordpad!CWordPadApp::InitInstance+0x521
1b 00000040`3607fc80 00007ff7`f37fb49e     MFC42u!AfxWinMain+0x97
1c 00000040`3607fcc0 00007ff8`0b03244d     wordpad!__wmainCRTStartup+0x1de
1d 00000040`3607fd80 00007ff8`0c18df78     KERNEL32!BaseThreadInitThunk+0x1d
1e 00000040`3607fdb0 00000000`00000000     ntdll!RtlUserThreadStart+0x28
 # Child-SP          RetAddr               Call Site
00 00000040`3627de98 00007fff`f3cce048     USER32!CreateWindowExW
01 00000040`3627dea0 00007fff`f3ccdf1f     urlmon!NotificationWindow::CreateHandle+0x68
02 00000040`3627df10 00007fff`f3c7cc9b     urlmon!CTransaction::GetNotificationWnd+0x13f
03 00000040`3627df50 00007fff`f3cb98ee     urlmon!GetTransactionObjects+0x2fb
04 00000040`3627dfd0 00007fff`f3cba81c     urlmon!CBinding::StartBinding+0x53e
05 00000040`3627e150 00007fff`f3cb9276     urlmon!CUrlMon::StartBinding+0x1b0
06 00000040`3627e220 00007fff`f1c05529     urlmon!CUrlMon::BindToStorage+0x96
07 00000040`3627e270 00007fff`f1c04acb     msxml6!URLMONStream::deferedOpen+0x159
08 00000040`3627e2d0 00007fff`f1c048df     msxml6!XMLParser::PushURL+0x1af
09 00000040`3627e350 00007fff`f1bbf1c6     msxml6!XMLParser::SetURL+0x7f
0a 00000040`3627e3a0 00007fff`f1bbed48     msxml6!Document::_load+0x162
0b 00000040`3627e420 00007fff`f1b87712     msxml6!Document::load+0x198
0c 00000040`3627e490 00007fff`7c46afae     msxml6!DOMDocumentWrapper::load+0x182
0d 00000040`3627e560 00007fff`7c414968     PrintConfig!PrintConfig::PrivateDevmodeMapFile::Parse+0x14e
0e 00000040`3627e6d0 00007fff`7c558d7b     PrintConfig!PrintConfigPlugIn::DevMode+0x458
0f 00000040`3627e7f0 00007fff`7c512b53     PrintConfig!BCalcTotalOEMDMSize+0xa3
10 00000040`3627e890 00007fff`7c5120ec     PrintConfig!UniDrvUI::LSimpleDocumentProperties+0x10f
11 00000040`3627e900 00007fff`7c40f54b     PrintConfig!UniDrvUI::DrvDocumentPropertySheets+0x60
12 00000040`3627edb0 00007fff`7c4249ad     PrintConfig!PrintConfig::DrvDocumentPropertySheets+0x123
13 00000040`3627ee20 00007fff`7c425543     PrintConfig!ExceptionBoundary<<lambda_ce41999e58d4ed069b2790d632f6e3b5> >+0x3d
14 00000040`3627ee70 00007fff`e032e9ba     PrintConfig!DrvDocumentPropertySheets+0x53
15 00000040`3627eee0 00007fff`e0315895     WINSPOOL!DocumentPropertySheets+0x3fa
16 00000040`3627ef30 00007fff`e0324aff     WINSPOOL!DocumentPropertiesWNative+0xe9
17 00000040`3627efc0 00007ff8`0a340fc8     WINSPOOL!DocumentPropertiesW+0xcf
18 00000040`3627f020 00007ff8`0a344d59     COMDLG32!PrintGetDevMode+0x8c
19 00000040`3627f060 00007ff8`0a34058b     COMDLG32!PrintReturnDefault+0x91
1a 00000040`3627f090 00007ff8`0a340407     COMDLG32!PrintDlgX+0x163
1b 00000040`3627f530 00007fff`7c7ff697     COMDLG32!PrintDlgW+0x47
1c 00000040`3627fa40 00007fff`7c7ff591     MFC42u!CWinApp::UpdatePrinterSelection+0x57
1d 00000040`3627fbe0 00007ff7`f383d742     MFC42u!CWinApp::GetPrinterDeviceDefaults+0x21
1e 00000040`3627fc10 00007ff7`f383d938     wordpad!CWordPadApp::CreateDevNames+0xb2
1f 00000040`3627fcf0 00007fff`7c7ffc1d     wordpad!CWordPadApp::DoDeferredInitialization+0x18
20 00000040`3627fd20 00007ff8`0a57e634     MFC42u!_AfxThreadEntry+0xdd
21 00000040`3627fde0 00007ff8`0a57e70c     msvcrt!_callthreadstartex+0x28
22 00000040`3627fe10 00007ff8`0b03244d     msvcrt!_threadstartex+0x7c
23 00000040`3627fe40 00007ff8`0c18df78     KERNEL32!BaseThreadInitThunk+0x1d
24 00000040`3627fe70 00000000`00000000     ntdll!RtlUserThreadStart+0x28
 # Child-SP          RetAddr               Call Site
00 00000040`3607d5c8 00007fff`7c7ec835     USER32!CreateWindowExW
01 00000040`3607d5d0 00007fff`7c7ec919     MFC42u!CWnd::CreateEx+0xf5
02 00000040`3607d690 00007fff`7c7f528e     MFC42u!CWnd::Create+0x99
03 00000040`3607d700 00007fff`7c7f4fc8     MFC42u!CDockBar::Create+0x7e
04 00000040`3607d780 00007ff7`f383529e     MFC42u!CFrameWnd::EnableDocking+0x88
05 00000040`3607d7b0 00007fff`7c7e8f27     wordpad!CMainFrame::OnCreate+0xbe
06 00000040`3607d800 00007ff7`f3836f5c     MFC42u!CWnd::OnWndMsg+0x387
07 00000040`3607d930 00007fff`7c7f0bff     wordpad!CMainFrame::OnWndMsg+0x14c
08 00000040`3607d9c0 00007fff`7c7eb42a     MFC42u!CWnd::WindowProc+0x4f
09 00000040`3607da00 00007fff`7c7ea76b     MFC42u!AfxCallWndProc+0x14a
0a 00000040`3607db00 00007ff8`0be38161     MFC42u!AfxWndProcBase+0x15b
0b 00000040`3607db70 00007ff8`0be37e1c     USER32!UserCallWinProcCheckWow+0x2d1
0c 00000040`3607dcd0 00007ff8`0be44edc     USER32!DispatchClientMessage+0x9c
0d 00000040`3607dd30 00007ff8`0c1d2db4     USER32!__fnINLPCREATESTRUCT+0xac
0e 00000040`3607dd90 00007ff8`097f2294     ntdll!KiUserCallbackDispatcherContinue
0f 00000040`3607ded8 00007ff8`0be2f6b0     win32u!NtUserCreateWindowEx+0x14
10 00000040`3607dee0 00007ff8`0be2f3cc     USER32!VerNtUserCreateWindowEx+0x210
11 00000040`3607e270 00007ff8`0be2f212     USER32!CreateWindowInternal+0x1a8
12 00000040`3607e3d0 00007fff`7c7ec835     USER32!CreateWindowExW+0x82
13 00000040`3607e460 00007fff`7c7ef6c5     MFC42u!CWnd::CreateEx+0xf5
14 00000040`3607e520 00007fff`7c7ef7b3     MFC42u!CFrameWnd::Create+0xe5
15 00000040`3607e590 00007fff`7c7ef0ea     MFC42u!CFrameWnd::LoadFrame+0xb3
16 00000040`3607e610 00007fff`7c7fa1ec     MFC42u!CDocTemplate::CreateNewFrame+0x7a
17 00000040`3607e680 00007ff7`f383e331     MFC42u!CSingleDocTemplate::OpenDocumentFile+0x8c
18 00000040`3607e6e0 00007fff`7c7efa57     wordpad!CWordPadApp::InitInstance+0x521
19 00000040`3607fc80 00007ff7`f37fb49e     MFC42u!AfxWinMain+0x97
1a 00000040`3607fcc0 00007ff8`0b03244d     wordpad!__wmainCRTStartup+0x1de
1b 00000040`3607fd80 00007ff8`0c18df78     KERNEL32!BaseThreadInitThunk+0x1d
1c 00000040`3607fdb0 00000000`00000000     ntdll!RtlUserThreadStart+0x28
 # Child-SP          RetAddr               Call Site
00 00000040`3607d5c8 00007fff`7c7ec835     USER32!CreateWindowExW
01 00000040`3607d5d0 00007fff`7c7ec919     MFC42u!CWnd::CreateEx+0xf5
02 00000040`3607d690 00007fff`7c7f528e     MFC42u!CWnd::Create+0x99
03 00000040`3607d700 00007fff`7c7f4fc8     MFC42u!CDockBar::Create+0x7e
04 00000040`3607d780 00007ff7`f383529e     MFC42u!CFrameWnd::EnableDocking+0x88
05 00000040`3607d7b0 00007fff`7c7e8f27     wordpad!CMainFrame::OnCreate+0xbe
06 00000040`3607d800 00007ff7`f3836f5c     MFC42u!CWnd::OnWndMsg+0x387
07 00000040`3607d930 00007fff`7c7f0bff     wordpad!CMainFrame::OnWndMsg+0x14c
08 00000040`3607d9c0 00007fff`7c7eb42a     MFC42u!CWnd::WindowProc+0x4f
09 00000040`3607da00 00007fff`7c7ea76b     MFC42u!AfxCallWndProc+0x14a
0a 00000040`3607db00 00007ff8`0be38161     MFC42u!AfxWndProcBase+0x15b
0b 00000040`3607db70 00007ff8`0be37e1c     USER32!UserCallWinProcCheckWow+0x2d1
```

```
0c 00000040`3607dcd0 00007ff8`0be44edc     USER32!DispatchClientMessage+0x9c
0d 00000040`3607dd30 00007ff8`0c1d2db4     USER32!__fnINLPCREATESTRUCT+0xac
0e 00000040`3607dd90 00007ff8`097f2294     ntdll!KiUserCallbackDispatcherContinue
0f 00000040`3607ded8 00007ff8`0be2f6b0     win32u!NtUserCreateWindowEx+0x14
10 00000040`3607dee0 00007ff8`0be2f3cc     USER32!VerNtUserCreateWindowEx+0x210
11 00000040`3607e270 00007ff8`0be2f212     USER32!CreateWindowInternal+0x1a8
12 00000040`3607e3d0 00007fff`7c7ec835     USER32!CreateWindowExW+0x82
13 00000040`3607e460 00007fff`7c7ef6c5     MFC42u!CWnd::CreateEx+0xf5
14 00000040`3607e520 00007fff`7c7ef7b3     MFC42u!CFrameWnd::Create+0xe5
15 00000040`3607e590 00007fff`7c7ef0ea     MFC42u!CFrameWnd::LoadFrame+0xb3
16 00000040`3607e610 00007fff`7c7fa1ec     MFC42u!CDocTemplate::CreateNewFrame+0x7a
17 00000040`3607e680 00007ff7`f383e331     MFC42u!CSingleDocTemplate::OpenDocumentFile+0x8c
18 00000040`3607e6e0 00007fff`7c7efa57     wordpad!CWordPadApp::InitInstance+0x521
19 00000040`3607fc80 00007ff7`f37fb49e     MFC42u!AfxWinMain+0x97
1a 00000040`3607fcc0 00007ff8`0b03244d     wordpad!__wmainCRTStartup+0x1de
1b 00000040`3607fd80 00007ff8`0c18df78     KERNEL32!BaseThreadInitThunk+0x1d
1c 00000040`3607fdb0 00000000`00000000     ntdll!RtlUserThreadStart+0x28
 # Child-SP          RetAddr               Call Site
00 00000040`3607d5c8 00007fff`7c7ec835     USER32!CreateWindowExW
01 00000040`3607d5d0 00007fff`7c7ec919     MFC42u!CWnd::CreateEx+0xf5
02 00000040`3607d690 00007fff`7c7f528e     MFC42u!CWnd::Create+0x99
03 00000040`3607d700 00007fff`7c7f4fc8     MFC42u!CDockBar::Create+0x7e
04 00000040`3607d780 00007ff7`f383529e     MFC42u!CFrameWnd::EnableDocking+0x88
05 00000040`3607d7b0 00007fff`7c7e8f27     wordpad!CMainFrame::OnCreate+0xbe
06 00000040`3607d800 00007ff7`f3836f5c     MFC42u!CWnd::OnWndMsg+0x387
07 00000040`3607d930 00007fff`7c7f0bff     wordpad!CMainFrame::OnWndMsg+0x14c
08 00000040`3607d9c0 00007fff`7c7eb42a     MFC42u!CWnd::WindowProc+0x4f
09 00000040`3607da00 00007fff`7c7ea76b     MFC42u!AfxCallWndProc+0x14a
0a 00000040`3607db00 00007ff8`0be38161     MFC42u!AfxWndProcBase+0x15b
0b 00000040`3607db70 00007ff8`0be37e1c     USER32!UserCallWinProcCheckWow+0x2d1
0c 00000040`3607dcd0 00007ff8`0be44edc     USER32!DispatchClientMessage+0x9c
0d 00000040`3607dd30 00007ff8`0c1d2db4     USER32!__fnINLPCREATESTRUCT+0xac
0e 00000040`3607dd90 00007ff8`097f2294     ntdll!KiUserCallbackDispatcherContinue
0f 00000040`3607ded8 00007ff8`0be2f6b0     win32u!NtUserCreateWindowEx+0x14
10 00000040`3607dee0 00007ff8`0be2f3cc     USER32!VerNtUserCreateWindowEx+0x210
11 00000040`3607e270 00007ff8`0be2f212     USER32!CreateWindowInternal+0x1a8
12 00000040`3607e3d0 00007fff`7c7ec835     USER32!CreateWindowExW+0x82
13 00000040`3607e460 00007fff`7c7ef6c5     MFC42u!CWnd::CreateEx+0xf5
14 00000040`3607e520 00007fff`7c7ef7b3     MFC42u!CFrameWnd::Create+0xe5
15 00000040`3607e590 00007fff`7c7ef0ea     MFC42u!CFrameWnd::LoadFrame+0xb3
16 00000040`3607e610 00007fff`7c7fa1ec     MFC42u!CDocTemplate::CreateNewFrame+0x7a
17 00000040`3607e680 00007ff7`f383e331     MFC42u!CSingleDocTemplate::OpenDocumentFile+0x8c
18 00000040`3607e6e0 00007fff`7c7efa57     wordpad!CWordPadApp::InitInstance+0x521
19 00000040`3607fc80 00007ff7`f37fb49e     MFC42u!AfxWinMain+0x97
1a 00000040`3607fcc0 00007ff8`0b03244d     wordpad!__wmainCRTStartup+0x1de
1b 00000040`3607fd80 00007ff8`0c18df78     KERNEL32!BaseThreadInitThunk+0x1d
1c 00000040`3607fdb0 00000000`00000000     ntdll!RtlUserThreadStart+0x28
 # Child-SP          RetAddr               Call Site
00 00000040`3627de98 00007fff`f3cce048     USER32!CreateWindowExW
01 00000040`3627dea0 00007fff`f3ccdf1f     urlmon!NotificationWindow::CreateHandle+0x68
02 00000040`3627df10 00007fff`f3c7cc9b     urlmon!CTransaction::GetNotificationWnd+0x13f
03 00000040`3627df50 00007fff`f3cb98ee     urlmon!GetTransactionObjects+0x2fb
04 00000040`3627dfd0 00007fff`f3cba81c     urlmon!CBinding::StartBinding+0x53e
05 00000040`3627e150 00007fff`f3cb9276     urlmon!CUrlMon::StartBinding+0x1b0
06 00000040`3627e220 00007fff`f1c05529     urlmon!CUrlMon::BindToStorage+0x96
07 00000040`3627e270 00007fff`f1c04acb     msxml6!URLMONStream::deferedOpen+0x159
08 00000040`3627e2d0 00007fff`f1c048df     msxml6!XMLParser::PushURL+0x1af
09 00000040`3627e350 00007fff`f1bbf1c6     msxml6!XMLParser::SetURL+0x7f
0a 00000040`3627e3a0 00007fff`f1bbed48     msxml6!Document::_load+0x162
0b 00000040`3627e420 00007fff`f1b87712     msxml6!Document::load+0x198
0c 00000040`3627e490 00007fff`7c46afae     msxml6!DOMDocumentWrapper::load+0x182
0d 00000040`3627e560 00007fff`7c414968     PrintConfig!PrintConfig::PrivateDevmodeMapFile::Parse+0x14e
0e 00000040`3627e6d0 00007fff`7c558d7b     PrintConfig!PrintConfigPlugIn::DevMode+0x458
0f 00000040`3627e7f0 00007fff`7c512b53     PrintConfig!BCalcTotalOEMDMSize+0xa3
10 00000040`3627e890 00007fff`7c5120ec     PrintConfig!UniDrvUI::LSimpleDocumentProperties+0x10f
11 00000040`3627e900 00007fff`7c40f54b     PrintConfig!UniDrvUI::DrvDocumentPropertySheets+0x60
12 00000040`3627edb0 00007fff`7c4249ad     PrintConfig!PrintConfig::DrvDocumentPropertySheets+0x123
13 00000040`3627ee20 00007fff`7c425543     PrintConfig!ExceptionBoundary<<lambda_ce41999e58d4ed069b2790d632f6e3b5> >+0x3d
14 00000040`3627ee70 00007fff`e032e9ba     PrintConfig!DrvDocumentPropertySheets+0x53
15 00000040`3627eee0 00007fff`e031593f     WINSPOOL!DocumentPropertySheets+0x3fa
16 00000040`3627ef30 00007fff`e0324aff     WINSPOOL!DocumentPropertiesWNative+0x193
17 00000040`3627efc0 00007ff8`0a340fc8     WINSPOOL!DocumentPropertiesW+0xcf
18 00000040`3627f020 00007ff8`0a344d59     COMDLG32!PrintGetDevMode+0x8c
19 00000040`3627f060 00007ff8`0a34058b     COMDLG32!PrintReturnDefault+0x91
1a 00000040`3627f090 00007ff8`0a340407     COMDLG32!PrintDlgX+0x163
1b 00000040`3627f530 00007fff`7c7ff697     COMDLG32!PrintDlgW+0x47
1c 00000040`3627fa40 00007fff`7c7ff591     MFC42u!CWinApp::UpdatePrinterSelection+0x57
1d 00000040`3627fbe0 00007ff7`f383d742     MFC42u!CWinApp::GetPrinterDeviceDefaults+0x21
1e 00000040`3627fc10 00007ff7`f383d938     wordpad!CWordPadApp::CreateDevNames+0xb2
1f 00000040`3627fcf0 00007fff`7c7ffc1d     wordpad!CWordPadApp::DoDeferredInitialization+0x18
20 00000040`3627fd20 00007ff8`0a57e634     MFC42u!_AfxThreadEntry+0xdd
21 00000040`3627fde0 00007ff8`0a57e70c     msvcrt!_callthreadstartex+0x28
22 00000040`3627fe10 00007ff8`0b03244d     msvcrt!_threadstartex+0x7c
23 00000040`3627fe40 00007ff8`0c18df78     KERNEL32!BaseThreadInitThunk+0x1d
24 00000040`3627fe70 00000000`00000000     ntdll!RtlUserThreadStart+0x28
 # Child-SP          RetAddr               Call Site
00 00000040`3607d5c8 00007fff`7c7ec835     USER32!CreateWindowExW
01 00000040`3607d5d0 00007fff`7c7ec919     MFC42u!CWnd::CreateEx+0xf5
02 00000040`3607d690 00007fff`7c7f528e     MFC42u!CWnd::Create+0x99
03 00000040`3607d700 00007fff`7c7f4fc8     MFC42u!CDockBar::Create+0x7e
04 00000040`3607d780 00007ff7`f383529e     MFC42u!CFrameWnd::EnableDocking+0x88
05 00000040`3607d7b0 00007fff`7c7e8f27     wordpad!CMainFrame::OnCreate+0xbe
06 00000040`3607d800 00007ff7`f3836f5c     MFC42u!CWnd::OnWndMsg+0x387
07 00000040`3607d930 00007fff`7c7f0bff     wordpad!CMainFrame::OnWndMsg+0x14c
08 00000040`3607d9c0 00007fff`7c7eb42a     MFC42u!CWnd::WindowProc+0x4f
09 00000040`3607da00 00007fff`7c7ea76b     MFC42u!AfxCallWndProc+0x14a
0a 00000040`3607db00 00007ff8`0be38161     MFC42u!AfxWndProcBase+0x15b
0b 00000040`3607db70 00007ff8`0be37e1c     USER32!UserCallWinProcCheckWow+0x2d1
```

```
0c 00000040`3607dcd0 00007ff8`0be44edc     USER32!DispatchClientMessage+0x9c
0d 00000040`3607dd30 00007ff8`0c1d2db4     USER32!__fnINLPCREATESTRUCT+0xac
0e 00000040`3607dd90 00007ff8`097f2294     ntdll!KiUserCallbackDispatcherContinue
0f 00000040`3607ded8 00007ff8`0be2f6b0     win32u!NtUserCreateWindowEx+0x14
10 00000040`3607dee0 00007ff8`0be2f3cc     USER32!VerNtUserCreateWindowEx+0x210
11 00000040`3607e270 00007ff8`0be2f212     USER32!CreateWindowInternal+0x1a8
12 00000040`3607e3d0 00007fff`7c7ec835     USER32!CreateWindowExW+0x82
13 00000040`3607e460 00007fff`7c7ef6c5     MFC42u!CWnd::CreateEx+0xf5
14 00000040`3607e520 00007fff`7c7ef7b3     MFC42u!CFrameWnd::Create+0xe5
15 00000040`3607e590 00007fff`7c7ef0ea     MFC42u!CFrameWnd::LoadFrame+0xb3
16 00000040`3607e610 00007fff`7c7fa1ec     MFC42u!CDocTemplate::CreateNewFrame+0x7a
17 00000040`3607e680 00007ff7`f383e331     MFC42u!CSingleDocTemplate::OpenDocumentFile+0x8c
18 00000040`3607e6e0 00007fff`7c7efa57     wordpad!CWordPadApp::InitInstance+0x521
19 00000040`3607fc80 00007ff7`f37fb49e     MFC42u!AfxWinMain+0x97
1a 00000040`3607fcc0 00007ff8`0b03244d     wordpad!__wmainCRTStartup+0x1de
1b 00000040`3607fd80 00007ff8`0c18df78     KERNEL32!BaseThreadInitThunk+0x1d
1c 00000040`3607fdb0 00000000`00000000     ntdll!RtlUserThreadStart+0x28
 # Child-SP          RetAddr               Call Site
00 00000040`3607d538 00007fff`7c7ec835     USER32!CreateWindowExW
01 00000040`3607d540 00007fff`7c7ec919     MFC42u!CWnd::CreateEx+0xf5
02 00000040`3607d600 00007ff7`f38389a4     MFC42u!CWnd::Create+0x99
03 00000040`3607d670 00007ff7`f38352d6     wordpad!CRulerBar::Create+0xa4
04 00000040`3607d7b0 00007fff`7c7e8f27     wordpad!CMainFrame::OnCreate+0xf6
05 00000040`3607d800 00007ff7`f3836f5c     MFC42u!CWnd::OnWndMsg+0x387
06 00000040`3607d930 00007fff`7c7f0bff     wordpad!CMainFrame::OnWndMsg+0x14c
07 00000040`3607d9c0 00007fff`7c7eb42a     MFC42u!CWnd::WindowProc+0x4f
08 00000040`3607da00 00007fff`7c7ea76b     MFC42u!AfxCallWndProc+0x14a
09 00000040`3607db00 00007ff8`0be38161     MFC42u!AfxWndProcBase+0x15b
0a 00000040`3607db70 00007ff8`0be37e1c     USER32!UserCallWinProcCheckWow+0x2d1
0b 00000040`3607dcd0 00007ff8`0be44edc     USER32!DispatchClientMessage+0x9c
0c 00000040`3607dd30 00007ff8`0c1d2db4     USER32!__fnINLPCREATESTRUCT+0xac
0d 00000040`3607dd90 00007ff8`097f2294     ntdll!KiUserCallbackDispatcherContinue
0e 00000040`3607ded8 00007ff8`0be2f6b0     win32u!NtUserCreateWindowEx+0x14
0f 00000040`3607dee0 00007ff8`0be2f3cc     USER32!VerNtUserCreateWindowEx+0x210
10 00000040`3607e270 00007ff8`0be2f212     USER32!CreateWindowInternal+0x1a8
11 00000040`3607e3d0 00007fff`7c7ec835     USER32!CreateWindowExW+0x82
12 00000040`3607e460 00007fff`7c7ef6c5     MFC42u!CWnd::CreateEx+0xf5
13 00000040`3607e520 00007fff`7c7ef7b3     MFC42u!CFrameWnd::Create+0xe5
14 00000040`3607e590 00007fff`7c7ef0ea     MFC42u!CFrameWnd::LoadFrame+0xb3
15 00000040`3607e610 00007fff`7c7fa1ec     MFC42u!CDocTemplate::CreateNewFrame+0x7a
16 00000040`3607e680 00007ff7`f383e331     MFC42u!CSingleDocTemplate::OpenDocumentFile+0x8c
17 00000040`3607e6e0 00007fff`7c7efa57     wordpad!CWordPadApp::InitInstance+0x521
18 00000040`3607fc80 00007ff7`f37fb49e     MFC42u!AfxWinMain+0x97
19 00000040`3607fcc0 00007ff8`0b03244d     wordpad!__wmainCRTStartup+0x1de
1a 00000040`3607fd80 00007ff8`0c18df78     KERNEL32!BaseThreadInitThunk+0x1d
1b 00000040`3607fdb0 00000000`00000000     ntdll!RtlUserThreadStart+0x28
 # Child-SP          RetAddr               Call Site
00 00000040`3627dd28 00007fff`f3cce048     USER32!CreateWindowExW
01 00000040`3627dd30 00007fff`f3ccdf1f     urlmon!NotificationWindow::CreateHandle+0x68
02 00000040`3627dda0 00007fff`f3c7cc9b     urlmon!CTransaction::GetNotificationWnd+0x13f
03 00000040`3627dde0 00007fff`f3cb98ee     urlmon!GetTransactionObjects+0x2fb
04 00000040`3627de60 00007fff`f3cba81c     urlmon!CBinding::StartBinding+0x53e
05 00000040`3627dfe0 00007fff`f3cb9276     urlmon!CUrlMon::StartBinding+0x1b0
06 00000040`3627e0b0 00007fff`f1c05529     urlmon!CUrlMon::BindToStorage+0x96
07 00000040`3627e100 00007fff`f1c04acb     msxml6!URLMONStream::deferedOpen+0x159
08 00000040`3627e160 00007fff`f1c048df     msxml6!XMLParser::PushURL+0x1af
09 00000040`3627e1e0 00007fff`f1bbf1c6     msxml6!XMLParser::SetURL+0x7f
0a 00000040`3627e230 00007fff`f1bbed48     msxml6!Document::_load+0x162
0b 00000040`3627e2b0 00007fff`f1b87712     msxml6!Document::load+0x198
0c 00000040`3627e320 00007fff`7c46afae     msxml6!DOMDocumentWrapper::load+0x182
0d 00000040`3627e3f0 00007fff`7c414813     PrintConfig!PrintConfig::PrivateDevmodeMapFile::Parse+0x14e
0e 00000040`3627e560 00007fff`7c5583f0     PrintConfig!PrintConfigPlugIn::DevMode+0x303
0f 00000040`3627e680 00007fff`7c55956a     PrintConfig!BCallOEMDevMode+0x84
10 00000040`3627e700 00007fff`7c514999     PrintConfig!PGetDefaultDevmodeWithOemPlugins+0x16a
11 00000040`3627e780 00007fff`7c5144fa     PrintConfig!UniDrvUI::BGetValidatedUserDefaultDevmode+0x81
12 00000040`3627e7f0 00007fff`7c512bb7     PrintConfig!UniDrvUI::BFillCommonInfoDevmode+0x82
13 00000040`3627e890 00007fff`7c5120ec     PrintConfig!UniDrvUI::LSimpleDocumentProperties+0x173
14 00000040`3627e900 00007fff`7c40f54b     PrintConfig!UniDrvUI::DrvDocumentPropertySheets+0x60
15 00000040`3627edb0 00007fff`7c4249ad     PrintConfig!PrintConfig::DrvDocumentPropertySheets+0x123
16 00000040`3627ee20 00007fff`7c425543     PrintConfig!ExceptionBoundary<<lambda_ce41999e58d4ed069b2790d632f6e3b5> >+0x3d
17 00000040`3627ee70 00007fff`e032e9ba     PrintConfig!DrvDocumentPropertySheets+0x53
18 00000040`3627eee0 00007fff`e031593f     WINSPOOL!DocumentPropertySheets+0x3fa
19 00000040`3627ef30 00007fff`e0324aff     WINSPOOL!DocumentPropertiesWNative+0x193
1a 00000040`3627efc0 00007ff8`0a340fc8     WINSPOOL!DocumentPropertiesW+0xcf
1b 00000040`3627f020 00007ff8`0a344d59     COMDLG32!PrintGetDevMode+0x8c
1c 00000040`3627f060 00007ff8`0a34058b     COMDLG32!PrintReturnDefault+0x91
1d 00000040`3627f090 00007ff8`0a340407     COMDLG32!PrintDlgX+0x163
1e 00000040`3627f530 00007fff`7c7ff697     COMDLG32!PrintDlgW+0x47
1f 00000040`3627fa40 00007fff`7c7ff591     MFC42u!CWinApp::UpdatePrinterSelection+0x57
20 00000040`3627fbe0 00007ff7`f383d742     MFC42u!CWinApp::GetPrinterDeviceDefaults+0x21
21 00000040`3627fc10 00007ff7`f383d938     wordpad!CWordPadApp::CreateDevNames+0xb2
22 00000040`3627fcf0 00007fff`7c7ffc1d     wordpad!CWordPadApp::DoDeferredInitialization+0x18
23 00000040`3627fd20 00007ff8`0a57e634     MFC42u!_AfxThreadEntry+0xdd
24 00000040`3627fde0 00007ff8`0a57e70c     msvcrt!_callthreadstartex+0x28
25 00000040`3627fe10 00007ff8`0b03244d     msvcrt!_threadstartex+0x7c
26 00000040`3627fe40 00007ff8`0c18df78     KERNEL32!BaseThreadInitThunk+0x1d
27 00000040`3627fe70 00000000`00000000     ntdll!RtlUserThreadStart+0x28
ModLoad: 00007ff8`08d30000 00007ff8`08d3e000   C:\WINDOWS\SYSTEM32\atlthunk.dll
 # Child-SP          RetAddr               Call Site
00 00000040`3607d3d8 00007fff`5cf1317c     USER32!CreateWindowExW
01 00000040`3607d3e0 00007fff`5cf243be     UIRibbon!IsolationAwareCreateWindowExW+0xe0
02 00000040`3607d460 00007fff`5cf1f3cd     UIRibbon!CreateWindowRegExW+0x9a
03 00000040`3607d4d0 00007fff`5cf20058     UIRibbon!SCM::FCreateWindow+0x51
04 00000040`3607d540 00007fff`5cf1fefa     UIRibbon!SCM::FRegisterDCMComp+0x44
05 00000040`3607d570 00007fff`5cf2b9e1     UIRibbon!SCM_MsoCompMgr::FRegisterComponent+0x1a
06 00000040`3607d5b0 00007fff`5cf2b7da     UIRibbon!TBComponent::FInit+0x71
07 00000040`3607d610 00007fff`5cf5668b     UIRibbon!FCreateTBComponent+0x32
08 00000040`3607d640 00007fff`5cf59500     UIRibbon!TBS::FInit+0x8f
```

```
09 00000040`3607d690 00007fff`5ceba402     UIRibbon!MsoFCreateToolbarSet+0x5c
0a 00000040`3607d6c0 00007fff`5ceaee0d     UIRibbon!COfficeSpaceUser::LoadUI+0x112
0b 00000040`3607d740 00007ff7`f3834e84     UIRibbon!CUIFramework::LoadUI+0x7d
0c 00000040`3607d780 00007ff7`f3835361     wordpad!CMainFrame::InitRibbonUI+0x180
0d 00000040`3607d7b0 00007fff`7c7e8f27     wordpad!CMainFrame::OnCreate+0x181
0e 00000040`3607d800 00007ff7`f3836f5c     MFC42u!CWnd::OnWndMsg+0x387
0f 00000040`3607d930 00007fff`7c7f0bff     wordpad!CMainFrame::OnWndMsg+0x14c
10 00000040`3607d9c0 00007fff`7c7eb42a     MFC42u!CWnd::WindowProc+0x4f
11 00000040`3607da00 00007fff`7c7ea76b     MFC42u!AfxCallWndProc+0x14a
12 00000040`3607db00 00007ff8`0be38161     MFC42u!AfxWndProcBase+0x15b
13 00000040`3607db70 00007ff8`0be37e1c     USER32!UserCallWinProcCheckWow+0x2d1
14 00000040`3607dcd0 00007ff8`0be44edc     USER32!DispatchClientMessage+0x9c
15 00000040`3607dd30 00007ff8`0c1d2db4     USER32!__fnINLPCREATESTRUCT+0xac
16 00000040`3607dd90 00007ff8`097f2294     ntdll!KiUserCallbackDispatcherContinue
17 00000040`3607ded8 00007ff8`0be2f6b0     win32u!NtUserCreateWindowEx+0x14
18 00000040`3607dee0 00007ff8`0be2f3cc     USER32!VerNtUserCreateWindowEx+0x210
19 00000040`3607e270 00007ff8`0be2f212     USER32!CreateWindowInternal+0x1a8
1a 00000040`3607e3d0 00007fff`7c7ec835     USER32!CreateWindowExW+0x82
1b 00000040`3607e460 00007fff`7c7ef6c5     MFC42u!CWnd::CreateEx+0xf5
1c 00000040`3607e520 00007fff`7c7ef7b3     MFC42u!CFrameWnd::Create+0xe5
1d 00000040`3607e590 00007fff`7c7ef0ea     MFC42u!CFrameWnd::LoadFrame+0xb3
1e 00000040`3607e610 00007fff`7c7fa1ec     MFC42u!CDocTemplate::CreateNewFrame+0x7a
1f 00000040`3607e680 00007ff7`f383e331     MFC42u!CSingleDocTemplate::OpenDocumentFile+0x8c
20 00000040`3607e6e0 00007fff`7c7efa57     wordpad!CWordPadApp::InitInstance+0x521
21 00000040`3607fc80 00007ff7`f37fb49e     MFC42u!AfxWinMain+0x97
22 00000040`3607fcc0 00007ff8`0b03244d     wordpad!__wmainCRTStartup+0x1de
23 00000040`3607fd80 00007ff8`0c18df78     KERNEL32!BaseThreadInitThunk+0x1d
24 00000040`3607fdb0 00000000`00000000     ntdll!RtlUserThreadStart+0x28
 # Child-SP          RetAddr               Call Site
00 00000040`3607d338 00007fff`5cf1317c     USER32!CreateWindowExW
01 00000040`3607d340 00007fff`5cf243be     UIRibbon!IsolationAwareCreateWindowExW+0xe0
02 00000040`3607d3c0 00007fff`5cf2555b     UIRibbon!CreateWindowRegExW+0x9a
03 00000040`3607d430 00007fff`5cf269af     UIRibbon!TBCWP::FCreatePane+0x7b
04 00000040`3607d4a0 00007fff`5cf75f9e     UIRibbon!TBCWP::FShow+0x4b3
05 00000040`3607d510 00007fff`5cf6c85c     UIRibbon!OfficeSpaceRibbonToolbarUser::FSetToolbarControls+0x29e
06 00000040`3607d5a0 00007fff`5cf6bf6e     UIRibbon!TB::FInitControls+0x50
07 00000040`3607d600 00007fff`5cf6c0cd     UIRibbon!TB::FEnsureControls+0x12
08 00000040`3607d630 00007fff`5cf350b6     UIRibbon!TB::FGetControl+0x1d
09 00000040`3607d660 00007fff`5ceba437     UIRibbon!OfficeSpace::Root::FCreate+0x1be
0a 00000040`3607d6c0 00007fff`5ceaee0d     UIRibbon!COfficeSpaceUser::LoadUI+0x147
0b 00000040`3607d740 00007ff7`f3834e84     UIRibbon!CUIFramework::LoadUI+0x7d
0c 00000040`3607d780 00007ff7`f3835361     wordpad!CMainFrame::InitRibbonUI+0x180
0d 00000040`3607d7b0 00007fff`7c7e8f27     wordpad!CMainFrame::OnCreate+0x181
0e 00000040`3607d800 00007ff7`f3836f5c     MFC42u!CWnd::OnWndMsg+0x387
0f 00000040`3607d930 00007fff`7c7f0bff     wordpad!CMainFrame::OnWndMsg+0x14c
10 00000040`3607d9c0 00007fff`7c7eb42a     MFC42u!CWnd::WindowProc+0x4f
11 00000040`3607da00 00007fff`7c7ea76b     MFC42u!AfxCallWndProc+0x14a
12 00000040`3607db00 00007ff8`0be38161     MFC42u!AfxWndProcBase+0x15b
13 00000040`3607db70 00007ff8`0be37e1c     USER32!UserCallWinProcCheckWow+0x2d1
14 00000040`3607dcd0 00007ff8`0be44edc     USER32!DispatchClientMessage+0x9c
15 00000040`3607dd30 00007ff8`0c1d2db4     USER32!__fnINLPCREATESTRUCT+0xac
16 00000040`3607dd90 00007ff8`097f2294     ntdll!KiUserCallbackDispatcherContinue
17 00000040`3607ded8 00007ff8`0be2f6b0     win32u!NtUserCreateWindowEx+0x14
18 00000040`3607dee0 00007ff8`0be2f3cc     USER32!VerNtUserCreateWindowEx+0x210
19 00000040`3607e270 00007ff8`0be2f212     USER32!CreateWindowInternal+0x1a8
1a 00000040`3607e3d0 00007fff`7c7ec835     USER32!CreateWindowExW+0x82
1b 00000040`3607e460 00007fff`7c7ef6c5     MFC42u!CWnd::CreateEx+0xf5
1c 00000040`3607e520 00007fff`7c7ef7b3     MFC42u!CFrameWnd::Create+0xe5
1d 00000040`3607e590 00007fff`7c7ef0ea     MFC42u!CFrameWnd::LoadFrame+0xb3
1e 00000040`3607e610 00007fff`7c7fa1ec     MFC42u!CDocTemplate::CreateNewFrame+0x7a
1f 00000040`3607e680 00007ff7`f383e331     MFC42u!CSingleDocTemplate::OpenDocumentFile+0x8c
20 00000040`3607e6e0 00007fff`7c7efa57     wordpad!CWordPadApp::InitInstance+0x521
21 00000040`3607fc80 00007ff7`f37fb49e     MFC42u!AfxWinMain+0x97
22 00000040`3607fcc0 00007ff8`0b03244d     wordpad!__wmainCRTStartup+0x1de
23 00000040`3607fd80 00007ff8`0c18df78     KERNEL32!BaseThreadInitThunk+0x1d
24 00000040`3607fdb0 00000000`00000000     ntdll!RtlUserThreadStart+0x28
ModLoad: 00007fff`d66e0000 00007fff`d6746000   C:\WINDOWS\SYSTEM32\Print.PrintSupport.Source.dll
 # Child-SP          RetAddr               Call Site
00 00000040`3607d1a8 00007fff`5cf1317c     USER32!CreateWindowExW
01 00000040`3607d1b0 00007fff`5cf243be     UIRibbon!IsolationAwareCreateWindowExW+0xe0
02 00000040`3607d230 00007fff`5cf29c8e     UIRibbon!CreateWindowRegExW+0x9a
03 00000040`3607d2a0 00007fff`5cf2a199     UIRibbon!NUIPaneHost::Create+0xa2
04 00000040`3607d310 00007fff`5cf3a835     UIRibbon!DefaultNetUIWorkPaneCallback+0x37d
05 00000040`3607d3c0 00007fff`5cf25593     UIRibbon!OfficeSpaceWorkPaneCallback+0xb5
06 00000040`3607d430 00007fff`5cf269af     UIRibbon!TBCWP::FCreatePane+0xb3
07 00000040`3607d4a0 00007fff`5cf75f9e     UIRibbon!TBCWP::FShow+0x4b3
08 00000040`3607d510 00007fff`5cf6c85c     UIRibbon!OfficeSpaceRibbonToolbarUser::FSetToolbarControls+0x29e
09 00000040`3607d5a0 00007fff`5cf6bf6e     UIRibbon!TB::FInitControls+0x50
0a 00000040`3607d600 00007fff`5cf6c0cd     UIRibbon!TB::FEnsureControls+0x12
0b 00000040`3607d630 00007fff`5cf350b6     UIRibbon!TB::FGetControl+0x1d
0c 00000040`3607d660 00007fff`5ceba437     UIRibbon!OfficeSpace::Root::FCreate+0x1be
0d 00000040`3607d6c0 00007fff`5ceaee0d     UIRibbon!COfficeSpaceUser::LoadUI+0x147
0e 00000040`3607d740 00007ff7`f3834e84     UIRibbon!CUIFramework::LoadUI+0x7d
0f 00000040`3607d780 00007ff7`f3835361     wordpad!CMainFrame::InitRibbonUI+0x180
10 00000040`3607d7b0 00007fff`7c7e8f27     wordpad!CMainFrame::OnCreate+0x181
11 00000040`3607d800 00007ff7`f3836f5c     MFC42u!CWnd::OnWndMsg+0x387
12 00000040`3607d930 00007fff`7c7f0bff     wordpad!CMainFrame::OnWndMsg+0x14c
13 00000040`3607d9c0 00007fff`7c7eb42a     MFC42u!CWnd::WindowProc+0x4f
14 00000040`3607da00 00007fff`7c7ea76b     MFC42u!AfxCallWndProc+0x14a
15 00000040`3607db00 00007ff8`0be38161     MFC42u!AfxWndProcBase+0x15b
16 00000040`3607db70 00007ff8`0be37e1c     USER32!UserCallWinProcCheckWow+0x2d1
17 00000040`3607dcd0 00007ff8`0be44edc     USER32!DispatchClientMessage+0x9c
18 00000040`3607dd30 00007ff8`0c1d2db4     USER32!__fnINLPCREATESTRUCT+0xac
19 00000040`3607dd90 00007ff8`097f2294     ntdll!KiUserCallbackDispatcherContinue
1a 00000040`3607ded8 00007ff8`0be2f6b0     win32u!NtUserCreateWindowEx+0x14
1b 00000040`3607dee0 00007ff8`0be2f3cc     USER32!VerNtUserCreateWindowEx+0x210
1c 00000040`3607e270 00007ff8`0be2f212     USER32!CreateWindowInternal+0x1a8
1d 00000040`3607e3d0 00007fff`7c7ec835     USER32!CreateWindowExW+0x82
```

```
1e 00000040`3607e460 00007fff`7c7ef6c5     MFC42u!CWnd::CreateEx+0xf5
1f 00000040`3607e520 00007fff`7c7ef7b3     MFC42u!CFrameWnd::Create+0xe5
20 00000040`3607e590 00007fff`7c7ef0ea     MFC42u!CFrameWnd::LoadFrame+0xb3
21 00000040`3607e610 00007fff`7c7fa1ec     MFC42u!CDocTemplate::CreateNewFrame+0x7a
22 00000040`3607e680 00007ff7`f383e331     MFC42u!CSingleDocTemplate::OpenDocumentFile+0x8c
23 00000040`3607e6e0 00007fff`7c7efa57     wordpad!CWordPadApp::InitInstance+0x521
24 00000040`3607fc80 00007ff7`f37fb49e     MFC42u!AfxWinMain+0x97
25 00000040`3607fcc0 00007ff8`0b03244d     wordpad!__wmainCRTStartup+0x1de
26 00000040`3607fd80 00007ff8`0c18df78     KERNEL32!BaseThreadInitThunk+0x1d
27 00000040`3607fdb0 00000000`00000000     ntdll!RtlUserThreadStart+0x28
ModLoad: 00007ff8`091a0000 00007ff8`091ef000   C:\WINDOWS\SYSTEM32\cfgmgr32.dll
ModLoad: 00000170`04570000 00000170`045bf000   C:\WINDOWS\SYSTEM32\cfgmgr32.dll
ModLoad: 00000170`045c0000 00000170`0460f000   C:\WINDOWS\SYSTEM32\cfgmgr32.dll
 # Child-SP          RetAddr               Call Site
00 00000040`3607d138 00007fff`5cf1317c     USER32!CreateWindowExW
01 00000040`3607d140 00007fff`5cf8c1d2     UIRibbon!IsolationAwareCreateWindowExW+0xe0
02 00000040`3607d1c0 00007fff`5cf8b2fc     UIRibbon!NetUI::HWNDElement::Initialize+0x222
03 00000040`3607d280 00007fff`5cfc290e     UIRibbon!NetUI::HWNDElement::Create+0x54
04 00000040`3607d2b0 00007fff`5cfc2106     UIRibbon!NetUI::NUIDocument::_CreateDocumentWindow+0x6e
05 00000040`3607d2e0 00007fff`5cf2a22a     UIRibbon!NetUI::NUIDocument::CreateDocumentWindow+0x92
06 00000040`3607d310 00007fff`5cf3a835     UIRibbon!DefaultNetUIWorkPaneCallback+0x40e
07 00000040`3607d3c0 00007fff`5cf25593     UIRibbon!OfficeSpaceWorkPaneCallback+0xb5
08 00000040`3607d430 00007fff`5cf269af     UIRibbon!TBCWP::FCreatePane+0xb3
09 00000040`3607d4a0 00007fff`5cf75f9e     UIRibbon!TBCWP::FShow+0x4b3
0a 00000040`3607d510 00007fff`5cf6c85c     UIRibbon!OfficeSpaceRibbonToolbarUser::FSetToolbarControls+0x29e
0b 00000040`3607d5a0 00007fff`5cf6bf6e     UIRibbon!TB::FInitControls+0x50
0c 00000040`3607d600 00007fff`5cf6c0cd     UIRibbon!TB::FEnsureControls+0x12
0d 00000040`3607d630 00007fff`5cf350b6     UIRibbon!TB::FGetControl+0x1d
0e 00000040`3607d660 00007fff`5ceba437     UIRibbon!OfficeSpace::Root::FCreate+0x1be
0f 00000040`3607d6c0 00007fff`5ceaee0d     UIRibbon!COfficeSpaceUser::LoadUI+0x147
10 00000040`3607d740 00007ff7`f3834e84     UIRibbon!CUIFramework::LoadUI+0x7d
11 00000040`3607d780 00007ff7`f3835361     wordpad!CMainFrame::InitRibbonUI+0x180
12 00000040`3607d7b0 00007fff`7c7e8f27     wordpad!CMainFrame::OnCreate+0x181
13 00000040`3607d800 00007ff7`f3836f5c     MFC42u!CWnd::OnWndMsg+0x387
14 00000040`3607d930 00007fff`7c7f0bff     wordpad!CMainFrame::OnWndMsg+0x14c
15 00000040`3607d9c0 00007fff`7c7eb42a     MFC42u!CWnd::WindowProc+0x4f
16 00000040`3607da00 00007fff`7c7ea76b     MFC42u!AfxCallWndProc+0x14a
17 00000040`3607db00 00007ff8`0be38161     MFC42u!AfxWndProcBase+0x15b
18 00000040`3607db70 00007ff8`0be37e1c     USER32!UserCallWinProcCheckWow+0x2d1
19 00000040`3607dcd0 00007ff8`0be44edc     USER32!DispatchClientMessage+0x9c
1a 00000040`3607dd30 00007ff8`0c1d2db4     USER32!__fnINLPCREATESTRUCT+0xac
1b 00000040`3607dd90 00007ff8`097f2294     ntdll!KiUserCallbackDispatcherContinue
1c 00000040`3607ded8 00007ff8`0be2f6b0     win32u!NtUserCreateWindowEx+0x14
1d 00000040`3607dee0 00007ff8`0be2f3cc     USER32!VerNtUserCreateWindowEx+0x210
1e 00000040`3607e270 00007ff8`0be2f212     USER32!CreateWindowInternal+0x1a8
1f 00000040`3607e3d0 00007fff`7c7ec835     USER32!CreateWindowExW+0x82
20 00000040`3607e460 00007fff`7c7ef6c5     MFC42u!CWnd::CreateEx+0xf5
21 00000040`3607e520 00007fff`7c7ef7b3     MFC42u!CFrameWnd::Create+0xe5
22 00000040`3607e590 00007fff`7c7ef0ea     MFC42u!CFrameWnd::LoadFrame+0xb3
23 00000040`3607e610 00007fff`7c7fa1ec     MFC42u!CDocTemplate::CreateNewFrame+0x7a
24 00000040`3607e680 00007ff7`f383e331     MFC42u!CSingleDocTemplate::OpenDocumentFile+0x8c
25 00000040`3607e6e0 00007fff`7c7efa57     wordpad!CWordPadApp::InitInstance+0x521
26 00000040`3607fc80 00007ff7`f37fb49e     MFC42u!AfxWinMain+0x97
27 00000040`3607fcc0 00007ff8`0b03244d     wordpad!__wmainCRTStartup+0x1de
28 00000040`3607fd80 00007ff8`0c18df78     KERNEL32!BaseThreadInitThunk+0x1d
29 00000040`3607fdb0 00000000`00000000     ntdll!RtlUserThreadStart+0x28
 # Child-SP          RetAddr               Call Site
00 00000040`3607cfe8 00007fff`5cf1317c     USER32!CreateWindowExW
01 00000040`3607cff0 00007fff`5d0356ff     UIRibbon!IsolationAwareCreateWindowExW+0xe0
02 00000040`3607d070 00007fff`5cffbb10     UIRibbon!CoreSC::CreateHiddenWindow+0x6f
03 00000040`3607d0e0 00007fff`5d04a5c4     UIRibbon!DuRootGadget::Build+0xd0
04 00000040`3607d110 00007fff`5cfd9aa5     UIRibbon!GdCreateHwndRootGadget+0x6c
05 00000040`3607d150 00007fff`5cf8c2f6     UIRibbon!CreateGadgetEx+0x221
06 00000040`3607d1c0 00007fff`5cf8b2fc     UIRibbon!NetUI::HWNDElement::Initialize+0x346
07 00000040`3607d280 00007fff`5cfc290e     UIRibbon!NetUI::HWNDElement::Create+0x54
08 00000040`3607d2b0 00007fff`5cfc2106     UIRibbon!NetUI::NUIDocument::_CreateDocumentWindow+0x6e
09 00000040`3607d2e0 00007fff`5cf2a22a     UIRibbon!NetUI::NUIDocument::CreateDocumentWindow+0x92
0a 00000040`3607d310 00007fff`5cf3a835     UIRibbon!DefaultNetUIWorkPaneCallback+0x40e
0b 00000040`3607d3c0 00007fff`5cf25593     UIRibbon!OfficeSpaceWorkPaneCallback+0xb5
0c 00000040`3607d430 00007fff`5cf269af     UIRibbon!TBCWP::FCreatePane+0xb3
0d 00000040`3607d4a0 00007fff`5cf75f9e     UIRibbon!TBCWP::FShow+0x4b3
0e 00000040`3607d510 00007fff`5cf6c85c     UIRibbon!OfficeSpaceRibbonToolbarUser::FSetToolbarControls+0x29e
0f 00000040`3607d5a0 00007fff`5cf6bf6e     UIRibbon!TB::FInitControls+0x50
10 00000040`3607d600 00007fff`5cf6c0cd     UIRibbon!TB::FEnsureControls+0x12
11 00000040`3607d630 00007fff`5cf350b6     UIRibbon!TB::FGetControl+0x1d
12 00000040`3607d660 00007fff`5ceba437     UIRibbon!OfficeSpace::Root::FCreate+0x1be
13 00000040`3607d6c0 00007fff`5ceaee0d     UIRibbon!COfficeSpaceUser::LoadUI+0x147
14 00000040`3607d740 00007ff7`f3834e84     UIRibbon!CUIFramework::LoadUI+0x7d
15 00000040`3607d780 00007ff7`f3835361     wordpad!CMainFrame::InitRibbonUI+0x180
16 00000040`3607d7b0 00007fff`7c7e8f27     wordpad!CMainFrame::OnCreate+0x181
17 00000040`3607d800 00007ff7`f3836f5c     MFC42u!CWnd::OnWndMsg+0x387
18 00000040`3607d930 00007fff`7c7f0bff     wordpad!CMainFrame::OnWndMsg+0x14c
19 00000040`3607d9c0 00007fff`7c7eb42a     MFC42u!CWnd::WindowProc+0x4f
1a 00000040`3607da00 00007fff`7c7ea76b     MFC42u!AfxCallWndProc+0x14a
1b 00000040`3607db00 00007ff8`0be38161     MFC42u!AfxWndProcBase+0x15b
1c 00000040`3607db70 00007ff8`0be37e1c     USER32!UserCallWinProcCheckWow+0x2d1
1d 00000040`3607dcd0 00007ff8`0be44edc     USER32!DispatchClientMessage+0x9c
1e 00000040`3607dd30 00007ff8`0c1d2db4     USER32!__fnINLPCREATESTRUCT+0xac
1f 00000040`3607dd90 00007ff8`097f2294     ntdll!KiUserCallbackDispatcherContinue
20 00000040`3607ded8 00007ff8`0be2f6b0     win32u!NtUserCreateWindowEx+0x14
21 00000040`3607dee0 00007ff8`0be2f3cc     USER32!VerNtUserCreateWindowEx+0x210
22 00000040`3607e270 00007ff8`0be2f212     USER32!CreateWindowInternal+0x1a8
23 00000040`3607e3d0 00007fff`7c7ec835     USER32!CreateWindowExW+0x82
24 00000040`3607e460 00007fff`7c7ef6c5     MFC42u!CWnd::CreateEx+0xf5
25 00000040`3607e520 00007fff`7c7ef7b3     MFC42u!CFrameWnd::Create+0xe5
26 00000040`3607e590 00007fff`7c7ef0ea     MFC42u!CFrameWnd::LoadFrame+0xb3
27 00000040`3607e610 00007fff`7c7fa1ec     MFC42u!CDocTemplate::CreateNewFrame+0x7a
28 00000040`3607e680 00007ff7`f383e331     MFC42u!CSingleDocTemplate::OpenDocumentFile+0x8c
```

```
29 00000040`3607e6e0 00007fff`7c7efa57     wordpad!CWordPadApp::InitInstance+0x521
2a 00000040`3607fc80 00007ff7`f37fb49e     MFC42u!AfxWinMain+0x97
2b 00000040`3607fcc0 00007ff8`0b03244d     wordpad!__wmainCRTStartup+0x1de
2c 00000040`3607fd80 00007ff8`0c18df78     KERNEL32!BaseThreadInitThunk+0x1d
2d 00000040`3607fdb0 00000000`00000000     ntdll!RtlUserThreadStart+0x28
ModLoad: 00007ff8`05e60000 00007ff8`05e8b000   C:\WINDOWS\system32\dwmapi.dll
 # Child-SP          RetAddr               Call Site
00 00000040`3607d578 00007fff`5cf1317c     USER32!CreateWindowExW
01 00000040`3607d580 00007fff`5cf243be     UIRibbon!IsolationAwareCreateWindowExW+0xe0
02 00000040`3607d600 00007fff`5cf58d91     UIRibbon!CreateWindowRegExW+0x9a
03 00000040`3607d670 00007fff`5ceb9c4f     UIRibbon!TBS::SetVisible+0xc1
04 00000040`3607d710 00007fff`5ceaee5c     UIRibbon!COfficeSpaceUser::InitializeUI+0x4f
05 00000040`3607d740 00007ff7`f3834e84     UIRibbon!CUIFramework::LoadUI+0xcc
06 00000040`3607d780 00007ff7`f3835361     wordpad!CMainFrame::InitRibbonUI+0x180
07 00000040`3607d7b0 00007fff`7c7e8f27     wordpad!CMainFrame::OnCreate+0x181
08 00000040`3607d800 00007ff7`f3836f5c     MFC42u!CWnd::OnWndMsg+0x387
09 00000040`3607d930 00007fff`7c7f0bff     wordpad!CMainFrame::OnWndMsg+0x14c
0a 00000040`3607d9c0 00007fff`7c7eb42a     MFC42u!CWnd::WindowProc+0x4f
0b 00000040`3607da00 00007fff`7c7ea76b     MFC42u!AfxCallWndProc+0x14a
0c 00000040`3607db00 00007ff8`0be38161     MFC42u!AfxWndProcBase+0x15b
0d 00000040`3607db70 00007ff8`0be37e1c     USER32!UserCallWinProcCheckWow+0x2d1
0e 00000040`3607dcd0 00007ff8`0be44edc     USER32!DispatchClientMessage+0x9c
0f 00000040`3607dd30 00007ff8`0c1d2db4     USER32!__fnINLPCREATESTRUCT+0xac
10 00000040`3607dd90 00007ff8`097f2294     ntdll!KiUserCallbackDispatcherContinue
11 00000040`3607ded8 00007ff8`0be2f6b0     win32u!NtUserCreateWindowEx+0x14
12 00000040`3607dee0 00007ff8`0be2f3cc     USER32!VerNtUserCreateWindowEx+0x210
13 00000040`3607e270 00007ff8`0be2f212     USER32!CreateWindowInternal+0x1a8
14 00000040`3607e3d0 00007fff`7c7ec835     USER32!CreateWindowExW+0x82
15 00000040`3607e460 00007fff`7c7ef6c5     MFC42u!CWnd::CreateEx+0xf5
16 00000040`3607e520 00007fff`7c7ef7b3     MFC42u!CFrameWnd::Create+0xe5
17 00000040`3607e590 00007fff`7c7ef0ea     MFC42u!CFrameWnd::LoadFrame+0xb3
18 00000040`3607e610 00007fff`7c7fa1ec     MFC42u!CDocTemplate::CreateNewFrame+0x7a
19 00000040`3607e680 00007ff7`f383e331     MFC42u!CSingleDocTemplate::OpenDocumentFile+0x8c
1a 00000040`3607e6e0 00007fff`7c7efa57     wordpad!CWordPadApp::InitInstance+0x521
1b 00000040`3607fc80 00007ff7`f37fb49e     MFC42u!AfxWinMain+0x97
1c 00000040`3607fcc0 00007ff8`0b03244d     wordpad!__wmainCRTStartup+0x1de
1d 00000040`3607fd80 00007ff8`0c18df78     KERNEL32!BaseThreadInitThunk+0x1d
1e 00000040`3607fdb0 00000000`00000000     ntdll!RtlUserThreadStart+0x28
 # Child-SP          RetAddr               Call Site
00 00000040`3607d578 00007fff`5cf1317c     USER32!CreateWindowExW
01 00000040`3607d580 00007fff`5cf243be     UIRibbon!IsolationAwareCreateWindowExW+0xe0
02 00000040`3607d600 00007fff`5cf58d91     UIRibbon!CreateWindowRegExW+0x9a
03 00000040`3607d670 00007fff`5ceb9c4f     UIRibbon!TBS::SetVisible+0xc1
04 00000040`3607d710 00007fff`5ceaee5c     UIRibbon!COfficeSpaceUser::InitializeUI+0x4f
05 00000040`3607d740 00007ff7`f3834e84     UIRibbon!CUIFramework::LoadUI+0xcc
06 00000040`3607d780 00007ff7`f3835361     wordpad!CMainFrame::InitRibbonUI+0x180
07 00000040`3607d7b0 00007fff`7c7e8f27     wordpad!CMainFrame::OnCreate+0x181
08 00000040`3607d800 00007ff7`f3836f5c     MFC42u!CWnd::OnWndMsg+0x387
09 00000040`3607d930 00007fff`7c7f0bff     wordpad!CMainFrame::OnWndMsg+0x14c
0a 00000040`3607d9c0 00007fff`7c7eb42a     MFC42u!CWnd::WindowProc+0x4f
0b 00000040`3607da00 00007fff`7c7ea76b     MFC42u!AfxCallWndProc+0x14a
0c 00000040`3607db00 00007ff8`0be38161     MFC42u!AfxWndProcBase+0x15b
0d 00000040`3607db70 00007ff8`0be37e1c     USER32!UserCallWinProcCheckWow+0x2d1
0e 00000040`3607dcd0 00007ff8`0be44edc     USER32!DispatchClientMessage+0x9c
0f 00000040`3607dd30 00007ff8`0c1d2db4     USER32!__fnINLPCREATESTRUCT+0xac
10 00000040`3607dd90 00007ff8`097f2294     ntdll!KiUserCallbackDispatcherContinue
11 00000040`3607ded8 00007ff8`0be2f6b0     win32u!NtUserCreateWindowEx+0x14
12 00000040`3607dee0 00007ff8`0be2f3cc     USER32!VerNtUserCreateWindowEx+0x210
13 00000040`3607e270 00007ff8`0be2f212     USER32!CreateWindowInternal+0x1a8
14 00000040`3607e3d0 00007fff`7c7ec835     USER32!CreateWindowExW+0x82
15 00000040`3607e460 00007fff`7c7ef6c5     MFC42u!CWnd::CreateEx+0xf5
16 00000040`3607e520 00007fff`7c7ef7b3     MFC42u!CFrameWnd::Create+0xe5
17 00000040`3607e590 00007fff`7c7ef0ea     MFC42u!CFrameWnd::LoadFrame+0xb3
18 00000040`3607e610 00007fff`7c7fa1ec     MFC42u!CDocTemplate::CreateNewFrame+0x7a
19 00000040`3607e680 00007ff7`f383e331     MFC42u!CSingleDocTemplate::OpenDocumentFile+0x8c
1a 00000040`3607e6e0 00007fff`7c7efa57     wordpad!CWordPadApp::InitInstance+0x521
1b 00000040`3607fc80 00007ff7`f37fb49e     MFC42u!AfxWinMain+0x97
1c 00000040`3607fcc0 00007ff8`0b03244d     wordpad!__wmainCRTStartup+0x1de
1d 00000040`3607fd80 00007ff8`0c18df78     KERNEL32!BaseThreadInitThunk+0x1d
1e 00000040`3607fdb0 00000000`00000000     ntdll!RtlUserThreadStart+0x28
 # Child-SP          RetAddr               Call Site
00 00000040`3607d578 00007fff`5cf1317c     USER32!CreateWindowExW
01 00000040`3607d580 00007fff`5cf243be     UIRibbon!IsolationAwareCreateWindowExW+0xe0
02 00000040`3607d600 00007fff`5cf58d91     UIRibbon!CreateWindowRegExW+0x9a
03 00000040`3607d670 00007fff`5ceb9c4f     UIRibbon!TBS::SetVisible+0xc1
04 00000040`3607d710 00007fff`5ceaee5c     UIRibbon!COfficeSpaceUser::InitializeUI+0x4f
05 00000040`3607d740 00007ff7`f3834e84     UIRibbon!CUIFramework::LoadUI+0xcc
06 00000040`3607d780 00007ff7`f3835361     wordpad!CMainFrame::InitRibbonUI+0x180
07 00000040`3607d7b0 00007fff`7c7e8f27     wordpad!CMainFrame::OnCreate+0x181
08 00000040`3607d800 00007ff7`f3836f5c     MFC42u!CWnd::OnWndMsg+0x387
09 00000040`3607d930 00007fff`7c7f0bff     wordpad!CMainFrame::OnWndMsg+0x14c
0a 00000040`3607d9c0 00007fff`7c7eb42a     MFC42u!CWnd::WindowProc+0x4f
0b 00000040`3607da00 00007fff`7c7ea76b     MFC42u!AfxCallWndProc+0x14a
0c 00000040`3607db00 00007ff8`0be38161     MFC42u!AfxWndProcBase+0x15b
0d 00000040`3607db70 00007ff8`0be37e1c     USER32!UserCallWinProcCheckWow+0x2d1
0e 00000040`3607dcd0 00007ff8`0be44edc     USER32!DispatchClientMessage+0x9c
0f 00000040`3607dd30 00007ff8`0c1d2db4     USER32!__fnINLPCREATESTRUCT+0xac
10 00000040`3607dd90 00007ff8`097f2294     ntdll!KiUserCallbackDispatcherContinue
11 00000040`3607ded8 00007ff8`0be2f6b0     win32u!NtUserCreateWindowEx+0x14
12 00000040`3607dee0 00007ff8`0be2f3cc     USER32!VerNtUserCreateWindowEx+0x210
13 00000040`3607e270 00007ff8`0be2f212     USER32!CreateWindowInternal+0x1a8
14 00000040`3607e3d0 00007fff`7c7ec835     USER32!CreateWindowExW+0x82
15 00000040`3607e460 00007fff`7c7ef6c5     MFC42u!CWnd::CreateEx+0xf5
16 00000040`3607e520 00007fff`7c7ef7b3     MFC42u!CFrameWnd::Create+0xe5
17 00000040`3607e590 00007fff`7c7ef0ea     MFC42u!CFrameWnd::LoadFrame+0xb3
18 00000040`3607e610 00007fff`7c7fa1ec     MFC42u!CDocTemplate::CreateNewFrame+0x7a
19 00000040`3607e680 00007ff7`f383e331     MFC42u!CSingleDocTemplate::OpenDocumentFile+0x8c
1a 00000040`3607e6e0 00007fff`7c7efa57     wordpad!CWordPadApp::InitInstance+0x521
```

```
1b 00000040`3607fc80 00007ff7`f37fb49e      MFC42u!AfxWinMain+0x97
1c 00000040`3607fcc0 00007ff8`0b03244d      wordpad!__wmainCRTStartup+0x1de
1d 00000040`3607fd80 00007ff8`0c18df78      KERNEL32!BaseThreadInitThunk+0x1d
1e 00000040`3607fdb0 00000000`00000000      ntdll!RtlUserThreadStart+0x28
 # Child-SP          RetAddr               Call Site
00 00000040`3607d578 00007fff`5cf1317c      USER32!CreateWindowExW
01 00000040`3607d580 00007fff`5cf243be      UIRibbon!IsolationAwareCreateWindowExW+0xe0
02 00000040`3607d600 00007fff`5cf58d91      UIRibbon!CreateWindowRegExW+0x9a
03 00000040`3607d670 00007fff`5ceb9c4f      UIRibbon!TBS::SetVisible+0xc1
04 00000040`3607d710 00007fff`5ceaee5c      UIRibbon!COfficeSpaceUser::InitializeUI+0x4f
05 00000040`3607d740 00007ff7`f3834e84      UIRibbon!CUIFramework::LoadUI+0xcc
06 00000040`3607d780 00007ff7`f3835361      wordpad!CMainFrame::InitRibbonUI+0x180
07 00000040`3607d7b0 00007fff`7c7e8f27      wordpad!CMainFrame::OnCreate+0x181
08 00000040`3607d800 00007ff7`f3836f5c      MFC42u!CWnd::OnWndMsg+0x387
09 00000040`3607d930 00007fff`7c7f0bff      wordpad!CMainFrame::OnWndMsg+0x14c
0a 00000040`3607d9c0 00007fff`7c7eb42a      MFC42u!CWnd::WindowProc+0x4f
0b 00000040`3607da00 00007fff`7c7ea76b      MFC42u!AfxCallWndProc+0x14a
0c 00000040`3607db00 00007ff8`0be38161      MFC42u!AfxWndProcBase+0x15b
0d 00000040`3607db70 00007ff8`0be37e1c      USER32!UserCallWinProcCheckWow+0x2d1
0e 00000040`3607dcd0 00007ff8`0be44edc      USER32!DispatchClientMessage+0x9c
0f 00000040`3607dd30 00007ff8`0c1d2db4      USER32!__fnINLPCREATESTRUCT+0xac
10 00000040`3607dd90 00007ff8`097f2294      ntdll!KiUserCallbackDispatcherContinue
11 00000040`3607ded8 00007ff8`0be2f6b0      win32u!NtUserCreateWindowEx+0x14
12 00000040`3607dee0 00007ff8`0be2f3cc      USER32!VerNtUserCreateWindowEx+0x210
13 00000040`3607e270 00007ff8`0be2f212      USER32!CreateWindowInternal+0x1a8
14 00000040`3607e3d0 00007fff`7c7ec835      USER32!CreateWindowExW+0x82
15 00000040`3607e460 00007fff`7c7ef6c5      MFC42u!CWnd::CreateEx+0xf5
16 00000040`3607e520 00007fff`7c7ef7b3      MFC42u!CFrameWnd::Create+0xe5
17 00000040`3607e590 00007fff`7c7ef0ea      MFC42u!CFrameWnd::LoadFrame+0xb3
18 00000040`3607e610 00007fff`7c7fa1ec      MFC42u!CDocTemplate::CreateNewFrame+0x7a
19 00000040`3607e680 00007ff7`f383e331      MFC42u!CSingleDocTemplate::OpenDocumentFile+0x8c
1a 00000040`3607e6e0 00007fff`7c7efa57      wordpad!CWordPadApp::InitInstance+0x521
1b 00000040`3607fc80 00007ff7`f37fb49e      MFC42u!AfxWinMain+0x97
1c 00000040`3607fcc0 00007ff8`0b03244d      wordpad!__wmainCRTStartup+0x1de
1d 00000040`3607fd80 00007ff8`0c18df78      KERNEL32!BaseThreadInitThunk+0x1d
1e 00000040`3607fdb0 00000000`00000000      ntdll!RtlUserThreadStart+0x28
 # Child-SP          RetAddr               Call Site
00 00000040`3627db88 00007fff`f3cce048      USER32!CreateWindowExW
01 00000040`3627db90 00007fff`f3ccdf1f      urlmon!NotificationWindow::CreateHandle+0x68
02 00000040`3627dc00 00007fff`f3c7cc9b      urlmon!CTransaction::GetNotificationWnd+0x13f
03 00000040`3627dc40 00007fff`f3cb98ee      urlmon!GetTransactionObjects+0x2fb
04 00000040`3627dcc0 00007fff`f3cba81c      urlmon!CBinding::StartBinding+0x53e
05 00000040`3627de40 00007fff`f3cb9276      urlmon!CUrlMon::StartBinding+0x1b0
06 00000040`3627df10 00007fff`f1c05529      urlmon!CUrlMon::BindToStorage+0x96
07 00000040`3627df60 00007fff`f1c04acb      msxml6!URLMONStream::deferedOpen+0x159
08 00000040`3627dfc0 00007fff`f1c048df      msxml6!XMLParser::PushURL+0x1af
09 00000040`3627e040 00007fff`f1bbf1c6      msxml6!XMLParser::SetURL+0x7f
0a 00000040`3627e090 00007fff`f1bbed48      msxml6!Document::_load+0x162
0b 00000040`3627e110 00007fff`f1b87712      msxml6!Document::load+0x198
0c 00000040`3627e180 00007fff`7c46afae      msxml6!DOMDocumentWrapper::load+0x182
0d 00000040`3627e250 00007fff`7c414968      PrintConfig!PrintConfig::PrivateDevmodeMapFile::Parse+0x14e
0e 00000040`3627e3c0 00007fff`7c558d7b      PrintConfig!PrintConfigPlugIn::DevMode+0x458
0f 00000040`3627e4e0 00007fff`7c55949b      PrintConfig!BCalcTotalOEMDMSize+0xa3
10 00000040`3627e580 00007fff`7c514999      PrintConfig!PGetDefaultDevmodeWithOemPlugins+0x9b
11 00000040`3627e600 00007fff`7c4fc131      PrintConfig!UniDrvUI::BGetValidatedUserDefaultDevmode+0x81
12 00000040`3627e670 00007fff`7c4f1696      PrintConfig!UniDrvUI::CPrintTicketProvider::GetValidatedDevmode+0x6d
13 00000040`3627e6c0 00007fff`7c51654d      PrintConfig!UniDrvUI::CPrintTicketProvider::ConvertDevModeToPrintTicket+0x36
14 00000040`3627e6f0 00007fff`7c514794      PrintConfig!UniDrvUI::PerformJScriptDevmodeValidation+0x479
15 00000040`3627e7f0 00007fff`7c512bb7      PrintConfig!UniDrvUI::BFillCommonInfoDevmode+0x31c
16 00000040`3627e890 00007fff`7c5120ec      PrintConfig!UniDrvUI::LSimpleDocumentProperties+0x173
17 00000040`3627e900 00007fff`7c40f54b      PrintConfig!UniDrvUI::DrvDocumentPropertySheets+0x60
18 00000040`3627edb0 00007fff`7c4249ad      PrintConfig!PrintConfig::DrvDocumentPropertySheets+0x123
19 00000040`3627ee20 00007fff`7c425543      PrintConfig!ExceptionBoundary<<lambda_ce41999e58d4ed069b2790d632f6e3b5> >+0x3d
1a 00000040`3627ee70 00007fff`e032e9ba      PrintConfig!DrvDocumentPropertySheets+0x53
1b 00000040`3627eee0 00007fff`e031593f      WINSPOOL!DocumentPropertySheets+0x3fa
1c 00000040`3627ef30 00007fff`e0324aff      WINSPOOL!DocumentPropertiesWNative+0x193
1d 00000040`3627efc0 00007ff8`0a340fc8      WINSPOOL!DocumentPropertiesW+0xcf
1e 00000040`3627f020 00007ff8`0a344d59      COMDLG32!PrintGetDevMode+0x8c
1f 00000040`3627f060 00007ff8`0a34058b      COMDLG32!PrintReturnDefault+0x91
20 00000040`3627f090 00007ff8`0a340407      COMDLG32!PrintDlgX+0x163
21 00000040`3627f530 00007fff`7c7ff697      COMDLG32!PrintDlgW+0x47
22 00000040`3627fa40 00007fff`7c7ff591      MFC42u!CWinApp::UpdatePrinterSelection+0x57
23 00000040`3627fbe0 00007ff7`f383d742      MFC42u!CWinApp::GetPrinterDeviceDefaults+0x21
24 00000040`3627fc10 00007ff7`f383d938      wordpad!CWordPadApp::CreateDevNames+0xb2
25 00000040`3627fcf0 00007fff`7c7ffc1d      wordpad!CWordPadApp::DoDeferredInitialization+0x18
26 00000040`3627fd20 00007ff8`0a57e634      MFC42u!_AfxThreadEntry+0xdd
27 00000040`3627fde0 00007ff8`0a57e70c      msvcrt!_callthreadstartex+0x28
28 00000040`3627fe10 00007ff8`0b03244d      msvcrt!_threadstartex+0x7c
29 00000040`3627fe40 00007ff8`0c18df78      KERNEL32!BaseThreadInitThunk+0x1d
2a 00000040`3627fe70 00000000`00000000      ntdll!RtlUserThreadStart+0x28
 # Child-SP          RetAddr               Call Site
00 00000040`3607d218 00007fff`5cf1317c      USER32!CreateWindowExW
01 00000040`3607d220 00007fff`5cf243be      UIRibbon!IsolationAwareCreateWindowExW+0xe0
02 00000040`3607d2a0 00007fff`5cf562e2      UIRibbon!CreateWindowRegExW+0x9a
03 00000040`3607d310 00007fff`5cf6e16d      UIRibbon!TBS::FGetHwnd+0x13a
04 00000040`3607d5f0 00007fff`5cf58f55      UIRibbon!TB::FShowTb+0xd5
05 00000040`3607d670 00007fff`5ceb9c4f      UIRibbon!TBS::SetVisible+0x285
06 00000040`3607d710 00007fff`5ceaee5c      UIRibbon!COfficeSpaceUser::InitializeUI+0x4f
07 00000040`3607d740 00007ff7`f3834e84      UIRibbon!CUIFramework::LoadUI+0xcc
08 00000040`3607d780 00007ff7`f3835361      wordpad!CMainFrame::InitRibbonUI+0x180
09 00000040`3607d7b0 00007fff`7c7e8f27      wordpad!CMainFrame::OnCreate+0x181
0a 00000040`3607d800 00007ff7`f3836f5c      MFC42u!CWnd::OnWndMsg+0x387
0b 00000040`3607d930 00007fff`7c7f0bff      wordpad!CMainFrame::OnWndMsg+0x14c
0c 00000040`3607d9c0 00007fff`7c7eb42a      MFC42u!CWnd::WindowProc+0x4f
0d 00000040`3607da00 00007fff`7c7ea76b      MFC42u!AfxCallWndProc+0x14a
0e 00000040`3607db00 00007ff8`0be38161      MFC42u!AfxWndProcBase+0x15b
0f 00000040`3607db70 00007ff8`0be37e1c      USER32!UserCallWinProcCheckWow+0x2d1
10 00000040`3607dcd0 00007ff8`0be44edc      USER32!DispatchClientMessage+0x9c
```

```
11 00000040`3607dd30 00007ff8`0c1d2db4     USER32!__fnINLPCREATESTRUCT+0xac
12 00000040`3607dd90 00007ff8`097f2294     ntdll!KiUserCallbackDispatcherContinue
13 00000040`3607ded8 00007ff8`0be2f6b0     win32u!NtUserCreateWindowEx+0x14
14 00000040`3607dee0 00007ff8`0be2f3cc     USER32!VerNtUserCreateWindowEx+0x210
15 00000040`3607e270 00007ff8`0be2f212     USER32!CreateWindowInternal+0x1a8
16 00000040`3607e3d0 00007fff`7c7ec835     USER32!CreateWindowExW+0x82
17 00000040`3607e460 00007fff`7c7ef6c5     MFC42u!CWnd::CreateEx+0xf5
18 00000040`3607e520 00007fff`7c7ef7b3     MFC42u!CFrameWnd::Create+0xe5
19 00000040`3607e590 00007fff`7c7ef0ea     MFC42u!CFrameWnd::LoadFrame+0xb3
1a 00000040`3607e610 00007fff`7c7fa1ec     MFC42u!CDocTemplate::CreateNewFrame+0x7a
1b 00000040`3607e680 00007ff7`f383e331     MFC42u!CSingleDocTemplate::OpenDocumentFile+0x8c
1c 00000040`3607e6e0 00007fff`7c7efa57     wordpad!CWordPadApp::InitInstance+0x521
1d 00000040`3607fc80 00007ff7`f37fb49e     MFC42u!AfxWinMain+0x97
1e 00000040`3607fcc0 00007ff8`0b03244d     wordpad!__wmainCRTStartup+0x1de
1f 00000040`3607fd80 00007ff8`0c18df78     KERNEL32!BaseThreadInitThunk+0x1d
20 00000040`3607fdb0 00000000`00000000     ntdll!RtlUserThreadStart+0x28
 # Child-SP          RetAddr               Call Site
00 00000040`3627dba8 00007fff`f3cce048     USER32!CreateWindowExW
01 00000040`3627dbb0 00007fff`f3ccdf1f     urlmon!NotificationWindow::CreateHandle+0x68
02 00000040`3627dc20 00007fff`f3c7cc9b     urlmon!CTransaction::GetNotificationWnd+0x13f
03 00000040`3627dc60 00007fff`f3cb98ee     urlmon!GetTransactionObjects+0x2fb
04 00000040`3627dce0 00007fff`f3cba81c     urlmon!CBinding::StartBinding+0x53e
05 00000040`3627de60 00007fff`f3cb9276     urlmon!CUrlMon::StartBinding+0x1b0
06 00000040`3627df30 00007fff`f1c05529     urlmon!CUrlMon::BindToStorage+0x96
07 00000040`3627df80 00007fff`f1c04acb     msxml6!URLMONStream::deferedOpen+0x159
08 00000040`3627dfe0 00007fff`f1c048df     msxml6!XMLParser::PushURL+0x1af
09 00000040`3627e060 00007fff`f1bbf1c6     msxml6!XMLParser::SetURL+0x7f
0a 00000040`3627e0b0 00007fff`f1bbed48     msxml6!Document::_load+0x162
0b 00000040`3627e130 00007fff`f1b87712     msxml6!Document::load+0x198
0c 00000040`3627e1a0 00007fff`7c46afae     msxml6!DOMDocumentWrapper::load+0x182
0d 00000040`3627e270 00007fff`7c414813     PrintConfig!PrintConfig::PrivateDevmodeMapFile::Parse+0x14e
0e 00000040`3627e3e0 00007fff`7c5583f0     PrintConfig!PrintConfigPlugIn::DevMode+0x303
0f 00000040`3627e500 00007fff`7c55956a     PrintConfig!BCallOEMDevMode+0x84
10 00000040`3627e580 00007fff`7c514999     PrintConfig!PGetDefaultDevmodeWithOemPlugins+0x16a
11 00000040`3627e600 00007fff`7c4fc131     PrintConfig!UniDrvUI::BGetValidatedUserDefaultDevmode+0x81
12 00000040`3627e670 00007fff`7c4f1696     PrintConfig!UniDrvUI::CPrintTicketProvider::GetValidatedDevmode+0x6d
13 00000040`3627e6c0 00007fff`7c51654d     PrintConfig!UniDrvUI::CPrintTicketProvider::ConvertDevModeToPrintTicket+0x36
14 00000040`3627e6f0 00007fff`7c514794     PrintConfig!UniDrvUI::PerformJScriptDevmodeValidation+0x479
15 00000040`3627e7f0 00007fff`7c512bb7     PrintConfig!UniDrvUI::BFillCommonInfoDevmode+0x31c
16 00000040`3627e890 00007fff`7c5120ec     PrintConfig!UniDrvUI::LSimpleDocumentProperties+0x173
17 00000040`3627e900 00007fff`7c40f54b     PrintConfig!UniDrvUI::DrvDocumentPropertySheets+0x60
18 00000040`3627edb0 00007fff`7c4249ad     PrintConfig!PrintConfig::DrvDocumentPropertySheets+0x123
19 00000040`3627ee20 00007fff`7c425543     PrintConfig!ExceptionBoundary<<lambda_ce41999e58d4ed069b2790d632f6e3b5> >+0x3d
1a 00000040`3627ee70 00007fff`e032e9ba     PrintConfig!DrvDocumentPropertySheets+0x53
1b 00000040`3627eee0 00007fff`e031593f     WINSPOOL!DocumentPropertySheets+0x3fa
1c 00000040`3627ef30 00007fff`e0324aff     WINSPOOL!DocumentPropertiesWNative+0x193
1d 00000040`3627efc0 00007ff8`0a340fc8     WINSPOOL!DocumentPropertiesW+0xcf
1e 00000040`3627f020 00007ff8`0a344d59     COMDLG32!PrintGetDevMode+0x8c
1f 00000040`3627f060 00007ff8`0a34058b     COMDLG32!PrintReturnDefault+0x91
20 00000040`3627f090 00007ff8`0a340407     COMDLG32!PrintDlgX+0x163
21 00000040`3627f530 00007fff`7c7ff697     COMDLG32!PrintDlgW+0x47
22 00000040`3627fa40 00007fff`7c7ff591     MFC42u!CWinApp::UpdatePrinterSelection+0x57
23 00000040`3627fbe0 00007ff7`f383d742     MFC42u!CWinApp::GetPrinterDeviceDefaults+0x21
24 00000040`3627fc10 00007ff7`f383d938     wordpad!CWordPadApp::CreateDevNames+0xb2
25 00000040`3627fcf0 00007fff`7c7ffc1d     wordpad!CWordPadApp::DoDeferredInitialization+0x18
26 00000040`3627fd20 00007ff8`0a57e634     MFC42u!_AfxThreadEntry+0xdd
27 00000040`3627fde0 00007ff8`0a57e70c     msvcrt!_callthreadstartex+0x28
28 00000040`3627fe10 00007ff8`0b03244d     msvcrt!_threadstartex+0x7c
29 00000040`3627fe40 00007ff8`0c18df78     KERNEL32!BaseThreadInitThunk+0x1d
2a 00000040`3627fe70 00000000`00000000     ntdll!RtlUserThreadStart+0x28
ModLoad: 00007ff8`043b0000 00007ff8`04560000   C:\WINDOWS\system32\windowscodecs.dll
ModLoad: 00007ff8`034b0000 00007ff8`034e8000   C:\WINDOWS\system32\fms.dll
ModLoad: 00007fff`d5b80000 00007fff`d5c56000   C:\Windows\System32\jscript.dll
ModLoad: 00007fff`e13e0000 00007fff`e13fd000   C:\WINDOWS\SYSTEM32\amsi.dll
ModLoad: 00007ff8`09490000 00007ff8`094b1000   C:\WINDOWS\SYSTEM32\profapi.dll
ModLoad: 00007fff`e0c20000 00007fff`e0c9b000   C:\ProgramData\Microsoft\Windows Defender\Platform\4.18.2210.4-0\MpOav.dll
ModLoad: 00007ff8`093d0000 00007ff8`09472000   C:\WINDOWS\SYSTEM32\sxs.dll
 # Child-SP          RetAddr               Call Site
00 00000040`3607ac78 00007fff`5cf1317c     USER32!CreateWindowExW
01 00000040`3607ac80 00007fff`5cfb02a0     UIRibbon!IsolationAwareCreateWindowExW+0xe0
02 00000040`3607ad00 00007fff`5cfb0881     UIRibbon!NetUI::HWNDHost::InitializeHWND+0x160
03 00000040`3607ad70 00007fff`5cf91f4a     UIRibbon!NetUI::HWNDHost::OnHosted+0x21
04 00000040`3607adc0 00007fff`5cf91f4a     UIRibbon!NetUI::Element::OnHosted+0xda
05 00000040`3607adf0 00007fff`5cf91f4a     UIRibbon!NetUI::Element::OnHosted+0xda
06 00000040`3607ae20 00007fff`5cf91f4a     UIRibbon!NetUI::Element::OnHosted+0xda
07 00000040`3607ae50 00007fff`5cf9b4cb     UIRibbon!NetUI::Element::OnHosted+0xda
08 00000040`3607ae80 00007fff`5cf9c40c     UIRibbon!NetUI::Element::OnElementOrderChanged+0x16f
09 00000040`3607aee0 00007fff`5cf9d7eb     UIRibbon!NetUI::Element::OnPropertyChanged+0xadc
0a 00000040`3607af80 00007fff`5cf88fb1     UIRibbon!NetUI::Node::_PostSourceChange+0x43b
0b 00000040`3607b000 00007fff`5cf8711a     UIRibbon!NetUI::Node::InsertNodes+0x231
0c 00000040`3607b0a0 00007fff`5cf8a3ed     UIRibbon!NetUI::Node::CreateStyleTreeFor+0x222
0d 00000040`3607b130 00007fff`5cf8a189     UIRibbon!NetUI::Node::TreeStyleChildrenInto+0x195
0e 00000040`3607b190 00007fff`5cf871ac     UIRibbon!NetUI::Node::TreeStyleChildrenBindingsInSubtree+0x5d
0f 00000040`3607b1e0 00007fff`5cf8a3ed     UIRibbon!NetUI::Node::CreateStyleTreeFor+0x2b4
10 00000040`3607b270 00007fff`5cf8a215     UIRibbon!NetUI::Node::TreeStyleChildrenInto+0x195
11 00000040`3607b2d0 00007fff`5cf871ac     UIRibbon!NetUI::Node::TreeStyleChildrenBindingsInSubtree+0xe9
12 00000040`3607b320 00007fff`5cf8a3ed     UIRibbon!NetUI::Node::CreateStyleTreeFor+0x2b4
13 00000040`3607b3b0 00007fff`5cf8a189     UIRibbon!NetUI::Node::TreeStyleChildrenInto+0x195
14 00000040`3607b410 00007fff`5cf871ac     UIRibbon!NetUI::Node::TreeStyleChildrenBindingsInSubtree+0x5d
15 00000040`3607b460 00007fff`5cf88b6e     UIRibbon!NetUI::Node::CreateStyleTreeFor+0x2b4
16 00000040`3607b4f0 00007fff`5cf86311     UIRibbon!NetUI::Node::IncrementalTreeStyle+0x1ba
17 00000040`3607b580 00007fff`5cf89404     UIRibbon!NetUI::Node::ApplyStyleTree+0x191
18 00000040`3607b5c0 00007fff`5cf9a30d     UIRibbon!NetUI::Node::OnGroupChanged+0x14
19 00000040`3607b5f0 00007fff`5cf9c5d9     UIRibbon!NetUI::Node::EndDefer+0x1e5
1a 00000040`3607b620 00007fff`5cf29fba     UIRibbon!NetUI::DeferCycle::PushDefer+0x6d
1b 00000040`3607b650 00007fff`5cf3a835     UIRibbon!DefaultNetUIWorkPaneCallback+0x19e
1c 00000040`3607b700 00007fff`5cf25ef4     UIRibbon!OfficeSpaceWorkPaneCallback+0xb5
1d 00000040`3607b770 00007fff`5cf25f73     UIRibbon!TBCWP::FGetMinMax+0xfc
```

```
1e 00000040`3607b810 00007fff`5cf26f81     UIRibbon!TBPUPANE::FGetMinMax+0x33
1f 00000040`3607b850 00007fff`5cf6a778     UIRibbon!TBCWP::QuerySize+0x81
20 00000040`3607b8c0 00007fff`5cf55274     UIRibbon!TB::CalcRectOrReflowToolbar+0x5d0
21 00000040`3607ba70 00007fff`5cf54409     UIRibbon!TBS::DxyReflowDockRow+0x94
22 00000040`3607baf0 00007fff`5cf58825     UIRibbon!TBS::DockReflowAsNeeded+0xf5
23 00000040`3607bb80 00007fff`5cf57cb7     UIRibbon!TBS::RedisplayDocks+0x3e9
24 00000040`3607bc20 00007fff`5ceb9d5b     UIRibbon!TBS::HrResizeBorderEx+0xf7
25 00000040`3607bc50 00007fff`5ceb8eb3     UIRibbon!COfficeSpaceUser::InvalidateLayout+0x8b
26 00000040`3607bca0 00007fff`5cebc534     UIRibbon!COfficeSpaceUser::FrameListenerWndProc+0x19f
27 00000040`3607bd00 00007fff`5d0154a7     UIRibbon!COfficeSpaceUser::s_FrameListenerWndProc+0xa4
28 00000040`3607bd50 00007ff8`08d310f2     UIRibbon!WndBridge::RawWndProc+0x107
29 00000040`3607bdc0 00007ff8`0be38161     atlthunk!AtlThunk_0x00+0x22
2a 00000040`3607be00 00007ff8`0be37e1c     USER32!UserCallWinProcCheckWow+0x2d1
2b 00000040`3607bf60 00007ff8`0be42cdd     USER32!DispatchClientMessage+0x9c
2c 00000040`3607bfc0 00007ff8`0c1d2db4     USER32!_fnDWORD+0x3d
2d 00000040`3607c020 00007ff8`097f1554     ntdll!KiUserCallbackDispatcherContinue
2e 00000040`3607c0a8 00007ff8`0be37798     win32u!NtUserMessageCall+0x14
2f 00000040`3607c0b0 00007ff8`0be3665b     USER32!SendMessageWorker+0x2e8
30 00000040`3607c160 00007ff8`0be36082     USER32!RealDefWindowProcWorker+0x23b
31 00000040`3607c240 00007ff8`05b56e26     USER32!RealDefWindowProcW+0x52
32 00000040`3607c280 00007ff8`05b55821     uxtheme!_ThemeDefWindowProc+0x15f6
33 00000040`3607c4a0 00007ff8`0be36205     uxtheme!ThemeDefWindowProcW+0x11
34 00000040`3607c4e0 00007ff8`0be38161     USER32!DefWindowProcW+0x135
35 00000040`3607c550 00007ff8`0be379ab     USER32!UserCallWinProcCheckWow+0x2d1
36 00000040`3607c6b0 00007fff`7c7f0c8e     USER32!CallWindowProcW+0x8b
37 00000040`3607c700 00007fff`7c7f0c28     MFC42u!CWnd::DefWindowProcW+0x3e
38 00000040`3607c740 00007fff`7c7eb42a     MFC42u!CWnd::WindowProc+0x78
39 00000040`3607c780 00007fff`7c7ea76b     MFC42u!AfxCallWndProc+0x14a
3a 00000040`3607c880 00007ff8`0be38161     MFC42u!AfxWndProcBase+0x15b
3b 00000040`3607c8f0 00007ff8`0be379ab     USER32!UserCallWinProcCheckWow+0x2d1
3c 00000040`3607ca50 00007fff`5d0154e9     USER32!CallWindowProcW+0x8b
3d 00000040`3607caa0 00007ff8`08d310f2     UIRibbon!WndBridge::RawWndProc+0x149
3e 00000040`3607cb10 00007ff8`0be38161     atlthunk!AtlThunk_0x00+0x22
3f 00000040`3607cb50 00007ff8`0be37e1c     USER32!UserCallWinProcCheckWow+0x2d1
40 00000040`3607ccb0 00007ff8`0be45d2a     USER32!DispatchClientMessage+0x9c
41 00000040`3607cd10 00007ff8`0c1d2db4     USER32!_fnINLPWINDOWPOS+0x3a
42 00000040`3607cd70 00007ff8`097f18d4     ntdll!KiUserCallbackDispatcherContinue
43 00000040`3607ce18 00007fff`5cf3c5eb     win32u!NtUserSetWindowPos+0x14
44 00000040`3607ce20 00007fff`5cf3c03e     UIRibbon!OfficeSpace::Root::ChromeMetricsUpdater::RecalcFrame+0x43
45 00000040`3607ce70 00007fff`5cfc7082     UIRibbon!OfficeSpace::Root::AppDataSourceListener::OnPropertyChanged+0x17e
46 00000040`3607ced0 00007fff`5cfc6e61     UIRibbon!FlexUI::PropertyChangeListenerManager::OnPropertyChanged+0x86
47 00000040`3607cf30 00007fff`5cfc7c8c     UIRibbon!FlexUI::DataSource::OnPropertyChanged+0x2d
48 00000040`3607cfb0 00007fff`5cf3cf83     UIRibbon!FlexUI::DataSource::SetValue+0x30c
49 00000040`3607d030 00007fff`5cf33cb1     UIRibbon!OfficeSpace::Root::RegisterAppFrame+0x1e3
4a 00000040`3607d090 00007fff`5cf33c08     UIRibbon!OfficeSpace::Root::EnsureContent+0x99
4b 00000040`3607d110 00007fff`5cf3a818     UIRibbon!OfficeSpace::Root::EnsureContent+0x70
4c 00000040`3607d150 00007fff`5cf25ef4     UIRibbon!OfficeSpaceWorkPaneCallback+0x98
4d 00000040`3607d1c0 00007fff`5cf25f73     UIRibbon!TBCWP::FGetMinMax+0xfc
4e 00000040`3607d260 00007fff`5cf26f81     UIRibbon!TBPUPANE::FGetMinMax+0x33
4f 00000040`3607d2a0 00007fff`5cf6a778     UIRibbon!TBCWP::QuerySize+0x81
50 00000040`3607d310 00007fff`5cf55274     UIRibbon!TB::CalcRectOrReflowToolbar+0x5d0
51 00000040`3607d4c0 00007fff`5cf54409     UIRibbon!TBS::DxyReflowDockRow+0x94
52 00000040`3607d540 00007fff`5cf58825     UIRibbon!TBS::DockReflowAsNeeded+0xf5
53 00000040`3607d5d0 00007fff`5cf59019     UIRibbon!TBS::RedisplayDocks+0x3e9
54 00000040`3607d670 00007fff`5ceb9c4f     UIRibbon!TBS::SetVisible+0x349
55 00000040`3607d710 00007fff`5ceaee5c     UIRibbon!COfficeSpaceUser::InitializeUI+0x4f
56 00000040`3607d740 00007ff7`f3834e84     UIRibbon!CUIFramework::LoadUI+0xcc
57 00000040`3607d780 00007ff7`f3835361     wordpad!CMainFrame::InitRibbonUI+0x180
58 00000040`3607d7b0 00007fff`7c7e8f27     wordpad!CMainFrame::OnCreate+0x181
59 00000040`3607d800 00007ff7`f3836f5c     MFC42u!CWnd::OnWndMsg+0x387
5a 00000040`3607d930 00007fff`7c7f0bff     wordpad!CMainFrame::OnWndMsg+0x14c
5b 00000040`3607d9c0 00007fff`7c7eb42a     MFC42u!CWnd::WindowProc+0x4f
5c 00000040`3607da00 00007fff`7c7ea76b     MFC42u!AfxCallWndProc+0x14a
5d 00000040`3607db00 00007ff8`0be38161     MFC42u!AfxWndProcBase+0x15b
5e 00000040`3607db70 00007ff8`0be37e1c     USER32!UserCallWinProcCheckWow+0x2d1
5f 00000040`3607dcd0 00007ff8`0be44edc     USER32!DispatchClientMessage+0x9c
60 00000040`3607dd30 00007ff8`0c1d2db4     USER32!__fnINLPCREATESTRUCT+0xac
61 00000040`3607dd90 00007ff8`097f2294     ntdll!KiUserCallbackDispatcherContinue
62 00000040`3607ded8 00007ff8`0be2f6b0     win32u!NtUserCreateWindowEx+0x14
63 00000040`3607dee0 00007ff8`0be2f3cc     USER32!VerNtUserCreateWindowEx+0x210
64 00000040`3607e270 00007ff8`0be2f212     USER32!CreateWindowInternal+0x1a8
65 00000040`3607e3d0 00007fff`7c7ec835     USER32!CreateWindowExW+0x82
66 00000040`3607e460 00007fff`7c7ef6c5     MFC42u!CWnd::CreateEx+0xf5
67 00000040`3607e520 00007fff`7c7ef7b3     MFC42u!CFrameWnd::Create+0xe5
68 00000040`3607e590 00007fff`7c7ef0ea     MFC42u!CFrameWnd::LoadFrame+0xb3
69 00000040`3607e610 00007fff`7c7fa1ec     MFC42u!CDocTemplate::CreateNewFrame+0x7a
6a 00000040`3607e680 00007ff7`f383e331     MFC42u!CSingleDocTemplate::OpenDocumentFile+0x8c
6b 00000040`3607e6e0 00007fff`7c7efa57     wordpad!CWordPadApp::InitInstance+0x521
6c 00000040`3607fc80 00007ff7`f37fb49e     MFC42u!AfxWinMain+0x97
6d 00000040`3607fcc0 00007ff8`0b03244d     wordpad!__wmainCRTStartup+0x1de
6e 00000040`3607fd80 00007ff8`0c18df78     KERNEL32!BaseThreadInitThunk+0x1d
6f 00000040`3607fdb0 00000000`00000000     ntdll!RtlUserThreadStart+0x28
 # Child-SP          RetAddr               Call Site
00 00000040`3607ac78 00007fff`5cf1317c     USER32!CreateWindowExW
01 00000040`3607ac80 00007fff`5cfb02a0     UIRibbon!IsolationAwareCreateWindowExW+0xe0
02 00000040`3607ad00 00007fff`5cfb0881     UIRibbon!NetUI::HWNDHost::InitializeHWND+0x160
03 00000040`3607ad70 00007fff`5cf91f4a     UIRibbon!NetUI::HWNDHost::OnHosted+0x21
04 00000040`3607adc0 00007fff`5cf91f4a     UIRibbon!NetUI::Element::OnHosted+0xda
05 00000040`3607adf0 00007fff`5cf91f4a     UIRibbon!NetUI::Element::OnHosted+0xda
06 00000040`3607ae20 00007fff`5cf91f4a     UIRibbon!NetUI::Element::OnHosted+0xda
07 00000040`3607ae50 00007fff`5cf9b4cb     UIRibbon!NetUI::Element::OnHosted+0xda
08 00000040`3607ae80 00007fff`5cf9c40c     UIRibbon!NetUI::Element::OnElementOrderChanged+0x16f
09 00000040`3607aee0 00007fff`5cf9d7eb     UIRibbon!NetUI::Element::OnPropertyChanged+0xadc
0a 00000040`3607af80 00007fff`5cf88fb1     UIRibbon!NetUI::Node::_PostSourceChange+0x43b
0b 00000040`3607b000 00007fff`5cf8711a     UIRibbon!NetUI::Node::InsertNodes+0x231
0c 00000040`3607b0a0 00007fff`5cf8a3ed     UIRibbon!NetUI::Node::CreateStyleTreeFor+0x222
0d 00000040`3607b130 00007fff`5cf8a189     UIRibbon!NetUI::Node::TreeStyleChildrenInto+0x195
0e 00000040`3607b190 00007fff`5cf871ac     UIRibbon!NetUI::Node::TreeStyleChildrenBindingsInSubtree+0x5d
```

```
0f 00000040`3607b1e0 00007fff`5cf8a3ed     UIRibbon!NetUI::Node::CreateStyleTreeFor+0x2b4
10 00000040`3607b270 00007fff`5cf8a215     UIRibbon!NetUI::Node::TreeStyleChildrenInto+0x195
11 00000040`3607b2d0 00007fff`5cf871ac     UIRibbon!NetUI::Node::TreeStyleChildrenBindingsInSubtree+0xe9
12 00000040`3607b320 00007fff`5cf8a3ed     UIRibbon!NetUI::Node::CreateStyleTreeFor+0x2b4
13 00000040`3607b3b0 00007fff`5cf8a189     UIRibbon!NetUI::Node::TreeStyleChildrenInto+0x195
14 00000040`3607b410 00007fff`5cf871ac     UIRibbon!NetUI::Node::TreeStyleChildrenBindingsInSubtree+0x5d
15 00000040`3607b460 00007fff`5cf88b6e     UIRibbon!NetUI::Node::CreateStyleTreeFor+0x2b4
16 00000040`3607b4f0 00007fff`5cf86311     UIRibbon!NetUI::Node::IncrementalTreeStyle+0x1ba
17 00000040`3607b580 00007fff`5cf89404     UIRibbon!NetUI::Node::ApplyStyleTree+0x191
18 00000040`3607b5c0 00007fff`5cf9a30d     UIRibbon!NetUI::Node::OnGroupChanged+0x14
19 00000040`3607b5f0 00007fff`5cf9c5d9     UIRibbon!NetUI::Node::EndDefer+0x1e5
1a 00000040`3607b620 00007fff`5cf29fba     UIRibbon!NetUI::DeferCycle::PushDefer+0x6d
1b 00000040`3607b650 00007fff`5cf3a835     UIRibbon!DefaultNetUIWorkPaneCallback+0x19e
1c 00000040`3607b700 00007fff`5cf25ef4     UIRibbon!OfficeSpaceWorkPaneCallback+0xb5
1d 00000040`3607b770 00007fff`5cf25f73     UIRibbon!TBCWP::FGetMinMax+0xfc
1e 00000040`3607b810 00007fff`5cf26f81     UIRibbon!TBPUPANE::FGetMinMax+0x33
1f 00000040`3607b850 00007fff`5cf6a778     UIRibbon!TBCWP::QuerySize+0x81
20 00000040`3607b8c0 00007fff`5cf55274     UIRibbon!TB::CalcRectOrReflowToolbar+0x5d0
21 00000040`3607ba70 00007fff`5cf54409     UIRibbon!TBS::DxyReflowDockRow+0x94
22 00000040`3607baf0 00007fff`5cf58825     UIRibbon!TBS::DockReflowAsNeeded+0xf5
23 00000040`3607bb80 00007fff`5cf57cb7     UIRibbon!TBS::RedisplayDocks+0x3e9
24 00000040`3607bc20 00007fff`5ceb9d5b     UIRibbon!TBS::HrResizeBorderEx+0xf7
25 00000040`3607bc50 00007fff`5ceb8eb3     UIRibbon!COfficeSpaceUser::InvalidateLayout+0x8b
26 00000040`3607bca0 00007fff`5cebc534     UIRibbon!COfficeSpaceUser::FrameListenerWndProc+0x19f
27 00000040`3607bd00 00007fff`5d0154a7     UIRibbon!COfficeSpaceUser::s_FrameListenerWndProc+0xa4
28 00000040`3607bd50 00007ff8`08d310f2     UIRibbon!WndBridge::RawWndProc+0x107
29 00000040`3607bdc0 00007ff8`0be38161     atlthunk!AtlThunk_0x00+0x22
2a 00000040`3607be00 00007ff8`0be37e1c     USER32!UserCallWinProcCheckWow+0x2d1
2b 00000040`3607bf60 00007ff8`0be42cdd     USER32!DispatchClientMessage+0x9c
2c 00000040`3607bfc0 00007ff8`0c1d2db4     USER32!_fnDWORD+0x3d
2d 00000040`3607c020 00007ff8`097f1554     ntdll!KiUserCallbackDispatcherContinue
2e 00000040`3607c0a8 00007ff8`0be37798     win32u!NtUserMessageCall+0x14
2f 00000040`3607c0b0 00007ff8`0be3665b     USER32!SendMessageWorker+0x2e8
30 00000040`3607c160 00007ff8`0be36082     USER32!RealDefWindowProcWorker+0x23b
31 00000040`3607c240 00007ff8`05b56e26     USER32!RealDefWindowProcW+0x52
32 00000040`3607c280 00007ff8`05b55821     uxtheme!_ThemeDefWindowProc+0x15f6
33 00000040`3607c4a0 00007ff8`0be36205     uxtheme!ThemeDefWindowProcW+0x11
34 00000040`3607c4e0 00007ff8`0be38161     USER32!DefWindowProcW+0x135
35 00000040`3607c550 00007ff8`0be379ab     USER32!UserCallWinProcCheckWow+0x2d1
36 00000040`3607c6b0 00007fff`7c7f0c8e     USER32!CallWindowProcW+0x8b
37 00000040`3607c700 00007fff`7c7f0c28     MFC42u!CWnd::DefWindowProcW+0x3e
38 00000040`3607c740 00007fff`7c7eb42a     MFC42u!CWnd::WindowProc+0x78
39 00000040`3607c780 00007fff`7c7ea76b     MFC42u!AfxCallWndProc+0x14a
3a 00000040`3607c880 00007ff8`0be38161     MFC42u!AfxWndProcBase+0x15b
3b 00000040`3607c8f0 00007ff8`0be379ab     USER32!UserCallWinProcCheckWow+0x2d1
3c 00000040`3607ca50 00007fff`5d0154e9     USER32!CallWindowProcW+0x8b
3d 00000040`3607caa0 00007ff8`08d310f2     UIRibbon!WndBridge::RawWndProc+0x149
3e 00000040`3607cb10 00007ff8`0be38161     atlthunk!AtlThunk_0x00+0x22
3f 00000040`3607cb50 00007ff8`0be37e1c     USER32!UserCallWinProcCheckWow+0x2d1
40 00000040`3607ccb0 00007ff8`0be45d2a     USER32!DispatchClientMessage+0x9c
41 00000040`3607cd10 00007ff8`0c1d2db4     USER32!_fnINLPWINDOWPOS+0x3a
42 00000040`3607cd70 00007ff8`097f18d4     ntdll!KiUserCallbackDispatcherContinue
43 00000040`3607ce18 00007fff`5cf3c5eb     win32u!NtUserSetWindowPos+0x14
44 00000040`3607ce20 00007fff`5cf3c03e     UIRibbon!OfficeSpace::Root::ChromeMetricsUpdater::RecalcFrame+0x43
45 00000040`3607ce70 00007fff`5cfc7082     UIRibbon!OfficeSpace::Root::AppDataSourceListener::OnPropertyChanged+0x17e
46 00000040`3607ced0 00007fff`5cfc6e61     UIRibbon!FlexUI::PropertyChangeListenerManager::OnPropertyChanged+0x86
47 00000040`3607cf30 00007fff`5cfc7c8c     UIRibbon!FlexUI::DataSource::OnPropertyChanged+0x2d
48 00000040`3607cfb0 00007fff`5cf3cf83     UIRibbon!FlexUI::DataSource::SetValue+0x30c
49 00000040`3607d030 00007fff`5cf33cb1     UIRibbon!OfficeSpace::Root::RegisterAppFrame+0x1e3
4a 00000040`3607d090 00007fff`5cf33c08     UIRibbon!OfficeSpace::Root::EnsureContent+0x99
4b 00000040`3607d110 00007fff`5cf3a818     UIRibbon!OfficeSpace::Root::EnsureContent+0x70
4c 00000040`3607d150 00007fff`5cf25ef4     UIRibbon!OfficeSpaceWorkPaneCallback+0x98
4d 00000040`3607d1c0 00007fff`5cf25f73     UIRibbon!TBCWP::FGetMinMax+0xfc
4e 00000040`3607d260 00007fff`5cf26f81     UIRibbon!TBPUPANE::FGetMinMax+0x33
4f 00000040`3607d2a0 00007fff`5cf6a778     UIRibbon!TBCWP::QuerySize+0x81
50 00000040`3607d310 00007fff`5cf55274     UIRibbon!TB::CalcRectOrReflowToolbar+0x5d0
51 00000040`3607d4c0 00007fff`5cf54409     UIRibbon!TBS::DxyReflowDockRow+0x94
52 00000040`3607d540 00007fff`5cf58825     UIRibbon!TBS::DockReflowAsNeeded+0xf5
53 00000040`3607d5d0 00007fff`5cf59019     UIRibbon!TBS::RedisplayDocks+0x3e9
54 00000040`3607d670 00007fff`5ceb9c4f     UIRibbon!TBS::SetVisible+0x349
55 00000040`3607d710 00007fff`5ceaee5c     UIRibbon!COfficeSpaceUser::InitializeUI+0x4f
56 00000040`3607d740 00007ff7`f3834e84     UIRibbon!CUIFramework::LoadUI+0xcc
57 00000040`3607d780 00007ff7`f3835361     wordpad!CMainFrame::InitRibbonUI+0x180
58 00000040`3607d7b0 00007fff`7c7e8f27     wordpad!CMainFrame::OnCreate+0x181
59 00000040`3607d800 00007ff7`f3836f5c     MFC42u!CWnd::OnWndMsg+0x387
5a 00000040`3607d930 00007fff`7c7f0bff     wordpad!CMainFrame::OnWndMsg+0x14c
5b 00000040`3607d9c0 00007fff`7c7eb42a     MFC42u!CWnd::WindowProc+0x4f
5c 00000040`3607da00 00007fff`7c7ea76b     MFC42u!AfxCallWndProc+0x14a
5d 00000040`3607db00 00007ff8`0be38161     MFC42u!AfxWndProcBase+0x15b
5e 00000040`3607db70 00007ff8`0be37e1c     USER32!UserCallWinProcCheckWow+0x2d1
5f 00000040`3607dcd0 00007ff8`0be44edc     USER32!DispatchClientMessage+0x9c
60 00000040`3607dd30 00007ff8`0c1d2db4     USER32!__fnINLPCREATESTRUCT+0xac
61 00000040`3607dd90 00007ff8`097f2294     ntdll!KiUserCallbackDispatcherContinue
62 00000040`3607ded8 00007ff8`0be2f6b0     win32u!NtUserCreateWindowEx+0x14
63 00000040`3607dee0 00007ff8`0be2f3cc     USER32!VerNtUserCreateWindowEx+0x210
64 00000040`3607e270 00007ff8`0be2f212     USER32!CreateWindowInternal+0x1a8
65 00000040`3607e3d0 00007fff`7c7ec835     USER32!CreateWindowExW+0x82
66 00000040`3607e460 00007fff`7c7ef6c5     MFC42u!CWnd::CreateEx+0xf5
67 00000040`3607e520 00007fff`7c7ef7b3     MFC42u!CFrameWnd::Create+0xe5
68 00000040`3607e590 00007fff`7c7ef0ea     MFC42u!CFrameWnd::LoadFrame+0xb3
69 00000040`3607e610 00007fff`7c7fa1ec     MFC42u!CDocTemplate::CreateNewFrame+0x7a
6a 00000040`3607e680 00007ff7`f383e331     MFC42u!CSingleDocTemplate::OpenDocumentFile+0x8c
6b 00000040`3607e6e0 00007fff`7c7efa57     wordpad!CWordPadApp::InitInstance+0x521
6c 00000040`3607fc80 00007ff7`f37fb49e     MFC42u!AfxWinMain+0x97
6d 00000040`3607fcc0 00007ff8`0b03244d     wordpad!__wmainCRTStartup+0x1de
6e 00000040`3607fd80 00007ff8`0c18df78     KERNEL32!BaseThreadInitThunk+0x1d
6f 00000040`3607fdb0 00000000`00000000     ntdll!RtlUserThreadStart+0x28
ModLoad: 00007ff8`0a3f0000 00007ff8`0a46d000   C:\WINDOWS\System32\coml2.dll
```

```
(1a28.623c): C++ EH exception - code e06d7363 (first chance)
(1a28.623c): C++ EH exception - code e06d7363 (first chance)
(1a28.623c): C++ EH exception - code e06d7363 (first chance)
ModLoad: 00007fff`e13e0000 00007fff`e13fd000   C:\WINDOWS\SYSTEM32\amsi.dll
 # Child-SP          RetAddr           Call Site
00 00000040`3627db58 00007fff`f3cce048 USER32!CreateWindowExW
01 00000040`3627db60 00007fff`f3ccdf1f urlmon!NotificationWindow::CreateHandle+0x68
02 00000040`3627dbd0 00007fff`f3c7cc9b urlmon!CTransaction::GetNotificationWnd+0x13f
03 00000040`3627dc10 00007fff`f3cb98ee urlmon!GetTransactionObjects+0x2fb
04 00000040`3627dc90 00007fff`f3cba81c urlmon!CBinding::StartBinding+0x53e
05 00000040`3627de10 00007fff`f3cb9276 urlmon!CUrlMon::StartBinding+0x1b0
06 00000040`3627dee0 00007fff`f1c05529 urlmon!CUrlMon::BindToStorage+0x96
07 00000040`3627df30 00007fff`f1c04acb msxml6!URLMONStream::deferedOpen+0x159
08 00000040`3627df90 00007fff`f1c048df msxml6!XMLParser::PushURL+0x1af
09 00000040`3627e010 00007fff`f1bbf1c6 msxml6!XMLParser::SetURL+0x7f
0a 00000040`3627e060 00007fff`f1bbed48 msxml6!Document::_load+0x162
0b 00000040`3627e0e0 00007fff`f1b87712 msxml6!Document::load+0x198
0c 00000040`3627e150 00007fff`7c46afae msxml6!DOMDocumentWrapper::load+0x182
0d 00000040`3627e220 00007fff`7c414813 PrintConfig!PrintConfig::PrivateDevmodeMapFile::Parse+0x14e
0e 00000040`3627e390 00007fff`7c5583f0 PrintConfig!PrintConfigPlugIn::DevMode+0x303
0f 00000040`3627e4b0 00007fff`7c55956a PrintConfig!BCallOEMDevMode+0x84
10 00000040`3627e530 00007fff`7c514999 PrintConfig!PGetDefaultDevmodeWithOemPlugins+0x16a
11 00000040`3627e5b0 00007fff`7c4fc131 PrintConfig!UniDrvUI::BGetValidatedUserDefaultDevmode+0x81
12 00000040`3627e620 00007fff`7c4f1c06 PrintConfig!UniDrvUI::CPrintTicketProvider::GetValidatedDevmode+0x6d
13 00000040`3627e670 00007fff`7c5165f9 PrintConfig!UniDrvUI::CPrintTicketProvider::ConvertPrintTicketToDevMode+0x76
14 00000040`3627e6f0 00007fff`7c514794 PrintConfig!UniDrvUI::PerformJScriptDevmodeValidation+0x525
15 00000040`3627e7f0 00007fff`7c512bb7 PrintConfig!UniDrvUI::BFillCommonInfoDevmode+0x31c
16 00000040`3627e890 00007fff`7c5120ec PrintConfig!UniDrvUI::LSimpleDocumentProperties+0x173
17 00000040`3627e900 00007fff`7c40f54b PrintConfig!UniDrvUI::DrvDocumentPropertySheets+0x60
18 00000040`3627edb0 00007fff`7c4249ad PrintConfig!PrintConfig::DrvDocumentPropertySheets+0x123
19 00000040`3627ee20 00007fff`7c425543 PrintConfig!ExceptionBoundary<<lambda_ce41999e58d4ed069b2790d632f6e3b5> >+0x3d
1a 00000040`3627ee70 00007fff`e032e9ba PrintConfig!DrvDocumentPropertySheets+0x53
1b 00000040`3627eee0 00007fff`e031593f WINSPOOL!DocumentPropertySheets+0x3fa
1c 00000040`3627ef30 00007fff`e0324aff WINSPOOL!DocumentPropertiesWNative+0x193
1d 00000040`3627efc0 00007ff8`0a340fc8 WINSPOOL!DocumentPropertiesW+0xcf
1e 00000040`3627f020 00007ff8`0a344d59 COMDLG32!PrintGetDevMode+0x8c
1f 00000040`3627f060 00007ff8`0a34058b COMDLG32!PrintReturnDefault+0x91
20 00000040`3627f090 00007ff8`0a340407 COMDLG32!PrintDlgX+0x163
21 00000040`3627f530 00007fff`7c7ff697 COMDLG32!PrintDlgW+0x47
22 00000040`3627fa40 00007fff`7c7ff591 MFC42u!CWinApp::UpdatePrinterSelection+0x57
23 00000040`3627fbe0 00007ff7`f383d742 MFC42u!CWinApp::GetPrinterDeviceDefaults+0x21
24 00000040`3627fc10 00007ff7`f383d938 wordpad!CWordPadApp::CreateDevNames+0xb2
25 00000040`3627fcf0 00007fff`7c7ffc1d wordpad!CWordPadApp::DoDeferredInitialization+0x18
26 00000040`3627fd20 00007ff8`0a57e634 MFC42u!_AfxThreadEntry+0xdd
27 00000040`3627fde0 00007ff8`0a57e70c msvcrt!_callthreadstartex+0x28
28 00000040`3627fe10 00007ff8`0b03244d msvcrt!_threadstartex+0x7c
29 00000040`3627fe40 00007ff8`0c18df78 KERNEL32!BaseThreadInitThunk+0x1d
2a 00000040`3627fe70 00000000`00000000 ntdll!RtlUserThreadStart+0x28
 # Child-SP          RetAddr           Call Site
00 00000040`3607bba8 00007ff8`0be2e567 USER32!CreateWindowExW
01 00000040`3607bbb0 00007ff8`0be3b5a0 USER32!CreateIMEUI+0xf7
02 00000040`3607be10 00007ff8`0be35e02 USER32!ImeSetContextHandler+0x2b4
03 00000040`3607bec0 00007ff8`0be35c2f USER32!ImeWndProcWorker+0x1a2
04 00000040`3607bf10 00007ff8`0be38161 USER32!ImeWndProcW+0x4f
05 00000040`3607bf50 00007ff8`0be379ab USER32!UserCallWinProcCheckWow+0x2d1
06 00000040`3607c0b0 00007fff`7c7e76bc USER32!CallWindowProcW+0x8b
07 00000040`3607c100 00007ff8`0be38161 MFC42u!_AfxActivationWndProc+0x12c
08 00000040`3607c1e0 00007ff8`0be37e1c USER32!UserCallWinProcCheckWow+0x2d1
09 00000040`3607c340 00007ff8`0be42cdd USER32!DispatchClientMessage+0x9c
0a 00000040`3607c3a0 00007ff8`0c1d2db4 USER32!_fnDWORD+0x3d
0b 00000040`3607c400 00007ff8`097f1554 ntdll!KiUserCallbackDispatcherContinue
0c 00000040`3607c488 00007ff8`0be37798 win32u!NtUserMessageCall+0x14
0d 00000040`3607c490 00007ff8`0be3679f USER32!SendMessageWorker+0x2e8
0e 00000040`3607c540 00007ff8`0be361d9 USER32!RealDefWindowProcWorker+0x37f
0f 00000040`3607c620 00007ff8`0be38161 USER32!DefWindowProcW+0x109
10 00000040`3607c690 00007ff8`0be379ab USER32!UserCallWinProcCheckWow+0x2d1
11 00000040`3607c7f0 00007fff`7c7f0c8e USER32!CallWindowProcW+0x8b
12 00000040`3607c840 00007fff`7c7f0c28 MFC42u!CWnd::DefWindowProcW+0x3e
13 00000040`3607c880 00007fff`7c7eb42a MFC42u!CWnd::WindowProc+0x78
14 00000040`3607c8c0 00007fff`7c7ea76b MFC42u!AfxCallWndProc+0x14a
15 00000040`3607c9c0 00007ff8`0be38161 MFC42u!AfxWndProcBase+0x15b
16 00000040`3607ca30 00007ff8`0be379ab USER32!UserCallWinProcCheckWow+0x2d1
17 00000040`3607cb90 00007fff`5d0154e9 USER32!CallWindowProcW+0x8b
18 00000040`3607cbe0 00007ff8`08d310f2 UIRibbon!WndBridge::RawWndProc+0x149
19 00000040`3607cc50 00007ff8`0be38161 atlthunk!AtlThunk_0x00+0x22
1a 00000040`3607cc90 00007ff8`0be37e1c USER32!UserCallWinProcCheckWow+0x2d1
1b 00000040`3607cdf0 00007ff8`0be42cdd USER32!DispatchClientMessage+0x9c
1c 00000040`3607ce50 00007ff8`0c1d2db4 USER32!_fnDWORD+0x3d
1d 00000040`3607ceb0 00007ff8`097f1554 ntdll!KiUserCallbackDispatcherContinue
1e 00000040`3607cf38 00007ff8`0be37798 win32u!NtUserMessageCall+0x14
1f 00000040`3607cf40 00007ff8`0be37447 USER32!SendMessageWorker+0x2e8
20 00000040`3607cff0 00007ff8`0a6d248e USER32!SendMessageW+0x137
21 00000040`3607d050 00007ff8`0be310ae IMM32!ImmSetActiveContext+0x72e
22 00000040`3607d0c0 00007ff8`0be322c3 USER32!FocusSetIMCContext+0x5e
23 00000040`3607d0f0 00007ff8`0be35dca USER32!ImeSystemHandler+0x77
24 00000040`3607d3b0 00007ff8`0be35c2f USER32!ImeWndProcWorker+0x16a
25 00000040`3607d400 00007ff8`0be38161 USER32!ImeWndProcW+0x4f
26 00000040`3607d440 00007ff8`0be379ab USER32!UserCallWinProcCheckWow+0x2d1
27 00000040`3607d5a0 00007fff`7c7e76bc USER32!CallWindowProcW+0x8b
28 00000040`3607d5f0 00007ff8`0be38161 MFC42u!_AfxActivationWndProc+0x12c
29 00000040`3607d6d0 00007ff8`0be37e1c USER32!UserCallWinProcCheckWow+0x2d1
2a 00000040`3607d830 00007ff8`0be42cdd USER32!DispatchClientMessage+0x9c
2b 00000040`3607d890 00007ff8`0c1d2db4 USER32!_fnDWORD+0x3d
2c 00000040`3607d8f0 00007ff8`097f1554 ntdll!KiUserCallbackDispatcherContinue
2d 00000040`3607d978 00007ff8`0be36575 win32u!NtUserMessageCall+0x14
2e 00000040`3607d980 00007ff8`0be361d9 USER32!RealDefWindowProcWorker+0x155
2f 00000040`3607da60 00007ff8`0be38161 USER32!DefWindowProcW+0x109
```

```
30 00000040`3607dad0 00007ff8`0be379ab     USER32!UserCallWinProcCheckWow+0x2d1
31 00000040`3607dc30 00007fff`7c7f0c8e     USER32!CallWindowProcW+0x8b
32 00000040`3607dc80 00007fff`7c7e7a09     MFC42u!CWnd::DefWindowProcW+0x3e
33 00000040`3607dcc0 00007fff`7c7e4cc9     MFC42u!CWnd::Default+0x39
34 00000040`3607dd00 00007fff`7c7e96fd     MFC42u!CFrameWnd::OnActivate+0x29
35 00000040`3607dd40 00007ff7`f3836f5c     MFC42u!CWnd::OnWndMsg+0xb5d
36 00000040`3607de70 00007fff`7c7f0bff     wordpad!CMainFrame::OnWndMsg+0x14c
37 00000040`3607df00 00007fff`7c7eb42a     MFC42u!CWnd::WindowProc+0x4f
38 00000040`3607df40 00007fff`7c7ea76b     MFC42u!AfxCallWndProc+0x14a
39 00000040`3607e040 00007ff8`0be38161     MFC42u!AfxWndProcBase+0x15b
3a 00000040`3607e0b0 00007ff8`0be379ab     USER32!UserCallWinProcCheckWow+0x2d1
3b 00000040`3607e210 00007fff`5d0154e9     USER32!CallWindowProcW+0x8b
3c 00000040`3607e260 00007ff8`08d310f2     UIRibbon!WndBridge::RawWndProc+0x149
3d 00000040`3607e2d0 00007ff8`0be38161     atlthunk!AtlThunk_0x00+0x22
3e 00000040`3607e310 00007ff8`0be37e1c     USER32!UserCallWinProcCheckWow+0x2d1
3f 00000040`3607e470 00007ff8`0be42cdd     USER32!DispatchClientMessage+0x9c
40 00000040`3607e4d0 00007ff8`0c1d2db4     USER32!_fnDWORD+0x3d
41 00000040`3607e530 00007ff8`097f2e94     ntdll!KiUserCallbackDispatcherContinue
42 00000040`3607e5b8 00007ff7`f38346a5     win32u!NtUserSetWindowPlacement+0x14
43 00000040`3607e5c0 00007fff`7c7e25de     wordpad!CMainFrame::ActivateFrame+0x75
44 00000040`3607e630 00007fff`7c7fa2e7     MFC42u!CFrameWnd::InitialUpdateFrame+0xfe
45 00000040`3607e680 00007ff7`f383e331     MFC42u!CSingleDocTemplate::OpenDocumentFile+0x187
46 00000040`3607e6e0 00007fff`7c7efa57     wordpad!CWordPadApp::InitInstance+0x521
47 00000040`3607fc80 00007ff7`f37fb49e     MFC42u!AfxWinMain+0x97
48 00000040`3607fcc0 00007ff8`0b03244d     wordpad!__wmainCRTStartup+0x1de
49 00000040`3607fd80 00007ff8`0c18df78     KERNEL32!BaseThreadInitThunk+0x1d
4a 00000040`3607fdb0 00000000`00000000     ntdll!RtlUserThreadStart+0x28
ModLoad: 00007fff`f5460000 00007fff`f55d4000   C:\Windows\System32\Windows.UI.dll
 # Child-SP          RetAddr               Call Site
00 00000040`3607cad8 00007fff`5cf1317c     USER32!CreateWindowExW
01 00000040`3607cae0 00007fff`5cf243be     UIRibbon!IsolationAwareCreateWindowExW+0xe0
02 00000040`3607cb60 00007fff`5cf2569e     UIRibbon!CreateWindowRegExW+0x9a
03 00000040`3607cbd0 00007fff`5cf2eced     UIRibbon!TBPUPANE::FCreateWindow+0x7e
04 00000040`3607cc40 00007fff`5cf255be     UIRibbon!TBCP::FCreateWindow+0x3d
05 00000040`3607cc80 00007fff`5cf2ee79     UIRibbon!TBCWP::FCreateWindow+0xe
06 00000040`3607ccb0 00007fff`5cf25d34     UIRibbon!TBCP::FDraw+0xe9
07 00000040`3607cd10 00007fff`5cf6bde4     UIRibbon!TBCWP::FDraw+0x24
08 00000040`3607cd40 00007fff`5cf71082     UIRibbon!TB::FDraw+0x6f0
09 00000040`3607cf00 00007ff8`0be38161     UIRibbon!TBWndProc+0x192
0a 00000040`3607cfe0 00007ff8`0be37e1c     USER32!UserCallWinProcCheckWow+0x2d1
0b 00000040`3607d140 00007ff8`0be42cdd     USER32!DispatchClientMessage+0x9c
0c 00000040`3607d1a0 00007ff8`0c1d2db4     USER32!_fnDWORD+0x3d
0d 00000040`3607d200 00007ff8`097fc7d4     ntdll!KiUserCallbackDispatcherContinue
0e 00000040`3607d288 00007fff`5cf544f8     win32u!NtUserUpdateWindow+0x14
0f 00000040`3607d290 00007fff`5cf58825     UIRibbon!TBS::DockReflowAsNeeded+0x1e4
10 00000040`3607d320 00007fff`5cf57cb7     UIRibbon!TBS::RedisplayDocks+0x3e9
11 00000040`3607d3c0 00007fff`5ceb9d5b     UIRibbon!TBS::HrResizeBorderEx+0xf7
12 00000040`3607d3f0 00007fff`5ceb8eb3     UIRibbon!COfficeSpaceUser::InvalidateLayout+0x8b
13 00000040`3607d440 00007fff`5cebc534     UIRibbon!COfficeSpaceUser::FrameListenerWndProc+0x19f
14 00000040`3607d4a0 00007fff`5d0154a7     UIRibbon!COfficeSpaceUser::s_FrameListenerWndProc+0xa4
15 00000040`3607d4f0 00007ff8`08d310f2     UIRibbon!WndBridge::RawWndProc+0x107
16 00000040`3607d560 00007ff8`0be38161     atlthunk!AtlThunk_0x00+0x22
17 00000040`3607d5a0 00007ff8`0be37e1c     USER32!UserCallWinProcCheckWow+0x2d1
18 00000040`3607d700 00007ff8`0be42cdd     USER32!DispatchClientMessage+0x9c
19 00000040`3607d760 00007ff8`0c1d2db4     USER32!_fnDWORD+0x3d
1a 00000040`3607d7c0 00007ff8`097f1554     ntdll!KiUserCallbackDispatcherContinue
1b 00000040`3607d848 00007ff8`0be37798     win32u!NtUserMessageCall+0x14
1c 00000040`3607d850 00007ff8`0be3665b     USER32!SendMessageWorker+0x2e8
1d 00000040`3607d900 00007ff8`0be36082     USER32!RealDefWindowProcWorker+0x23b
1e 00000040`3607d9e0 00007ff8`05b56e26     USER32!RealDefWindowProcW+0x52
1f 00000040`3607da20 00007ff8`05b55821     uxtheme!_ThemeDefWindowProc+0x15f6
20 00000040`3607dc40 00007ff8`0be36205     uxtheme!ThemeDefWindowProcW+0x11
21 00000040`3607dc80 00007ff8`0be38161     USER32!DefWindowProcW+0x135
22 00000040`3607dcf0 00007ff8`0be379ab     USER32!UserCallWinProcCheckWow+0x2d1
23 00000040`3607de50 00007fff`7c7f0c8e     USER32!CallWindowProcW+0x8b
24 00000040`3607dea0 00007fff`7c7f0c28     MFC42u!CWnd::DefWindowProcW+0x3e
25 00000040`3607dee0 00007fff`7c7eb42a     MFC42u!CWnd::WindowProc+0x78
26 00000040`3607df20 00007fff`7c7ea76b     MFC42u!AfxCallWndProc+0x14a
27 00000040`3607e020 00007ff8`0be38161     MFC42u!AfxWndProcBase+0x15b
28 00000040`3607e090 00007ff8`0be379ab     USER32!UserCallWinProcCheckWow+0x2d1
29 00000040`3607e1f0 00007fff`5d0154e9     USER32!CallWindowProcW+0x8b
2a 00000040`3607e240 00007ff8`08d310f2     UIRibbon!WndBridge::RawWndProc+0x149
2b 00000040`3607e2b0 00007ff8`0be38161     atlthunk!AtlThunk_0x00+0x22
2c 00000040`3607e2f0 00007ff8`0be37e1c     USER32!UserCallWinProcCheckWow+0x2d1
2d 00000040`3607e450 00007ff8`0be45d2a     USER32!DispatchClientMessage+0x9c
2e 00000040`3607e4b0 00007ff8`0c1d2db4     USER32!_fnINLPWINDOWPOS+0x3a
2f 00000040`3607e510 00007ff8`097f2e94     ntdll!KiUserCallbackDispatcherContinue
30 00000040`3607e5b8 00007ff7`f38346a5     win32u!NtUserSetWindowPlacement+0x14
31 00000040`3607e5c0 00007fff`7c7e25de     wordpad!CMainFrame::ActivateFrame+0x75
32 00000040`3607e630 00007fff`7c7fa2e7     MFC42u!CFrameWnd::InitialUpdateFrame+0xfe
33 00000040`3607e680 00007ff7`f383e331     MFC42u!CSingleDocTemplate::OpenDocumentFile+0x187
34 00000040`3607e6e0 00007fff`7c7efa57     wordpad!CWordPadApp::InitInstance+0x521
35 00000040`3607fc80 00007ff7`f37fb49e     MFC42u!AfxWinMain+0x97
36 00000040`3607fcc0 00007ff8`0b03244d     wordpad!__wmainCRTStartup+0x1de
37 00000040`3607fd80 00007ff8`0c18df78     KERNEL32!BaseThreadInitThunk+0x1d
38 00000040`3607fdb0 00000000`00000000     ntdll!RtlUserThreadStart+0x28
 # Child-SP          RetAddr               Call Site
00 00000040`3627d888 00007fff`f3cce048     USER32!CreateWindowExW
01 00000040`3627d890 00007fff`f3ccdf1f     urlmon!NotificationWindow::CreateHandle+0x68
02 00000040`3627d900 00007fff`f3c7cc9b     urlmon!CTransaction::GetNotificationWnd+0x13f
03 00000040`3627d940 00007fff`f3cb98ee     urlmon!GetTransactionObjects+0x2fb
04 00000040`3627d9c0 00007fff`f3cba81c     urlmon!CBinding::StartBinding+0x53e
05 00000040`3627db40 00007fff`f3cb9276     urlmon!CUrlMon::StartBinding+0x1b0
06 00000040`3627dc10 00007fff`f1c05529     urlmon!CUrlMon::BindToStorage+0x96
07 00000040`3627dc60 00007fff`f1c04acb     msxml6!URLMONStream::deferedOpen+0x159
08 00000040`3627dcc0 00007fff`f1c048df     msxml6!XMLParser::PushURL+0x1af
09 00000040`3627dd40 00007fff`f1bbf1c6     msxml6!XMLParser::SetURL+0x7f
0a 00000040`3627dd90 00007fff`f1bbed48     msxml6!Document::_load+0x162
```

```
0b 00000040`3627de10 00007fff`f1b87712     msxml6!Document::load+0x198
0c 00000040`3627de80 00007fff`7c46afae     msxml6!DOMDocumentWrapper::load+0x182
0d 00000040`3627df50 00007fff`7c414968     PrintConfig!PrintConfig::PrivateDevmodeMapFile::Parse+0x14e
0e 00000040`3627e0c0 00007fff`7c558d7b     PrintConfig!PrintConfigPlugIn::DevMode+0x458
0f 00000040`3627e1e0 00007fff`7c55949b     PrintConfig!BCalcTotalOEMDMSize+0xa3
10 00000040`3627e280 00007fff`7c514999     PrintConfig!PGetDefaultDevmodeWithOemPlugins+0x9b
11 00000040`3627e300 00007fff`7c5144fa     PrintConfig!UniDrvUI::BGetValidatedUserDefaultDevmode+0x81
12 00000040`3627e370 00007fff`7c50d112     PrintConfig!UniDrvUI::BFillCommonInfoDevmode+0x82
13 00000040`3627e410 00007fff`7c50ce44     PrintConfig!UniDrvUI::DwDeviceCapabilities+0xc6
14 00000040`3627e4c0 00007fff`7c42514f     PrintConfig!UniDrvUI::DrvDeviceCapabilities+0x18
15 00000040`3627e500 00007fff`e03150a1     PrintConfig!DrvDeviceCapabilities+0x11f
16 00000040`3627e580 00007fff`e0324a04     WINSPOOL!DeviceCapabilitiesWNative+0x105
17 00000040`3627e5e0 00007fff`7c4feafb     WINSPOOL!DeviceCapabilitiesW+0x54
18 00000040`3627e620 00007fff`7c516637     PrintConfig!publicdm::JobTicketToPublicDevmode+0x9f
19 00000040`3627e6f0 00007fff`7c514794     PrintConfig!UniDrvUI::PerformJScriptDevmodeValidation+0x563
1a 00000040`3627e7f0 00007fff`7c512bb7     PrintConfig!UniDrvUI::BFillCommonInfoDevmode+0x31c
1b 00000040`3627e890 00007fff`7c5120ec     PrintConfig!UniDrvUI::LSimpleDocumentProperties+0x173
1c 00000040`3627e900 00007fff`7c40f54b     PrintConfig!UniDrvUI::DrvDocumentPropertySheets+0x60
1d 00000040`3627edb0 00007fff`7c4249ad     PrintConfig!PrintConfig::DrvDocumentPropertySheets+0x123
1e 00000040`3627ee20 00007fff`7c425543     PrintConfig!ExceptionBoundary<<lambda_ce41999e58d4ed069b2790d632f6e3b5> >+0x3d
1f 00000040`3627ee70 00007fff`e032e9ba     PrintConfig!DrvDocumentPropertySheets+0x53
20 00000040`3627eee0 00007fff`e031593f     WINSPOOL!DocumentPropertySheets+0x3fa
21 00000040`3627ef30 00007fff`e0324aff     WINSPOOL!DocumentPropertiesWNative+0x193
22 00000040`3627efc0 00007ff8`0a340fc8     WINSPOOL!DocumentPropertiesW+0xcf
23 00000040`3627f020 00007ff8`0a344d59     COMDLG32!PrintGetDevMode+0x8c
24 00000040`3627f060 00007ff8`0a34058b     COMDLG32!PrintReturnDefault+0x91
25 00000040`3627f090 00007ff8`0a340407     COMDLG32!PrintDlgX+0x163
26 00000040`3627f530 00007fff`7c7ff697     COMDLG32!PrintDlgW+0x47
27 00000040`3627fa40 00007fff`7c7ff591     MFC42u!CWinApp::UpdatePrinterSelection+0x57
28 00000040`3627fbe0 00007ff7`f383d742     MFC42u!CWinApp::GetPrinterDeviceDefaults+0x21
29 00000040`3627fc10 00007ff7`f383d938     wordpad!CWordPadApp::CreateDevNames+0xb2
2a 00000040`3627fcf0 00007fff`7c7ffc1d     wordpad!CWordPadApp::DoDeferredInitialization+0x18
2b 00000040`3627fd20 00007ff8`0a57e634     MFC42u!_AfxThreadEntry+0xdd
2c 00000040`3627fde0 00007ff8`0a57e70c     msvcrt!_callthreadstartex+0x28
2d 00000040`3627fe10 00007ff8`0b03244d     msvcrt!_threadstartex+0x7c
2e 00000040`3627fe40 00007ff8`0c18df78     KERNEL32!BaseThreadInitThunk+0x1d
2f 00000040`3627fe70 00000000`00000000     ntdll!RtlUserThreadStart+0x28
 # Child-SP          RetAddr               Call Site
00 00000040`3627d8a8 00007fff`f3cce048     USER32!CreateWindowExW
01 00000040`3627d8b0 00007fff`f3ccdf1f     urlmon!NotificationWindow::CreateHandle+0x68
02 00000040`3627d920 00007fff`f3c7cc9b     urlmon!CTransaction::GetNotificationWnd+0x13f
03 00000040`3627d960 00007fff`f3cb98ee     urlmon!GetTransactionObjects+0x2fb
04 00000040`3627d9e0 00007fff`f3cba81c     urlmon!CBinding::StartBinding+0x53e
05 00000040`3627db60 00007fff`f3cb9276     urlmon!CUrlMon::StartBinding+0x1b0
06 00000040`3627dc30 00007fff`f1c05529     urlmon!CUrlMon::BindToStorage+0x96
07 00000040`3627dc80 00007fff`f1c04acb     msxml6!URLMONStream::deferedOpen+0x159
08 00000040`3627dce0 00007fff`f1c048df     msxml6!XMLParser::PushURL+0x1af
09 00000040`3627dd60 00007fff`f1bbf1c6     msxml6!XMLParser::SetURL+0x7f
0a 00000040`3627ddb0 00007fff`f1bbed48     msxml6!Document::_load+0x162
0b 00000040`3627de30 00007fff`f1b87712     msxml6!Document::load+0x198
0c 00000040`3627dea0 00007fff`7c46afae     msxml6!DOMDocumentWrapper::load+0x182
0d 00000040`3627df70 00007fff`7c414813     PrintConfig!PrintConfig::PrivateDevmodeMapFile::Parse+0x14e
0e 00000040`3627e0e0 00007fff`7c5583f0     PrintConfig!PrintConfigPlugIn::DevMode+0x303
0f 00000040`3627e200 00007fff`7c55956a     PrintConfig!BCallOEMDevMode+0x84
10 00000040`3627e280 00007fff`7c514999     PrintConfig!PGetDefaultDevmodeWithOemPlugins+0x16a
11 00000040`3627e300 00007fff`7c5144fa     PrintConfig!UniDrvUI::BGetValidatedUserDefaultDevmode+0x81
12 00000040`3627e370 00007fff`7c50d112     PrintConfig!UniDrvUI::BFillCommonInfoDevmode+0x82
13 00000040`3627e410 00007fff`7c50ce44     PrintConfig!UniDrvUI::DwDeviceCapabilities+0xc6
14 00000040`3627e4c0 00007fff`7c42514f     PrintConfig!UniDrvUI::DrvDeviceCapabilities+0x18
15 00000040`3627e500 00007fff`e03150a1     PrintConfig!DrvDeviceCapabilities+0x11f
16 00000040`3627e580 00007fff`e0324a04     WINSPOOL!DeviceCapabilitiesWNative+0x105
17 00000040`3627e5e0 00007fff`7c4feafb     WINSPOOL!DeviceCapabilitiesW+0x54
18 00000040`3627e620 00007fff`7c516637     PrintConfig!publicdm::JobTicketToPublicDevmode+0x9f
19 00000040`3627e6f0 00007fff`7c514794     PrintConfig!UniDrvUI::PerformJScriptDevmodeValidation+0x563
1a 00000040`3627e7f0 00007fff`7c512bb7     PrintConfig!UniDrvUI::BFillCommonInfoDevmode+0x31c
1b 00000040`3627e890 00007fff`7c5120ec     PrintConfig!UniDrvUI::LSimpleDocumentProperties+0x173
1c 00000040`3627e900 00007fff`7c40f54b     PrintConfig!UniDrvUI::DrvDocumentPropertySheets+0x60
1d 00000040`3627edb0 00007fff`7c4249ad     PrintConfig!PrintConfig::DrvDocumentPropertySheets+0x123
1e 00000040`3627ee20 00007fff`7c425543     PrintConfig!ExceptionBoundary<<lambda_ce41999e58d4ed069b2790d632f6e3b5> >+0x3d
1f 00000040`3627ee70 00007fff`e032e9ba     PrintConfig!DrvDocumentPropertySheets+0x53
20 00000040`3627eee0 00007fff`e031593f     WINSPOOL!DocumentPropertySheets+0x3fa
21 00000040`3627ef30 00007fff`e0324aff     WINSPOOL!DocumentPropertiesWNative+0x193
22 00000040`3627efc0 00007ff8`0a340fc8     WINSPOOL!DocumentPropertiesW+0xcf
23 00000040`3627f020 00007ff8`0a344d59     COMDLG32!PrintGetDevMode+0x8c
24 00000040`3627f060 00007ff8`0a34058b     COMDLG32!PrintReturnDefault+0x91
25 00000040`3627f090 00007ff8`0a340407     COMDLG32!PrintDlgX+0x163
26 00000040`3627f530 00007fff`7c7ff697     COMDLG32!PrintDlgW+0x47
27 00000040`3627fa40 00007fff`7c7ff591     MFC42u!CWinApp::UpdatePrinterSelection+0x57
28 00000040`3627fbe0 00007ff7`f383d742     MFC42u!CWinApp::GetPrinterDeviceDefaults+0x21
29 00000040`3627fc10 00007ff7`f383d938     wordpad!CWordPadApp::CreateDevNames+0xb2
2a 00000040`3627fcf0 00007fff`7c7ffc1d     wordpad!CWordPadApp::DoDeferredInitialization+0x18
2b 00000040`3627fd20 00007ff8`0a57e634     MFC42u!_AfxThreadEntry+0xdd
2c 00000040`3627fde0 00007ff8`0a57e70c     msvcrt!_callthreadstartex+0x28
2d 00000040`3627fe10 00007ff8`0b03244d     msvcrt!_threadstartex+0x7c
2e 00000040`3627fe40 00007ff8`0c18df78     KERNEL32!BaseThreadInitThunk+0x1d
2f 00000040`3627fe70 00000000`00000000     ntdll!RtlUserThreadStart+0x28
 # Child-SP          RetAddr               Call Site
00 00000040`3607b118 00007fff`f3cce048     USER32!CreateWindowExW
01 00000040`3607b120 00007fff`f3ccdf1f     urlmon!NotificationWindow::CreateHandle+0x68
02 00000040`3607b190 00007fff`f3c7cc9b     urlmon!CTransaction::GetNotificationWnd+0x13f
03 00000040`3607b1d0 00007fff`f3cb98ee     urlmon!GetTransactionObjects+0x2fb
04 00000040`3607b250 00007fff`f3cba81c     urlmon!CBinding::StartBinding+0x53e
```

```
05 00000040`3607b3d0 00007fff`f3cb9276     urlmon!CUrlMon::StartBinding+0x1b0
06 00000040`3607b4a0 00007fff`f1c05529     urlmon!CUrlMon::BindToStorage+0x96
07 00000040`3607b4f0 00007fff`f1c04acb     msxml6!URLMONStream::deferedOpen+0x159
08 00000040`3607b550 00007fff`f1c048df     msxml6!XMLParser::PushURL+0x1af
09 00000040`3607b5d0 00007fff`f1bbf1c6     msxml6!XMLParser::SetURL+0x7f
0a 00000040`3607b620 00007fff`f1bbed48     msxml6!Document::_load+0x162
0b 00000040`3607b6a0 00007fff`f1b87712     msxml6!Document::load+0x198
0c 00000040`3607b710 00007fff`7c46afae     msxml6!DOMDocumentWrapper::load+0x182
0d 00000040`3607b7e0 00007fff`7c414968     PrintConfig!PrintConfig::PrivateDevmodeMapFile::Parse+0x14e
0e 00000040`3607b950 00007fff`7c558d7b     PrintConfig!PrintConfigPlugIn::DevMode+0x458
0f 00000040`3607ba70 00007fff`7c512b53     PrintConfig!BCalcTotalOEMDMSize+0xa3
10 00000040`3607bb10 00007fff`7c5120ec     PrintConfig!UniDrvUI::LSimpleDocumentProperties+0x10f
11 00000040`3607bb80 00007fff`7c40f54b     PrintConfig!UniDrvUI::DrvDocumentPropertySheets+0x60
12 00000040`3607c030 00007fff`7c4249ad     PrintConfig!PrintConfig::DrvDocumentPropertySheets+0x123
13 00000040`3607c0a0 00007fff`7c425543     PrintConfig!ExceptionBoundary<<lambda_ce41999e58d4ed069b2790d632f6e3b5> >+0x3d
14 00000040`3607c0f0 00007fff`e032e9ba     PrintConfig!DrvDocumentPropertySheets+0x53
15 00000040`3607c160 00007fff`e031593f     WINSPOOL!DocumentPropertySheets+0x3fa
16 00000040`3607c1b0 00007fff`e0324aff     WINSPOOL!DocumentPropertiesWNative+0x193
17 00000040`3607c240 00007fff`a4de735d     WINSPOOL!DocumentPropertiesW+0xcf
18 00000040`3607c2a0 00007fff`a4ddb5d0     mxdwdrv!edocs::PDev::PDev+0x1d9
19 00000040`3607c7b0 00007ff8`09876470     mxdwdrv!DrvEnablePDEV+0x170
1a 00000040`3607d0f0 00007ff8`0bea991a     gdi32full!GdiPrinterThunk+0x370
1b 00000040`3607d1c0 00007ff8`0c1d2db4     USER32!_ClientPrinterThunk+0x3a
1c 00000040`3607da40 00007ff8`097f2d94     ntdll!KiUserCallbackDispatcherContinue
1d 00000040`3607db48 00007ff8`0983dbd2     win32u!NtGdiOpenDCW+0x14
1e 00000040`3607db50 00007ff8`0a5f166d     gdi32full!hdcCreateDCW+0xb2
1f 00000040`3607dea0 00007ff8`0a5f1545     GDI32!bCreateDCW+0xed
20 00000040`3607df10 00007fff`7c7ff0c7     GDI32!CreateDCW+0x55
21 00000040`3607df50 00007fff`7c7ff020     MFC42u!AfxCreateDC+0x67
22 00000040`3607df80 00007ff7`f3852407     MFC42u!CWinApp::CreatePrinterDC+0x20
23 00000040`3607dfb0 00007fff`7c7e9920     wordpad!CWordPadView::OnPrinterChangedMsg+0x37
24 00000040`3607e000 00007fff`7c7f0bff     MFC42u!CWnd::OnWndMsg+0xd80
25 00000040`3607e130 00007fff`7c7eb42a     MFC42u!CWnd::WindowProc+0x4f
26 00000040`3607e170 00007fff`7c7ea76b     MFC42u!AfxCallWndProc+0x14a
27 00000040`3607e270 00007ff8`0be38161     MFC42u!AfxWndProcBase+0x15b
28 00000040`3607e2e0 00007ff8`0be37e1c     USER32!UserCallWinProcCheckWow+0x2d1
29 00000040`3607e440 00007ff8`0be42cdd     USER32!DispatchClientMessage+0x9c
2a 00000040`3607e4a0 00007ff8`0c1d2db4     USER32!_fnDWORD+0x3d
2b 00000040`3607e500 00007ff8`097f1ef4     ntdll!KiUserCallbackDispatcherContinue
2c 00000040`3607e588 00007fff`7c7e21cd     win32u!NtUserShowWindow+0x14
2d 00000040`3607e590 00007ff7`f38346b6     MFC42u!CFrameWnd::ActivateFrame+0x2d
2e 00000040`3607e5c0 00007fff`7c7e25de     wordpad!CMainFrame::ActivateFrame+0x86
2f 00000040`3607e630 00007fff`7c7fa2e7     MFC42u!CFrameWnd::InitialUpdateFrame+0xfe
30 00000040`3607e680 00007ff7`f383e331     MFC42u!CSingleDocTemplate::OpenDocumentFile+0x187
31 00000040`3607e6e0 00007fff`7c7efa57     wordpad!CWordPadApp::InitInstance+0x521
32 00000040`3607fc80 00007ff7`f37fb49e     MFC42u!AfxWinMain+0x97
33 00000040`3607fcc0 00007ff8`0b03244d     wordpad!__wmainCRTStartup+0x1de
34 00000040`3607fd80 00007ff8`0c18df78     KERNEL32!BaseThreadInitThunk+0x1d
35 00000040`3607fdb0 00000000`00000000     ntdll!RtlUserThreadStart+0x28
 # Child-SP          RetAddr               Call Site
00 00000040`3607b118 00007fff`f3cce048     USER32!CreateWindowExW
01 00000040`3607b120 00007fff`f3ccdf1f     urlmon!NotificationWindow::CreateHandle+0x68
02 00000040`3607b190 00007fff`f3c7cc9b     urlmon!CTransaction::GetNotificationWnd+0x13f
03 00000040`3607b1d0 00007fff`f3cb98ee     urlmon!GetTransactionObjects+0x2fb
04 00000040`3607b250 00007fff`f3cba81c     urlmon!CBinding::StartBinding+0x53e
05 00000040`3607b3d0 00007fff`f3cb9276     urlmon!CUrlMon::StartBinding+0x1b0
06 00000040`3607b4a0 00007fff`f1c05529     urlmon!CUrlMon::BindToStorage+0x96
07 00000040`3607b4f0 00007fff`f1c04acb     msxml6!URLMONStream::deferedOpen+0x159
08 00000040`3607b550 00007fff`f1c048df     msxml6!XMLParser::PushURL+0x1af
09 00000040`3607b5d0 00007fff`f1bbf1c6     msxml6!XMLParser::SetURL+0x7f
0a 00000040`3607b620 00007fff`f1bbed48     msxml6!Document::_load+0x162
0b 00000040`3607b6a0 00007fff`f1b87712     msxml6!Document::load+0x198
0c 00000040`3607b710 00007fff`7c46afae     msxml6!DOMDocumentWrapper::load+0x182
0d 00000040`3607b7e0 00007fff`7c414968     PrintConfig!PrintConfig::PrivateDevmodeMapFile::Parse+0x14e
0e 00000040`3607b950 00007fff`7c558d7b     PrintConfig!PrintConfigPlugIn::DevMode+0x458
0f 00000040`3607ba70 00007fff`7c512b53     PrintConfig!BCalcTotalOEMDMSize+0xa3
10 00000040`3607bb10 00007fff`7c5120ec     PrintConfig!UniDrvUI::LSimpleDocumentProperties+0x10f
11 00000040`3607bb80 00007fff`7c40f54b     PrintConfig!UniDrvUI::DrvDocumentPropertySheets+0x60
12 00000040`3607c030 00007fff`7c4249ad     PrintConfig!PrintConfig::DrvDocumentPropertySheets+0x123
13 00000040`3607c0a0 00007fff`7c425543     PrintConfig!ExceptionBoundary<<lambda_ce41999e58d4ed069b2790d632f6e3b5> >+0x3d
14 00000040`3607c0f0 00007fff`e032e9ba     PrintConfig!DrvDocumentPropertySheets+0x53
15 00000040`3607c160 00007fff`e0315895     WINSPOOL!DocumentPropertySheets+0x3fa
16 00000040`3607c1b0 00007fff`e0324aff     WINSPOOL!DocumentPropertiesWNative+0xe9
17 00000040`3607c240 00007fff`a4de746f     WINSPOOL!DocumentPropertiesW+0xcf
18 00000040`3607c2a0 00007fff`a4ddb5d0     mxdwdrv!edocs::PDev::PDev+0x2eb
19 00000040`3607c7b0 00007ff8`09876470     mxdwdrv!DrvEnablePDEV+0x170
1a 00000040`3607d0f0 00007ff8`0bea991a     gdi32full!GdiPrinterThunk+0x370
1b 00000040`3607d1c0 00007ff8`0c1d2db4     USER32!_ClientPrinterThunk+0x3a
1c 00000040`3607da40 00007ff8`097f2d94     ntdll!KiUserCallbackDispatcherContinue
1d 00000040`3607db48 00007ff8`0983dbd2     win32u!NtGdiOpenDCW+0x14
1e 00000040`3607db50 00007ff8`0a5f166d     gdi32full!hdcCreateDCW+0xb2
1f 00000040`3607dea0 00007ff8`0a5f1545     GDI32!bCreateDCW+0xed
20 00000040`3607df10 00007fff`7c7ff0c7     GDI32!CreateDCW+0x55
21 00000040`3607df50 00007fff`7c7ff020     MFC42u!AfxCreateDC+0x67
22 00000040`3607df80 00007ff7`f3852407     MFC42u!CWinApp::CreatePrinterDC+0x20
23 00000040`3607dfb0 00007fff`7c7e9920     wordpad!CWordPadView::OnPrinterChangedMsg+0x37
24 00000040`3607e000 00007fff`7c7f0bff     MFC42u!CWnd::OnWndMsg+0xd80
25 00000040`3607e130 00007fff`7c7eb42a     MFC42u!CWnd::WindowProc+0x4f
26 00000040`3607e170 00007fff`7c7ea76b     MFC42u!AfxCallWndProc+0x14a
27 00000040`3607e270 00007ff8`0be38161     MFC42u!AfxWndProcBase+0x15b
28 00000040`3607e2e0 00007ff8`0be37e1c     USER32!UserCallWinProcCheckWow+0x2d1
29 00000040`3607e440 00007ff8`0be42cdd     USER32!DispatchClientMessage+0x9c
2a 00000040`3607e4a0 00007ff8`0c1d2db4     USER32!_fnDWORD+0x3d
2b 00000040`3607e500 00007ff8`097f1ef4     ntdll!KiUserCallbackDispatcherContinue
2c 00000040`3607e588 00007fff`7c7e21cd     win32u!NtUserShowWindow+0x14
2d 00000040`3607e590 00007ff7`f38346b6     MFC42u!CFrameWnd::ActivateFrame+0x2d
2e 00000040`3607e5c0 00007fff`7c7e25de     wordpad!CMainFrame::ActivateFrame+0x86
2f 00000040`3607e630 00007fff`7c7fa2e7     MFC42u!CFrameWnd::InitialUpdateFrame+0xfe
```

```
30 00000040`3607e680 00007ff7`f383e331     MFC42u!CSingleDocTemplate::OpenDocumentFile+0x187
31 00000040`3607e6e0 00007fff`7c7efa57     wordpad!CWordPadApp::InitInstance+0x521
32 00000040`3607fc80 00007ff7`f37fb49e     MFC42u!AfxWinMain+0x97
33 00000040`3607fcc0 00007ff8`0b03244d     wordpad!__wmainCRTStartup+0x1de
34 00000040`3607fd80 00007ff8`0c18df78     KERNEL32!BaseThreadInitThunk+0x1d
35 00000040`3607fdb0 00000000`00000000     ntdll!RtlUserThreadStart+0x28
 # Child-SP          RetAddr               Call Site
00 00000040`3607b118 00007fff`f3cce048     USER32!CreateWindowExW
01 00000040`3607b120 00007fff`f3ccdf1f     urlmon!NotificationWindow::CreateHandle+0x68
02 00000040`3607b190 00007fff`f3c7cc9b     urlmon!CTransaction::GetNotificationWnd+0x13f
03 00000040`3607b1d0 00007fff`f3cb98ee     urlmon!GetTransactionObjects+0x2fb
04 00000040`3607b250 00007fff`f3cba81c     urlmon!CBinding::StartBinding+0x53e
05 00000040`3607b3d0 00007fff`f3cb9276     urlmon!CUrlMon::StartBinding+0x1b0
06 00000040`3607b4a0 00007fff`f1c05529     urlmon!CUrlMon::BindToStorage+0x96
07 00000040`3607b4f0 00007fff`f1c04acb     msxml6!URLMONStream::deferedOpen+0x159
08 00000040`3607b550 00007fff`f1c048df     msxml6!XMLParser::PushURL+0x1af
09 00000040`3607b5d0 00007fff`f1bbf1c6     msxml6!XMLParser::SetURL+0x7f
0a 00000040`3607b620 00007fff`f1bbed48     msxml6!Document::_load+0x162
0b 00000040`3607b6a0 00007fff`f1b87712     msxml6!Document::load+0x198
0c 00000040`3607b710 00007fff`7c46afae     msxml6!DOMDocumentWrapper::load+0x182
0d 00000040`3607b7e0 00007fff`7c414968     PrintConfig!PrintConfig::PrivateDevmodeMapFile::Parse+0x14e
0e 00000040`3607b950 00007fff`7c558d7b     PrintConfig!PrintConfigPlugIn::DevMode+0x458
0f 00000040`3607ba70 00007fff`7c512b53     PrintConfig!BCalcTotalOEMDMSize+0xa3
10 00000040`3607bb10 00007fff`7c5120ec     PrintConfig!UniDrvUI::LSimpleDocumentProperties+0x10f
11 00000040`3607bb80 00007fff`7c40f54b     PrintConfig!UniDrvUI::DrvDocumentPropertySheets+0x60
12 00000040`3607c030 00007fff`7c4249ad     PrintConfig!PrintConfig::DrvDocumentPropertySheets+0x123
13 00000040`3607c0a0 00007fff`7c425543     PrintConfig!ExceptionBoundary<<lambda_ce41999e58d4ed069b2790d632f6e3b5> >+0x3d
14 00000040`3607c0f0 00007fff`e032e9ba     PrintConfig!DrvDocumentPropertySheets+0x53
15 00000040`3607c160 00007fff`e031593f     WINSPOOL!DocumentPropertySheets+0x3fa
16 00000040`3607c1b0 00007fff`e0324aff     WINSPOOL!DocumentPropertiesWNative+0x193
17 00000040`3607c240 00007fff`a4de746f     WINSPOOL!DocumentPropertiesW+0xcf
18 00000040`3607c2a0 00007fff`a4ddb5d0     mxdwdrv!edocs::PDev::PDev+0x2eb
19 00000040`3607c7b0 00007ff8`09876470     mxdwdrv!DrvEnablePDEV+0x170
1a 00000040`3607d0f0 00007ff8`0bea991a     gdi32full!GdiPrinterThunk+0x370
1b 00000040`3607d1c0 00007ff8`0c1d2db4     USER32!_ClientPrinterThunk+0x3a
1c 00000040`3607da40 00007ff8`097f2d94     ntdll!KiUserCallbackDispatcherContinue
1d 00000040`3607db48 00007ff8`0983dbd2     win32u!NtGdiOpenDCW+0x14
1e 00000040`3607db50 00007ff8`0a5f166d     gdi32full!hdcCreateDCW+0xb2
1f 00000040`3607dea0 00007ff8`0a5f1545     GDI32!bCreateDCW+0xed
20 00000040`3607df10 00007fff`7c7ff0c7     GDI32!CreateDCW+0x55
21 00000040`3607df50 00007fff`7c7ff020     MFC42u!AfxCreateDC+0x67
22 00000040`3607df80 00007ff7`f3852407     MFC42u!CWinApp::CreatePrinterDC+0x20
23 00000040`3607dfb0 00007fff`7c7e9920     wordpad!CWordPadView::OnPrinterChangedMsg+0x37
24 00000040`3607e000 00007fff`7c7f0bff     MFC42u!CWnd::OnWndMsg+0xd80
25 00000040`3607e130 00007fff`7c7eb42a     MFC42u!CWnd::WindowProc+0x4f
26 00000040`3607e170 00007fff`7c7ea76b     MFC42u!AfxCallWndProc+0x14a
27 00000040`3607e270 00007ff8`0be38161     MFC42u!AfxWndProcBase+0x15b
28 00000040`3607e2e0 00007ff8`0be37e1c     USER32!UserCallWinProcCheckWow+0x2d1
29 00000040`3607e440 00007ff8`0be42cdd     USER32!DispatchClientMessage+0x9c
2a 00000040`3607e4a0 00007ff8`0c1d2db4     USER32!_fnDWORD+0x3d
2b 00000040`3607e500 00007ff8`097f1ef4     ntdll!KiUserCallbackDispatcherContinue
2c 00000040`3607e588 00007fff`7c7e21cd     win32u!NtUserShowWindow+0x14
2d 00000040`3607e590 00007ff7`f38346b6     MFC42u!CFrameWnd::ActivateFrame+0x2d
2e 00000040`3607e5c0 00007fff`7c7e25de     wordpad!CMainFrame::ActivateFrame+0x86
2f 00000040`3607e630 00007fff`7c7fa2e7     MFC42u!CFrameWnd::InitialUpdateFrame+0xfe
30 00000040`3607e680 00007ff7`f383e331     MFC42u!CSingleDocTemplate::OpenDocumentFile+0x187
31 00000040`3607e6e0 00007fff`7c7efa57     wordpad!CWordPadApp::InitInstance+0x521
32 00000040`3607fc80 00007ff7`f37fb49e     MFC42u!AfxWinMain+0x97
33 00000040`3607fcc0 00007ff8`0b03244d     wordpad!__wmainCRTStartup+0x1de
34 00000040`3607fd80 00007ff8`0c18df78     KERNEL32!BaseThreadInitThunk+0x1d
35 00000040`3607fdb0 00000000`00000000     ntdll!RtlUserThreadStart+0x28
 # Child-SP          RetAddr               Call Site
00 00000040`3607afa8 00007fff`f3cce048     USER32!CreateWindowExW
01 00000040`3607afb0 00007fff`f3ccdf1f     urlmon!NotificationWindow::CreateHandle+0x68
02 00000040`3607b020 00007fff`f3c7cc9b     urlmon!CTransaction::GetNotificationWnd+0x13f
03 00000040`3607b060 00007fff`f3cb98ee     urlmon!GetTransactionObjects+0x2fb
04 00000040`3607b0e0 00007fff`f3cba81c     urlmon!CBinding::StartBinding+0x53e
05 00000040`3607b260 00007fff`f3cb9276     urlmon!CUrlMon::StartBinding+0x1b0
06 00000040`3607b330 00007fff`f1c05529     urlmon!CUrlMon::BindToStorage+0x96
07 00000040`3607b380 00007fff`f1c04acb     msxml6!URLMONStream::deferedOpen+0x159
08 00000040`3607b3e0 00007fff`f1c048df     msxml6!XMLParser::PushURL+0x1af
09 00000040`3607b460 00007fff`f1bbf1c6     msxml6!XMLParser::SetURL+0x7f
0a 00000040`3607b4b0 00007fff`f1bbed48     msxml6!Document::_load+0x162
0b 00000040`3607b530 00007fff`f1b87712     msxml6!Document::load+0x198
0c 00000040`3607b5a0 00007fff`7c46afae     msxml6!DOMDocumentWrapper::load+0x182
0d 00000040`3607b670 00007fff`7c414813     PrintConfig!PrintConfig::PrivateDevmodeMapFile::Parse+0x14e
0e 00000040`3607b7e0 00007fff`7c5583f0     PrintConfig!PrintConfigPlugIn::DevMode+0x303
0f 00000040`3607b900 00007fff`7c55956a     PrintConfig!BCallOEMDevMode+0x84
10 00000040`3607b980 00007fff`7c514999     PrintConfig!PGetDefaultDevmodeWithOemPlugins+0x16a
11 00000040`3607ba00 00007fff`7c5144fa     PrintConfig!UniDrvUI::BGetValidatedUserDefaultDevmode+0x81
12 00000040`3607ba70 00007fff`7c512bb7     PrintConfig!UniDrvUI::BFillCommonInfoDevmode+0x82
13 00000040`3607bb10 00007fff`7c5120ec     PrintConfig!UniDrvUI::LSimpleDocumentProperties+0x173
14 00000040`3607bb80 00007fff`7c40f54b     PrintConfig!UniDrvUI::DrvDocumentPropertySheets+0x60
15 00000040`3607c030 00007fff`7c4249ad     PrintConfig!PrintConfig::DrvDocumentPropertySheets+0x123
16 00000040`3607c0a0 00007fff`7c425543     PrintConfig!ExceptionBoundary<<lambda_ce41999e58d4ed069b2790d632f6e3b5> >+0x3d
17 00000040`3607c0f0 00007fff`e032e9ba     PrintConfig!DrvDocumentPropertySheets+0x53
18 00000040`3607c160 00007fff`e031593f     WINSPOOL!DocumentPropertySheets+0x3fa
19 00000040`3607c1b0 00007fff`e0324aff     WINSPOOL!DocumentPropertiesWNative+0x193
1a 00000040`3607c240 00007fff`a4de746f     WINSPOOL!DocumentPropertiesW+0xcf
1b 00000040`3607c2a0 00007fff`a4ddb5d0     mxdwdrv!edocs::PDev::PDev+0x2eb
1c 00000040`3607c7b0 00007ff8`09876470     mxdwdrv!DrvEnablePDEV+0x170
1d 00000040`3607d0f0 00007ff8`0bea991a     gdi32full!GdiPrinterThunk+0x370
1e 00000040`3607d1c0 00007ff8`0c1d2db4     USER32!_ClientPrinterThunk+0x3a
1f 00000040`3607da40 00007ff8`097f2d94     ntdll!KiUserCallbackDispatcherContinue
20 00000040`3607db48 00007ff8`0983dbd2     win32u!NtGdiOpenDCW+0x14
21 00000040`3607db50 00007ff8`0a5f166d     gdi32full!hdcCreateDCW+0xb2
22 00000040`3607dea0 00007ff8`0a5f1545     GDI32!bCreateDCW+0xed
23 00000040`3607df10 00007fff`7c7ff0c7     GDI32!CreateDCW+0x55
```

```
24 00000040`3607df50 00007fff`7c7ff020       MFC42u!AfxCreateDC+0x67
25 00000040`3607df80 00007ff7`f3852407       MFC42u!CWinApp::CreatePrinterDC+0x20
26 00000040`3607dfb0 00007fff`7c7e9920       wordpad!CWordPadView::OnPrinterChangedMsg+0x37
27 00000040`3607e000 00007fff`7c7f0bff       MFC42u!CWnd::OnWndMsg+0xd80
28 00000040`3607e130 00007fff`7c7eb42a       MFC42u!CWnd::WindowProc+0x4f
29 00000040`3607e170 00007fff`7c7ea76b       MFC42u!AfxCallWndProc+0x14a
2a 00000040`3607e270 00007ff8`0be38161       MFC42u!AfxWndProcBase+0x15b
2b 00000040`3607e2e0 00007ff8`0be37e1c       USER32!UserCallWinProcCheckWow+0x2d1
2c 00000040`3607e440 00007ff8`0be42cdd       USER32!DispatchClientMessage+0x9c
2d 00000040`3607e4a0 00007ff8`0c1d2db4       USER32!_fnDWORD+0x3d
2e 00000040`3607e500 00007ff8`097f1ef4       ntdll!KiUserCallbackDispatcherContinue
2f 00000040`3607e588 00007fff`7c7e21cd       win32u!NtUserShowWindow+0x14
30 00000040`3607e590 00007ff7`f38346b6       MFC42u!CFrameWnd::ActivateFrame+0x2d
31 00000040`3607e5c0 00007fff`7c7e25de       wordpad!CMainFrame::ActivateFrame+0x86
32 00000040`3607e630 00007fff`7c7fa2e7       MFC42u!CFrameWnd::InitialUpdateFrame+0xfe
33 00000040`3607e680 00007ff7`f383e331       MFC42u!CSingleDocTemplate::OpenDocumentFile+0x187
34 00000040`3607e6e0 00007fff`7c7efa57       wordpad!CWordPadApp::InitInstance+0x521
35 00000040`3607fc80 00007ff7`f37fb49e       MFC42u!AfxWinMain+0x97
36 00000040`3607fcc0 00007ff8`0b03244d       wordpad!__wmainCRTStartup+0x1de
37 00000040`3607fd80 00007ff8`0c18df78       KERNEL32!BaseThreadInitThunk+0x1d
38 00000040`3607fdb0 00000000`00000000       ntdll!RtlUserThreadStart+0x28
ModLoad: 00007fff`aff10000 00007fff`aff33000   C:\WINDOWS\system32\FontSub.dll
 # Child-SP          RetAddr               Call Site
00 00000040`3607aaa8 00007fff`f3cce048       USER32!CreateWindowExW
01 00000040`3607aab0 00007fff`f3ccdf1f       urlmon!NotificationWindow::CreateHandle+0x68
02 00000040`3607ab20 00007fff`f3c7cc9b       urlmon!CTransaction::GetNotificationWnd+0x13f
03 00000040`3607ab60 00007fff`f3cb98ee       urlmon!GetTransactionObjects+0x2fb
04 00000040`3607abe0 00007fff`f3cba81c       urlmon!CBinding::StartBinding+0x53e
05 00000040`3607ad60 00007fff`f3cb9276       urlmon!CUrlMon::StartBinding+0x1b0
06 00000040`3607ae30 00007fff`f1c05529       urlmon!CUrlMon::BindToStorage+0x96
07 00000040`3607ae80 00007fff`f1c04acb       msxml6!URLMONStream::deferedOpen+0x159
08 00000040`3607aee0 00007fff`f1c048df       msxml6!XMLParser::PushURL+0x1af
09 00000040`3607af60 00007fff`f1bbf1c6       msxml6!XMLParser::SetURL+0x7f
0a 00000040`3607afb0 00007fff`f1bbed48       msxml6!Document::_load+0x162
0b 00000040`3607b030 00007fff`f1b87712       msxml6!Document::load+0x198
0c 00000040`3607b0a0 00007fff`7c46afae       msxml6!DOMDocumentWrapper::load+0x182
0d 00000040`3607b170 00007fff`7c414968       PrintConfig!PrintConfig::PrivateDevmodeMapFile::Parse+0x14e
0e 00000040`3607b2e0 00007fff`7c558d7b       PrintConfig!PrintConfigPlugIn::DevMode+0x458
0f 00000040`3607b400 00007fff`7c55949b       PrintConfig!BCalcTotalOEMDMSize+0xa3
10 00000040`3607b4a0 00007fff`7c514999       PrintConfig!PGetDefaultDevmodeWithOemPlugins+0x9b
11 00000040`3607b520 00007fff`7c4fc131       PrintConfig!UniDrvUI::BGetValidatedUserDefaultDevmode+0x81
12 00000040`3607b590 00007fff`7c4f1696       PrintConfig!UniDrvUI::CPrintTicketProvider::GetValidatedDevmode+0x6d
13 00000040`3607b5e0 00007fff`7c412e2d       PrintConfig!UniDrvUI::CPrintTicketProvider::ConvertDevModeToPrintTicket+0x36
14 00000040`3607b610 00007fff`d5863f96       PrintConfig!PrintConfig::PrintConfigPTProvider::ConvertDevModeToPrintTicket+0x7d
15 00000040`3607b670 00007fff`d5853aff       prntvpt!PTConvertDevModeToPrintTicketImp+0x156
16 00000040`3607b730 00007fff`d5857941       prntvpt!CPrintTicketServerBase::ConvertDevModeToPrintTicket+0x1f
17 00000040`3607b780 00007fff`7c48ebe8       prntvpt!PTConvertDevModeToPrintTicket+0x91
18 00000040`3607b800 00007fff`7c48eb69       PrintConfig!PrintConfig::CSystemPrintTicketService::ConvertDevmodeToPrintTicket+0x68
19 00000040`3607b840 00007fff`7c4bfa7a       PrintConfig!PrintConfig::CPrintTicketServiceDecorator::ConvertDevmodeToPrintTicket+0x29
1a 00000040`3607b880 00007fff`7c4c012a       PrintConfig!PrintConfig::WithPrintTicketServiceDo<<lambda_2a2b4feca19798ea4352c64d6d699e02> >+0x82
1b 00000040`3607b8c0 00007fff`7c4c37b3       PrintConfig!PrintConfig::CPrintTicketService2::ConvertDevmodeToPrintTicket+0x1ea
1c 00000040`3607b950 00007fff`7c4c38ec       PrintConfig!PrintConfig::CPrintTicket::EnsureXmlDocument+0x8b
1d 00000040`3607b990 00007fff`7c418a3f       PrintConfig!PrintConfig::CPrintTicket::GetCapabilities+0xfc
1e 00000040`3607ba00 00007fff`7c4248a4       PrintConfig!PrintConfig::V4MxdcGetPDEVAdjustment+0x9ef
1f 00000040`3607bcc0 00007fff`7c424c33       PrintConfig!ExceptionBoundary<<lambda_4bfcdea1f8082feba17903613e8e1079> >+0x54
20 00000040`3607bd30 00007fff`a4de9cc7       PrintConfig!MxdcGetPDEVAdjustment+0x93
21 00000040`3607bdd0 00007fff`a4dead1f       mxdwdrv!edocs::PDev::InvokeMxdcGetPDEVAdjustment+0xb3
22 00000040`3607c2b0 00007fff`a4ddb5f0       mxdwdrv!edocs::PDev::ValidateForm+0x1fb
23 00000040`3607c7b0 00007ff8`09876470       mxdwdrv!DrvEnablePDEV+0x190
24 00000040`3607d0f0 00007ff8`0bea991a       gdi32full!GdiPrinterThunk+0x370
25 00000040`3607d1c0 00007ff8`0c1d2db4       USER32!_ClientPrinterThunk+0x3a
26 00000040`3607da40 00007ff8`097f2d94       ntdll!KiUserCallbackDispatcherContinue
27 00000040`3607db48 00007ff8`0983dbd2       win32u!NtGdiOpenDCW+0x14
28 00000040`3607db50 00007ff8`0a5f166d       gdi32full!hdcCreateDCW+0xb2
29 00000040`3607dea0 00007ff8`0a5f1545       GDI32!bCreateDCW+0xed
2a 00000040`3607df10 00007fff`7c7ff0c7       GDI32!CreateDCW+0x55
2b 00000040`3607df50 00007fff`7c7ff020       MFC42u!AfxCreateDC+0x67
2c 00000040`3607df80 00007ff7`f3852407       MFC42u!CWinApp::CreatePrinterDC+0x20
2d 00000040`3607dfb0 00007fff`7c7e9920       wordpad!CWordPadView::OnPrinterChangedMsg+0x37
2e 00000040`3607e000 00007fff`7c7f0bff       MFC42u!CWnd::OnWndMsg+0xd80
2f 00000040`3607e130 00007fff`7c7eb42a       MFC42u!CWnd::WindowProc+0x4f
30 00000040`3607e170 00007fff`7c7ea76b       MFC42u!AfxCallWndProc+0x14a
31 00000040`3607e270 00007ff8`0be38161       MFC42u!AfxWndProcBase+0x15b
32 00000040`3607e2e0 00007ff8`0be37e1c       USER32!UserCallWinProcCheckWow+0x2d1
33 00000040`3607e440 00007ff8`0be42cdd       USER32!DispatchClientMessage+0x9c
34 00000040`3607e4a0 00007ff8`0c1d2db4       USER32!_fnDWORD+0x3d
35 00000040`3607e500 00007ff8`097f1ef4       ntdll!KiUserCallbackDispatcherContinue
36 00000040`3607e588 00007fff`7c7e21cd       win32u!NtUserShowWindow+0x14
37 00000040`3607e590 00007ff7`f38346b6       MFC42u!CFrameWnd::ActivateFrame+0x2d
38 00000040`3607e5c0 00007fff`7c7e25de       wordpad!CMainFrame::ActivateFrame+0x86
39 00000040`3607e630 00007fff`7c7fa2e7       MFC42u!CFrameWnd::InitialUpdateFrame+0xfe
3a 00000040`3607e680 00007ff7`f383e331       MFC42u!CSingleDocTemplate::OpenDocumentFile+0x187
3b 00000040`3607e6e0 00007fff`7c7efa57       wordpad!CWordPadApp::InitInstance+0x521
3c 00000040`3607fc80 00007ff7`f37fb49e       MFC42u!AfxWinMain+0x97
3d 00000040`3607fcc0 00007ff8`0b03244d       wordpad!__wmainCRTStartup+0x1de
3e 00000040`3607fd80 00007ff8`0c18df78       KERNEL32!BaseThreadInitThunk+0x1d
3f 00000040`3607fdb0 00000000`00000000       ntdll!RtlUserThreadStart+0x28
 # Child-SP          RetAddr               Call Site
00 00000040`3607aac8 00007fff`f3cce048       USER32!CreateWindowExW
01 00000040`3607aad0 00007fff`f3ccdf1f       urlmon!NotificationWindow::CreateHandle+0x68
02 00000040`3607ab40 00007fff`f3c7cc9b       urlmon!CTransaction::GetNotificationWnd+0x13f
03 00000040`3607ab80 00007fff`f3cb98ee       urlmon!GetTransactionObjects+0x2fb
04 00000040`3607ac00 00007fff`f3cba81c       urlmon!CBinding::StartBinding+0x53e
05 00000040`3607ad80 00007fff`f3cb9276       urlmon!CUrlMon::StartBinding+0x1b0
06 00000040`3607ae50 00007fff`f1c05529       urlmon!CUrlMon::BindToStorage+0x96
07 00000040`3607aea0 00007fff`f1c04acb       msxml6!URLMONStream::deferedOpen+0x159
08 00000040`3607af00 00007fff`f1c048df       msxml6!XMLParser::PushURL+0x1af
09 00000040`3607af80 00007fff`f1bbf1c6       msxml6!XMLParser::SetURL+0x7f
```

```
0a 00000040`3607afd0 00007fff`f1bbed48     msxml6!Document::_load+0x162
0b 00000040`3607b050 00007fff`f1b87712     msxml6!Document::load+0x198
0c 00000040`3607b0c0 00007fff`7c46afae     msxml6!DOMDocumentWrapper::load+0x182
0d 00000040`3607b190 00007fff`7c414813     PrintConfig!PrintConfig::PrivateDevmodeMapFile::Parse+0x14e
0e 00000040`3607b300 00007fff`7c5583f0     PrintConfig!PrintConfigPlugIn::DevMode+0x303
0f 00000040`3607b420 00007fff`7c55956a     PrintConfig!BCallOEMDevMode+0x84
10 00000040`3607b4a0 00007fff`7c514999     PrintConfig!PGetDefaultDevmodeWithOemPlugins+0x16a
11 00000040`3607b520 00007fff`7c4fc131     PrintConfig!UniDrvUI::BGetValidatedUserDefaultDevmode+0x81
12 00000040`3607b590 00007fff`7c4f1696     PrintConfig!UniDrvUI::CPrintTicketProvider::GetValidatedDevmode+0x6d
13 00000040`3607b5e0 00007fff`7c412e2d     PrintConfig!UniDrvUI::CPrintTicketProvider::ConvertDevModeToPrintTicket+0x36
14 00000040`3607b610 00007fff`d5863f96     PrintConfig!PrintConfig::PrintConfigPTProvider::ConvertDevModeToPrintTicket+0x7d
15 00000040`3607b670 00007fff`d5853aff     prntvpt!PTConvertDevModeToPrintTicketImp+0x156
16 00000040`3607b730 00007fff`d5857941     prntvpt!CPrintTicketServerBase::ConvertDevModeToPrintTicket+0x1f
17 00000040`3607b780 00007fff`7c48ebe8     prntvpt!PTConvertDevModeToPrintTicket+0x91
18 00000040`3607b800 00007fff`7c48eb69     PrintConfig!PrintConfig::CSystemPrintTicketService::ConvertDevmodeToPrintTicket+0x68
19 00000040`3607b840 00007fff`7c4bfa7a     PrintConfig!PrintConfig::CPrintTicketServiceDecorator::ConvertDevmodeToPrintTicket+0x29
1a 00000040`3607b880 00007fff`7c4c012a     PrintConfig!PrintConfig::WithPrintTicketServiceDo<<lambda_2a2b4feca19798ea4352c64d6d699e02> >+0x82
1b 00000040`3607b8c0 00007fff`7c4c37b3     PrintConfig!PrintConfig::CPrintTicketService2::ConvertDevmodeToPrintTicket+0x1ea
1c 00000040`3607b950 00007fff`7c4c38ec     PrintConfig!PrintConfig::CPrintTicket::EnsureXmlDocument+0x8b
1d 00000040`3607b990 00007fff`7c418a3f     PrintConfig!PrintConfig::CPrintTicket::GetCapabilities+0xfc
1e 00000040`3607ba00 00007fff`7c4248a4     PrintConfig!PrintConfig::V4MxdcGetPDEVAdjustment+0x9ef
1f 00000040`3607bcc0 00007fff`7c424c33     PrintConfig!ExceptionBoundary<<lambda_4bfcdea1f8082feba17903613e8e1079> >+0x54
20 00000040`3607bd30 00007fff`a4de9cc7     PrintConfig!MxdcGetPDEVAdjustment+0x93
21 00000040`3607bdd0 00007fff`a4dead1f     mxdwdrv!edocs::PDev::InvokeMxdcGetPDEVAdjustment+0xb3
22 00000040`3607c2b0 00007fff`a4ddb5f0     mxdwdrv!edocs::PDev::ValidateForm+0x1fb
23 00000040`3607c7b0 00007ff8`09876470     mxdwdrv!DrvEnablePDEV+0x190
24 00000040`3607d0f0 00007ff8`0bea991a     gdi32full!GdiPrinterThunk+0x370
25 00000040`3607d1c0 00007ff8`0c1d2db4     USER32!_ClientPrinterThunk+0x3a
26 00000040`3607da40 00007ff8`097f2d94     ntdll!KiUserCallbackDispatcherContinue
27 00000040`3607db48 00007ff8`0983dbd2     win32u!NtGdiOpenDCW+0x14
28 00000040`3607db50 00007ff8`0a5f166d     gdi32full!hdcCreateDCW+0xb2
29 00000040`3607dea0 00007ff8`0a5f1545     GDI32!bCreateDCW+0xed
2a 00000040`3607df10 00007fff`7c7ff0c7     GDI32!CreateDCW+0x55
2b 00000040`3607df50 00007fff`7c7ff020     MFC42u!AfxCreateDC+0x67
2c 00000040`3607df80 00007ff7`f3852407     MFC42u!CWinApp::CreatePrinterDC+0x20
2d 00000040`3607dfb0 00007fff`7c7e9920     wordpad!CWordPadView::OnPrinterChangedMsg+0x37
2e 00000040`3607e000 00007fff`7c7f0bff     MFC42u!CWnd::OnWndMsg+0xd80
2f 00000040`3607e130 00007fff`7c7eb42a     MFC42u!CWnd::WindowProc+0x4f
30 00000040`3607e170 00007fff`7c7ea76b     MFC42u!AfxCallWndProc+0x14a
31 00000040`3607e270 00007ff8`0be38161     MFC42u!AfxWndProcBase+0x15b
32 00000040`3607e2e0 00007ff8`0be37e1c     USER32!UserCallWinProcCheckWow+0x2d1
33 00000040`3607e440 00007ff8`0be42cdd     USER32!DispatchClientMessage+0x9c
34 00000040`3607e4a0 00007ff8`0c1d2db4     USER32!_fnDWORD+0x3d
35 00000040`3607e500 00007ff8`097f1ef4     ntdll!KiUserCallbackDispatcherContinue
36 00000040`3607e588 00007fff`7c7e21cd     win32u!NtUserShowWindow+0x14
37 00000040`3607e590 00007ff7`f38346b6     MFC42u!CFrameWnd::ActivateFrame+0x2d
38 00000040`3607e5c0 00007fff`7c7e25de     wordpad!CMainFrame::ActivateFrame+0x86
39 00000040`3607e630 00007fff`7c7fa2e7     MFC42u!CFrameWnd::InitialUpdateFrame+0xfe
3a 00000040`3607e680 00007ff7`f383e331     MFC42u!CSingleDocTemplate::OpenDocumentFile+0x187
3b 00000040`3607e6e0 00007fff`7c7efa57     wordpad!CWordPadApp::InitInstance+0x521
3c 00000040`3607fc80 00007ff7`f37fb49e     MFC42u!AfxWinMain+0x97
3d 00000040`3607fcc0 00007ff8`0b03244d     wordpad!__wmainCRTStartup+0x1de
3e 00000040`3607fd80 00007ff8`0c18df78     KERNEL32!BaseThreadInitThunk+0x1d
3f 00000040`3607fdb0 00000000`00000000     ntdll!RtlUserThreadStart+0x28
ModLoad: 00007fff`e13e0000 00007fff`e13fd000   C:\WINDOWS\SYSTEM32\amsi.dll
(1a28.d24): C++ EH exception - code e06d7363 (first chance)
(1a28.d24): C++ EH exception - code e06d7363 (first chance)
(1a28.d24): C++ EH exception - code e06d7363 (first chance)
 # Child-SP          RetAddr               Call Site
00 00000040`36079f38 00007fff`f3cce048     USER32!CreateWindowExW
01 00000040`36079f40 00007fff`f3ccdf1f     urlmon!NotificationWindow::CreateHandle+0x68
02 00000040`36079fb0 00007fff`f3c7cc9b     urlmon!CTransaction::GetNotificationWnd+0x13f
03 00000040`36079ff0 00007fff`f3cb98ee     urlmon!GetTransactionObjects+0x2fb
04 00000040`3607a070 00007fff`f3cba81c     urlmon!CBinding::StartBinding+0x53e
05 00000040`3607a1f0 00007fff`f3cb9276     urlmon!CUrlMon::StartBinding+0x1b0
06 00000040`3607a2c0 00007fff`f1c05529     urlmon!CUrlMon::BindToStorage+0x96
07 00000040`3607a310 00007fff`f1c04acb     msxml6!URLMONStream::deferedOpen+0x159
08 00000040`3607a370 00007fff`f1c048df     msxml6!XMLParser::PushURL+0x1af
09 00000040`3607a3f0 00007fff`f1bbf1c6     msxml6!XMLParser::SetURL+0x7f
0a 00000040`3607a440 00007fff`f1bbed48     msxml6!Document::_load+0x162
0b 00000040`3607a4c0 00007fff`f1b87712     msxml6!Document::load+0x198
0c 00000040`3607a530 00007fff`7c46afae     msxml6!DOMDocumentWrapper::load+0x182
0d 00000040`3607a600 00007fff`7c414968     PrintConfig!PrintConfig::PrivateDevmodeMapFile::Parse+0x14e
0e 00000040`3607a770 00007fff`7c558d7b     PrintConfig!PrintConfigPlugIn::DevMode+0x458
0f 00000040`3607a890 00007fff`7c55949b     PrintConfig!BCalcTotalOEMDMSize+0xa3
10 00000040`3607a930 00007fff`7c514999     PrintConfig!PGetDefaultDevmodeWithOemPlugins+0x9b
11 00000040`3607a9b0 00007fff`7c4fc131     PrintConfig!UniDrvUI::BGetValidatedUserDefaultDevmode+0x81
12 00000040`3607aa20 00007fff`7c4f60a0     PrintConfig!UniDrvUI::CPrintTicketProvider::GetValidatedDevmode+0x6d
13 00000040`3607aa70 00007fff`7c4f98a8     PrintConfig!UniDrvUI::CPrintTicketProvider::GenerateCoreDriverPrintCapabilities+0x84
14 00000040`3607b3b0 00007fff`7c412fc9     PrintConfig!UniDrvUI::CPrintTicketProvider::GetPrintCapabilities+0x98
15 00000040`3607b470 00007fff`d586443f     PrintConfig!PrintConfig::PrintConfigPTProvider::GetPrintCapabilities+0x69
16 00000040`3607b4c0 00007fff`d586c501     prntvpt!PTGetPrintCapabilitiesImp+0x14b
17 00000040`3607b550 00007fff`d5864400     prntvpt!TProviderInfo::GetSupportedFeaturesList+0x61
18 00000040`3607b5c0 00007fff`d58541fb     prntvpt!PTGetPrintCapabilitiesImp+0x10c
19 00000040`3607b650 00007fff`d5857bd4     prntvpt!CPrintTicketServerBase::GetPrintCapabilities+0x5b
1a 00000040`3607b690 00007fff`7c48f310     prntvpt!PTGetPrintCapabilities+0x94
1b 00000040`3607b710 00007fff`7c48f29f     PrintConfig!PrintConfig::CSystemPrintTicketService::GetPrintCapabilities+0x60
1c 00000040`3607b740 00007fff`7c4bfc00     PrintConfig!PrintConfig::CPrintTicketServiceDecorator::GetPrintCapabilities+0x1f
1d 00000040`3607b780 00007fff`7c4c0aad     PrintConfig!PrintConfig::WithPrintTicketServiceDo<<lambda_49af5cf54d102c592666395395ebbfac> >+0x78
1e 00000040`3607b7c0 00007fff`7c4cc122     PrintConfig!PrintConfig::CPrintTicketService2::GetPrintCapabilities+0x21d
1f 00000040`3607b850 00007fff`7c4cc01d     PrintConfig!PrintConfig::CPrintCapabilities::EnsureXmlDocument+0x4a
20 00000040`3607b8a0 00007fff`7c4cd552     PrintConfig!PrintConfig::CPrintCapabilities::EnsurePrintSchema+0x35
21 00000040`3607b8f0 00007fff`7c417389     PrintConfig!PrintConfig::CPrintCapabilities::get_PageImageableSize+0x82
22 00000040`3607b9b0 00007fff`7c418a54     PrintConfig!PrintConfig::ExtractImageableArea+0x39
23 00000040`3607ba00 00007fff`7c4248a4     PrintConfig!PrintConfig::V4MxdcGetPDEVAdjustment+0xa04
24 00000040`3607bcc0 00007fff`7c424c33     PrintConfig!ExceptionBoundary<<lambda_4bfcdea1f8082feba17903613e8e1079> >+0x54
25 00000040`3607bd30 00007fff`a4de9cc7     PrintConfig!MxdcGetPDEVAdjustment+0x93
26 00000040`3607bdd0 00007fff`a4dead1f     mxdwdrv!edocs::PDev::InvokeMxdcGetPDEVAdjustment+0xb3
```

```
27 00000040`3607c2b0 00007fff`a4ddb5f0     mxdwdrv!edocs::PDev::ValidateForm+0x1fb
28 00000040`3607c7b0 00007ff8`09876470     mxdwdrv!DrvEnablePDEV+0x190
29 00000040`3607d0f0 00007ff8`0bea991a     gdi32full!GdiPrinterThunk+0x370
2a 00000040`3607d1c0 00007ff8`0c1d2db4     USER32!_ClientPrinterThunk+0x3a
2b 00000040`3607da40 00007ff8`097f2d94     ntdll!KiUserCallbackDispatcherContinue
2c 00000040`3607db48 00007ff8`0983dbd2     win32u!NtGdiOpenDCW+0x14
2d 00000040`3607db50 00007ff8`0a5f166d     gdi32full!hdcCreateDCW+0xb2
2e 00000040`3607dea0 00007ff8`0a5f1545     GDI32!bCreateDCW+0xed
2f 00000040`3607df10 00007fff`7c7ff0c7     GDI32!CreateDCW+0x55
30 00000040`3607df50 00007fff`7c7ff020     MFC42u!AfxCreateDC+0x67
31 00000040`3607df80 00007ff7`f3852407     MFC42u!CWinApp::CreatePrinterDC+0x20
32 00000040`3607dfb0 00007fff`7c7e9920     wordpad!CWordPadView::OnPrinterChangedMsg+0x37
33 00000040`3607e000 00007fff`7c7f0bff     MFC42u!CWnd::OnWndMsg+0xd80
34 00000040`3607e130 00007fff`7c7eb42a     MFC42u!CWnd::WindowProc+0x4f
35 00000040`3607e170 00007fff`7c7ea76b     MFC42u!AfxCallWndProc+0x14a
36 00000040`3607e270 00007ff8`0be38161     MFC42u!AfxWndProcBase+0x15b
37 00000040`3607e2e0 00007ff8`0be37e1c     USER32!UserCallWinProcCheckWow+0x2d1
38 00000040`3607e440 00007ff8`0be42cdd     USER32!DispatchClientMessage+0x9c
39 00000040`3607e4a0 00007ff8`0c1d2db4     USER32!_fnDWORD+0x3d
3a 00000040`3607e500 00007ff8`097f1ef4     ntdll!KiUserCallbackDispatcherContinue
3b 00000040`3607e588 00007fff`7c7e21cd     win32u!NtUserShowWindow+0x14
3c 00000040`3607e590 00007ff7`f38346b6     MFC42u!CFrameWnd::ActivateFrame+0x2d
3d 00000040`3607e5c0 00007fff`7c7e25de     wordpad!CMainFrame::ActivateFrame+0x86
3e 00000040`3607e630 00007fff`7c7fa2e7     MFC42u!CFrameWnd::InitialUpdateFrame+0xfe
3f 00000040`3607e680 00007ff7`f383e331     MFC42u!CSingleDocTemplate::OpenDocumentFile+0x187
40 00000040`3607e6e0 00007fff`7c7efa57     wordpad!CWordPadApp::InitInstance+0x521
41 00000040`3607fc80 00007ff7`f37fb49e     MFC42u!AfxWinMain+0x97
42 00000040`3607fcc0 00007ff8`0b03244d     wordpad!__wmainCRTStartup+0x1de
43 00000040`3607fd80 00007ff8`0c18df78     KERNEL32!BaseThreadInitThunk+0x1d
44 00000040`3607fdb0 00000000`00000000     ntdll!RtlUserThreadStart+0x28
 # Child-SP          RetAddr               Call Site
00 00000040`36079f58 00007fff`f3cce048     USER32!CreateWindowExW
01 00000040`36079f60 00007fff`f3ccdf1f     urlmon!NotificationWindow::CreateHandle+0x68
02 00000040`36079fd0 00007fff`f3c7cc9b     urlmon!CTransaction::GetNotificationWnd+0x13f
03 00000040`3607a010 00007fff`f3cb98ee     urlmon!GetTransactionObjects+0x2fb
04 00000040`3607a090 00007fff`f3cba81c     urlmon!CBinding::StartBinding+0x53e
05 00000040`3607a210 00007fff`f3cb9276     urlmon!CUrlMon::StartBinding+0x1b0
06 00000040`3607a2e0 00007fff`f1c05529     urlmon!CUrlMon::BindToStorage+0x96
07 00000040`3607a330 00007fff`f1c04acb     msxml6!URLMONStream::deferedOpen+0x159
08 00000040`3607a390 00007fff`f1c048df     msxml6!XMLParser::PushURL+0x1af
09 00000040`3607a410 00007fff`f1bbf1c6     msxml6!XMLParser::SetURL+0x7f
0a 00000040`3607a460 00007fff`f1bbed48     msxml6!Document::_load+0x162
0b 00000040`3607a4e0 00007fff`f1b87712     msxml6!Document::load+0x198
0c 00000040`3607a550 00007fff`7c46afae     msxml6!DOMDocumentWrapper::load+0x182
0d 00000040`3607a620 00007fff`7c414813     PrintConfig!PrintConfig::PrivateDevmodeMapFile::Parse+0x14e
0e 00000040`3607a790 00007fff`7c5583f0     PrintConfig!PrintConfigPlugIn::DevMode+0x303
0f 00000040`3607a8b0 00007fff`7c55956a     PrintConfig!BCallOEMDevMode+0x84
10 00000040`3607a930 00007fff`7c514999     PrintConfig!PGetDefaultDevmodeWithOemPlugins+0x16a
11 00000040`3607a9b0 00007fff`7c4fc131     PrintConfig!UniDrvUI::BGetValidatedUserDefaultDevmode+0x81
12 00000040`3607aa20 00007fff`7c4f60a0     PrintConfig!UniDrvUI::CPrintTicketProvider::GetValidatedDevmode+0x6d
13 00000040`3607aa70 00007fff`7c4f98a8     PrintConfig!UniDrvUI::CPrintTicketProvider::GenerateCoreDriverPrintCapabilities+0x84
14 00000040`3607b3b0 00007fff`7c412fc9     PrintConfig!UniDrvUI::CPrintTicketProvider::GetPrintCapabilities+0x98
15 00000040`3607b470 00007fff`d586443f     PrintConfig!PrintConfig::PrintConfigPTProvider::GetPrintCapabilities+0x69
16 00000040`3607b4c0 00007fff`d586c501     prntvpt!PTGetPrintCapabilitiesImp+0x14b
17 00000040`3607b550 00007fff`d5864400     prntvpt!TProviderInfo::GetSupportedFeaturesList+0x61
18 00000040`3607b5c0 00007fff`d58541fb     prntvpt!PTGetPrintCapabilitiesImp+0x10c
19 00000040`3607b650 00007fff`d5857bd4     prntvpt!CPrintTicketServerBase::GetPrintCapabilities+0x5b
1a 00000040`3607b690 00007fff`7c48f310     prntvpt!PTGetPrintCapabilities+0x94
1b 00000040`3607b710 00007fff`7c48f29f     PrintConfig!PrintConfig::CSystemPrintTicketService::GetPrintCapabilities+0x60
1c 00000040`3607b740 00007fff`7c4bfc00     PrintConfig!PrintConfig::CPrintTicketServiceDecorator::GetPrintCapabilities+0x1f
1d 00000040`3607b780 00007fff`7c4c0aad     PrintConfig!PrintConfig::WithPrintTicketServiceDo<<lambda_49af5cf54d102c592666395395ebbfac> >+0x78
1e 00000040`3607b7c0 00007fff`7c4cc122     PrintConfig!PrintConfig::CPrintTicketService2::GetPrintCapabilities+0x21d
1f 00000040`3607b850 00007fff`7c4cc01d     PrintConfig!PrintConfig::CPrintCapabilities::EnsureXmlDocument+0x4a
20 00000040`3607b8a0 00007fff`7c4cd552     PrintConfig!PrintConfig::CPrintCapabilities::EnsurePrintSchema+0x35
21 00000040`3607b8f0 00007fff`7c417389     PrintConfig!PrintConfig::CPrintCapabilities::get_PageImageableSize+0x82
22 00000040`3607b9b0 00007fff`7c418a54     PrintConfig!PrintConfig::ExtractImageableArea+0x39
23 00000040`3607ba00 00007fff`7c4248a4     PrintConfig!PrintConfig::V4MxdcGetPDEVAdjustment+0xa04
24 00000040`3607bcc0 00007fff`7c424c33     PrintConfig!ExceptionBoundary<<lambda_4bfcdea1f8082feba17903613e8e1079> >+0x54
25 00000040`3607bd30 00007fff`a4de9cc7     PrintConfig!MxdcGetPDEVAdjustment+0x93
26 00000040`3607bdd0 00007fff`a4dead1f     mxdwdrv!edocs::PDev::InvokeMxdcGetPDEVAdjustment+0xb3
27 00000040`3607c2b0 00007fff`a4ddb5f0     mxdwdrv!edocs::PDev::ValidateForm+0x1fb
28 00000040`3607c7b0 00007ff8`09876470     mxdwdrv!DrvEnablePDEV+0x190
29 00000040`3607d0f0 00007ff8`0bea991a     gdi32full!GdiPrinterThunk+0x370
2a 00000040`3607d1c0 00007ff8`0c1d2db4     USER32!_ClientPrinterThunk+0x3a
2b 00000040`3607da40 00007ff8`097f2d94     ntdll!KiUserCallbackDispatcherContinue
2c 00000040`3607db48 00007ff8`0983dbd2     win32u!NtGdiOpenDCW+0x14
2d 00000040`3607db50 00007ff8`0a5f166d     gdi32full!hdcCreateDCW+0xb2
2e 00000040`3607dea0 00007ff8`0a5f1545     GDI32!bCreateDCW+0xed
2f 00000040`3607df10 00007fff`7c7ff0c7     GDI32!CreateDCW+0x55
30 00000040`3607df50 00007fff`7c7ff020     MFC42u!AfxCreateDC+0x67
31 00000040`3607df80 00007ff7`f3852407     MFC42u!CWinApp::CreatePrinterDC+0x20
32 00000040`3607dfb0 00007fff`7c7e9920     wordpad!CWordPadView::OnPrinterChangedMsg+0x37
33 00000040`3607e000 00007fff`7c7f0bff     MFC42u!CWnd::OnWndMsg+0xd80
34 00000040`3607e130 00007fff`7c7eb42a     MFC42u!CWnd::WindowProc+0x4f
35 00000040`3607e170 00007fff`7c7ea76b     MFC42u!AfxCallWndProc+0x14a
36 00000040`3607e270 00007ff8`0be38161     MFC42u!AfxWndProcBase+0x15b
37 00000040`3607e2e0 00007ff8`0be37e1c     USER32!UserCallWinProcCheckWow+0x2d1
38 00000040`3607e440 00007ff8`0be42cdd     USER32!DispatchClientMessage+0x9c
39 00000040`3607e4a0 00007ff8`0c1d2db4     USER32!_fnDWORD+0x3d
3a 00000040`3607e500 00007ff8`097f1ef4     ntdll!KiUserCallbackDispatcherContinue
3b 00000040`3607e588 00007fff`7c7e21cd     win32u!NtUserShowWindow+0x14
3c 00000040`3607e590 00007ff7`f38346b6     MFC42u!CFrameWnd::ActivateFrame+0x2d
3d 00000040`3607e5c0 00007fff`7c7e25de     wordpad!CMainFrame::ActivateFrame+0x86
3e 00000040`3607e630 00007fff`7c7fa2e7     MFC42u!CFrameWnd::InitialUpdateFrame+0xfe
3f 00000040`3607e680 00007ff7`f383e331     MFC42u!CSingleDocTemplate::OpenDocumentFile+0x187
40 00000040`3607e6e0 00007fff`7c7efa57     wordpad!CWordPadApp::InitInstance+0x521
41 00000040`3607fc80 00007ff7`f37fb49e     MFC42u!AfxWinMain+0x97
42 00000040`3607fcc0 00007ff8`0b03244d     wordpad!__wmainCRTStartup+0x1de
```

```
43 00000040`3607fd80 00007ff8`0c18df78     KERNEL32!BaseThreadInitThunk+0x1d
44 00000040`3607fdb0 00000000`00000000     ntdll!RtlUserThreadStart+0x28
ModLoad: 00007fff`aef80000 00007fff`aefa3000   C:\WINDOWS\system32\compstui.dll
ModLoad: 00007fff`e1470000 00007fff`e1477000   C:\WINDOWS\SYSTEM32\MSIMG32.dll
ModLoad: 00007fff`e13e0000 00007fff`e13fd000   C:\WINDOWS\SYSTEM32\amsi.dll
 # Child-SP          RetAddr               Call Site
00 00000040`3607a058 00007fff`f3cce048     USER32!CreateWindowExW
01 00000040`3607a060 00007fff`f3ccdf1f     urlmon!NotificationWindow::CreateHandle+0x68
02 00000040`3607a0d0 00007fff`f3c7cc9b     urlmon!CTransaction::GetNotificationWnd+0x13f
03 00000040`3607a110 00007fff`f3cb98ee     urlmon!GetTransactionObjects+0x2fb
04 00000040`3607a190 00007fff`f3cba81c     urlmon!CBinding::StartBinding+0x53e
05 00000040`3607a310 00007fff`f3cb9276     urlmon!CUrlMon::StartBinding+0x1b0
06 00000040`3607a3e0 00007fff`f1c05529     urlmon!CUrlMon::BindToStorage+0x96
07 00000040`3607a430 00007fff`f1c04acb     msxml6!URLMONStream::deferedOpen+0x159
08 00000040`3607a490 00007fff`f1c048df     msxml6!XMLParser::PushURL+0x1af
09 00000040`3607a510 00007fff`f1bbf1c6     msxml6!XMLParser::SetURL+0x7f
0a 00000040`3607a560 00007fff`f1bbed48     msxml6!Document::_load+0x162
0b 00000040`3607a5e0 00007fff`f1b87712     msxml6!Document::load+0x198
0c 00000040`3607a650 00007fff`7c46afae     msxml6!DOMDocumentWrapper::load+0x182
0d 00000040`3607a720 00007fff`7c414813     PrintConfig!PrintConfig::PrivateDevmodeMapFile::Parse+0x14e
0e 00000040`3607a890 00007fff`7c5583f0     PrintConfig!PrintConfigPlugIn::DevMode+0x303
0f 00000040`3607a9b0 00007fff`7c55956a     PrintConfig!BCallOEMDevMode+0x84
10 00000040`3607aa30 00007fff`7c514999     PrintConfig!PGetDefaultDevmodeWithOemPlugins+0x16a
11 00000040`3607aab0 00007fff`7c4fc131     PrintConfig!UniDrvUI::BGetValidatedUserDefaultDevmode+0x81
12 00000040`3607ab20 00007fff`7c4f60a0     PrintConfig!UniDrvUI::CPrintTicketProvider::GetValidatedDevmode+0x6d
13 00000040`3607ab70 00007fff`7c4f98a8     PrintConfig!UniDrvUI::CPrintTicketProvider::GenerateCoreDriverPrintCapabilities+0x84
14 00000040`3607b4b0 00007fff`7c412fc9     PrintConfig!UniDrvUI::CPrintTicketProvider::GetPrintCapabilities+0x98
15 00000040`3607b570 00007fff`d586443f     PrintConfig!PrintConfig::PrintConfigPTProvider::GetPrintCapabilities+0x69
16 00000040`3607b5c0 00007fff`d58541fb     prntvpt!PTGetPrintCapabilitiesImp+0x14b
17 00000040`3607b650 00007fff`d5857bd4     prntvpt!CPrintTicketServerBase::GetPrintCapabilities+0x5b
18 00000040`3607b690 00007fff`7c48f310     prntvpt!PTGetPrintCapabilities+0x94
19 00000040`3607b710 00007fff`7c48f29f     PrintConfig!PrintConfig::CSystemPrintTicketService::GetPrintCapabilities+0x60
1a 00000040`3607b740 00007fff`7c4bfc00     PrintConfig!PrintConfig::CPrintTicketServiceDecorator::GetPrintCapabilities+0x1f
1b 00000040`3607b780 00007fff`7c4c0aad     PrintConfig!PrintConfig::WithPrintTicketServiceDo<<lambda_49af5cf54d102c592666395395ebbfac> >+0x78
1c 00000040`3607b7c0 00007fff`7c4cc122     PrintConfig!PrintConfig::CPrintTicketService2::GetPrintCapabilities+0x21d
1d 00000040`3607b850 00007fff`7c4cc01d     PrintConfig!PrintConfig::CPrintCapabilities::EnsureXmlDocument+0x4a
1e 00000040`3607b8a0 00007fff`7c4cd552     PrintConfig!PrintConfig::CPrintCapabilities::EnsurePrintSchema+0x35
1f 00000040`3607b8f0 00007fff`7c417389     PrintConfig!PrintConfig::CPrintCapabilities::get_PageImageableSize+0x82
20 00000040`3607b9b0 00007fff`7c418a54     PrintConfig!PrintConfig::ExtractImageableArea+0x39
21 00000040`3607ba00 00007fff`7c4248a4     PrintConfig!PrintConfig::V4MxdcGetPDEVAdjustment+0xa04
22 00000040`3607bcc0 00007fff`7c424c33     PrintConfig!ExceptionBoundary<<lambda_4bfcdea1f8082feba17903613e8e1079> >+0x54
23 00000040`3607bd30 00007fff`a4de9cc7     PrintConfig!MxdcGetPDEVAdjustment+0x93
24 00000040`3607bdd0 00007fff`a4dead1f     mxdwdrv!edocs::PDev::InvokeMxdcGetPDEVAdjustment+0xb3
25 00000040`3607c2b0 00007fff`a4ddb5f0     mxdwdrv!edocs::PDev::ValidateForm+0x1fb
26 00000040`3607c7b0 00007ff8`09876470     mxdwdrv!DrvEnablePDEV+0x190
27 00000040`3607d0f0 00007ff8`0bea991a     gdi32full!GdiPrinterThunk+0x370
28 00000040`3607d1c0 00007ff8`0c1d2db4     USER32!_ClientPrinterThunk+0x3a
29 00000040`3607da40 00007ff8`097f2d94     ntdll!KiUserCallbackDispatcherContinue
2a 00000040`3607db48 00007ff8`0983dbd2     win32u!NtGdiOpenDCW+0x14
2b 00000040`3607db50 00007ff8`0a5f166d     gdi32full!hdcCreateDCW+0xb2
2c 00000040`3607dea0 00007ff8`0a5f1545     GDI32!bCreateDCW+0xed
2d 00000040`3607df10 00007fff`7c7ff0c7     GDI32!CreateDCW+0x55
2e 00000040`3607df50 00007fff`7c7ff020     MFC42u!AfxCreateDC+0x67
2f 00000040`3607df80 00007ff7`f3852407     MFC42u!CWinApp::CreatePrinterDC+0x20
30 00000040`3607dfb0 00007fff`7c7e9920     wordpad!CWordPadView::OnPrinterChangedMsg+0x37
31 00000040`3607e000 00007fff`7c7f0bff     MFC42u!CWnd::OnWndMsg+0xd80
32 00000040`3607e130 00007fff`7c7eb42a     MFC42u!CWnd::WindowProc+0x4f
33 00000040`3607e170 00007fff`7c7ea76b     MFC42u!AfxCallWndProc+0x14a
34 00000040`3607e270 00007ff8`0be38161     MFC42u!AfxWndProcBase+0x15b
35 00000040`3607e2e0 00007ff8`0be37e1c     USER32!UserCallWinProcCheckWow+0x2d1
36 00000040`3607e440 00007ff8`0be42cdd     USER32!DispatchClientMessage+0x9c
37 00000040`3607e4a0 00007ff8`0c1d2db4     USER32!_fnDWORD+0x3d
38 00000040`3607e500 00007ff8`097f1ef4     ntdll!KiUserCallbackDispatcherContinue
39 00000040`3607e588 00007fff`7c7e21cd     win32u!NtUserShowWindow+0x14
3a 00000040`3607e590 00007ff7`f38346b6     MFC42u!CFrameWnd::ActivateFrame+0x2d
3b 00000040`3607e5c0 00007fff`7c7e25de     wordpad!CMainFrame::ActivateFrame+0x86
3c 00000040`3607e630 00007fff`7c7fa2e7     MFC42u!CFrameWnd::InitialUpdateFrame+0xfe
3d 00000040`3607e680 00007ff7`f383e331     MFC42u!CSingleDocTemplate::OpenDocumentFile+0x187
3e 00000040`3607e6e0 00007fff`7c7efa57     wordpad!CWordPadApp::InitInstance+0x521
3f 00000040`3607fc80 00007ff7`f37fb49e     MFC42u!AfxWinMain+0x97
40 00000040`3607fcc0 00007ff8`0b03244d     wordpad!__wmainCRTStartup+0x1de
41 00000040`3607fd80 00007ff8`0c18df78     KERNEL32!BaseThreadInitThunk+0x1d
42 00000040`3607fdb0 00000000`00000000     ntdll!RtlUserThreadStart+0x28
ModLoad: 00007fff`e13e0000 00007fff`e13fd000   C:\WINDOWS\SYSTEM32\amsi.dll
 # Child-SP          RetAddr               Call Site
00 00000040`36079d08 00007fff`f3cce048     USER32!CreateWindowExW
01 00000040`36079d10 00007fff`f3ccdf1f     urlmon!NotificationWindow::CreateHandle+0x68
02 00000040`36079d80 00007fff`f3c7cc9b     urlmon!CTransaction::GetNotificationWnd+0x13f
03 00000040`36079dc0 00007fff`f3cb98ee     urlmon!GetTransactionObjects+0x2fb
04 00000040`36079e40 00007fff`f3cba81c     urlmon!CBinding::StartBinding+0x53e
05 00000040`36079fc0 00007fff`f3cb9276     urlmon!CUrlMon::StartBinding+0x1b0
06 00000040`3607a090 00007fff`f1c05529     urlmon!CUrlMon::BindToStorage+0x96
07 00000040`3607a0e0 00007fff`f1c04acb     msxml6!URLMONStream::deferedOpen+0x159
08 00000040`3607a140 00007fff`f1c048df     msxml6!XMLParser::PushURL+0x1af
09 00000040`3607a1c0 00007fff`f1bbf1c6     msxml6!XMLParser::SetURL+0x7f
0a 00000040`3607a210 00007fff`f1bbed48     msxml6!Document::_load+0x162
0b 00000040`3607a290 00007fff`f1b87712     msxml6!Document::load+0x198
0c 00000040`3607a300 00007fff`7c46afae     msxml6!DOMDocumentWrapper::load+0x182
0d 00000040`3607a3d0 00007fff`7c414968     PrintConfig!PrintConfig::PrivateDevmodeMapFile::Parse+0x14e
0e 00000040`3607a540 00007fff`7c558d7b     PrintConfig!PrintConfigPlugIn::DevMode+0x458
0f 00000040`3607a660 00007fff`7c55949b     PrintConfig!BCalcTotalOEMDMSize+0xa3
10 00000040`3607a700 00007fff`7c514999     PrintConfig!PGetDefaultDevmodeWithOemPlugins+0x9b
11 00000040`3607a780 00007fff`7c5144fa     PrintConfig!UniDrvUI::BGetValidatedUserDefaultDevmode+0x81
12 00000040`3607a7f0 00007fff`7c50d112     PrintConfig!UniDrvUI::BFillCommonInfoDevmode+0x82
13 00000040`3607a890 00007fff`7c50ce44     PrintConfig!UniDrvUI::DwDeviceCapabilities+0xc6
14 00000040`3607a940 00007fff`7c42514f     PrintConfig!UniDrvUI::DrvDeviceCapabilities+0x18
15 00000040`3607a980 00007fff`e03150a1     PrintConfig!DrvDeviceCapabilities+0x11f
16 00000040`3607aa00 00007fff`e0324a04     WINSPOOL!DeviceCapabilitiesWNative+0x105
```

```
17 00000040`3607aa60 00007fff`7c4feafb     WINSPOOL!DeviceCapabilitiesW+0x54
18 00000040`3607aaa0 00007fff`7c4f61e3     PrintConfig!publicdm::JobTicketToPublicDevmode+0x9f
19 00000040`3607ab70 00007fff`7c4f98a8     PrintConfig!UniDrvUI::CPrintTicketProvider::GenerateCoreDriverPrintCapabilities+0x1c7
1a 00000040`3607b4b0 00007fff`7c412fc9     PrintConfig!UniDrvUI::CPrintTicketProvider::GetPrintCapabilities+0x98
1b 00000040`3607b570 00007fff`d586443f     PrintConfig!PrintConfig::PrintConfigPTProvider::GetPrintCapabilities+0x69
1c 00000040`3607b5c0 00007fff`d58541fb     prntvpt!PTGetPrintCapabilitiesImp+0x14b
1d 00000040`3607b650 00007fff`d5857bd4     prntvpt!CPrintTicketServerBase::GetPrintCapabilities+0x5b
1e 00000040`3607b690 00007fff`7c48f310     prntvpt!PTGetPrintCapabilities+0x94
1f 00000040`3607b710 00007fff`7c48f29f     PrintConfig!PrintConfig::CSystemPrintTicketService::GetPrintCapabilities+0x60
20 00000040`3607b740 00007fff`7c4bfc00     PrintConfig!PrintConfig::CPrintTicketServiceDecorator::GetPrintCapabilities+0x1f
21 00000040`3607b780 00007fff`7c4c0aad     PrintConfig!PrintConfig::WithPrintTicketServiceDo<<lambda_49af5cf54d102c592666395395ebbfac> >+0x78
22 00000040`3607b7c0 00007fff`7c4cc122     PrintConfig!PrintConfig::CPrintTicketService2::GetPrintCapabilities+0x21d
23 00000040`3607b850 00007fff`7c4cc01d     PrintConfig!PrintConfig::CPrintCapabilities::EnsureXmlDocument+0x4a
24 00000040`3607b8a0 00007fff`7c4cd552     PrintConfig!PrintConfig::CPrintCapabilities::EnsurePrintSchema+0x35
25 00000040`3607b8f0 00007fff`7c417389     PrintConfig!PrintConfig::CPrintCapabilities::get_PageImageableSize+0x82
26 00000040`3607b9b0 00007fff`7c418a54     PrintConfig!PrintConfig::ExtractImageableArea+0x39
27 00000040`3607ba00 00007fff`7c4248a4     PrintConfig!PrintConfig::V4MxdcGetPDEVAdjustment+0xa04
28 00000040`3607bcc0 00007fff`7c424c33     PrintConfig!ExceptionBoundary<<lambda_4bfcdea1f8082feba17903613e8e1079> >+0x54
29 00000040`3607bd30 00007fff`a4de9cc7     PrintConfig!MxdcGetPDEVAdjustment+0x93
2a 00000040`3607bdd0 00007fff`a4dead1f     mxdwdrv!edocs::PDev::InvokeMxdcGetPDEVAdjustment+0xb3
2b 00000040`3607c2b0 00007fff`a4ddb5f0     mxdwdrv!edocs::PDev::ValidateForm+0x1fb
2c 00000040`3607c7b0 00007ff8`09876470     mxdwdrv!DrvEnablePDEV+0x190
2d 00000040`3607d0f0 00007ff8`0bea991a     gdi32full!GdiPrinterThunk+0x370
2e 00000040`3607d1c0 00007ff8`0c1d2db4     USER32!_ClientPrinterThunk+0x3a
2f 00000040`3607da40 00007ff8`097f2d94     ntdll!KiUserCallbackDispatcherContinue
30 00000040`3607db48 00007ff8`0983dbd2     win32u!NtGdiOpenDCW+0x14
31 00000040`3607db50 00007ff8`0a5f166d     gdi32full!hdcCreateDCW+0xb2
32 00000040`3607dea0 00007ff8`0a5f1545     GDI32!bCreateDCW+0xed
33 00000040`3607df10 00007fff`7c7ff0c7     GDI32!CreateDCW+0x55
34 00000040`3607df50 00007fff`7c7ff020     MFC42u!AfxCreateDC+0x67
35 00000040`3607df80 00007ff7`f3852407     MFC42u!CWinApp::CreatePrinterDC+0x20
36 00000040`3607dfb0 00007fff`7c7e9920     wordpad!CWordPadView::OnPrinterChangedMsg+0x37
37 00000040`3607e000 00007fff`7c7f0bff     MFC42u!CWnd::OnWndMsg+0xd80
38 00000040`3607e130 00007fff`7c7eb42a     MFC42u!CWnd::WindowProc+0x4f
39 00000040`3607e170 00007fff`7c7ea76b     MFC42u!AfxCallWndProc+0x14a
3a 00000040`3607e270 00007ff8`0be38161     MFC42u!AfxWndProcBase+0x15b
3b 00000040`3607e2e0 00007ff8`0be37e1c     USER32!UserCallWinProcCheckWow+0x2d1
3c 00000040`3607e440 00007ff8`0be42cdd     USER32!DispatchClientMessage+0x9c
3d 00000040`3607e4a0 00007ff8`0c1d2db4     USER32!_fnDWORD+0x3d
3e 00000040`3607e500 00007ff8`097f1ef4     ntdll!KiUserCallbackDispatcherContinue
3f 00000040`3607e588 00007fff`7c7e21cd     win32u!NtUserShowWindow+0x14
40 00000040`3607e590 00007ff7`f38346b6     MFC42u!CFrameWnd::ActivateFrame+0x2d
41 00000040`3607e5c0 00007fff`7c7e25de     wordpad!CMainFrame::ActivateFrame+0x86
42 00000040`3607e630 00007fff`7c7fa2e7     MFC42u!CFrameWnd::InitialUpdateFrame+0xfe
43 00000040`3607e680 00007ff7`f383e331     MFC42u!CSingleDocTemplate::OpenDocumentFile+0x187
44 00000040`3607e6e0 00007fff`7c7efa57     wordpad!CWordPadApp::InitInstance+0x521
45 00000040`3607fc80 00007ff7`f37fb49e     MFC42u!AfxWinMain+0x97
46 00000040`3607fcc0 00007ff8`0b03244d     wordpad!__wmainCRTStartup+0x1de
47 00000040`3607fd80 00007ff8`0c18df78     KERNEL32!BaseThreadInitThunk+0x1d
48 00000040`3607fdb0 00000000`00000000     ntdll!RtlUserThreadStart+0x28
 # Child-SP          RetAddr               Call Site
00 00000040`36079d28 00007fff`f3cce048     USER32!CreateWindowExW
01 00000040`36079d30 00007fff`f3ccdf1f     urlmon!NotificationWindow::CreateHandle+0x68
02 00000040`36079da0 00007fff`f3c7cc9b     urlmon!CTransaction::GetNotificationWnd+0x13f
03 00000040`36079de0 00007fff`f3cb98ee     urlmon!GetTransactionObjects+0x2fb
04 00000040`36079e60 00007fff`f3cba81c     urlmon!CBinding::StartBinding+0x53e
05 00000040`36079fe0 00007fff`f3cb9276     urlmon!CUrlMon::StartBinding+0x1b0
06 00000040`3607a0b0 00007fff`f1c05529     urlmon!CUrlMon::BindToStorage+0x96
07 00000040`3607a100 00007fff`f1c04acb     msxml6!URLMONStream::deferedOpen+0x159
08 00000040`3607a160 00007fff`f1c048df     msxml6!XMLParser::PushURL+0x1af
09 00000040`3607a1e0 00007fff`f1bbf1c6     msxml6!XMLParser::SetURL+0x7f
0a 00000040`3607a230 00007fff`f1bbed48     msxml6!Document::_load+0x162
0b 00000040`3607a2b0 00007fff`f1b87712     msxml6!Document::load+0x198
0c 00000040`3607a320 00007fff`7c46afae     msxml6!DOMDocumentWrapper::load+0x182
0d 00000040`3607a3f0 00007fff`7c414813     PrintConfig!PrintConfig::PrivateDevmodeMapFile::Parse+0x14e
0e 00000040`3607a560 00007fff`7c5583f0     PrintConfig!PrintConfigPlugIn::DevMode+0x303
0f 00000040`3607a680 00007fff`7c55956a     PrintConfig!BCallOEMDevMode+0x84
10 00000040`3607a700 00007fff`7c514999     PrintConfig!PGetDefaultDevmodeWithOemPlugins+0x16a
11 00000040`3607a780 00007fff`7c5144fa     PrintConfig!UniDrvUI::BGetValidatedUserDefaultDevmode+0x81
12 00000040`3607a7f0 00007fff`7c50d112     PrintConfig!UniDrvUI::BFillCommonInfoDevmode+0x82
13 00000040`3607a890 00007fff`7c50ce44     PrintConfig!UniDrvUI::DwDeviceCapabilities+0xc6
14 00000040`3607a940 00007fff`7c42514f     PrintConfig!UniDrvUI::DrvDeviceCapabilities+0x18
15 00000040`3607a980 00007fff`e03150a1     PrintConfig!DrvDeviceCapabilities+0x11f
16 00000040`3607aa00 00007fff`e0324a04     WINSPOOL!DeviceCapabilitiesWNative+0x105
17 00000040`3607aa60 00007fff`7c4feafb     WINSPOOL!DeviceCapabilitiesW+0x54
18 00000040`3607aaa0 00007fff`7c4f61e3     PrintConfig!publicdm::JobTicketToPublicDevmode+0x9f
19 00000040`3607ab70 00007fff`7c4f98a8     PrintConfig!UniDrvUI::CPrintTicketProvider::GenerateCoreDriverPrintCapabilities+0x1c7
1a 00000040`3607b4b0 00007fff`7c412fc9     PrintConfig!UniDrvUI::CPrintTicketProvider::GetPrintCapabilities+0x98
1b 00000040`3607b570 00007fff`d586443f     PrintConfig!PrintConfig::PrintConfigPTProvider::GetPrintCapabilities+0x69
1c 00000040`3607b5c0 00007fff`d58541fb     prntvpt!PTGetPrintCapabilitiesImp+0x14b
1d 00000040`3607b650 00007fff`d5857bd4     prntvpt!CPrintTicketServerBase::GetPrintCapabilities+0x5b
1e 00000040`3607b690 00007fff`7c48f310     prntvpt!PTGetPrintCapabilities+0x94
1f 00000040`3607b710 00007fff`7c48f29f     PrintConfig!PrintConfig::CSystemPrintTicketService::GetPrintCapabilities+0x60
20 00000040`3607b740 00007fff`7c4bfc00     PrintConfig!PrintConfig::CPrintTicketServiceDecorator::GetPrintCapabilities+0x1f
21 00000040`3607b780 00007fff`7c4c0aad     PrintConfig!PrintConfig::WithPrintTicketServiceDo<<lambda_49af5cf54d102c592666395395ebbfac> >+0x78
22 00000040`3607b7c0 00007fff`7c4cc122     PrintConfig!PrintConfig::CPrintTicketService2::GetPrintCapabilities+0x21d
23 00000040`3607b850 00007fff`7c4cc01d     PrintConfig!PrintConfig::CPrintCapabilities::EnsureXmlDocument+0x4a
24 00000040`3607b8a0 00007fff`7c4cd552     PrintConfig!PrintConfig::CPrintCapabilities::EnsurePrintSchema+0x35
25 00000040`3607b8f0 00007fff`7c417389     PrintConfig!PrintConfig::CPrintCapabilities::get_PageImageableSize+0x82
26 00000040`3607b9b0 00007fff`7c418a54     PrintConfig!PrintConfig::ExtractImageableArea+0x39
27 00000040`3607ba00 00007fff`7c4248a4     PrintConfig!PrintConfig::V4MxdcGetPDEVAdjustment+0xa04
28 00000040`3607bcc0 00007fff`7c424c33     PrintConfig!ExceptionBoundary<<lambda_4bfcdea1f8082feba17903613e8e1079> >+0x54
29 00000040`3607bd30 00007fff`a4de9cc7     PrintConfig!MxdcGetPDEVAdjustment+0x93
2a 00000040`3607bdd0 00007fff`a4dead1f     mxdwdrv!edocs::PDev::InvokeMxdcGetPDEVAdjustment+0xb3
2b 00000040`3607c2b0 00007fff`a4ddb5f0     mxdwdrv!edocs::PDev::ValidateForm+0x1fb
2c 00000040`3607c7b0 00007ff8`09876470     mxdwdrv!DrvEnablePDEV+0x190
2d 00000040`3607d0f0 00007ff8`0bea991a     gdi32full!GdiPrinterThunk+0x370
2e 00000040`3607d1c0 00007ff8`0c1d2db4     USER32!_ClientPrinterThunk+0x3a
```

```
2f 00000040`3607da40 00007ff8`097f2d94     ntdll!KiUserCallbackDispatcherContinue
30 00000040`3607db48 00007ff8`0983dbd2     win32u!NtGdiOpenDCW+0x14
31 00000040`3607db50 00007ff8`0a5f166d     gdi32full!hdcCreateDCW+0xb2
32 00000040`3607dea0 00007ff8`0a5f1545     GDI32!bCreateDCW+0xed
33 00000040`3607df10 00007fff`7c7ff0c7     GDI32!CreateDCW+0x55
34 00000040`3607df50 00007fff`7c7ff020     MFC42u!AfxCreateDC+0x67
35 00000040`3607df80 00007ff7`f3852407     MFC42u!CWinApp::CreatePrinterDC+0x20
36 00000040`3607dfb0 00007fff`7c7e9920     wordpad!CWordPadView::OnPrinterChangedMsg+0x37
37 00000040`3607e000 00007fff`7c7f0bff     MFC42u!CWnd::OnWndMsg+0xd80
38 00000040`3607e130 00007fff`7c7eb42a     MFC42u!CWnd::WindowProc+0x4f
39 00000040`3607e170 00007fff`7c7ea76b     MFC42u!AfxCallWndProc+0x14a
3a 00000040`3607e270 00007ff8`0be38161     MFC42u!AfxWndProcBase+0x15b
3b 00000040`3607e2e0 00007ff8`0be37e1c     USER32!UserCallWinProcCheckWow+0x2d1
3c 00000040`3607e440 00007ff8`0be42cdd     USER32!DispatchClientMessage+0x9c
3d 00000040`3607e4a0 00007ff8`0c1d2db4     USER32!_fnDWORD+0x3d
3e 00000040`3607e500 00007ff8`097f1ef4     ntdll!KiUserCallbackDispatcherContinue
3f 00000040`3607e588 00007fff`7c7e21cd     win32u!NtUserShowWindow+0x14
40 00000040`3607e590 00007ff7`f38346b6     MFC42u!CFrameWnd::ActivateFrame+0x2d
41 00000040`3607e5c0 00007fff`7c7e25de     wordpad!CMainFrame::ActivateFrame+0x86
42 00000040`3607e630 00007fff`7c7fa2e7     MFC42u!CFrameWnd::InitialUpdateFrame+0xfe
43 00000040`3607e680 00007ff7`f383e331     MFC42u!CSingleDocTemplate::OpenDocumentFile+0x187
44 00000040`3607e6e0 00007fff`7c7efa57     wordpad!CWordPadApp::InitInstance+0x521
45 00000040`3607fc80 00007ff7`f37fb49e     MFC42u!AfxWinMain+0x97
46 00000040`3607fcc0 00007ff8`0b03244d     wordpad!__wmainCRTStartup+0x1de
47 00000040`3607fd80 00007ff8`0c18df78     KERNEL32!BaseThreadInitThunk+0x1d
48 00000040`3607fdb0 00000000`00000000     ntdll!RtlUserThreadStart+0x28
ModLoad: 00007fff`aef80000 00007fff`aefa3000   C:\WINDOWS\system32\compstui.dll
ModLoad: 00007fff`e1470000 00007fff`e1477000   C:\WINDOWS\SYSTEM32\MSIMG32.dll
ModLoad: 00007fff`e13e0000 00007fff`e13fd000   C:\WINDOWS\SYSTEM32\amsi.dll
 # Child-SP          RetAddr               Call Site
00 00000040`3607c9c8 00007fff`f3cce048     USER32!CreateWindowExW
01 00000040`3607c9d0 00007fff`f3ccdf1f     urlmon!NotificationWindow::CreateHandle+0x68
02 00000040`3607ca40 00007fff`f3c7cc9b     urlmon!CTransaction::GetNotificationWnd+0x13f
03 00000040`3607ca80 00007fff`f3cb98ee     urlmon!GetTransactionObjects+0x2fb
04 00000040`3607cb00 00007fff`f3cba81c     urlmon!CBinding::StartBinding+0x53e
05 00000040`3607cc80 00007fff`f3cb9276     urlmon!CUrlMon::StartBinding+0x1b0
06 00000040`3607cd50 00007fff`f1c05529     urlmon!CUrlMon::BindToStorage+0x96
07 00000040`3607cda0 00007fff`f1c04acb     msxml6!URLMONStream::deferedOpen+0x159
08 00000040`3607ce00 00007fff`f1c048df     msxml6!XMLParser::PushURL+0x1af
09 00000040`3607ce80 00007fff`f1bbf1c6     msxml6!XMLParser::SetURL+0x7f
0a 00000040`3607ced0 00007fff`f1bbed48     msxml6!Document::_load+0x162
0b 00000040`3607cf50 00007fff`f1b87712     msxml6!Document::load+0x198
0c 00000040`3607cfc0 00007fff`7c46afae     msxml6!DOMDocumentWrapper::load+0x182
0d 00000040`3607d090 00007fff`7c414968     PrintConfig!PrintConfig::PrivateDevmodeMapFile::Parse+0x14e
0e 00000040`3607d200 00007fff`7c558d7b     PrintConfig!PrintConfigPlugIn::DevMode+0x458
0f 00000040`3607d320 00007fff`7c512b53     PrintConfig!BCalcTotalOEMDMSize+0xa3
10 00000040`3607d3c0 00007fff`7c5120ec     PrintConfig!UniDrvUI::LSimpleDocumentProperties+0x10f
11 00000040`3607d430 00007fff`7c40f54b     PrintConfig!UniDrvUI::DrvDocumentPropertySheets+0x60
12 00000040`3607d8e0 00007fff`7c4249ad     PrintConfig!PrintConfig::DrvDocumentPropertySheets+0x123
13 00000040`3607d950 00007fff`7c425543     PrintConfig!ExceptionBoundary<<lambda_ce41999e58d4ed069b2790d632f6e3b5> >+0x3d
14 00000040`3607d9a0 00007fff`e032e9ba     PrintConfig!DrvDocumentPropertySheets+0x53
15 00000040`3607da10 00007fff`e031593f     WINSPOOL!DocumentPropertySheets+0x3fa
16 00000040`3607da60 00007fff`e0324aff     WINSPOOL!DocumentPropertiesWNative+0x193
17 00000040`3607daf0 00007ff8`098585d8     WINSPOOL!DocumentPropertiesW+0xcf
18 00000040`3607db50 00007ff8`0a5f166d     gdi32full!hdcCreateDCW+0x1aab8
19 00000040`3607dea0 00007ff8`0a5f1545     GDI32!bCreateDCW+0xed
1a 00000040`3607df10 00007fff`7c7ff0c7     GDI32!CreateDCW+0x55
1b 00000040`3607df50 00007fff`7c7ff020     MFC42u!AfxCreateDC+0x67
1c 00000040`3607df80 00007ff7`f3852407     MFC42u!CWinApp::CreatePrinterDC+0x20
1d 00000040`3607dfb0 00007fff`7c7e9920     wordpad!CWordPadView::OnPrinterChangedMsg+0x37
1e 00000040`3607e000 00007fff`7c7f0bff     MFC42u!CWnd::OnWndMsg+0xd80
1f 00000040`3607e130 00007fff`7c7eb42a     MFC42u!CWnd::WindowProc+0x4f
20 00000040`3607e170 00007fff`7c7ea76b     MFC42u!AfxCallWndProc+0x14a
21 00000040`3607e270 00007ff8`0be38161     MFC42u!AfxWndProcBase+0x15b
22 00000040`3607e2e0 00007ff8`0be37e1c     USER32!UserCallWinProcCheckWow+0x2d1
23 00000040`3607e440 00007ff8`0be42cdd     USER32!DispatchClientMessage+0x9c
24 00000040`3607e4a0 00007ff8`0c1d2db4     USER32!_fnDWORD+0x3d
25 00000040`3607e500 00007ff8`097f1ef4     ntdll!KiUserCallbackDispatcherContinue
26 00000040`3607e588 00007fff`7c7e21cd     win32u!NtUserShowWindow+0x14
27 00000040`3607e590 00007ff7`f38346b6     MFC42u!CFrameWnd::ActivateFrame+0x2d
28 00000040`3607e5c0 00007fff`7c7e25de     wordpad!CMainFrame::ActivateFrame+0x86
29 00000040`3607e630 00007fff`7c7fa2e7     MFC42u!CFrameWnd::InitialUpdateFrame+0xfe
2a 00000040`3607e680 00007ff7`f383e331     MFC42u!CSingleDocTemplate::OpenDocumentFile+0x187
2b 00000040`3607e6e0 00007fff`7c7efa57     wordpad!CWordPadApp::InitInstance+0x521
2c 00000040`3607fc80 00007ff7`f37fb49e     MFC42u!AfxWinMain+0x97
2d 00000040`3607fcc0 00007ff8`0b03244d     wordpad!__wmainCRTStartup+0x1de
2e 00000040`3607fd80 00007ff8`0c18df78     KERNEL32!BaseThreadInitThunk+0x1d
2f 00000040`3607fdb0 00000000`00000000     ntdll!RtlUserThreadStart+0x28
 # Child-SP          RetAddr               Call Site
00 00000040`3607c9c8 00007fff`f3cce048     USER32!CreateWindowExW
01 00000040`3607c9d0 00007fff`f3ccdf1f     urlmon!NotificationWindow::CreateHandle+0x68
02 00000040`3607ca40 00007fff`f3c7cc9b     urlmon!CTransaction::GetNotificationWnd+0x13f
03 00000040`3607ca80 00007fff`f3cb98ee     urlmon!GetTransactionObjects+0x2fb
04 00000040`3607cb00 00007fff`f3cba81c     urlmon!CBinding::StartBinding+0x53e
05 00000040`3607cc80 00007fff`f3cb9276     urlmon!CUrlMon::StartBinding+0x1b0
06 00000040`3607cd50 00007fff`f1c05529     urlmon!CUrlMon::BindToStorage+0x96
07 00000040`3607cda0 00007fff`f1c04acb     msxml6!URLMONStream::deferedOpen+0x159
08 00000040`3607ce00 00007fff`f1c048df     msxml6!XMLParser::PushURL+0x1af
09 00000040`3607ce80 00007fff`f1bbf1c6     msxml6!XMLParser::SetURL+0x7f
0a 00000040`3607ced0 00007fff`f1bbed48     msxml6!Document::_load+0x162
0b 00000040`3607cf50 00007fff`f1b87712     msxml6!Document::load+0x198
0c 00000040`3607cfc0 00007fff`7c46afae     msxml6!DOMDocumentWrapper::load+0x182
0d 00000040`3607d090 00007fff`7c414968     PrintConfig!PrintConfig::PrivateDevmodeMapFile::Parse+0x14e
0e 00000040`3607d200 00007fff`7c558d7b     PrintConfig!PrintConfigPlugIn::DevMode+0x458
0f 00000040`3607d320 00007fff`7c512b53     PrintConfig!BCalcTotalOEMDMSize+0xa3
10 00000040`3607d3c0 00007fff`7c5120ec     PrintConfig!UniDrvUI::LSimpleDocumentProperties+0x10f
11 00000040`3607d430 00007fff`7c40f54b     PrintConfig!UniDrvUI::DrvDocumentPropertySheets+0x60
12 00000040`3607d8e0 00007fff`7c4249ad     PrintConfig!PrintConfig::DrvDocumentPropertySheets+0x123
```

```
13 00000040`3607d950 00007fff`7c425543     PrintConfig!ExceptionBoundary<<lambda_ce41999e58d4ed069b2790d632f6e3b5> >+0x3d
14 00000040`3607d9a0 00007fff`e032e9ba     PrintConfig!DrvDocumentPropertySheets+0x53
15 00000040`3607da10 00007fff`e0315895     WINSPOOL!DocumentPropertySheets+0x3fa
16 00000040`3607da60 00007fff`e0324aff     WINSPOOL!DocumentPropertiesWNative+0xe9
17 00000040`3607daf0 00007ff8`0985865c     WINSPOOL!DocumentPropertiesW+0xcf
18 00000040`3607db50 00007ff8`0a5f166d     gdi32full!hdcCreateDCW+0x1ab3c
19 00000040`3607dea0 00007ff8`0a5f1545     GDI32!bCreateDCW+0xed
1a 00000040`3607df10 00007fff`7c7ff0c7     GDI32!CreateDCW+0x55
1b 00000040`3607df50 00007fff`7c7ff020     MFC42u!AfxCreateDC+0x67
1c 00000040`3607df80 00007ff7`f3852407     MFC42u!CWinApp::CreatePrinterDC+0x20
1d 00000040`3607dfb0 00007fff`7c7e9920     wordpad!CWordPadView::OnPrinterChangedMsg+0x37
1e 00000040`3607e000 00007fff`7c7f0bff     MFC42u!CWnd::OnWndMsg+0xd80
1f 00000040`3607e130 00007fff`7c7eb42a     MFC42u!CWnd::WindowProc+0x4f
20 00000040`3607e170 00007fff`7c7ea76b     MFC42u!AfxCallWndProc+0x14a
21 00000040`3607e270 00007ff8`0be38161     MFC42u!AfxWndProcBase+0x15b
22 00000040`3607e2e0 00007ff8`0be37e1c     USER32!UserCallWinProcCheckWow+0x2d1
23 00000040`3607e440 00007ff8`0be42cdd     USER32!DispatchClientMessage+0x9c
24 00000040`3607e4a0 00007ff8`0c1d2db4     USER32!_fnDWORD+0x3d
25 00000040`3607e500 00007ff8`097f1ef4     ntdll!KiUserCallbackDispatcherContinue
26 00000040`3607e588 00007fff`7c7e21cd     win32u!NtUserShowWindow+0x14
27 00000040`3607e590 00007ff7`f38346b6     MFC42u!CFrameWnd::ActivateFrame+0x2d
28 00000040`3607e5c0 00007fff`7c7e25de     wordpad!CMainFrame::ActivateFrame+0x86
29 00000040`3607e630 00007fff`7c7fa2e7     MFC42u!CFrameWnd::InitialUpdateFrame+0xfe
2a 00000040`3607e680 00007ff7`f383e331     MFC42u!CSingleDocTemplate::OpenDocumentFile+0x187
2b 00000040`3607e6e0 00007fff`7c7efa57     wordpad!CWordPadApp::InitInstance+0x521
2c 00000040`3607fc80 00007ff7`f37fb49e     MFC42u!AfxWinMain+0x97
2d 00000040`3607fcc0 00007ff8`0b03244d     wordpad!__wmainCRTStartup+0x1de
2e 00000040`3607fd80 00007ff8`0c18df78     KERNEL32!BaseThreadInitThunk+0x1d
2f 00000040`3607fdb0 00000000`00000000     ntdll!RtlUserThreadStart+0x28
 # Child-SP          RetAddr               Call Site
00 00000040`3607c9c8 00007fff`f3cce048     USER32!CreateWindowExW
01 00000040`3607c9d0 00007fff`f3ccdf1f     urlmon!NotificationWindow::CreateHandle+0x68
02 00000040`3607ca40 00007fff`f3c7cc9b     urlmon!CTransaction::GetNotificationWnd+0x13f
03 00000040`3607ca80 00007fff`f3cb98ee     urlmon!GetTransactionObjects+0x2fb
04 00000040`3607cb00 00007fff`f3cba81c     urlmon!CBinding::StartBinding+0x53e
05 00000040`3607cc80 00007fff`f3cb9276     urlmon!CUrlMon::StartBinding+0x1b0
06 00000040`3607cd50 00007fff`f1c05529     urlmon!CUrlMon::BindToStorage+0x96
07 00000040`3607cda0 00007fff`f1c04acb     msxml6!URLMONStream::deferedOpen+0x159
08 00000040`3607ce00 00007fff`f1c048df     msxml6!XMLParser::PushURL+0x1af
09 00000040`3607ce80 00007fff`f1bbf1c6     msxml6!XMLParser::SetURL+0x7f
0a 00000040`3607ced0 00007fff`f1bbed48     msxml6!Document::_load+0x162
0b 00000040`3607cf50 00007fff`f1b87712     msxml6!Document::load+0x198
0c 00000040`3607cfc0 00007fff`7c46afae     msxml6!DOMDocumentWrapper::load+0x182
0d 00000040`3607d090 00007fff`7c414968     PrintConfig!PrintConfig::PrivateDevmodeMapFile::Parse+0x14e
0e 00000040`3607d200 00007fff`7c558d7b     PrintConfig!PrintConfigPlugIn::DevMode+0x458
0f 00000040`3607d320 00007fff`7c512b53     PrintConfig!BCalcTotalOEMDMSize+0xa3
10 00000040`3607d3c0 00007fff`7c5120ec     PrintConfig!UniDrvUI::LSimpleDocumentProperties+0x10f
11 00000040`3607d430 00007fff`7c40f54b     PrintConfig!UniDrvUI::DrvDocumentPropertySheets+0x60
12 00000040`3607d8e0 00007fff`7c4249ad     PrintConfig!PrintConfig::DrvDocumentPropertySheets+0x123
13 00000040`3607d950 00007fff`7c425543     PrintConfig!ExceptionBoundary<<lambda_ce41999e58d4ed069b2790d632f6e3b5> >+0x3d
14 00000040`3607d9a0 00007fff`e032e9ba     PrintConfig!DrvDocumentPropertySheets+0x53
15 00000040`3607da10 00007fff`e031593f     WINSPOOL!DocumentPropertySheets+0x3fa
16 00000040`3607da60 00007fff`e0324aff     WINSPOOL!DocumentPropertiesWNative+0x193
17 00000040`3607daf0 00007ff8`0985865c     WINSPOOL!DocumentPropertiesW+0xcf
18 00000040`3607db50 00007ff8`0a5f166d     gdi32full!hdcCreateDCW+0x1ab3c
19 00000040`3607dea0 00007ff8`0a5f1545     GDI32!bCreateDCW+0xed
1a 00000040`3607df10 00007fff`7c7ff0c7     GDI32!CreateDCW+0x55
1b 00000040`3607df50 00007fff`7c7ff020     MFC42u!AfxCreateDC+0x67
1c 00000040`3607df80 00007ff7`f3852407     MFC42u!CWinApp::CreatePrinterDC+0x20
1d 00000040`3607dfb0 00007fff`7c7e9920     wordpad!CWordPadView::OnPrinterChangedMsg+0x37
1e 00000040`3607e000 00007fff`7c7f0bff     MFC42u!CWnd::OnWndMsg+0xd80
1f 00000040`3607e130 00007fff`7c7eb42a     MFC42u!CWnd::WindowProc+0x4f
20 00000040`3607e170 00007fff`7c7ea76b     MFC42u!AfxCallWndProc+0x14a
21 00000040`3607e270 00007ff8`0be38161     MFC42u!AfxWndProcBase+0x15b
22 00000040`3607e2e0 00007ff8`0be37e1c     USER32!UserCallWinProcCheckWow+0x2d1
23 00000040`3607e440 00007ff8`0be42cdd     USER32!DispatchClientMessage+0x9c
24 00000040`3607e4a0 00007ff8`0c1d2db4     USER32!_fnDWORD+0x3d
25 00000040`3607e500 00007ff8`097f1ef4     ntdll!KiUserCallbackDispatcherContinue
26 00000040`3607e588 00007fff`7c7e21cd     win32u!NtUserShowWindow+0x14
27 00000040`3607e590 00007ff7`f38346b6     MFC42u!CFrameWnd::ActivateFrame+0x2d
28 00000040`3607e5c0 00007fff`7c7e25de     wordpad!CMainFrame::ActivateFrame+0x86
29 00000040`3607e630 00007fff`7c7fa2e7     MFC42u!CFrameWnd::InitialUpdateFrame+0xfe
2a 00000040`3607e680 00007ff7`f383e331     MFC42u!CSingleDocTemplate::OpenDocumentFile+0x187
2b 00000040`3607e6e0 00007fff`7c7efa57     wordpad!CWordPadApp::InitInstance+0x521
2c 00000040`3607fc80 00007ff7`f37fb49e     MFC42u!AfxWinMain+0x97
2d 00000040`3607fcc0 00007ff8`0b03244d     wordpad!__wmainCRTStartup+0x1de
2e 00000040`3607fd80 00007ff8`0c18df78     KERNEL32!BaseThreadInitThunk+0x1d
2f 00000040`3607fdb0 00000000`00000000     ntdll!RtlUserThreadStart+0x28
 # Child-SP          RetAddr               Call Site
00 00000040`3607c858 00007fff`f3cce048     USER32!CreateWindowExW
01 00000040`3607c860 00007fff`f3ccdf1f     urlmon!NotificationWindow::CreateHandle+0x68
02 00000040`3607c8d0 00007fff`f3c7cc9b     urlmon!CTransaction::GetNotificationWnd+0x13f
03 00000040`3607c910 00007fff`f3cb98ee     urlmon!GetTransactionObjects+0x2fb
04 00000040`3607c990 00007fff`f3cba81c     urlmon!CBinding::StartBinding+0x53e
05 00000040`3607cb10 00007fff`f3cb9276     urlmon!CUrlMon::StartBinding+0x1b0
06 00000040`3607cbe0 00007fff`f1c05529     urlmon!CUrlMon::BindToStorage+0x96
07 00000040`3607cc30 00007fff`f1c04acb     msxml6!URLMONStream::deferedOpen+0x159
08 00000040`3607cc90 00007fff`f1c048df     msxml6!XMLParser::PushURL+0x1af
09 00000040`3607cd10 00007fff`f1bbf1c6     msxml6!XMLParser::SetURL+0x7f
0a 00000040`3607cd60 00007fff`f1bbed48     msxml6!Document::_load+0x162
0b 00000040`3607cde0 00007fff`f1b87712     msxml6!Document::load+0x198
0c 00000040`3607ce50 00007fff`7c46afae     msxml6!DOMDocumentWrapper::load+0x182
0d 00000040`3607cf20 00007fff`7c414813     PrintConfig!PrintConfig::PrivateDevmodeMapFile::Parse+0x14e
0e 00000040`3607d090 00007fff`7c5583f0     PrintConfig!PrintConfigPlugIn::DevMode+0x303
0f 00000040`3607d1b0 00007fff`7c55956a     PrintConfig!BCallOEMDevMode+0x84
10 00000040`3607d230 00007fff`7c514999     PrintConfig!PGetDefaultDevmodeWithOemPlugins+0x16a
11 00000040`3607d2b0 00007fff`7c5144fa     PrintConfig!UniDrvUI::BGetValidatedUserDefaultDevmode+0x81
12 00000040`3607d320 00007fff`7c512bb7     PrintConfig!UniDrvUI::BFillCommonInfoDevmode+0x82
```

```
13 00000040`3607d3c0 00007fff`7c5120ec     PrintConfig!UniDrvUI::LSimpleDocumentProperties+0x173
14 00000040`3607d430 00007fff`7c40f54b     PrintConfig!UniDrvUI::DrvDocumentPropertySheets+0x60
15 00000040`3607d8e0 00007fff`7c4249ad     PrintConfig!PrintConfig::DrvDocumentPropertySheets+0x123
16 00000040`3607d950 00007fff`7c425543     PrintConfig!ExceptionBoundary<<lambda_ce41999e58d4ed069b2790d632f6e3b5> >+0x3d
17 00000040`3607d9a0 00007fff`e032e9ba     PrintConfig!DrvDocumentPropertySheets+0x53
18 00000040`3607da10 00007fff`e031593f     WINSPOOL!DocumentPropertySheets+0x3fa
19 00000040`3607da60 00007fff`e0324aff     WINSPOOL!DocumentPropertiesWNative+0x193
1a 00000040`3607daf0 00007ff8`0985865c     WINSPOOL!DocumentPropertiesW+0xcf
1b 00000040`3607db50 00007ff8`0a5f166d     gdi32full!hdcCreateDCW+0x1ab3c
1c 00000040`3607dea0 00007ff8`0a5f1545     GDI32!bCreateDCW+0xed
1d 00000040`3607df10 00007fff`7c7ff0c7     GDI32!CreateDCW+0x55
1e 00000040`3607df50 00007fff`7c7ff020     MFC42u!AfxCreateDC+0x67
1f 00000040`3607df80 00007ff7`f3852407     MFC42u!CWinApp::CreatePrinterDC+0x20
20 00000040`3607dfb0 00007fff`7c7e9920     wordpad!CWordPadView::OnPrinterChangedMsg+0x37
21 00000040`3607e000 00007fff`7c7f0bff     MFC42u!CWnd::OnWndMsg+0xd80
22 00000040`3607e130 00007fff`7c7eb42a     MFC42u!CWnd::WindowProc+0x4f
23 00000040`3607e170 00007fff`7c7ea76b     MFC42u!AfxCallWndProc+0x14a
24 00000040`3607e270 00007ff8`0be38161     MFC42u!AfxWndProcBase+0x15b
25 00000040`3607e2e0 00007ff8`0be37e1c     USER32!UserCallWinProcCheckWow+0x2d1
26 00000040`3607e440 00007ff8`0be42cdd     USER32!DispatchClientMessage+0x9c
27 00000040`3607e4a0 00007ff8`0c1d2db4     USER32!_fnDWORD+0x3d
28 00000040`3607e500 00007ff8`097f1ef4     ntdll!KiUserCallbackDispatcherContinue
29 00000040`3607e588 00007fff`7c7e21cd     win32u!NtUserShowWindow+0x14
2a 00000040`3607e590 00007ff7`f38346b6     MFC42u!CFrameWnd::ActivateFrame+0x2d
2b 00000040`3607e5c0 00007fff`7c7e25de     wordpad!CMainFrame::ActivateFrame+0x86
2c 00000040`3607e630 00007fff`7c7fa2e7     MFC42u!CFrameWnd::InitialUpdateFrame+0xfe
2d 00000040`3607e680 00007ff7`f383e331     MFC42u!CSingleDocTemplate::OpenDocumentFile+0x187
2e 00000040`3607e6e0 00007fff`7c7efa57     wordpad!CWordPadApp::InitInstance+0x521
2f 00000040`3607fc80 00007ff7`f37fb49e     MFC42u!AfxWinMain+0x97
30 00000040`3607fcc0 00007ff8`0b03244d     wordpad!__wmainCRTStartup+0x1de
31 00000040`3607fd80 00007ff8`0c18df78     KERNEL32!BaseThreadInitThunk+0x1d
32 00000040`3607fdb0 00000000`00000000     ntdll!RtlUserThreadStart+0x28
 # Child-SP          RetAddr               Call Site
00 00000040`3607c6b8 00007fff`f3cce048     USER32!CreateWindowExW
01 00000040`3607c6c0 00007fff`f3ccdf1f     urlmon!NotificationWindow::CreateHandle+0x68
02 00000040`3607c730 00007fff`f3c7cc9b     urlmon!CTransaction::GetNotificationWnd+0x13f
03 00000040`3607c770 00007fff`f3cb98ee     urlmon!GetTransactionObjects+0x2fb
04 00000040`3607c7f0 00007fff`f3cba81c     urlmon!CBinding::StartBinding+0x53e
05 00000040`3607c970 00007fff`f3cb9276     urlmon!CUrlMon::StartBinding+0x1b0
06 00000040`3607ca40 00007fff`f1c05529     urlmon!CUrlMon::BindToStorage+0x96
07 00000040`3607ca90 00007fff`f1c04acb     msxml6!URLMONStream::deferedOpen+0x159
08 00000040`3607caf0 00007fff`f1c048df     msxml6!XMLParser::PushURL+0x1af
09 00000040`3607cb70 00007fff`f1bbf1c6     msxml6!XMLParser::SetURL+0x7f
0a 00000040`3607cbc0 00007fff`f1bbed48     msxml6!Document::_load+0x162
0b 00000040`3607cc40 00007fff`f1b87712     msxml6!Document::load+0x198
0c 00000040`3607ccb0 00007fff`7c46afae     msxml6!DOMDocumentWrapper::load+0x182
0d 00000040`3607cd80 00007fff`7c414968     PrintConfig!PrintConfig::PrivateDevmodeMapFile::Parse+0x14e
0e 00000040`3607cef0 00007fff`7c558d7b     PrintConfig!PrintConfigPlugIn::DevMode+0x458
0f 00000040`3607d010 00007fff`7c55949b     PrintConfig!BCalcTotalOEMDMSize+0xa3
10 00000040`3607d0b0 00007fff`7c514999     PrintConfig!PGetDefaultDevmodeWithOemPlugins+0x9b
11 00000040`3607d130 00007fff`7c4fc131     PrintConfig!UniDrvUI::BGetValidatedUserDefaultDevmode+0x81
12 00000040`3607d1a0 00007fff`7c4f1696     PrintConfig!UniDrvUI::CPrintTicketProvider::GetValidatedDevmode+0x6d
13 00000040`3607d1f0 00007fff`7c51654d     PrintConfig!UniDrvUI::CPrintTicketProvider::ConvertDevModeToPrintTicket+0x36
14 00000040`3607d220 00007fff`7c514794     PrintConfig!UniDrvUI::PerformJScriptDevmodeValidation+0x479
15 00000040`3607d320 00007fff`7c512bb7     PrintConfig!UniDrvUI::BFillCommonInfoDevmode+0x31c
16 00000040`3607d3c0 00007fff`7c5120ec     PrintConfig!UniDrvUI::LSimpleDocumentProperties+0x173
17 00000040`3607d430 00007fff`7c40f54b     PrintConfig!UniDrvUI::DrvDocumentPropertySheets+0x60
18 00000040`3607d8e0 00007fff`7c4249ad     PrintConfig!PrintConfig::DrvDocumentPropertySheets+0x123
19 00000040`3607d950 00007fff`7c425543     PrintConfig!ExceptionBoundary<<lambda_ce41999e58d4ed069b2790d632f6e3b5> >+0x3d
1a 00000040`3607d9a0 00007fff`e032e9ba     PrintConfig!DrvDocumentPropertySheets+0x53
1b 00000040`3607da10 00007fff`e031593f     WINSPOOL!DocumentPropertySheets+0x3fa
1c 00000040`3607da60 00007fff`e0324aff     WINSPOOL!DocumentPropertiesWNative+0x193
1d 00000040`3607daf0 00007ff8`0985865c     WINSPOOL!DocumentPropertiesW+0xcf
1e 00000040`3607db50 00007ff8`0a5f166d     gdi32full!hdcCreateDCW+0x1ab3c
1f 00000040`3607dea0 00007ff8`0a5f1545     GDI32!bCreateDCW+0xed
20 00000040`3607df10 00007fff`7c7ff0c7     GDI32!CreateDCW+0x55
21 00000040`3607df50 00007fff`7c7ff020     MFC42u!AfxCreateDC+0x67
22 00000040`3607df80 00007ff7`f3852407     MFC42u!CWinApp::CreatePrinterDC+0x20
23 00000040`3607dfb0 00007fff`7c7e9920     wordpad!CWordPadView::OnPrinterChangedMsg+0x37
24 00000040`3607e000 00007fff`7c7f0bff     MFC42u!CWnd::OnWndMsg+0xd80
25 00000040`3607e130 00007fff`7c7eb42a     MFC42u!CWnd::WindowProc+0x4f
26 00000040`3607e170 00007fff`7c7ea76b     MFC42u!AfxCallWndProc+0x14a
27 00000040`3607e270 00007ff8`0be38161     MFC42u!AfxWndProcBase+0x15b
28 00000040`3607e2e0 00007ff8`0be37e1c     USER32!UserCallWinProcCheckWow+0x2d1
29 00000040`3607e440 00007ff8`0be42cdd     USER32!DispatchClientMessage+0x9c
2a 00000040`3607e4a0 00007ff8`0c1d2db4     USER32!_fnDWORD+0x3d
2b 00000040`3607e500 00007ff8`097f1ef4     ntdll!KiUserCallbackDispatcherContinue
2c 00000040`3607e588 00007fff`7c7e21cd     win32u!NtUserShowWindow+0x14
2d 00000040`3607e590 00007ff7`f38346b6     MFC42u!CFrameWnd::ActivateFrame+0x2d
2e 00000040`3607e5c0 00007fff`7c7e25de     wordpad!CMainFrame::ActivateFrame+0x86
2f 00000040`3607e630 00007fff`7c7fa2e7     MFC42u!CFrameWnd::InitialUpdateFrame+0xfe
30 00000040`3607e680 00007ff7`f383e331     MFC42u!CSingleDocTemplate::OpenDocumentFile+0x187
31 00000040`3607e6e0 00007fff`7c7efa57     wordpad!CWordPadApp::InitInstance+0x521
32 00000040`3607fc80 00007ff7`f37fb49e     MFC42u!AfxWinMain+0x97
33 00000040`3607fcc0 00007ff8`0b03244d     wordpad!__wmainCRTStartup+0x1de
34 00000040`3607fd80 00007ff8`0c18df78     KERNEL32!BaseThreadInitThunk+0x1d
35 00000040`3607fdb0 00000000`00000000     ntdll!RtlUserThreadStart+0x28
 # Child-SP          RetAddr               Call Site
00 00000040`3607c6d8 00007fff`f3cce048     USER32!CreateWindowExW
01 00000040`3607c6e0 00007fff`f3ccdf1f     urlmon!NotificationWindow::CreateHandle+0x68
02 00000040`3607c750 00007fff`f3c7cc9b     urlmon!CTransaction::GetNotificationWnd+0x13f
03 00000040`3607c790 00007fff`f3cb98ee     urlmon!GetTransactionObjects+0x2fb
04 00000040`3607c810 00007fff`f3cba81c     urlmon!CBinding::StartBinding+0x53e
05 00000040`3607c990 00007fff`f3cb9276     urlmon!CUrlMon::StartBinding+0x1b0
06 00000040`3607ca60 00007fff`f1c05529     urlmon!CUrlMon::BindToStorage+0x96
07 00000040`3607cab0 00007fff`f1c04acb     msxml6!URLMONStream::deferedOpen+0x159
08 00000040`3607cb10 00007fff`f1c048df     msxml6!XMLParser::PushURL+0x1af
09 00000040`3607cb90 00007fff`f1bbf1c6     msxml6!XMLParser::SetURL+0x7f
```

```
0a 00000040`3607cbe0 00007fff`f1bbed48     msxml6!Document::_load+0x162
0b 00000040`3607cc60 00007fff`f1b87712     msxml6!Document::load+0x198
0c 00000040`3607ccd0 00007fff`7c46afae     msxml6!DOMDocumentWrapper::load+0x182
0d 00000040`3607cda0 00007fff`7c414813     PrintConfig!PrintConfig::PrivateDevmodeMapFile::Parse+0x14e
0e 00000040`3607cf10 00007fff`7c5583f0     PrintConfig!PrintConfigPlugIn::DevMode+0x303
0f 00000040`3607d030 00007fff`7c55956a     PrintConfig!BCallOEMDevMode+0x84
10 00000040`3607d0b0 00007fff`7c514999     PrintConfig!PGetDefaultDevmodeWithOemPlugins+0x16a
11 00000040`3607d130 00007fff`7c4fc131     PrintConfig!UniDrvUI::BGetValidatedUserDefaultDevmode+0x81
12 00000040`3607d1a0 00007fff`7c4f1696     PrintConfig!UniDrvUI::CPrintTicketProvider::GetValidatedDevmode+0x6d
13 00000040`3607d1f0 00007fff`7c51654d     PrintConfig!UniDrvUI::CPrintTicketProvider::ConvertDevModeToPrintTicket+0x36
14 00000040`3607d220 00007fff`7c514794     PrintConfig!UniDrvUI::PerformJScriptDevmodeValidation+0x479
15 00000040`3607d320 00007fff`7c512bb7     PrintConfig!UniDrvUI::BFillCommonInfoDevmode+0x31c
16 00000040`3607d3c0 00007fff`7c5120ec     PrintConfig!UniDrvUI::LSimpleDocumentProperties+0x173
17 00000040`3607d430 00007fff`7c40f54b     PrintConfig!UniDrvUI::DrvDocumentPropertySheets+0x60
18 00000040`3607d8e0 00007fff`7c4249ad     PrintConfig!PrintConfig::DrvDocumentPropertySheets+0x123
19 00000040`3607d950 00007fff`7c425543     PrintConfig!ExceptionBoundary<<lambda_ce41999e58d4ed069b2790d632f6e3b5> >+0x3d
1a 00000040`3607d9a0 00007fff`e032e9ba     PrintConfig!DrvDocumentPropertySheets+0x53
1b 00000040`3607da10 00007fff`e031593f     WINSPOOL!DocumentPropertySheets+0x3fa
1c 00000040`3607da60 00007fff`e0324aff     WINSPOOL!DocumentPropertiesWNative+0x193
1d 00000040`3607daf0 00007ff8`0985865c     WINSPOOL!DocumentPropertiesW+0xcf
1e 00000040`3607db50 00007ff8`0a5f166d     gdi32full!hdcCreateDCW+0x1ab3c
1f 00000040`3607dea0 00007ff8`0a5f1545     GDI32!bCreateDCW+0xed
20 00000040`3607df10 00007fff`7c7ff0c7     GDI32!CreateDCW+0x55
21 00000040`3607df50 00007fff`7c7ff020     MFC42u!AfxCreateDC+0x67
22 00000040`3607df80 00007ff7`f3852407     MFC42u!CWinApp::CreatePrinterDC+0x20
23 00000040`3607dfb0 00007fff`7c7e9920     wordpad!CWordPadView::OnPrinterChangedMsg+0x37
24 00000040`3607e000 00007fff`7c7f0bff     MFC42u!CWnd::OnWndMsg+0xd80
25 00000040`3607e130 00007fff`7c7eb42a     MFC42u!CWnd::WindowProc+0x4f
26 00000040`3607e170 00007fff`7c7ea76b     MFC42u!AfxCallWndProc+0x14a
27 00000040`3607e270 00007ff8`0be38161     MFC42u!AfxWndProcBase+0x15b
28 00000040`3607e2e0 00007ff8`0be37e1c     USER32!UserCallWinProcCheckWow+0x2d1
29 00000040`3607e440 00007ff8`0be42cdd     USER32!DispatchClientMessage+0x9c
2a 00000040`3607e4a0 00007ff8`0c1d2db4     USER32!_fnDWORD+0x3d
2b 00000040`3607e500 00007ff8`097f1ef4     ntdll!KiUserCallbackDispatcherContinue
2c 00000040`3607e588 00007fff`7c7e21cd     win32u!NtUserShowWindow+0x14
2d 00000040`3607e590 00007ff7`f38346b6     MFC42u!CFrameWnd::ActivateFrame+0x2d
2e 00000040`3607e5c0 00007fff`7c7e25de     wordpad!CMainFrame::ActivateFrame+0x86
2f 00000040`3607e630 00007fff`7c7fa2e7     MFC42u!CFrameWnd::InitialUpdateFrame+0xfe
30 00000040`3607e680 00007ff7`f383e331     MFC42u!CSingleDocTemplate::OpenDocumentFile+0x187
31 00000040`3607e6e0 00007ff7`7c7efa57     wordpad!CWordPadApp::InitInstance+0x521
32 00000040`3607fc80 00007ff7`f37fb49e     MFC42u!AfxWinMain+0x97
33 00000040`3607fcc0 00007ff8`0b03244d     wordpad!__wmainCRTStartup+0x1de
34 00000040`3607fd80 00007ff8`0c18df78     KERNEL32!BaseThreadInitThunk+0x1d
35 00000040`3607fdb0 00000000`00000000     ntdll!RtlUserThreadStart+0x28
ModLoad: 00007fff`e13e0000 00007fff`e13fd000   C:\WINDOWS\SYSTEM32\amsi.dll
(1a28.d24): C++ EH exception - code e06d7363 (first chance)
(1a28.d24): C++ EH exception - code e06d7363 (first chance)
(1a28.d24): C++ EH exception - code e06d7363 (first chance)
ModLoad: 00007fff`e13e0000 00007fff`e13fd000   C:\WINDOWS\SYSTEM32\amsi.dll
 # Child-SP          RetAddr               Call Site
00 00000040`3607c688 00007fff`f3cce048     USER32!CreateWindowExW
01 00000040`3607c690 00007fff`f3ccdf1f     urlmon!NotificationWindow::CreateHandle+0x68
02 00000040`3607c700 00007fff`f3c7cc9b     urlmon!CTransaction::GetNotificationWnd+0x13f
03 00000040`3607c740 00007fff`f3cb98ee     urlmon!GetTransactionObjects+0x2fb
04 00000040`3607c7c0 00007fff`f3cba81c     urlmon!CBinding::StartBinding+0x53e
05 00000040`3607c940 00007fff`f3cb9276     urlmon!CUrlMon::StartBinding+0x1b0
06 00000040`3607ca10 00007fff`f1c05529     urlmon!CUrlMon::BindToStorage+0x96
07 00000040`3607ca60 00007fff`f1c04acb     msxml6!URLMONStream::deferedOpen+0x159
08 00000040`3607cac0 00007fff`f1c048df     msxml6!XMLParser::PushURL+0x1af
09 00000040`3607cb40 00007fff`f1bbf1c6     msxml6!XMLParser::SetURL+0x7f
0a 00000040`3607cb90 00007fff`f1bbed48     msxml6!Document::_load+0x162
0b 00000040`3607cc10 00007fff`f1b87712     msxml6!Document::load+0x198
0c 00000040`3607cc80 00007fff`7c46afae     msxml6!DOMDocumentWrapper::load+0x182
0d 00000040`3607cd50 00007fff`7c414813     PrintConfig!PrintConfig::PrivateDevmodeMapFile::Parse+0x14e
0e 00000040`3607cec0 00007fff`7c5583f0     PrintConfig!PrintConfigPlugIn::DevMode+0x303
0f 00000040`3607cfe0 00007fff`7c55956a     PrintConfig!BCallOEMDevMode+0x84
10 00000040`3607d060 00007fff`7c514999     PrintConfig!PGetDefaultDevmodeWithOemPlugins+0x16a
11 00000040`3607d0e0 00007fff`7c4fc131     PrintConfig!UniDrvUI::BGetValidatedUserDefaultDevmode+0x81
12 00000040`3607d150 00007fff`7c4f1c06     PrintConfig!UniDrvUI::CPrintTicketProvider::GetValidatedDevmode+0x6d
13 00000040`3607d1a0 00007fff`7c5165f9     PrintConfig!UniDrvUI::CPrintTicketProvider::ConvertPrintTicketToDevMode+0x76
14 00000040`3607d220 00007fff`7c514794     PrintConfig!UniDrvUI::PerformJScriptDevmodeValidation+0x525
15 00000040`3607d320 00007fff`7c512bb7     PrintConfig!UniDrvUI::BFillCommonInfoDevmode+0x31c
16 00000040`3607d3c0 00007fff`7c5120ec     PrintConfig!UniDrvUI::LSimpleDocumentProperties+0x173
17 00000040`3607d430 00007fff`7c40f54b     PrintConfig!UniDrvUI::DrvDocumentPropertySheets+0x60
18 00000040`3607d8e0 00007fff`7c4249ad     PrintConfig!PrintConfig::DrvDocumentPropertySheets+0x123
19 00000040`3607d950 00007fff`7c425543     PrintConfig!ExceptionBoundary<<lambda_ce41999e58d4ed069b2790d632f6e3b5> >+0x3d
1a 00000040`3607d9a0 00007fff`e032e9ba     PrintConfig!DrvDocumentPropertySheets+0x53
1b 00000040`3607da10 00007fff`e031593f     WINSPOOL!DocumentPropertySheets+0x3fa
1c 00000040`3607da60 00007fff`e0324aff     WINSPOOL!DocumentPropertiesWNative+0x193
1d 00000040`3607daf0 00007ff8`0985865c     WINSPOOL!DocumentPropertiesW+0xcf
1e 00000040`3607db50 00007ff8`0a5f166d     gdi32full!hdcCreateDCW+0x1ab3c
1f 00000040`3607dea0 00007ff8`0a5f1545     GDI32!bCreateDCW+0xed
20 00000040`3607df10 00007fff`7c7ff0c7     GDI32!CreateDCW+0x55
21 00000040`3607df50 00007fff`7c7ff020     MFC42u!AfxCreateDC+0x67
22 00000040`3607df80 00007ff7`f3852407     MFC42u!CWinApp::CreatePrinterDC+0x20
23 00000040`3607dfb0 00007fff`7c7e9920     wordpad!CWordPadView::OnPrinterChangedMsg+0x37
24 00000040`3607e000 00007fff`7c7f0bff     MFC42u!CWnd::OnWndMsg+0xd80
25 00000040`3607e130 00007fff`7c7eb42a     MFC42u!CWnd::WindowProc+0x4f
26 00000040`3607e170 00007fff`7c7ea76b     MFC42u!AfxCallWndProc+0x14a
27 00000040`3607e270 00007ff8`0be38161     MFC42u!AfxWndProcBase+0x15b
28 00000040`3607e2e0 00007ff8`0be37e1c     USER32!UserCallWinProcCheckWow+0x2d1
29 00000040`3607e440 00007ff8`0be42cdd     USER32!DispatchClientMessage+0x9c
2a 00000040`3607e4a0 00007ff8`0c1d2db4     USER32!_fnDWORD+0x3d
2b 00000040`3607e500 00007ff8`097f1ef4     ntdll!KiUserCallbackDispatcherContinue
2c 00000040`3607e588 00007fff`7c7e21cd     win32u!NtUserShowWindow+0x14
2d 00000040`3607e590 00007ff7`f38346b6     MFC42u!CFrameWnd::ActivateFrame+0x2d
2e 00000040`3607e5c0 00007fff`7c7e25de     wordpad!CMainFrame::ActivateFrame+0x86
2f 00000040`3607e630 00007fff`7c7fa2e7     MFC42u!CFrameWnd::InitialUpdateFrame+0xfe
```

```
30 00000040`3607e680 00007ff7`f383e331     MFC42u!CSingleDocTemplate::OpenDocumentFile+0x187
31 00000040`3607e6e0 00007fff`7c7efa57     wordpad!CWordPadApp::InitInstance+0x521
32 00000040`3607fc80 00007ff7`f37fb49e     MFC42u!AfxWinMain+0x97
33 00000040`3607fcc0 00007ff8`0b03244d     wordpad!__wmainCRTStartup+0x1de
34 00000040`3607fd80 00007ff8`0c18df78     KERNEL32!BaseThreadInitThunk+0x1d
35 00000040`3607fdb0 00000000`00000000     ntdll!RtlUserThreadStart+0x28
ModLoad: 00007fff`e13e0000 00007fff`e13fd000   C:\WINDOWS\SYSTEM32\amsi.dll
 # Child-SP          RetAddr               Call Site
00 00000040`3607c3b8 00007fff`f3cce048     USER32!CreateWindowExW
01 00000040`3607c3c0 00007fff`f3ccdf1f     urlmon!NotificationWindow::CreateHandle+0x68
02 00000040`3607c430 00007fff`f3c7cc9b     urlmon!CTransaction::GetNotificationWnd+0x13f
03 00000040`3607c470 00007fff`f3cb98ee     urlmon!GetTransactionObjects+0x2fb
04 00000040`3607c4f0 00007fff`f3cba81c     urlmon!CBinding::StartBinding+0x53e
05 00000040`3607c670 00007fff`f3cb9276     urlmon!CUrlMon::StartBinding+0x1b0
06 00000040`3607c740 00007fff`f1c05529     urlmon!CUrlMon::BindToStorage+0x96
07 00000040`3607c790 00007fff`f1c04acb     msxml6!URLMONStream::deferedOpen+0x159
08 00000040`3607c7f0 00007fff`f1c048df     msxml6!XMLParser::PushURL+0x1af
09 00000040`3607c870 00007fff`f1bbf1c6     msxml6!XMLParser::SetURL+0x7f
0a 00000040`3607c8c0 00007fff`f1bbed48     msxml6!Document::_load+0x162
0b 00000040`3607c940 00007fff`f1b87712     msxml6!Document::load+0x198
0c 00000040`3607c9b0 00007fff`7c46afae     msxml6!DOMDocumentWrapper::load+0x182
0d 00000040`3607ca80 00007fff`7c414968     PrintConfig!PrintConfig::PrivateDevmodeMapFile::Parse+0x14e
0e 00000040`3607cbf0 00007fff`7c558d7b     PrintConfig!PrintConfigPlugIn::DevMode+0x458
0f 00000040`3607cd10 00007fff`7c55949b     PrintConfig!BCalcTotalOEMDMSize+0xa3
10 00000040`3607cdb0 00007fff`7c514999     PrintConfig!PGetDefaultDevmodeWithOemPlugins+0x9b
11 00000040`3607ce30 00007fff`7c5144fa     PrintConfig!UniDrvUI::BGetValidatedUserDefaultDevmode+0x81
12 00000040`3607cea0 00007fff`7c50d112     PrintConfig!UniDrvUI::BFillCommonInfoDevmode+0x82
13 00000040`3607cf40 00007fff`7c50ce44     PrintConfig!UniDrvUI::DwDeviceCapabilities+0xc6
14 00000040`3607cff0 00007fff`7c42514f     PrintConfig!UniDrvUI::DrvDeviceCapabilities+0x18
15 00000040`3607d030 00007fff`e03150a1     PrintConfig!DrvDeviceCapabilities+0x11f
16 00000040`3607d0b0 00007fff`e0324a04     WINSPOOL!DeviceCapabilitiesWNative+0x105
17 00000040`3607d110 00007fff`7c4feafb     WINSPOOL!DeviceCapabilitiesW+0x54
18 00000040`3607d150 00007fff`7c516637     PrintConfig!publicdm::JobTicketToPublicDevmode+0x9f
19 00000040`3607d220 00007fff`7c514794     PrintConfig!UniDrvUI::PerformJScriptDevmodeValidation+0x563
1a 00000040`3607d320 00007fff`7c512bb7     PrintConfig!UniDrvUI::BFillCommonInfoDevmode+0x31c
1b 00000040`3607d3c0 00007fff`7c5120ec     PrintConfig!UniDrvUI::LSimpleDocumentProperties+0x173
1c 00000040`3607d430 00007fff`7c40f54b     PrintConfig!UniDrvUI::DrvDocumentPropertySheets+0x60
1d 00000040`3607d8e0 00007fff`7c4249ad     PrintConfig!PrintConfig::DrvDocumentPropertySheets+0x123
1e 00000040`3607d950 00007fff`7c425543     PrintConfig!ExceptionBoundary<<lambda_ce41999e58d4ed069b2790d632f6e3b5> >+0x3d
1f 00000040`3607d9a0 00007fff`e032e9ba     PrintConfig!DrvDocumentPropertySheets+0x53
20 00000040`3607da10 00007fff`e031593f     WINSPOOL!DocumentPropertySheets+0x3fa
21 00000040`3607da60 00007fff`e0324aff     WINSPOOL!DocumentPropertiesWNative+0x193
22 00000040`3607daf0 00007ff8`0985865c     WINSPOOL!DocumentPropertiesW+0xcf
23 00000040`3607db50 00007ff8`0a5f166d     gdi32full!hdcCreateDCW+0x1ab3c
24 00000040`3607dea0 00007ff8`0a5f1545     GDI32!bCreateDCW+0xed
25 00000040`3607df10 00007fff`7c7ff0c7     GDI32!CreateDCW+0x55
26 00000040`3607df50 00007fff`7c7ff020     MFC42u!AfxCreateDC+0x67
27 00000040`3607df80 00007ff7`f3852407     MFC42u!CWinApp::CreatePrinterDC+0x20
28 00000040`3607dfb0 00007fff`7c7e9920     wordpad!CWordPadView::OnPrinterChangedMsg+0x37
29 00000040`3607e000 00007fff`7c7f0bff     MFC42u!CWnd::OnWndMsg+0xd80
2a 00000040`3607e130 00007fff`7c7eb42a     MFC42u!CWnd::WindowProc+0x4f
2b 00000040`3607e170 00007fff`7c7ea76b     MFC42u!AfxCallWndProc+0x14a
2c 00000040`3607e270 00007ff8`0be38161     MFC42u!AfxWndProcBase+0x15b
2d 00000040`3607e2e0 00007ff8`0be37e1c     USER32!UserCallWinProcCheckWow+0x2d1
2e 00000040`3607e440 00007ff8`0be42cdd     USER32!DispatchClientMessage+0x9c
2f 00000040`3607e4a0 00007ff8`0c1d2db4     USER32!_fnDWORD+0x3d
30 00000040`3607e500 00007ff8`097f1ef4     ntdll!KiUserCallbackDispatcherContinue
31 00000040`3607e588 00007fff`7c7e21cd     win32u!NtUserShowWindow+0x14
32 00000040`3607e590 00007ff7`f38346b6     MFC42u!CFrameWnd::ActivateFrame+0x2d
33 00000040`3607e5c0 00007fff`7c7e25de     wordpad!CMainFrame::ActivateFrame+0x86
34 00000040`3607e630 00007fff`7c7fa2e7     MFC42u!CFrameWnd::InitialUpdateFrame+0xfe
35 00000040`3607e680 00007ff7`f383e331     MFC42u!CSingleDocTemplate::OpenDocumentFile+0x187
36 00000040`3607e6e0 00007fff`7c7efa57     wordpad!CWordPadApp::InitInstance+0x521
37 00000040`3607fc80 00007ff7`f37fb49e     MFC42u!AfxWinMain+0x97
38 00000040`3607fcc0 00007ff8`0b03244d     wordpad!__wmainCRTStartup+0x1de
39 00000040`3607fd80 00007ff8`0c18df78     KERNEL32!BaseThreadInitThunk+0x1d
3a 00000040`3607fdb0 00000000`00000000     ntdll!RtlUserThreadStart+0x28
 # Child-SP          RetAddr               Call Site
00 00000040`3607c3d8 00007fff`f3cce048     USER32!CreateWindowExW
01 00000040`3607c3e0 00007fff`f3ccdf1f     urlmon!NotificationWindow::CreateHandle+0x68
02 00000040`3607c450 00007fff`f3c7cc9b     urlmon!CTransaction::GetNotificationWnd+0x13f
03 00000040`3607c490 00007fff`f3cb98ee     urlmon!GetTransactionObjects+0x2fb
04 00000040`3607c510 00007fff`f3cba81c     urlmon!CBinding::StartBinding+0x53e
05 00000040`3607c690 00007fff`f3cb9276     urlmon!CUrlMon::StartBinding+0x1b0
06 00000040`3607c760 00007fff`f1c05529     urlmon!CUrlMon::BindToStorage+0x96
07 00000040`3607c7b0 00007fff`f1c04acb     msxml6!URLMONStream::deferedOpen+0x159
08 00000040`3607c810 00007fff`f1c048df     msxml6!XMLParser::PushURL+0x1af
09 00000040`3607c890 00007fff`f1bbf1c6     msxml6!XMLParser::SetURL+0x7f
0a 00000040`3607c8e0 00007fff`f1bbed48     msxml6!Document::_load+0x162
0b 00000040`3607c960 00007fff`f1b87712     msxml6!Document::load+0x198
0c 00000040`3607c9d0 00007fff`7c46afae     msxml6!DOMDocumentWrapper::load+0x182
0d 00000040`3607caa0 00007fff`7c414813     PrintConfig!PrintConfig::PrivateDevmodeMapFile::Parse+0x14e
0e 00000040`3607cc10 00007fff`7c5583f0     PrintConfig!PrintConfigPlugIn::DevMode+0x303
0f 00000040`3607cd30 00007fff`7c55956a     PrintConfig!BCallOEMDevMode+0x84
10 00000040`3607cdb0 00007fff`7c514999     PrintConfig!PGetDefaultDevmodeWithOemPlugins+0x16a
11 00000040`3607ce30 00007fff`7c5144fa     PrintConfig!UniDrvUI::BGetValidatedUserDefaultDevmode+0x81
12 00000040`3607cea0 00007fff`7c50d112     PrintConfig!UniDrvUI::BFillCommonInfoDevmode+0x82
13 00000040`3607cf40 00007fff`7c50ce44     PrintConfig!UniDrvUI::DwDeviceCapabilities+0xc6
14 00000040`3607cff0 00007fff`7c42514f     PrintConfig!UniDrvUI::DrvDeviceCapabilities+0x18
15 00000040`3607d030 00007fff`e03150a1     PrintConfig!DrvDeviceCapabilities+0x11f
16 00000040`3607d0b0 00007fff`e0324a04     WINSPOOL!DeviceCapabilitiesWNative+0x105
17 00000040`3607d110 00007fff`7c4feafb     WINSPOOL!DeviceCapabilitiesW+0x54
18 00000040`3607d150 00007fff`7c516637     PrintConfig!publicdm::JobTicketToPublicDevmode+0x9f
19 00000040`3607d220 00007fff`7c514794     PrintConfig!UniDrvUI::PerformJScriptDevmodeValidation+0x563
1a 00000040`3607d320 00007fff`7c512bb7     PrintConfig!UniDrvUI::BFillCommonInfoDevmode+0x31c
1b 00000040`3607d3c0 00007fff`7c5120ec     PrintConfig!UniDrvUI::LSimpleDocumentProperties+0x173
1c 00000040`3607d430 00007fff`7c40f54b     PrintConfig!UniDrvUI::DrvDocumentPropertySheets+0x60
1d 00000040`3607d8e0 00007fff`7c4249ad     PrintConfig!PrintConfig::DrvDocumentPropertySheets+0x123
```

```
1e 00000040`3607d950 00007fff`7c425543     PrintConfig!ExceptionBoundary<<lambda_ce41999e58d4ed069b2790d632f6e3b5> >+0x3d
1f 00000040`3607d9a0 00007fff`e032e9ba     PrintConfig!DrvDocumentPropertySheets+0x53
20 00000040`3607da10 00007fff`e031593f     WINSPOOL!DocumentPropertySheets+0x3fa
21 00000040`3607da60 00007fff`e0324aff     WINSPOOL!DocumentPropertiesWNative+0x193
22 00000040`3607daf0 00007ff8`0985865c     WINSPOOL!DocumentPropertiesW+0xcf
23 00000040`3607db50 00007ff8`0a5f166d     gdi32full!hdcCreateDCW+0x1ab3c
24 00000040`3607dea0 00007ff8`0a5f1545     GDI32!bCreateDCW+0xed
25 00000040`3607df10 00007fff`7c7ff0c7     GDI32!CreateDCW+0x55
26 00000040`3607df50 00007fff`7c7ff020     MFC42u!AfxCreateDC+0x67
27 00000040`3607df80 00007ff7`f3852407     MFC42u!CWinApp::CreatePrinterDC+0x20
28 00000040`3607dfb0 00007fff`7c7e9920     wordpad!CWordPadView::OnPrinterChangedMsg+0x37
29 00000040`3607e000 00007fff`7c7f0bff     MFC42u!CWnd::OnWndMsg+0xd80
2a 00000040`3607e130 00007fff`7c7eb42a     MFC42u!CWnd::WindowProc+0x4f
2b 00000040`3607e170 00007fff`7c7ea76b     MFC42u!AfxCallWndProc+0x14a
2c 00000040`3607e270 00007ff8`0be38161     MFC42u!AfxWndProcBase+0x15b
2d 00000040`3607e2e0 00007ff8`0be37e1c     USER32!UserCallWinProcCheckWow+0x2d1
2e 00000040`3607e440 00007ff8`0be42cdd     USER32!DispatchClientMessage+0x9c
2f 00000040`3607e4a0 00007ff8`0c1d2db4     USER32!_fnDWORD+0x3d
30 00000040`3607e500 00007ff8`097f1ef4     ntdll!KiUserCallbackDispatcherContinue
31 00000040`3607e588 00007fff`7c7e21cd     win32u!NtUserShowWindow+0x14
32 00000040`3607e590 00007ff7`f38346b6     MFC42u!CFrameWnd::ActivateFrame+0x2d
33 00000040`3607e5c0 00007fff`7c7e25de     wordpad!CMainFrame::ActivateFrame+0x86
34 00000040`3607e630 00007fff`7c7fa2e7     MFC42u!CFrameWnd::InitialUpdateFrame+0xfe
35 00000040`3607e680 00007ff7`f383e331     MFC42u!CSingleDocTemplate::OpenDocumentFile+0x187
36 00000040`3607e6e0 00007fff`7c7efa57     wordpad!CWordPadApp::InitInstance+0x521
37 00000040`3607fc80 00007ff7`f37fb49e     MFC42u!AfxWinMain+0x97
38 00000040`3607fcc0 00007ff8`0b03244d     wordpad!__wmainCRTStartup+0x1de
39 00000040`3607fd80 00007ff8`0c18df78     KERNEL32!BaseThreadInitThunk+0x1d
3a 00000040`3607fdb0 00000000`00000000     ntdll!RtlUserThreadStart+0x28
 # Child-SP          RetAddr               Call Site
00 00000040`3607b088 00007fff`f3cce048     USER32!CreateWindowExW
01 00000040`3607b090 00007fff`f3ccdf1f     urlmon!NotificationWindow::CreateHandle+0x68
02 00000040`3607b100 00007fff`f3c7cc9b     urlmon!CTransaction::GetNotificationWnd+0x13f
03 00000040`3607b140 00007fff`f3cb98ee     urlmon!GetTransactionObjects+0x2fb
04 00000040`3607b1c0 00007fff`f3cba81c     urlmon!CBinding::StartBinding+0x53e
05 00000040`3607b340 00007fff`f3cb9276     urlmon!CUrlMon::StartBinding+0x1b0
06 00000040`3607b410 00007fff`f1c05529     urlmon!CUrlMon::BindToStorage+0x96
07 00000040`3607b460 00007fff`f1c04acb     msxml6!URLMONStream::deferedOpen+0x159
08 00000040`3607b4c0 00007fff`f1c048df     msxml6!XMLParser::PushURL+0x1af
09 00000040`3607b540 00007fff`f1bbf1c6     msxml6!XMLParser::SetURL+0x7f
0a 00000040`3607b590 00007fff`f1bbed48     msxml6!Document::_load+0x162
0b 00000040`3607b610 00007fff`f1b87712     msxml6!Document::load+0x198
0c 00000040`3607b680 00007fff`7c46afae     msxml6!DOMDocumentWrapper::load+0x182
0d 00000040`3607b750 00007fff`7c414968     PrintConfig!PrintConfig::PrivateDevmodeMapFile::Parse+0x14e
0e 00000040`3607b8c0 00007fff`7c558d7b     PrintConfig!PrintConfigPlugIn::DevMode+0x458
0f 00000040`3607b9e0 00007fff`7c512b53     PrintConfig!BCalcTotalOEMDMSize+0xa3
10 00000040`3607ba80 00007fff`7c5120ec     PrintConfig!UniDrvUI::LSimpleDocumentProperties+0x10f
11 00000040`3607baf0 00007fff`7c40f54b     PrintConfig!UniDrvUI::DrvDocumentPropertySheets+0x60
12 00000040`3607bfa0 00007fff`7c4249ad     PrintConfig!PrintConfig::DrvDocumentPropertySheets+0x123
13 00000040`3607c010 00007fff`7c425543     PrintConfig!ExceptionBoundary<<lambda_ce41999e58d4ed069b2790d632f6e3b5> >+0x3d
14 00000040`3607c060 00007fff`e032e9ba     PrintConfig!DrvDocumentPropertySheets+0x53
15 00000040`3607c0d0 00007fff`e031593f     WINSPOOL!DocumentPropertySheets+0x3fa
16 00000040`3607c120 00007fff`e0324aff     WINSPOOL!DocumentPropertiesWNative+0x193
17 00000040`3607c1b0 00007fff`a4de735d     WINSPOOL!DocumentPropertiesW+0xcf
18 00000040`3607c210 00007fff`a4ddb5d0     mxdwdrv!edocs::PDev::PDev+0x1d9
19 00000040`3607c720 00007ff8`09876470     mxdwdrv!DrvEnablePDEV+0x170
1a 00000040`3607d060 00007ff8`0bea991a     gdi32full!GdiPrinterThunk+0x370
1b 00000040`3607d130 00007ff8`0c1d2db4     USER32!_ClientPrinterThunk+0x3a
1c 00000040`3607d9b0 00007ff8`097f2d94     ntdll!KiUserCallbackDispatcherContinue
1d 00000040`3607dab8 00007ff8`0983dbd2     win32u!NtGdiOpenDCW+0x14
1e 00000040`3607dac0 00007ff8`0a5f166d     gdi32full!hdcCreateDCW+0xb2
1f 00000040`3607de10 00007ff8`0a5f1545     GDI32!bCreateDCW+0xed
20 00000040`3607de80 00007fff`7c7ff0c7     GDI32!CreateDCW+0x55
21 00000040`3607dec0 00007fff`7c7ff020     MFC42u!AfxCreateDC+0x67
22 00000040`3607def0 00007ff7`f37f8a92     MFC42u!CWinApp::CreatePrinterDC+0x20
23 00000040`3607df20 00007ff7`f3854634     wordpad!CRichEdit2View::WrapChanged+0x62
24 00000040`3607df50 00007ff7`f37f615f     wordpad!CWordPadView::WrapChanged+0x84
25 00000040`3607df80 00007ff7`f385242e     wordpad!CRichEdit2View::OnPrinterChanged+0x11f
26 00000040`3607dfb0 00007fff`7c7e9920     wordpad!CWordPadView::OnPrinterChangedMsg+0x5e
27 00000040`3607e000 00007fff`7c7f0bff     MFC42u!CWnd::OnWndMsg+0xd80
28 00000040`3607e130 00007fff`7c7eb42a     MFC42u!CWnd::WindowProc+0x4f
29 00000040`3607e170 00007fff`7c7ea76b     MFC42u!AfxCallWndProc+0x14a
2a 00000040`3607e270 00007ff8`0be38161     MFC42u!AfxWndProcBase+0x15b
2b 00000040`3607e2e0 00007ff8`0be37e1c     USER32!UserCallWinProcCheckWow+0x2d1
2c 00000040`3607e440 00007ff8`0be42cdd     USER32!DispatchClientMessage+0x9c
2d 00000040`3607e4a0 00007ff8`0c1d2db4     USER32!_fnDWORD+0x3d
2e 00000040`3607e500 00007ff8`097f1ef4     ntdll!KiUserCallbackDispatcherContinue
2f 00000040`3607e588 00007fff`7c7e21cd     win32u!NtUserShowWindow+0x14
30 00000040`3607e590 00007ff7`f38346b6     MFC42u!CFrameWnd::ActivateFrame+0x2d
31 00000040`3607e5c0 00007fff`7c7e25de     wordpad!CMainFrame::ActivateFrame+0x86
32 00000040`3607e630 00007fff`7c7fa2e7     MFC42u!CFrameWnd::InitialUpdateFrame+0xfe
33 00000040`3607e680 00007ff7`f383e331     MFC42u!CSingleDocTemplate::OpenDocumentFile+0x187
34 00000040`3607e6e0 00007fff`7c7efa57     wordpad!CWordPadApp::InitInstance+0x521
35 00000040`3607fc80 00007ff7`f37fb49e     MFC42u!AfxWinMain+0x97
36 00000040`3607fcc0 00007ff8`0b03244d     wordpad!__wmainCRTStartup+0x1de
37 00000040`3607fd80 00007ff8`0c18df78     KERNEL32!BaseThreadInitThunk+0x1d
38 00000040`3607fdb0 00000000`00000000     ntdll!RtlUserThreadStart+0x28
 # Child-SP          RetAddr               Call Site
00 00000040`3607b088 00007fff`f3cce048     USER32!CreateWindowExW
01 00000040`3607b090 00007fff`f3ccdf1f     urlmon!NotificationWindow::CreateHandle+0x68
02 00000040`3607b100 00007fff`f3c7cc9b     urlmon!CTransaction::GetNotificationWnd+0x13f
03 00000040`3607b140 00007fff`f3cb98ee     urlmon!GetTransactionObjects+0x2fb
04 00000040`3607b1c0 00007fff`f3cba81c     urlmon!CBinding::StartBinding+0x53e
05 00000040`3607b340 00007fff`f3cb9276     urlmon!CUrlMon::StartBinding+0x1b0
06 00000040`3607b410 00007fff`f1c05529     urlmon!CUrlMon::BindToStorage+0x96
07 00000040`3607b460 00007fff`f1c04acb     msxml6!URLMONStream::deferedOpen+0x159
08 00000040`3607b4c0 00007fff`f1c048df     msxml6!XMLParser::PushURL+0x1af
09 00000040`3607b540 00007fff`f1bbf1c6     msxml6!XMLParser::SetURL+0x7f
```

```
0a 00000040`3607b590 00007fff`f1bbed48     msxml6!Document::_load+0x162
0b 00000040`3607b610 00007fff`f1b87712     msxml6!Document::load+0x198
0c 00000040`3607b680 00007fff`7c46afae     msxml6!DOMDocumentWrapper::load+0x182
0d 00000040`3607b750 00007fff`7c414968     PrintConfig!PrintConfig::PrivateDevmodeMapFile::Parse+0x14e
0e 00000040`3607b8c0 00007fff`7c558d7b     PrintConfig!PrintConfigPlugIn::DevMode+0x458
0f 00000040`3607b9e0 00007fff`7c512b53     PrintConfig!BCalcTotalOEMDMSize+0xa3
10 00000040`3607ba80 00007fff`7c5120ec     PrintConfig!UniDrvUI::LSimpleDocumentProperties+0x10f
11 00000040`3607baf0 00007fff`7c40f54b     PrintConfig!UniDrvUI::DrvDocumentPropertySheets+0x60
12 00000040`3607bfa0 00007fff`7c4249ad     PrintConfig!PrintConfig::DrvDocumentPropertySheets+0x123
13 00000040`3607c010 00007fff`7c425543     PrintConfig!ExceptionBoundary<<lambda_ce41999e58d4ed069b2790d632f6e3b5> >+0x3d
14 00000040`3607c060 00007fff`e032e9ba     PrintConfig!DrvDocumentPropertySheets+0x53
15 00000040`3607c0d0 00007fff`e0315895     WINSPOOL!DocumentPropertySheets+0x3fa
16 00000040`3607c120 00007fff`e0324aff     WINSPOOL!DocumentPropertiesWNative+0xe9
17 00000040`3607c1b0 00007fff`a4de746f     WINSPOOL!DocumentPropertiesW+0xcf
18 00000040`3607c210 00007fff`a4ddb5d0     mxdwdrv!edocs::PDev::PDev+0x2eb
19 00000040`3607c720 00007ff8`09876470     mxdwdrv!DrvEnablePDEV+0x170
1a 00000040`3607d060 00007ff8`0bea991a     gdi32full!GdiPrinterThunk+0x370
1b 00000040`3607d130 00007ff8`0c1d2db4     USER32!_ClientPrinterThunk+0x3a
1c 00000040`3607d9b0 00007ff8`097f2d94     ntdll!KiUserCallbackDispatcherContinue
1d 00000040`3607dab8 00007ff8`0983dbd2     win32u!NtGdiOpenDCW+0x14
1e 00000040`3607dac0 00007ff8`0a5f166d     gdi32full!hdcCreateDCW+0xb2
1f 00000040`3607de10 00007ff8`0a5f1545     GDI32!bCreateDCW+0xed
20 00000040`3607de80 00007fff`7c7ff0c7     GDI32!CreateDCW+0x55
21 00000040`3607dec0 00007fff`7c7ff020     MFC42u!AfxCreateDC+0x67
22 00000040`3607def0 00007ff7`f37f8a92     MFC42u!CWinApp::CreatePrinterDC+0x20
23 00000040`3607df20 00007ff7`f3854634     wordpad!CRichEdit2View::WrapChanged+0x62
24 00000040`3607df50 00007ff7`f37f615f     wordpad!CWordPadView::WrapChanged+0x84
25 00000040`3607df80 00007ff7`f385242e     wordpad!CRichEdit2View::OnPrinterChanged+0x11f
26 00000040`3607dfb0 00007fff`7c7e9920     wordpad!CWordPadView::OnPrinterChangedMsg+0x5e
27 00000040`3607e000 00007fff`7c7f0bff     MFC42u!CWnd::OnWndMsg+0xd80
28 00000040`3607e130 00007fff`7c7eb42a     MFC42u!CWnd::WindowProc+0x4f
29 00000040`3607e170 00007fff`7c7ea76b     MFC42u!AfxCallWndProc+0x14a
2a 00000040`3607e270 00007ff8`0be38161     MFC42u!AfxWndProcBase+0x15b
2b 00000040`3607e2e0 00007ff8`0be37e1c     USER32!UserCallWinProcCheckWow+0x2d1
2c 00000040`3607e440 00007ff8`0be42cdd     USER32!DispatchClientMessage+0x9c
2d 00000040`3607e4a0 00007ff8`0c1d2db4     USER32!_fnDWORD+0x3d
2e 00000040`3607e500 00007ff8`097f1ef4     ntdll!KiUserCallbackDispatcherContinue
2f 00000040`3607e588 00007fff`7c7e21cd     win32u!NtUserShowWindow+0x14
30 00000040`3607e590 00007ff7`f38346b6     MFC42u!CFrameWnd::ActivateFrame+0x2d
31 00000040`3607e5c0 00007fff`7c7e25de     wordpad!CMainFrame::ActivateFrame+0x86
32 00000040`3607e630 00007fff`7c7fa2e7     MFC42u!CFrameWnd::InitialUpdateFrame+0xfe
33 00000040`3607e680 00007ff7`f383e331     MFC42u!CSingleDocTemplate::OpenDocumentFile+0x187
34 00000040`3607e6e0 00007fff`7c7efa57     wordpad!CWordPadApp::InitInstance+0x521
35 00000040`3607fc80 00007ff7`f37fb49e     MFC42u!AfxWinMain+0x97
36 00000040`3607fcc0 00007ff8`0b03244d     wordpad!__wmainCRTStartup+0x1de
37 00000040`3607fd80 00007ff8`0c18df78     KERNEL32!BaseThreadInitThunk+0x1d
38 00000040`3607fdb0 00000000`00000000     ntdll!RtlUserThreadStart+0x28
 # Child-SP          RetAddr               Call Site
00 00000040`3607b088 00007fff`f3cce048     USER32!CreateWindowExW
01 00000040`3607b090 00007fff`f3ccdf1f     urlmon!NotificationWindow::CreateHandle+0x68
02 00000040`3607b100 00007fff`f3c7cc9b     urlmon!CTransaction::GetNotificationWnd+0x13f
03 00000040`3607b140 00007fff`f3cb98ee     urlmon!GetTransactionObjects+0x2fb
04 00000040`3607b1c0 00007fff`f3cba81c     urlmon!CBinding::StartBinding+0x53e
05 00000040`3607b340 00007fff`f3cb9276     urlmon!CUrlMon::StartBinding+0x1b0
06 00000040`3607b410 00007fff`f1c05529     urlmon!CUrlMon::BindToStorage+0x96
07 00000040`3607b460 00007fff`f1c04acb     msxml6!URLMONStream::deferedOpen+0x159
08 00000040`3607b4c0 00007fff`f1c048df     msxml6!XMLParser::PushURL+0x1af
09 00000040`3607b540 00007fff`f1bbf1c6     msxml6!XMLParser::SetURL+0x7f
0a 00000040`3607b590 00007fff`f1bbed48     msxml6!Document::_load+0x162
0b 00000040`3607b610 00007fff`f1b87712     msxml6!Document::load+0x198
0c 00000040`3607b680 00007fff`7c46afae     msxml6!DOMDocumentWrapper::load+0x182
0d 00000040`3607b750 00007fff`7c414968     PrintConfig!PrintConfig::PrivateDevmodeMapFile::Parse+0x14e
0e 00000040`3607b8c0 00007fff`7c558d7b     PrintConfig!PrintConfigPlugIn::DevMode+0x458
0f 00000040`3607b9e0 00007fff`7c512b53     PrintConfig!BCalcTotalOEMDMSize+0xa3
10 00000040`3607ba80 00007fff`7c5120ec     PrintConfig!UniDrvUI::LSimpleDocumentProperties+0x10f
11 00000040`3607baf0 00007fff`7c40f54b     PrintConfig!UniDrvUI::DrvDocumentPropertySheets+0x60
12 00000040`3607bfa0 00007fff`7c4249ad     PrintConfig!PrintConfig::DrvDocumentPropertySheets+0x123
13 00000040`3607c010 00007fff`7c425543     PrintConfig!ExceptionBoundary<<lambda_ce41999e58d4ed069b2790d632f6e3b5> >+0x3d
14 00000040`3607c060 00007fff`e032e9ba     PrintConfig!DrvDocumentPropertySheets+0x53
15 00000040`3607c0d0 00007fff`e031593f     WINSPOOL!DocumentPropertySheets+0x3fa
16 00000040`3607c120 00007fff`e0324aff     WINSPOOL!DocumentPropertiesWNative+0x193
17 00000040`3607c1b0 00007fff`a4de746f     WINSPOOL!DocumentPropertiesW+0xcf
18 00000040`3607c210 00007fff`a4ddb5d0     mxdwdrv!edocs::PDev::PDev+0x2eb
19 00000040`3607c720 00007ff8`09876470     mxdwdrv!DrvEnablePDEV+0x170
1a 00000040`3607d060 00007ff8`0bea991a     gdi32full!GdiPrinterThunk+0x370
1b 00000040`3607d130 00007ff8`0c1d2db4     USER32!_ClientPrinterThunk+0x3a
1c 00000040`3607d9b0 00007ff8`097f2d94     ntdll!KiUserCallbackDispatcherContinue
1d 00000040`3607dab8 00007ff8`0983dbd2     win32u!NtGdiOpenDCW+0x14
1e 00000040`3607dac0 00007ff8`0a5f166d     gdi32full!hdcCreateDCW+0xb2
1f 00000040`3607de10 00007ff8`0a5f1545     GDI32!bCreateDCW+0xed
20 00000040`3607de80 00007fff`7c7ff0c7     GDI32!CreateDCW+0x55
21 00000040`3607dec0 00007fff`7c7ff020     MFC42u!AfxCreateDC+0x67
22 00000040`3607def0 00007ff7`f37f8a92     MFC42u!CWinApp::CreatePrinterDC+0x20
23 00000040`3607df20 00007ff7`f3854634     wordpad!CRichEdit2View::WrapChanged+0x62
24 00000040`3607df50 00007ff7`f37f615f     wordpad!CWordPadView::WrapChanged+0x84
25 00000040`3607df80 00007ff7`f385242e     wordpad!CRichEdit2View::OnPrinterChanged+0x11f
26 00000040`3607dfb0 00007fff`7c7e9920     wordpad!CWordPadView::OnPrinterChangedMsg+0x5e
27 00000040`3607e000 00007fff`7c7f0bff     MFC42u!CWnd::OnWndMsg+0xd80
28 00000040`3607e130 00007fff`7c7eb42a     MFC42u!CWnd::WindowProc+0x4f
29 00000040`3607e170 00007fff`7c7ea76b     MFC42u!AfxCallWndProc+0x14a
2a 00000040`3607e270 00007ff8`0be38161     MFC42u!AfxWndProcBase+0x15b
2b 00000040`3607e2e0 00007ff8`0be37e1c     USER32!UserCallWinProcCheckWow+0x2d1
2c 00000040`3607e440 00007ff8`0be42cdd     USER32!DispatchClientMessage+0x9c
2d 00000040`3607e4a0 00007ff8`0c1d2db4     USER32!_fnDWORD+0x3d
2e 00000040`3607e500 00007ff8`097f1ef4     ntdll!KiUserCallbackDispatcherContinue
2f 00000040`3607e588 00007fff`7c7e21cd     win32u!NtUserShowWindow+0x14
30 00000040`3607e590 00007ff7`f38346b6     MFC42u!CFrameWnd::ActivateFrame+0x2d
31 00000040`3607e5c0 00007fff`7c7e25de     wordpad!CMainFrame::ActivateFrame+0x86
```

```
32 00000040`3607e630 00007fff`7c7fa2e7     MFC42u!CFrameWnd::InitialUpdateFrame+0xfe
33 00000040`3607e680 00007ff7`f383e331     MFC42u!CSingleDocTemplate::OpenDocumentFile+0x187
34 00000040`3607e6e0 00007fff`7c7efa57     wordpad!CWordPadApp::InitInstance+0x521
35 00000040`3607fc80 00007ff7`f37fb49e     MFC42u!AfxWinMain+0x97
36 00000040`3607fcc0 00007ff8`0b03244d     wordpad!__wmainCRTStartup+0x1de
37 00000040`3607fd80 00007ff8`0c18df78     KERNEL32!BaseThreadInitThunk+0x1d
38 00000040`3607fdb0 00000000`00000000     ntdll!RtlUserThreadStart+0x28
 # Child-SP          RetAddr               Call Site
00 00000040`3607af18 00007fff`f3cce048     USER32!CreateWindowExW
01 00000040`3607af20 00007fff`f3ccdf1f     urlmon!NotificationWindow::CreateHandle+0x68
02 00000040`3607af90 00007fff`f3c7cc9b     urlmon!CTransaction::GetNotificationWnd+0x13f
03 00000040`3607afd0 00007fff`f3cb98ee     urlmon!GetTransactionObjects+0x2fb
04 00000040`3607b050 00007fff`f3cba81c     urlmon!CBinding::StartBinding+0x53e
05 00000040`3607b1d0 00007fff`f3cb9276     urlmon!CUrlMon::StartBinding+0x1b0
06 00000040`3607b2a0 00007fff`f1c05529     urlmon!CUrlMon::BindToStorage+0x96
07 00000040`3607b2f0 00007fff`f1c04acb     msxml6!URLMONStream::deferedOpen+0x159
08 00000040`3607b350 00007fff`f1c048df     msxml6!XMLParser::PushURL+0x1af
09 00000040`3607b3d0 00007fff`f1bbf1c6     msxml6!XMLParser::SetURL+0x7f
0a 00000040`3607b420 00007fff`f1bbed48     msxml6!Document::_load+0x162
0b 00000040`3607b4a0 00007fff`f1b87712     msxml6!Document::load+0x198
0c 00000040`3607b510 00007fff`7c46afae     msxml6!DOMDocumentWrapper::load+0x182
0d 00000040`3607b5e0 00007fff`7c414813     PrintConfig!PrintConfig::PrivateDevmodeMapFile::Parse+0x14e
0e 00000040`3607b750 00007fff`7c5583f0     PrintConfig!PrintConfigPlugIn::DevMode+0x303
0f 00000040`3607b870 00007fff`7c55956a     PrintConfig!BCallOEMDevMode+0x84
10 00000040`3607b8f0 00007fff`7c514999     PrintConfig!PGetDefaultDevmodeWithOemPlugins+0x16a
11 00000040`3607b970 00007fff`7c5144fa     PrintConfig!UniDrvUI::BGetValidatedUserDefaultDevmode+0x81
12 00000040`3607b9e0 00007fff`7c512bb7     PrintConfig!UniDrvUI::BFillCommonInfoDevmode+0x82
13 00000040`3607ba80 00007fff`7c5120ec     PrintConfig!UniDrvUI::LSimpleDocumentProperties+0x173
14 00000040`3607baf0 00007fff`7c40f54b     PrintConfig!UniDrvUI::DrvDocumentPropertySheets+0x60
15 00000040`3607bfa0 00007fff`7c4249ad     PrintConfig!PrintConfig::DrvDocumentPropertySheets+0x123
16 00000040`3607c010 00007fff`7c425543     PrintConfig!ExceptionBoundary<<lambda_ce41999e58d4ed069b2790d632f6e3b5> >+0x3d
17 00000040`3607c060 00007fff`e032e9ba     PrintConfig!DrvDocumentPropertySheets+0x53
18 00000040`3607c0d0 00007fff`e031593f     WINSPOOL!DocumentPropertySheets+0x3fa
19 00000040`3607c120 00007fff`e0324aff     WINSPOOL!DocumentPropertiesWNative+0x193
1a 00000040`3607c1b0 00007fff`a4de746f     WINSPOOL!DocumentPropertiesW+0xcf
1b 00000040`3607c210 00007fff`a4ddb5d0     mxdwdrv!edocs::PDev::PDev+0x2eb
1c 00000040`3607c720 00007ff8`09876470     mxdwdrv!DrvEnablePDEV+0x170
1d 00000040`3607d060 00007ff8`0bea991a     gdi32full!GdiPrinterThunk+0x370
1e 00000040`3607d130 00007ff8`0c1d2db4     USER32!_ClientPrinterThunk+0x3a
1f 00000040`3607d9b0 00007ff8`097f2d94     ntdll!KiUserCallbackDispatcherContinue
20 00000040`3607dab8 00007ff8`0983dbd2     win32u!NtGdiOpenDCW+0x14
21 00000040`3607dac0 00007ff8`0a5f166d     gdi32full!hdcCreateDCW+0xb2
22 00000040`3607de10 00007ff8`0a5f1545     GDI32!bCreateDCW+0xed
23 00000040`3607de80 00007fff`7c7ff0c7     GDI32!CreateDCW+0x55
24 00000040`3607dec0 00007fff`7c7ff020     MFC42u!AfxCreateDC+0x67
25 00000040`3607def0 00007ff7`f37f8a92     MFC42u!CWinApp::CreatePrinterDC+0x20
26 00000040`3607df20 00007ff7`f3854634     wordpad!CRichEdit2View::WrapChanged+0x62
27 00000040`3607df50 00007ff7`f37f615f     wordpad!CWordPadView::WrapChanged+0x84
28 00000040`3607df80 00007ff7`f385242e     wordpad!CRichEdit2View::OnPrinterChanged+0x11f
29 00000040`3607dfb0 00007fff`7c7e9920     wordpad!CWordPadView::OnPrinterChangedMsg+0x5e
2a 00000040`3607e000 00007fff`7c7f0bff     MFC42u!CWnd::OnWndMsg+0xd80
2b 00000040`3607e130 00007fff`7c7eb42a     MFC42u!CWnd::WindowProc+0x4f
2c 00000040`3607e170 00007fff`7c7ea76b     MFC42u!AfxCallWndProc+0x14a
2d 00000040`3607e270 00007ff8`0be38161     MFC42u!AfxWndProcBase+0x15b
2e 00000040`3607e2e0 00007ff8`0be37e1c     USER32!UserCallWinProcCheckWow+0x2d1
2f 00000040`3607e440 00007ff8`0be42cdd     USER32!DispatchClientMessage+0x9c
30 00000040`3607e4a0 00007ff8`0c1d2db4     USER32!_fnDWORD+0x3d
31 00000040`3607e500 00007ff8`097f1ef4     ntdll!KiUserCallbackDispatcherContinue
32 00000040`3607e588 00007fff`7c7e21cd     win32u!NtUserShowWindow+0x14
33 00000040`3607e590 00007ff7`f38346b6     MFC42u!CFrameWnd::ActivateFrame+0x2d
34 00000040`3607e5c0 00007fff`7c7e25de     wordpad!CMainFrame::ActivateFrame+0x86
35 00000040`3607e630 00007fff`7c7fa2e7     MFC42u!CFrameWnd::InitialUpdateFrame+0xfe
36 00000040`3607e680 00007ff7`f383e331     MFC42u!CSingleDocTemplate::OpenDocumentFile+0x187
37 00000040`3607e6e0 00007fff`7c7efa57     wordpad!CWordPadApp::InitInstance+0x521
38 00000040`3607fc80 00007ff7`f37fb49e     MFC42u!AfxWinMain+0x97
39 00000040`3607fcc0 00007ff8`0b03244d     wordpad!__wmainCRTStartup+0x1de
3a 00000040`3607fd80 00007ff8`0c18df78     KERNEL32!BaseThreadInitThunk+0x1d
3b 00000040`3607fdb0 00000000`00000000     ntdll!RtlUserThreadStart+0x28
 # Child-SP          RetAddr               Call Site
00 00000040`3607c938 00007fff`f3cce048     USER32!CreateWindowExW
01 00000040`3607c940 00007fff`f3ccdf1f     urlmon!NotificationWindow::CreateHandle+0x68
02 00000040`3607c9b0 00007fff`f3c7cc9b     urlmon!CTransaction::GetNotificationWnd+0x13f
03 00000040`3607c9f0 00007fff`f3cb98ee     urlmon!GetTransactionObjects+0x2fb
04 00000040`3607ca70 00007fff`f3cba81c     urlmon!CBinding::StartBinding+0x53e
05 00000040`3607cbf0 00007fff`f3cb9276     urlmon!CUrlMon::StartBinding+0x1b0
06 00000040`3607ccc0 00007fff`f1c05529     urlmon!CUrlMon::BindToStorage+0x96
07 00000040`3607cd10 00007fff`f1c04acb     msxml6!URLMONStream::deferedOpen+0x159
08 00000040`3607cd70 00007fff`f1c048df     msxml6!XMLParser::PushURL+0x1af
09 00000040`3607cdf0 00007fff`f1bbf1c6     msxml6!XMLParser::SetURL+0x7f
0a 00000040`3607ce40 00007fff`f1bbed48     msxml6!Document::_load+0x162
0b 00000040`3607cec0 00007fff`f1b87712     msxml6!Document::load+0x198
0c 00000040`3607cf30 00007fff`7c46afae     msxml6!DOMDocumentWrapper::load+0x182
0d 00000040`3607d000 00007fff`7c414968     PrintConfig!PrintConfig::PrivateDevmodeMapFile::Parse+0x14e
0e 00000040`3607d170 00007fff`7c558d7b     PrintConfig!PrintConfigPlugIn::DevMode+0x458
0f 00000040`3607d290 00007fff`7c512b53     PrintConfig!BCalcTotalOEMDMSize+0xa3
10 00000040`3607d330 00007fff`7c5120ec     PrintConfig!UniDrvUI::LSimpleDocumentProperties+0x10f
11 00000040`3607d3a0 00007fff`7c40f54b     PrintConfig!UniDrvUI::DrvDocumentPropertySheets+0x60
12 00000040`3607d850 00007fff`7c4249ad     PrintConfig!PrintConfig::DrvDocumentPropertySheets+0x123
13 00000040`3607d8c0 00007fff`7c425543     PrintConfig!ExceptionBoundary<<lambda_ce41999e58d4ed069b2790d632f6e3b5> >+0x3d
14 00000040`3607d910 00007fff`e032e9ba     PrintConfig!DrvDocumentPropertySheets+0x53
15 00000040`3607d980 00007fff`e031593f     WINSPOOL!DocumentPropertySheets+0x3fa
16 00000040`3607d9d0 00007fff`e0324aff     WINSPOOL!DocumentPropertiesWNative+0x193
17 00000040`3607da60 00007ff8`098585d8     WINSPOOL!DocumentPropertiesW+0xcf
18 00000040`3607dac0 00007ff8`0a5f166d     gdi32full!hdcCreateDCW+0x1aab8
19 00000040`3607de10 00007ff8`0a5f1545     GDI32!bCreateDCW+0xed
1a 00000040`3607de80 00007fff`7c7ff0c7     GDI32!CreateDCW+0x55
1b 00000040`3607dec0 00007fff`7c7ff020     MFC42u!AfxCreateDC+0x67
1c 00000040`3607def0 00007ff7`f37f8a92     MFC42u!CWinApp::CreatePrinterDC+0x20
```

```
1d 00000040`3607df20 00007ff7`f3854634     wordpad!CRichEdit2View::WrapChanged+0x62
1e 00000040`3607df50 00007ff7`f37f615f     wordpad!CWordPadView::WrapChanged+0x84
1f 00000040`3607df80 00007ff7`f385242e     wordpad!CRichEdit2View::OnPrinterChanged+0x11f
20 00000040`3607dfb0 00007fff`7c7e9920     wordpad!CWordPadView::OnPrinterChangedMsg+0x5e
21 00000040`3607e000 00007fff`7c7f0bff     MFC42u!CWnd::OnWndMsg+0xd80
22 00000040`3607e130 00007fff`7c7eb42a     MFC42u!CWnd::WindowProc+0x4f
23 00000040`3607e170 00007fff`7c7ea76b     MFC42u!AfxCallWndProc+0x14a
24 00000040`3607e270 00007ff8`0be38161     MFC42u!AfxWndProcBase+0x15b
25 00000040`3607e2e0 00007ff8`0be37e1c     USER32!UserCallWinProcCheckWow+0x2d1
26 00000040`3607e440 00007ff8`0be42cdd     USER32!DispatchClientMessage+0x9c
27 00000040`3607e4a0 00007ff8`0c1d2db4     USER32!_fnDWORD+0x3d
28 00000040`3607e500 00007ff8`097f1ef4     ntdll!KiUserCallbackDispatcherContinue
29 00000040`3607e588 00007fff`7c7e21cd     win32u!NtUserShowWindow+0x14
2a 00000040`3607e590 00007ff7`f38346b6     MFC42u!CFrameWnd::ActivateFrame+0x2d
2b 00000040`3607e5c0 00007fff`7c7e25de     wordpad!CMainFrame::ActivateFrame+0x86
2c 00000040`3607e630 00007fff`7c7fa2e7     MFC42u!CFrameWnd::InitialUpdateFrame+0xfe
2d 00000040`3607e680 00007ff7`f383e331     MFC42u!CSingleDocTemplate::OpenDocumentFile+0x187
2e 00000040`3607e6e0 00007fff`7c7efa57     wordpad!CWordPadApp::InitInstance+0x521
2f 00000040`3607fc80 00007ff7`f37fb49e     MFC42u!AfxWinMain+0x97
30 00000040`3607fcc0 00007ff8`0b03244d     wordpad!__wmainCRTStartup+0x1de
31 00000040`3607fd80 00007ff8`0c18df78     KERNEL32!BaseThreadInitThunk+0x1d
32 00000040`3607fdb0 00000000`00000000     ntdll!RtlUserThreadStart+0x28
 # Child-SP          RetAddr               Call Site
00 00000040`3607c938 00007fff`f3cce048     USER32!CreateWindowExW
01 00000040`3607c940 00007fff`f3ccdf1f     urlmon!NotificationWindow::CreateHandle+0x68
02 00000040`3607c9b0 00007fff`f3c7cc9b     urlmon!CTransaction::GetNotificationWnd+0x13f
03 00000040`3607c9f0 00007fff`f3cb98ee     urlmon!GetTransactionObjects+0x2fb
04 00000040`3607ca70 00007fff`f3cba81c     urlmon!CBinding::StartBinding+0x53e
05 00000040`3607cbf0 00007fff`f3cb9276     urlmon!CUrlMon::StartBinding+0x1b0
06 00000040`3607ccc0 00007fff`f1c05529     urlmon!CUrlMon::BindToStorage+0x96
07 00000040`3607cd10 00007fff`f1c04acb     msxml6!URLMONStream::deferedOpen+0x159
08 00000040`3607cd70 00007fff`f1c048df     msxml6!XMLParser::PushURL+0x1af
09 00000040`3607cdf0 00007fff`f1bbf1c6     msxml6!XMLParser::SetURL+0x7f
0a 00000040`3607ce40 00007fff`f1bbed48     msxml6!Document::_load+0x162
0b 00000040`3607cec0 00007fff`f1b87712     msxml6!Document::load+0x198
0c 00000040`3607cf30 00007fff`7c46afae     msxml6!DOMDocumentWrapper::load+0x182
0d 00000040`3607d000 00007fff`7c414968     PrintConfig!PrintConfig::PrivateDevmodeMapFile::Parse+0x14e
0e 00000040`3607d170 00007fff`7c558d7b     PrintConfig!PrintConfigPlugIn::DevMode+0x458
0f 00000040`3607d290 00007fff`7c512b53     PrintConfig!BCalcTotalOEMDMSize+0xa3
10 00000040`3607d330 00007fff`7c5120ec     PrintConfig!UniDrvUI::LSimpleDocumentProperties+0x10f
11 00000040`3607d3a0 00007fff`7c40f54b     PrintConfig!UniDrvUI::DrvDocumentPropertySheets+0x60
12 00000040`3607d850 00007fff`7c4249ad     PrintConfig!PrintConfig::DrvDocumentPropertySheets+0x123
13 00000040`3607d8c0 00007fff`7c425543     PrintConfig!ExceptionBoundary<<lambda_ce41999e58d4ed069b2790d632f6e3b5> >+0x3d
14 00000040`3607d910 00007fff`e032e9ba     PrintConfig!DrvDocumentPropertySheets+0x53
15 00000040`3607d980 00007fff`e0315895     WINSPOOL!DocumentPropertySheets+0x3fa
16 00000040`3607d9d0 00007fff`e0324aff     WINSPOOL!DocumentPropertiesWNative+0xe9
17 00000040`3607da60 00007ff8`0985865c     WINSPOOL!DocumentPropertiesW+0xcf
18 00000040`3607dac0 00007ff8`0a5f166d     gdi32full!hdcCreateDCW+0x1ab3c
19 00000040`3607de10 00007ff8`0a5f1545     GDI32!bCreateDCW+0xed
1a 00000040`3607de80 00007fff`7c7ff0c7     GDI32!CreateDCW+0x55
1b 00000040`3607dec0 00007fff`7c7ff020     MFC42u!AfxCreateDC+0x67
1c 00000040`3607def0 00007ff7`f37f8a92     MFC42u!CWinApp::CreatePrinterDC+0x20
1d 00000040`3607df20 00007ff7`f3854634     wordpad!CRichEdit2View::WrapChanged+0x62
1e 00000040`3607df50 00007ff7`f37f615f     wordpad!CWordPadView::WrapChanged+0x84
1f 00000040`3607df80 00007ff7`f385242e     wordpad!CRichEdit2View::OnPrinterChanged+0x11f
20 00000040`3607dfb0 00007fff`7c7e9920     wordpad!CWordPadView::OnPrinterChangedMsg+0x5e
21 00000040`3607e000 00007fff`7c7f0bff     MFC42u!CWnd::OnWndMsg+0xd80
22 00000040`3607e130 00007fff`7c7eb42a     MFC42u!CWnd::WindowProc+0x4f
23 00000040`3607e170 00007fff`7c7ea76b     MFC42u!AfxCallWndProc+0x14a
24 00000040`3607e270 00007ff8`0be38161     MFC42u!AfxWndProcBase+0x15b
25 00000040`3607e2e0 00007ff8`0be37e1c     USER32!UserCallWinProcCheckWow+0x2d1
26 00000040`3607e440 00007ff8`0be42cdd     USER32!DispatchClientMessage+0x9c
27 00000040`3607e4a0 00007ff8`0c1d2db4     USER32!_fnDWORD+0x3d
28 00000040`3607e500 00007ff8`097f1ef4     ntdll!KiUserCallbackDispatcherContinue
29 00000040`3607e588 00007fff`7c7e21cd     win32u!NtUserShowWindow+0x14
2a 00000040`3607e590 00007ff7`f38346b6     MFC42u!CFrameWnd::ActivateFrame+0x2d
2b 00000040`3607e5c0 00007fff`7c7e25de     wordpad!CMainFrame::ActivateFrame+0x86
2c 00000040`3607e630 00007fff`7c7fa2e7     MFC42u!CFrameWnd::InitialUpdateFrame+0xfe
2d 00000040`3607e680 00007ff7`f383e331     MFC42u!CSingleDocTemplate::OpenDocumentFile+0x187
2e 00000040`3607e6e0 00007fff`7c7efa57     wordpad!CWordPadApp::InitInstance+0x521
2f 00000040`3607fc80 00007ff7`f37fb49e     MFC42u!AfxWinMain+0x97
30 00000040`3607fcc0 00007ff8`0b03244d     wordpad!__wmainCRTStartup+0x1de
31 00000040`3607fd80 00007ff8`0c18df78     KERNEL32!BaseThreadInitThunk+0x1d
32 00000040`3607fdb0 00000000`00000000     ntdll!RtlUserThreadStart+0x28
...
...
...
ModLoad: 00007ff8`06660000 00007ff8`06f17000   C:\WINDOWS\SYSTEM32\windows.storage.dll
ModLoad: 00007ff8`06520000 00007ff8`0665d000   C:\WINDOWS\SYSTEM32\wintypes.dll
```

Note: We keep these stack traces for reference in later exercises and slide descriptions.

9. We close logging before exiting WinDbg:

```
0:000> .logclose
Closing open log file C:\AWAPI-Dumps\W3.log
```

API Sequences (Prescriptive)

- CreateThread, ..., CloseHandle
- RegisterClass, CreateWindowEx
- GetMessage, TranslateMessage, DispatchMessage
- BeginPaint, ..., EndPaint
- GetDC, ..., ReleaseDC

By prescriptive API sequences, we mean how to use API functions and in what sequence. It is like a traditional approach to language grammar teaching how to use a language. The slide shows some iconic prescriptive sequences we talk about when we discuss appropriate API classes.

API Sequences (Descriptive)

- Horizontal
  - Code disassembly
  - Traces and logs (Thread of Activity analysis pattern)
- Vertical
  - Stack trace
  - Traces and logs (Fiber Bundle analysis pattern)

Descriptive sequences are sequences that we actually see in memory dumps, traces, and logs, and also during live debugging. They may deviate from prescriptive sequences, which should trigger an investigation if detected. We can discover and analyze horizontal API sequences by disassembling code or looking at trace and log messages corresponding to particular threads, and vertical API sequences by looking at stack traces from problem threads during postmortem debugging, API usage during live debugging sessions as we did in the previous exercise, or looking at stack traces associated with particular trace and log messages, the so-called **Fiber Bundle** trace and log analysis pattern.

Thread of Activity
https://www.dumpanalysis.org/blog/index.php/2009/08/03/trace-analysis-patterns-part-7/

Fiber Bundle
https://www.dumpanalysis.org/blog/index.php/2012/09/26/trace-analysis-patterns-part-52/

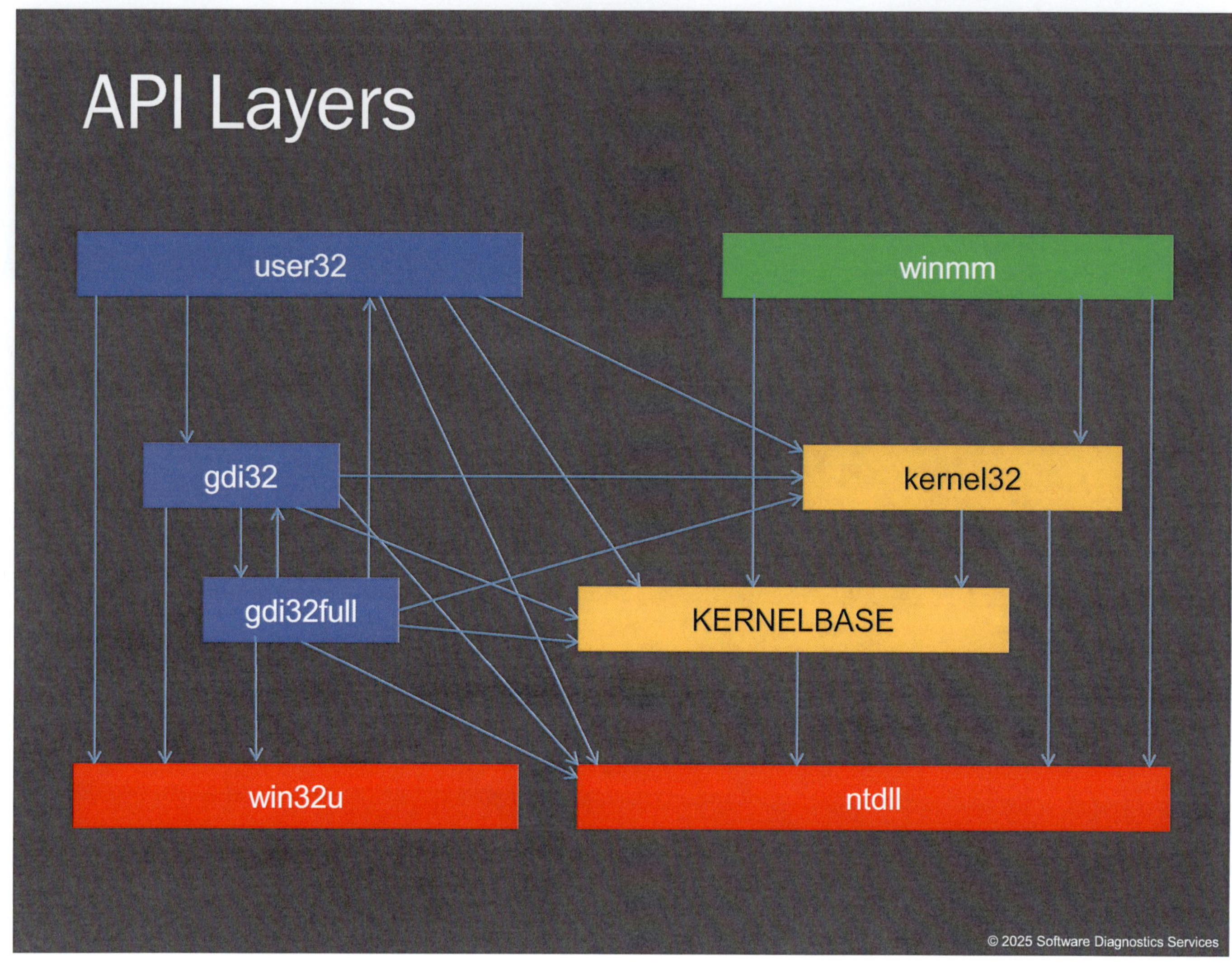

Suppose you start looking at the IAT of the *user32.dll* module and another non-GUI module and look at imports. If you continue looking at referenced modules, you may get this picture. Here, I chose the *winmm* module because I used to work for many years with multimedia API and soundcards. Note that some modules are also linked to the C runtime library, such as *ucrtbase*. We didn't include it in the diagram. If we look at IAT dependencies, we find 3 clusters that we call layers. The bottom layer that talks to OS is shown in red color, and it consists of *ntdll* and *win32u* modules. All other modules use them. Apart from them, there are 2 groups of modules: kernel-related ones and *user/gdi* ones. The former group consists of *kerne32* and the recently refactored *KERNELBASE* to which the old *kernel32* redirects. All other modules use their services. The latter group consists of modules related to windowing and graphics. They are *user32*, *gdi32*, and the refactored *gdi32full*. Usually, UI applications use the latter group of modules. We also didn't include *advapi32.dll* – we look at it in the next exercise.

API Internals

- Memory analysis patterns:
  - Hooked Functions (User Space)
  - Module patterns
    - Hooked Modules
- Malware analysis patterns:
  - Patched Code

WinDbg Commands

```
0:000> .chkimg

0:000> !for_each_module

0:000> u fname

0:000> uf /c fname
```


Windows API functions can be patched by value-adding code and malware, and modules in memory can be checked by comparing code sections from modules downloaded from the Microsoft symbol server since code sections are supposed to be execute-readonly. There are a few WinDbg commands we will try in the next exercise covering ADDR patterns, such as **Function Skeleton** and **Call Path**.

Hooked Functions (User Space)
https://www.dumpanalysis.org/blog/index.php/2007/11/22/crash-dump-analysis-patterns-part-38/

Module patterns
https://www.dumpanalysis.org/blog/index.php/2012/07/15/module-patterns/

Hooked Modules
https://www.dumpanalysis.org/blog/index.php/2008/09/19/hooked-modules/

Patched Code
https://www.dumpanalysis.org/blog/index.php/2013/02/09/malware-analysis-patterns-part-21/

Exercise W4

- **Goal:** Explore API layers and internals of specific API functions
- **ADDR Patterns:** Function Skeleton; Call Path
- \AWAPI-Dumps\Exercise-W4.pdf

Exercise W4

Goal: Explore API layers and internals of specific API functions.

ADDR Patterns: Function Skeleton; Call Path.

1. Launch WinDbg.

2. Open \AWAPI-Dumps\Process\Notepad.DMP

3. We get the dump file loaded:

```
Microsoft (R) Windows Debugger Version 10.0.27725.1000 AMD64
Copyright (c) Microsoft Corporation. All rights reserved.


Loading Dump File [C:\AWAPI-Dumps\Process\Notepad.DMP]
User Mini Dump File with Full Memory: Only application data is available


************* Path validation summary **************
Response                         Time (ms)     Location
Deferred                                       srv*
Symbol search path is: srv*
Executable search path is:
Windows 10 Version 22000 MP (2 procs) Free x64
Product: WinNt, suite: SingleUserTS Personal
Edition build lab: 22000.1.amd64fre.co_release.210604-1628
Debug session time: Sat Oct 15 20:05:58.000 2022 (UTC + 0:00)
System Uptime: 0 days 0:18:47.338
Process Uptime: 0 days 0:01:01.000
................................................................
....................................
Loading unloaded module list
.
For analysis of this file, run !analyze -v
win32u!NtUserGetMessage+0x14:
00007ffc`f3811414 ret
```

4. Open a log file using **.logopen**:

```
0:000> .logopen C:\AWAPI-Dumps\W4.log
Opened log file 'C:\AWAPI-Dumps\W4.log'
```

5. Let's look at the current thread stack trace:

```
0:000> k
 # Child-SP          RetAddr               Call Site
00 00000088`716ffcb8 00007ffc`f586464e     win32u!NtUserGetMessage+0x14
01 00000088`716ffcc0 00007ff7`0b4394e7     user32!GetMessageW+0x2e
02 00000088`716ffd20 00007ff7`0b465dba     Notepad+0x194e7
03 00000088`716ffdf0 00007ffc`f4fd54e0     Notepad+0x45dba
04 00000088`716ffe30 00007ffc`f62a485b     kernel32!BaseThreadInitThunk+0x10
05 00000088`716ffe60 00000000`00000000     ntdll!RtlUserThreadStart+0x2b
```

Note: We don't have symbols for the Notepad module, but since it uses *GetMessageW*, the latter could be a part of the message loop processing API sequence. We look at **Function Skeleton** (the **/c** option also translates indirect calls to their destination addresses):

```
0:000> uf /c Notepad+0x194e7
Notepad+0x19010 (00007ff7`0b439010)
  Notepad+0x1904f (00007ff7`0b43904f):
    call to kernel32!GetCommandLineWStub (00007ffc`f4fdd660)
  Notepad+0x1905f (00007ff7`0b43905f):
    call to combase!CoCreateGuid (00007ffc`f5d8ef50) [onecore\com\combase\class\cocrguid.cxx @ 49]
  Notepad+0x190b2 (00007ff7`0b4390b2):
    call to ntdll!EtwEventRegister (00007ffc`f62b59f0)
  Notepad+0x190dd (00007ff7`0b4390dd):
    call to Notepad+0xb350 (00007ff7`0b42b350)
  Notepad+0x190fe (00007ff7`0b4390fe):
    call to ntdll!EtwEventSetInformation (00007ffc`f62b5520)
  Notepad+0x19147 (00007ff7`0b439147):
    call to ntdll!EtwEventRegister (00007ffc`f62b59f0)
  Notepad+0x19172 (00007ff7`0b439172):
    call to Notepad+0xb350 (00007ff7`0b42b350)
  Notepad+0x19190 (00007ff7`0b439190):
    call to ntdll!EtwEventSetInformation (00007ffc`f62b5520)
  Notepad+0x19196 (00007ff7`0b439196):
    call to Notepad+0x34870 (00007ff7`0b454870)
  Notepad+0x1919b (00007ff7`0b43919b):
    call to msvcp140!_Query_perf_frequency (00007ffc`b1c239c0) [d:\a01\_work\3\s\src\vctools\crt\github\stl\src\xtime.cpp @ 89]
  Notepad+0x191a4 (00007ff7`0b4391a4):
    call to msvcp140!_Query_perf_counter (00007ffc`b1c239a0) [d:\a01\_work\3\s\src\vctools\crt\github\stl\src\xtime.cpp @ 83]
  Notepad+0x191e0 (00007ff7`0b4391e0):
    call to Notepad+0x59f0 (00007ff7`0b4259f0)
  Notepad+0x191f8 (00007ff7`0b4391f8):
    call to kernel32!HeapSetInformationStub (00007ffc`f4fde9c0)
  Notepad+0x19205 (00007ff7`0b439205):
    call to combase!CoInitializeEx (00007ffc`f5da0f00) [onecore\com\combase\class\compobj.cxx @ 3734]
  Notepad+0x1923e (00007ff7`0b43923e):
    call to user32!CharNextWStub (00007ffc`f586b130)
  Notepad+0x1926e (00007ff7`0b43926e):
    call to Notepad+0x2ddc0 (00007ff7`0b44ddc0)
  Notepad+0x19287 (00007ff7`0b439287):
    call to user32!GetMessageW (00007ffc`f5864620)
  Notepad+0x192b4 (00007ff7`0b4392b4):
    unresolvable call: call    qword ptr [rax+30h]
  Notepad+0x192c5 (00007ff7`0b4392c5):
    call to user32!PostMessageW (00007ffc`f5857070)
  Notepad+0x192e0 (00007ff7`0b4392e0):
    unresolvable call: call    qword ptr [rax+30h]
  Notepad+0x192ed (00007ff7`0b4392ed):
    call to user32!TranslateAcceleratorW (00007ffc`f5864ea0)
  Notepad+0x1931b (00007ff7`0b43931b):
    call to Notepad+0x45980 (00007ff7`0b465980)
  Notepad+0x19341 (00007ff7`0b439341):
    call to Notepad+0x45920 (00007ff7`0b465920)
  Notepad+0x19351 (00007ff7`0b439351):
    call to Notepad+0x3f330 (00007ff7`0b45f330)
  Notepad+0x1936e (00007ff7`0b43936e):
    call to Notepad+0xed80 (00007ff7`0b42ed80)
  Notepad+0x1939f (00007ff7`0b43939f):
    call to Notepad+0xc4b0 (00007ff7`0b42c4b0)
  Notepad+0x193d3 (00007ff7`0b4393d3):
    call to user32!GetFocus (00007ffc`f5859480)
  Notepad+0x193e4 (00007ff7`0b4393e4):
    call to user32!GetKeyState (00007ffc`f585ace0)
  Notepad+0x193f6 (00007ff7`0b4393f6):
    call to user32!GetKeyState (00007ffc`f585ace0)
  Notepad+0x19433 (00007ff7`0b439433):
    call to Notepad+0x3fc90 (00007ff7`0b45fc90)
  Notepad+0x19451 (00007ff7`0b439451):
    call to Notepad+0x45980 (00007ff7`0b465980)
  Notepad+0x19477 (00007ff7`0b439477):
    call to Notepad+0x45920 (00007ff7`0b465920)
  Notepad+0x19487 (00007ff7`0b439487):
    call to Notepad+0x3f1b0 (00007ff7`0b45f1b0)
  Notepad+0x194a5 (00007ff7`0b4394a5):
    unresolvable call: call    qword ptr [rax+30h]
  Notepad+0x194b2 (00007ff7`0b4394b2):
    call to user32!TranslateAcceleratorW (00007ffc`f5864ea0)
  Notepad+0x194c5 (00007ff7`0b4394c5):
    call to user32!TranslateMessage (00007ffc`f58563e0)
  Notepad+0x194cf (00007ff7`0b4394cf):
    call to user32!DispatchMessageW (00007ffc`f5850bf0)
```

```
Notepad+0x194e1 (00007ff7`0b4394e1):
  call to user32!GetMessageW (00007ffc`f5864620)
Notepad+0x194fd (00007ff7`0b4394fd):
  unresolvable call: call    qword ptr [rax+20h]
Notepad+0x19524 (00007ff7`0b439524):
  call to Notepad+0x45980 (00007ff7`0b465980)
Notepad+0x19541 (00007ff7`0b439541):
  call to Notepad+0x46424 (00007ff7`0b466424)
Notepad+0x1954d (00007ff7`0b43954d):
  call to Notepad+0x45920 (00007ff7`0b465920)
Notepad+0x1956e (00007ff7`0b43956e):
  unresolvable call: call    qword ptr [rax+60h]
Notepad+0x19577 (00007ff7`0b439577):
  unresolvable call: call    rbx
Notepad+0x1957b (00007ff7`0b43957b):
  call to Notepad+0x59f0 (00007ff7`0b4259f0)
Notepad+0x19591 (00007ff7`0b439591):
  call to Notepad+0x35ab0 (00007ff7`0b455ab0)
Notepad+0x195a5 (00007ff7`0b4395a5):
  unresolvable call: call    qword ptr [rax+10h]
Notepad+0x195bb (00007ff7`0b4395bb):
  call to kernel32!GlobalFreeStub (00007ffc`f4fd6b30)
Notepad+0x195cd (00007ff7`0b4395cd):
  call to kernel32!GlobalFreeStub (00007ffc`f4fd6b30)
Notepad+0x1960a (00007ff7`0b43960a):
  call to combase!WindowsCreateStringReference (00007ffc`f5d70ac0) [onecore\com\combase\winrt\string\string.cpp @ 70]
Notepad+0x19628 (00007ff7`0b439628):
  unresolvable call: call    qword ptr [rax+10h]
Notepad+0x1963a (00007ff7`0b43963a):
  call to combase!RoGetActivationFactory (00007ffc`f5da6520) [onecore\com\combase\winrtbase\winrtbase.cpp @ 1060]
Notepad+0x19657 (00007ff7`0b439657):
  call to Notepad+0xb350 (00007ff7`0b42b350)
Notepad+0x19669 (00007ff7`0b439669):
  unresolvable call: call    qword ptr [rax+68h]
Notepad+0x1967d (00007ff7`0b43967d):
  unresolvable call: call    qword ptr [rax+10h]
Notepad+0x19696 (00007ff7`0b439696):
  unresolvable call: call    qword ptr [rax+0A8h]
Notepad+0x1969c (00007ff7`0b43969c):
  call to combase!CoUninitialize (00007ffc`f5da15f0) [onecore\com\combase\class\compobj.cxx @ 3793]
Notepad+0x196b7 (00007ff7`0b4396b7):
  call to ntdll!EtwEventUnregister (00007ffc`f62a65e0)
Notepad+0x196d2 (00007ff7`0b4396d2):
  call to ntdll!EtwEventUnregister (00007ffc`f62a65e0)
Notepad+0x196e2 (00007ff7`0b4396e2):
  call to Notepad+0x45380 (00007ff7`0b465380)
Notepad+0x19704 (00007ff7`0b439704):
  call to Notepad+0x1adf0 (00007ff7`0b43adf0)
Notepad+0x1979a (00007ff7`0b43979a):
  call to Notepad+0x45980 (00007ff7`0b465980)
Notepad+0x197b7 (00007ff7`0b4397b7):
  call to Notepad+0x46424 (00007ff7`0b466424)
Notepad+0x197c3 (00007ff7`0b4397c3):
  call to Notepad+0x45920 (00007ff7`0b465920)
Notepad+0x197e7 (00007ff7`0b4397e7):
  unresolvable call: call    qword ptr [r8+0F8h]
Notepad+0x19808 (00007ff7`0b439808):
  call to Notepad+0x1dda0 (00007ff7`0b43dda0)
Notepad+0x19810 (00007ff7`0b439810):
  call to apphelp!MbHook_IsWindowVisible (00007ffc`f09d3160)
Notepad+0x19836 (00007ff7`0b439836):
  call to Notepad+0x1dac0 (00007ff7`0b43dac0)
Notepad+0x1985c (00007ff7`0b43985c):
  call to user32!GetWindowRect (00007ffc`f58503b0)
Notepad+0x19899 (00007ff7`0b439899):
  call to user32!NtUserGetGUIThreadInfo (00007ffc`f5872320)
Notepad+0x198b8 (00007ff7`0b4398b8):
  call to user32!ClientToScreen (00007ffc`f585d6f0)
Notepad+0x1991f (00007ff7`0b43991f):
  call to user32!ScreenToClient (00007ffc`f584c780)
Notepad+0x19933 (00007ff7`0b439933):
  call to Notepad+0x107c0 (00007ff7`0b4307c0)
Notepad+0x19951 (00007ff7`0b439951):
  call to user32!GetActiveWindow (00007ffc`f586a2b0)
Notepad+0x19983 (00007ff7`0b439983):
  call to Notepad+0x45980 (00007ff7`0b465980)
Notepad+0x199a0 (00007ff7`0b4399a0):
  call to Notepad+0x46424 (00007ff7`0b466424)
Notepad+0x199ac (00007ff7`0b4399ac):
  call to Notepad+0x45920 (00007ff7`0b465920)
Notepad+0x199d6 (00007ff7`0b4399d6):
  call to user32!GetCursorPos (00007ffc`f5868520)
Notepad+0x199e4 (00007ff7`0b4399e4):
  call to user32!ScreenToClient (00007ffc`f584c780)
```

```
  Notepad+0x199ef (00007ff7`0b4399ef):
    call to user32!GetKeyState (00007ffc`f585ace0)
  Notepad+0x199fd (00007ff7`0b4399fd):
    call to user32!GetKeyState (00007ffc`f585ace0)
  Notepad+0x19a3c (00007ff7`0b439a3c):
    call to user32!SendMessageW (00007ffc`f5850600)
  Notepad+0x19a5d (00007ff7`0b439a5d):
    call to comctl32!DefSubclassProc (00007ffc`e1914750)
  Notepad+0x19a6a (00007ff7`0b439a6a):
    call to Notepad+0x45380 (00007ff7`0b465380)
```

6. We see that the Notepad function used a *PostMessageW* API call that is supposed to be non-blocking. Let's check its internals:

```
0:000> uf /c user32!PostMessageW
user32!PostMessageW (00007ffc`f5857070)
  user32!PostMessageW+0x4d (00007ffc`f58570bd):
    call to win32u!NtUserPostMessage (00007ffc`f3811520)
  user32!PostMessageW+0x27be8 (00007ffc`f587ec58):
    call to win32u!NtUserQueryWindow (00007ffc`f3811540)
  user32!PostMessageW+0x27c06 (00007ffc`f587ec76):
    call to kernel32!GlobalSize (00007ffc`f4fd7c50)
  user32!PostMessageW+0x27c20 (00007ffc`f587ec90):
    call to user32!SendMessageW (00007ffc`f5850600)
  user32!PostMessageW+0x27c55 (00007ffc`f587ecc5):
    call to user32!ForwardTouchMessage (00007ffc`f5890118)
  user32!PostMessageW+0x27c81 (00007ffc`f587ecf1):
    call to user32!ForwardGestureMessage (00007ffc`f588d184)
```

Note: We see a *SendMessageW* call which may be potentially blocking.

7. Let's now look at Notepad IAT:

```
0:000> !dh Notepad

File Type: EXECUTABLE IMAGE
FILE HEADER VALUES
    8664 machine (X64)
       6 number of sections
63176285 time date stamp Tue Sep  6 16:08:53 2022

       0 file pointer to symbol table
       0 number of symbols
      F0 size of optional header
      22 characteristics
            Executable
            App can handle >2gb addresses

OPTIONAL HEADER VALUES
     20B magic #
   14.33 linker version
   49C00 size of code
   37800 size of initialized data
       0 size of uninitialized data
   45E28 address of entry point
    1000 base of code
         ----- new -----
00007ff70b420000 image base
    1000 section alignment
     200 file alignment
       2 subsystem (Windows GUI)
```

```
    6.00 operating system version
    0.00 image version
    6.00 subsystem version
   84000 size of image
     400 size of headers
       0 checksum
0000000000100000 size of stack reserve
0000000000001000 size of stack commit
0000000000100000 size of heap reserve
0000000000001000 size of heap commit
    8160  DLL characteristics
            High entropy VA supported
            Dynamic base
            NX compatible
            Terminal server aware
       0 [        0] address [size] of Export Directory
  612E8 [      258] address [size] of Import Directory
   75000 [     D930] address [size] of Resource Directory
   72000 [     2DCC] address [size] of Exception Directory
       0 [        0] address [size] of Security Directory
   83000 [      998] address [size] of Base Relocation Directory
   51610 [       70] address [size] of Debug Directory
       0 [        0] address [size] of Description Directory
       0 [        0] address [size] of Special Directory
   51680 [       28] address [size] of Thread Storage Directory
   514D0 [      140] address [size] of Load Configuration Directory
       0 [        0] address [size] of Bound Import Directory
   4B000 [      D68] address [size] of Import Address Table Directory
       0 [        0] address [size] of Delay Import Directory
       0 [        0] address [size] of COR20 Header Directory
       0 [        0] address [size] of Reserved Directory


SECTION HEADER #1
   .text name
   49AF7 virtual size
    1000 virtual address
   49C00 size of raw data
     400 file pointer to raw data
       0 file pointer to relocation table
       0 file pointer to line numbers
       0 number of relocations
       0 number of line numbers
60000020 flags
         Code
         (no align specified)
         Execute Read

SECTION HEADER #2
  .rdata name
   1974E virtual size
   4B000 virtual address
   19800 size of raw data
   4A000 file pointer to raw data
       0 file pointer to relocation table
       0 file pointer to line numbers
       0 number of relocations
       0 number of line numbers
40000040 flags
```

```
         Initialized Data
         (no align specified)
         Read Only


Debug Directories(4)
       Type       Size     Address  Pointer
       cv           47       59b10    58b10    Format: RSDS, guid, 1,
D:\a\_work\1\b\Release\x64\Notepad\Notepad.pdb
       (   12)      14       59b58    58b58
       (   13)     400       59b6c    58b6c
       (   14)       0           0        0

SECTION HEADER #3
   .data name
    CC50 virtual size
   65000 virtual address
    B200 size of raw data
   63800 file pointer to raw data
       0 file pointer to relocation table
       0 file pointer to line numbers
       0 number of relocations
       0 number of line numbers
C0000040 flags
         Initialized Data
         (no align specified)
         Read Write

SECTION HEADER #4
  .pdata name
    2DCC virtual size
   72000 virtual address
    2E00 size of raw data
   6EA00 file pointer to raw data
       0 file pointer to relocation table
       0 file pointer to line numbers
       0 number of relocations
       0 number of line numbers
40000040 flags
         Initialized Data
         (no align specified)
         Read Only

SECTION HEADER #5
   .rsrc name
    D930 virtual size
   75000 virtual address
    DA00 size of raw data
   71800 file pointer to raw data
       0 file pointer to relocation table
       0 file pointer to line numbers
       0 number of relocations
       0 number of line numbers
40000040 flags
         Initialized Data
         (no align specified)
         Read Only

SECTION HEADER #6
```

```
   .reloc name
      998 virtual size
    83000 virtual address
      A00 size of raw data
    7F200 file pointer to raw data
        0 file pointer to relocation table
        0 file pointer to line numbers
        0 number of relocations
        0 number of line numbers
 42000040 flags
          Initialized Data
          Discardable
          (no align specified)
          Read Only

0:000> dps 00007ff70b420000 + 4B000 LD68/8
00007ff7`0b46b000  00007ffc`f526d000 advapi32!DecryptFileW
00007ff7`0b46b008  00007ffc`f5246800 advapi32!RegOpenKeyExWStub
00007ff7`0b46b010  00007ffc`f5248460 advapi32!RegCreateKeyW
00007ff7`0b46b018  00007ffc`f5246b20 advapi32!RegCloseKeyStub
00007ff7`0b46b020  00007ffc`f5246750 advapi32!RegQueryValueExWStub
00007ff7`0b46b028  00007ffc`f526d090 advapi32!DuplicateEncryptionInfoFile
00007ff7`0b46b030  00007ffc`f5247680 advapi32!RegCreateKeyExWStub
00007ff7`0b46b038  00007ffc`f5248170 advapi32!RegSetValueExWStub
00007ff7`0b46b040  00007ffc`f525f950 advapi32!RegDeleteKeyExWStub
00007ff7`0b46b048  00007ffc`f5246900 advapi32!RegQueryInfoKeyWStub
00007ff7`0b46b050  00007ffc`f5246d50 advapi32!RegEnumValueWStub
00007ff7`0b46b058  00007ffc`f52468e0 advapi32!GetTokenInformationStub
00007ff7`0b46b060  00007ffc`f62a4f40 ntdll!EtwEventWriteTransfer
00007ff7`0b46b068  00007ffc`f62a65e0 ntdll!EtwEventUnregister
00007ff7`0b46b070  00007ffc`f62b59f0 ntdll!EtwEventRegister
00007ff7`0b46b078  00007ffc`f62b5520 ntdll!EtwEventSetInformation
00007ff7`0b46b080  00007ffc`f5246b40 advapi32!IsTextUnicode
00007ff7`0b46b088  00000000`00000000
00007ff7`0b46b090  00007ffc`e1907530 comctl32!SetWindowSubclass
00007ff7`0b46b098  00007ffc`e18da3e0 comctl32!CreateStatusWindowW
00007ff7`0b46b0a0  00007ffc`e1914750 comctl32!DefSubclassProc
00007ff7`0b46b0a8  00000000`00000000
00007ff7`0b46b0b0  00007ffc`f59f1440 comdlg32!GetFileTitleW
00007ff7`0b46b0b8  00007ffc`f5a815c0 comdlg32!PrintDlgExW
00007ff7`0b46b0c0  00007ffc`f5a532d0 comdlg32!PageSetupDlgW
00007ff7`0b46b0c8  00007ffc`f5a41b80 comdlg32!CommDlgExtendedError
00007ff7`0b46b0d0  00000000`00000000
00007ff7`0b46b0d8  00007ffc`f41fd2b0 gdi32!AbortDoc
00007ff7`0b46b0e0  00007ffc`f41f1c70 gdi32!DeleteObject
00007ff7`0b46b0e8  00007ffc`f41f5a80 gdi32!EndPage
00007ff7`0b46b0f0  00007ffc`f41fe290 gdi32!StartDocW
00007ff7`0b46b0f8  00007ffc`f41f5a30 gdi32!StartPage
00007ff7`0b46b100  00007ffc`f41f5cf0 gdi32!EndDoc
00007ff7`0b46b108  00007ffc`f41f3d60 gdi32!SetBkMode
00007ff7`0b46b110  00007ffc`f41f3a90 gdi32!SelectObject
00007ff7`0b46b118  00007ffc`f41f1350 gdi32!CreateFontIndirectW
00007ff7`0b46b120  00007ffc`f41f4c50 gdi32!CreateSolidBrushStub
00007ff7`0b46b128  00007ffc`f41f73b0 gdi32!EnumFontsW
00007ff7`0b46b130  00007ffc`f41f3ba0 gdi32!GetStockObjectStub
00007ff7`0b46b138  00007ffc`f41f3fb0 gdi32!GetTextMetricsWStub
00007ff7`0b46b140  00007ffc`f41f45c0 gdi32!LPtoDPStub
00007ff7`0b46b148  00007ffc`f41fc620 gdi32!SetWindowExtExStub
00007ff7`0b46b150  00007ffc`f41fc5b0 gdi32!SetViewportExtExStub
00007ff7`0b46b158  00007ffc`f41f16e0 gdi32!SetMapModeStub
00007ff7`0b46b160  00007ffc`f41f33d0 gdi32!GetDeviceCaps
00007ff7`0b46b168  00007ffc`f41f12d0 gdi32!GetTextExtentPoint32WStub
00007ff7`0b46b170  00007ffc`f41f7610 gdi32!TextOutW
00007ff7`0b46b178  00007ffc`f41f13a0 gdi32!CreateDCW
00007ff7`0b46b180  00007ffc`f41f2b40 gdi32!CreateRectRgn
00007ff7`0b46b188  00007ffc`f41f3c80 gdi32!BitBltStub
00007ff7`0b46b190  00007ffc`f41fa9d0 gdi32!SetAbortProc
00007ff7`0b46b198  00007ffc`f41f3e60 gdi32!SetTextColor
00007ff7`0b46b1a0  00007ffc`f41f40a0 gdi32!SetBkColorStub
00007ff7`0b46b1a8  00007ffc`f41f3eb0 gdi32!CreateCompatibleDCStub
00007ff7`0b46b1b0  00007ffc`f41f2ef0 gdi32!DeleteDC
00007ff7`0b46b1b8  00007ffc`f41f7310 gdi32!EnumFontFamiliesExW
00007ff7`0b46b1c0  00007ffc`f41f2a40 gdi32!CreateDIBSectionStub
00007ff7`0b46b1c8  00000000`00000000
00007ff7`0b46b1d0  00007ffc`f5c858e0 imm32!ImmReleaseContext
00007ff7`0b46b1d8  00007ffc`f5c8d240 imm32!ImmSetCompositionStringW
00007ff7`0b46b1e0  00007ffc`f5c85840 imm32!ImmIsIME
00007ff7`0b46b1e8  00007ffc`f5c91d50 imm32!ImmGetProperty
00007ff7`0b46b1f0  00007ffc`f5c81160 imm32!ImmSetCompositionWindow
00007ff7`0b46b1f8  00007ffc`f5c83030 imm32!ImmGetContext
00007ff7`0b46b200  00007ffc`f5c81d40 imm32!ImmGetCompositionWindow
00007ff7`0b46b208  00000000`00000000
00007ff7`0b46b210  00007ffc`f4fe2bd0 kernel32!GetCurrentProcess
00007ff7`0b46b218  00007ffc`f4fd5ef0 kernel32!HeapFreeStub
00007ff7`0b46b220  00007ffc`f4fe2df0 kernel32!SetEvent
00007ff7`0b46b228  00007ffc`f4fe2cc0 kernel32!CreateEventExW
00007ff7`0b46b230  00007ffc`f4fe2da0 kernel32!OpenSemaphoreW
00007ff7`0b46b238  00007ffc`f4fdb7d0 kernel32!IsProcessorFeaturePresentStub
```

```
00007ff7`0b46b240  00007ffc`f4fd93b0 kernel32!GetProcAddressStub
00007ff7`0b46b248  00007ffc`f4fe2d00 kernel32!CreateMutexExW
00007ff7`0b46b250  00007ffc`f4fd7b20 kernel32!CompareStringOrdinalStub
00007ff7`0b46b258  00007ffc`f4fe2e50 kernel32!WaitForSingleObjectEx
00007ff7`0b46b260  00007ffc`f4fe3270 kernel32!ReadFile
00007ff7`0b46b268  00007ffc`f4fe2fb0 kernel32!FindFirstFileW
00007ff7`0b46b270  00007ffc`f4fe2dc0 kernel32!ReleaseMutex
00007ff7`0b46b278  00007ffc`f4fde090 kernel32!GetModuleHandleExWStub
00007ff7`0b46b280  00007ffc`f4fe2dd0 kernel32!ReleaseSemaphore
00007ff7`0b46b288  00007ffc`f4fe2d20 kernel32!CreateSemaphoreExW
00007ff7`0b46b290  00007ffc`f4fe2be0 kernel32!GetCurrentProcessId
00007ff7`0b46b298  00007ffc`f62da4e0 ntdll!RtlEnterCriticalSection
00007ff7`0b46b2a0  00007ffc`f4fe33a0 kernel32!MulDiv
00007ff7`0b46b2a8  00007ffc`f4fd6030 kernel32!GlobalUnlock
00007ff7`0b46b2b0  00007ffc`f62db4d0 ntdll!RtlLeaveCriticalSection
00007ff7`0b46b2b8  00007ffc`f4fd6100 kernel32!GlobalLock
00007ff7`0b46b2c0  00007ffc`f4fdb790 kernel32!GetModuleHandleWStub
00007ff7`0b46b2c8  00007ffc`f4fe2d50 kernel32!InitializeCriticalSectionAndSpinCount
00007ff7`0b46b2d0  00007ffc`f62be080 ntdll!RtlDeleteCriticalSection
00007ff7`0b46b2d8  00007ffc`f4fe2de0 kernel32!ResetEvent
00007ff7`0b46b2e0  00007ffc`f4fdc070 kernel32!UnmapViewOfFileStub
00007ff7`0b46b2e8  00007ffc`f3b4be90 KERNELBASE!GetCurrentPackageFullName
00007ff7`0b46b2f0  00007ffc`f3b9fc20 KERNELBASE!ParseApplicationUserModelId
00007ff7`0b46b2f8  00007ffc`f3b3c250 KERNELBASE!GetCurrentApplicationUserModelId
00007ff7`0b46b300  00007ffc`f4fdb9e0 kernel32!MapViewOfFileStub
00007ff7`0b46b308  00007ffc`f4fda7c0 kernel32!CreateFileMappingWStub
00007ff7`0b46b310  00007ffc`f4fde8e0 kernel32!LocalReAllocStub
00007ff7`0b46b318  00007ffc`f4fde920 kernel32!GetCurrentDirectoryWStub
00007ff7`0b46b320  00007ffc`f4fe0230 kernel32!RegisterApplicationRestartStub
00007ff7`0b46b328  00007ffc`f4fd5fc0 kernel32!MultiByteToWideCharStub
00007ff7`0b46b330  00007ffc`f4fe3060 kernel32!GetDiskFreeSpaceExW
00007ff7`0b46b338  00007ffc`f4fe2cd0 kernel32!CreateEventW
00007ff7`0b46b340  00007ffc`f4fe01a0 kernel32!SetCurrentDirectoryWStub
00007ff7`0b46b348  00007ffc`f4fe30c0 kernel32!GetFileAttributesExW
00007ff7`0b46b350  00007ffc`f4fe59f0 kernel32!LocalUnlockStub
00007ff7`0b46b358  00007ffc`f4fd6670 kernel32!QueryPerformanceCounterStub
00007ff7`0b46b360  00007ffc`f4fe32c0 kernel32!SetEndOfFile
00007ff7`0b46b368  00007ffc`f4fde050 kernel32!GetACPStub
00007ff7`0b46b370  00007ffc`f62c8ac0 ntdll!RtlAllocateHeap
00007ff7`0b46b378  00007ffc`f4fe2ea0 kernel32!CreateDirectoryW
00007ff7`0b46b380  00007ffc`f4fd6b30 kernel32!GlobalFreeStub
00007ff7`0b46b388  00007ffc`f4fd9370 kernel32!GlobalAllocStub
00007ff7`0b46b390  00007ffc`f4fe2a00 kernel32!RtlCaptureContext
00007ff7`0b46b398  00007ffc`f4fe3360 kernel32!WriteFile
00007ff7`0b46b3a0  00007ffc`f4fd6010 kernel32!WideCharToMultiByteStub
00007ff7`0b46b3a8  00007ffc`f4fd7a90 kernel32!GetSystemTimeAsFileTimeStub
00007ff7`0b46b3b0  00007ffc`f631fda0 ntdll!VerSetConditionMask
00007ff7`0b46b3b8  00007ffc`f4fd6690 kernel32!VerifyVersionInfoW
00007ff7`0b46b3c0  00007ffc`f63155b0 ntdll!RtlInitializeSListHead
00007ff7`0b46b3c8  00007ffc`f4fe2f30 kernel32!FindClose
00007ff7`0b46b3d0  00007ffc`f4fe30e0 kernel32!GetFileInformationByHandle
00007ff7`0b46b3d8  00007ffc`f4fe23f0 kernel32!K32GetModuleFileNameExWStub
00007ff7`0b46b3e0  00007ffc`f4fdbfa0 kernel32!GetModuleFileNameWStub
00007ff7`0b46b3e8  00007ffc`f4fe2c50 kernel32!CloseHandle
00007ff7`0b46b3f0  00007ffc`f4fdd660 kernel32!GetCommandLineWStub
00007ff7`0b46b3f8  00007ffc`f4fde9c0 kernel32!HeapSetInformationStub
00007ff7`0b46b400  00007ffc`f4fe30d0 kernel32!GetFileAttributesW
00007ff7`0b46b408  00007ffc`f4fe2f00 kernel32!DeleteFileW
00007ff7`0b46b410  00007ffc`f4fd9330 kernel32!LocalAllocStub
00007ff7`0b46b418  00007ffc`f4fe2ed0 kernel32!CreateFileW
00007ff7`0b46b420  00007ffc`f6333740 ntdll!_C_specific_handler
00007ff7`0b46b428  00007ffc`f4ff8630 kernel32!DebugBreakStub
00007ff7`0b46b430  00007ffc`f4fd6360 kernel32!SetLastErrorStub
00007ff7`0b46b438  00007ffc`f62a6c60 ntdll!RtlInterlockedPushEntrySList
00007ff7`0b46b440  00007ffc`f4fda650 kernel32!FreeLibraryStub
00007ff7`0b46b448  00007ffc`f62b9860 ntdll!RtlAcquireSRWLockExclusive
00007ff7`0b46b450  00007ffc`f62bb270 ntdll!RtlReleaseSRWLockExclusive
00007ff7`0b46b458  00007ffc`f4fd62e0 kernel32!GetLastErrorStub
00007ff7`0b46b460  00007ffc`f4fde730 kernel32!IsDebuggerPresentStub
00007ff7`0b46b468  00007ffc`f4fdbf80 kernel32!FormatMessageWStub
00007ff7`0b46b470  00007ffc`f4fdbe40 kernel32!RaiseExceptionStub
00007ff7`0b46b478  00007ffc`f4fc6170 kernel32!GetCurrentThreadId
00007ff7`0b46b480  00007ffc`f4fdaf10 kernel32!OutputDebugStringWStub
00007ff7`0b46b488  00007ffc`f4fd82d0 kernel32!LocalFreeStub
00007ff7`0b46b490  00007ffc`f4fe0160 kernel32!TrySubmitThreadpoolCallbackStub
00007ff7`0b46b498  00007ffc`f4fe5840 kernel32!FindNLSStringStub
00007ff7`0b46b4a0  00007ffc`f4fd6340 kernel32!GetProcessHeapStub
00007ff7`0b46b4a8  00007ffc`f4fdedf0 kernel32!GetLocaleInfoWStub
00007ff7`0b46b4b0  00007ffc`f4fdf9e0 kernel32!GetUserDefaultUILanguageStub
00007ff7`0b46b4b8  00007ffc`f4fde680 kernel32!GetLocalTimeStub
00007ff7`0b46b4c0  00007ffc`f4fdf4a0 kernel32!GetDateFormatWStub
00007ff7`0b46b4c8  00007ffc`f4fdd6a0 kernel32!GetTimeFormatWStub
00007ff7`0b46b4d0  00007ffc`f4fe0b20 kernel32!RtlLookupFunctionEntryStub
00007ff7`0b46b4d8  00007ffc`f4fe5ab0 kernel32!RtlVirtualUnwindStub
00007ff7`0b46b4e0  00007ffc`f4ffa370 kernel32!UnhandledExceptionFilterStub
00007ff7`0b46b4e8  00007ffc`f4fde6d0 kernel32!SetUnhandledExceptionFilterStub
00007ff7`0b46b4f0  00007ffc`f4fdba00 kernel32!GetStartupInfoWStub
00007ff7`0b46b4f8  00007ffc`f4fde880 kernel32!LoadLibraryWStub
00007ff7`0b46b500  00007ffc`f4fdf800 kernel32!TerminateProcessStub
00007ff7`0b46b508  00007ffc`f4fe59d0 kernel32!LocalLockStub
00007ff7`0b46b510  00007ffc`f4fe3160 kernel32!GetFullPathNameW
00007ff7`0b46b518  00007ffc`f4fdd600 kernel32!GetModuleFileNameAStub
00007ff7`0b46b520  00007ffc`f4fe2e40 kernel32!WaitForSingleObject
00007ff7`0b46b528  00000000`00000000
00007ff7`0b46b530  00007ffc`b1c239a0 msvcp140!_Query_perf_counter [d:\a01\_work\3\s\src\vctools\crt\github\stl\src\xtime.cpp @ 83]
00007ff7`0b46b538  00007ffc`b1c239c0 msvcp140!_Query_perf_frequency [d:\a01\_work\3\s\src\vctools\crt\github\stl\src\xtime.cpp @ 89]
00007ff7`0b46b540  00007ffc`b1c5d4c0 msvcp140!std::basic_istream<wchar_t,std::char_traits<wchar_t> >::ignore
[d:\a01\_work\3\s\src\vctools\crt\github\stl\inc\istream @ 482]
```

```
00007ff7`0b46b548  00007ffc`b1c4b170 msvcp140!std::_Xout_of_range [d:\a01\_work\3\s\src\vctools\crt\github\stl\src\xthrow.cpp @ 24]
00007ff7`0b46b550  00007ffc`b1c124c0 msvcp140!__ExceptionPtrCreate [d:\a01\_work\3\s\src\vctools\crt\github\stl\src\excptptr.cpp @ 427]
00007ff7`0b46b558  00007ffc`b1c12550 msvcp140!__ExceptionPtrDestroy [d:\a01\_work\3\s\src\vctools\crt\github\stl\src\excptptr.cpp @ 431]
00007ff7`0b46b560  00007ffc`b1c230c0 msvcp140!_Thrd_yield [d:\a01\_work\3\s\src\vctools\crt\github\stl\src\cthread.cpp @ 81]
00007ff7`0b46b568  00007ffc`b1c12440 msvcp140!__ExceptionPtrAssign [d:\a01\_work\3\s\src\vctools\crt\github\stl\src\excptptr.cpp @ 439]
00007ff7`0b46b570  00007ffc`b1c12460 msvcp140!__ExceptionPtrCopy [d:\a01\_work\3\s\src\vctools\crt\github\stl\src\excptptr.cpp @ 435]
00007ff7`0b46b578  00007ffc`b1c124d0 msvcp140!__ExceptionPtrCurrentException [d:\a01\_work\3\s\src\vctools\crt\github\stl\src\excptptr.cpp @ 459]
00007ff7`0b46b580  00007ffc`b1c12560 msvcp140!__ExceptionPtrRethrow [d:\a01\_work\3\s\src\vctools\crt\github\stl\src\excptptr.cpp @ 476]
00007ff7`0b46b588  00007ffc`b1c12490 msvcp140!__ExceptionPtrCopyException [d:\a01\_work\3\s\src\vctools\crt\github\stl\src\excptptr.cpp @ 539]
00007ff7`0b46b590  00007ffc`b1c14a10 msvcp140!std::basic_ios<char,std::char_traits<char> >::~basic_ios<char,std::char_traits<char> >
[d:\a01\_work\3\s\src\vctools\crt\github\stl\inc\ios @ 37]
00007ff7`0b46b598  00007ffc`b1c4b140 msvcp140!std::_Xlength_error [d:\a01\_work\3\s\src\vctools\crt\github\stl\src\xthrow.cpp @ 20]
00007ff7`0b46b5a0  00007ffc`b1c25880 msvcp140!std::basic_streambuf<wchar_t,std::char_traits<wchar_t> >::~basic_streambuf<wchar_t,std::char_traits<wchar_t> >
[d:\a01\_work\3\s\src\vctools\crt\github\stl\inc\streambuf @ 68]
00007ff7`0b46b5a8  00007ffc`b1c25cd0 msvcp140!std::basic_istream<wchar_t,std::char_traits<wchar_t> >::~basic_istream<wchar_t,std::char_traits<wchar_t> >
[d:\a01\_work\3\s\src\vctools\crt\github\stl\inc\istream @ 71]
00007ff7`0b46b5b0  00007ffc`b1c53f60 msvcp140!std::basic_istream<wchar_t,std::char_traits<wchar_t> >::operator>>
[d:\a01\_work\3\s\src\vctools\crt\github\stl\inc\istream @ 263]
00007ff7`0b46b5b8  00007ffc`b1c3f2f0 msvcp140!std::basic_streambuf<unsigned short,std::char_traits<unsigned short> >::sbumpc
[d:\a01\_work\3\s\src\vctools\crt\github\stl\inc\streambuf @ 125]
00007ff7`0b46b5c0  00007ffc`b1c24f00 msvcp140!std::basic_streambuf<unsigned short,std::char_traits<unsigned short> >::uflow
[d:\a01\_work\3\s\src\vctools\crt\github\stl\inc\streambuf @ 299]
00007ff7`0b46b5c8  00007ffc`b1c122f0 msvcp140!`anonymous namespace'::_ExceptionPtr_static<std::bad_alloc>::_Delete_this
[d:\a01\_work\3\s\src\vctools\crt\github\stl\src\excptptr.cpp @ 221]
00007ff7`0b46b5d0  00007ffc`b1c122f0 msvcp140!`anonymous namespace'::_ExceptionPtr_static<std::bad_alloc>::_Delete_this
[d:\a01\_work\3\s\src\vctools\crt\github\stl\src\excptptr.cpp @ 221]
00007ff7`0b46b5d8  00007ffc`b1c3f330 msvcp140!std::basic_streambuf<unsigned short,std::char_traits<unsigned short> >::sgetc
[d:\a01\_work\3\s\src\vctools\crt\github\stl\inc\streambuf @ 129]
00007ff7`0b46b5e0  00007ffc`b1c123a0 msvcp140!std::_Ref_count_base::_Get_deleter [d:\a01\_work\3\s\src\vctools\crt\github\stl\inc\memory @ 1117]
00007ff7`0b46b5e8  00007ffc`b1c123a0 msvcp140!std::_Ref_count_base::_Get_deleter [d:\a01\_work\3\s\src\vctools\crt\github\stl\inc\memory @ 1117]
00007ff7`0b46b5f0  00007ffc`b1c25050 msvcp140!std::basic_streambuf<unsigned short,std::char_traits<unsigned short> >::xsgetn
[d:\a01\_work\3\s\src\vctools\crt\github\stl\inc\streambuf @ 303]
00007ff7`0b46b5f8  00007ffc`b1c25130 msvcp140!std::basic_streambuf<unsigned short,std::char_traits<unsigned short> >::xsputn
[d:\a01\_work\3\s\src\vctools\crt\github\stl\inc\streambuf @ 332]
00007ff7`0b46b600  00007ffc`b1c11d60 msvcp140!std::_Init_locks::operator=
00007ff7`0b46b608  00007ffc`b1c122f0 msvcp140!`anonymous namespace'::_ExceptionPtr_static<std::bad_alloc>::_Delete_this
[d:\a01\_work\3\s\src\vctools\crt\github\stl\src\excptptr.cpp @ 221]
00007ff7`0b46b610  00007ffc`b1c61000 msvcp140!std::basic_streambuf<unsigned short,std::char_traits<unsigned short> >::snextc
[d:\a01\_work\3\s\src\vctools\crt\github\stl\inc\streambuf @ 138]
00007ff7`0b46b618  00007ffc`b1c25bb0 msvcp140!std::basic_ios<wchar_t,std::char_traits<wchar_t> >::widen [d:\a01\_work\3\s\src\vctools\crt\github\stl\inc\ios @
113]
00007ff7`0b46b620  00007ffc`b1c25700 msvcp140!std::basic_streambuf<wchar_t,std::char_traits<wchar_t> >::basic_streambuf<wchar_t,std::char_traits<wchar_t> >
[d:\a01\_work\3\s\src\vctools\crt\github\stl\inc\streambuf @ 24]
00007ff7`0b46b628  00007ffc`b1c5a0c0 msvcp140!std::basic_istream<wchar_t,std::char_traits<wchar_t> >::_Ipfx
[d:\a01\_work\3\s\src\vctools\crt\github\stl\inc\istream @ 114]
00007ff7`0b46b630  00007ffc`b1c25c40 msvcp140!std::basic_istream<wchar_t,std::char_traits<wchar_t> >::basic_istream<wchar_t,std::char_traits<wchar_t> >
[d:\a01\_work\3\s\src\vctools\crt\github\stl\inc\istream @ 40]
00007ff7`0b46b638  00007ffc`b1c25650 msvcp140!std::basic_ios<wchar_t,std::char_traits<wchar_t> >::basic_ios<wchar_t,std::char_traits<wchar_t> >
[d:\a01\_work\3\s\src\vctools\crt\github\stl\inc\ios @ 160]
00007ff7`0b46b640  00007ffc`b1c24480 msvcp140!std::basic_streambuf<unsigned short,std::char_traits<unsigned short> >::_Pninc
[d:\a01\_work\3\s\src\vctools\crt\github\stl\inc\streambuf @ 252]
00007ff7`0b46b648  00007ffc`b1c18d20 msvcp140!std::basic_ios<char,std::char_traits<char> >::setstate [d:\a01\_work\3\s\src\vctools\crt\github\stl\inc\ios @
51]
00007ff7`0b46b650  00007ffc`b1c1b420 msvcp140!std::basic_ios<char,std::char_traits<char> >::rdbuf [d:\a01\_work\3\s\src\vctools\crt\github\stl\inc\ios @ 78]
00007ff7`0b46b658  00000000`00000000
00007ff7`0b46b660  00007ffc`f5b629e0 oleaut32!SetErrorInfo
00007ff7`0b46b668  00007ffc`f5b629d0 oleaut32!GetErrorInfo
00007ff7`0b46b670  00007ffc`f5b57300 oleaut32!SysStringLen
00007ff7`0b46b678  00007ffc`f5b4a940 oleaut32!SysFreeString
00007ff7`0b46b680  00007ffc`f5b4a7a0 oleaut32!SysAllocString
00007ff7`0b46b688  00007ffc`f5b4a6c0 oleaut32!SysAllocStringLen
00007ff7`0b46b690  00000000`00000000
00007ff7`0b46b698  00007ffc`f1514820 propsys!PropVariantToStringVectorAlloc
00007ff7`0b46b6a0  00007ffc`f14fa310 propsys!PSGetPropertyDescriptionListFromString
00007ff7`0b46b6a8  00000000`00000000
00007ff7`0b46b6b0  00007ffc`f473f0e0 shell32!DragAcceptFiles
00007ff7`0b46b6b8  00007ffc`f48f51b0 shell32!ShellExecuteW
00007ff7`0b46b6c0  00007ffc`f47d3a20 shell32!SHGetKnownFolderPathStub
00007ff7`0b46b6c8  00007ffc`f48c8900 shell32!DragQueryFileW
00007ff7`0b46b6d0  00007ffc`f4794610 shell32!SHCreateItemFromParsingName
00007ff7`0b46b6d8  00007ffc`f48c85e0 shell32!DragFinish
00007ff7`0b46b6e0  00007ffc`f47f9380 shell32!SHAddToRecentDocs
00007ff7`0b46b6e8  00000000`00000000
00007ff7`0b46b6f0  00007ffc`f5c216f0 shlwapi!SHStrDupWStub
00007ff7`0b46b6f8  00007ffc`f5c27eb0 shlwapi!PathFindExtensionWStub
00007ff7`0b46b700  00007ffc`f5c28b30 shlwapi!PathIsNetworkPathWStub
00007ff7`0b46b708  00007ffc`f5c31280 shlwapi!PathIsFileSpecWStub
00007ff7`0b46b710  00007ffc`f5c28d30 shlwapi!PathFileExistsWStub
00007ff7`0b46b718  00000000`00000000
00007ff7`0b46b720  00007ffc`f5864ee0 user32!SystemParametersInfoW
00007ff7`0b46b728  00007ffc`f5852210 user32!IsZoomed
00007ff7`0b46b730  00007ffc`f58657f0 user32!CopyRect
00007ff7`0b46b738  00007ffc`f58501e0 user32!SetWindowLongPtrW
00007ff7`0b46b740  00007ffc`f5854bf0 user32!SetScrollInfo
00007ff7`0b46b748  00007ffc`f5848030 user32!CreateWindowExW
00007ff7`0b46b750  00007ffc`f58531e0 user32!GetWindow
00007ff7`0b46b758  00007ffc`f5842690 user32!GetNextDlgTabItem
00007ff7`0b46b760  00007ffc`f58723c0 user32!NtUserGetKeyboardState
00007ff7`0b46b768  00007ffc`f58413f0 user32!SendDlgItemMessageW
00007ff7`0b46b770  00007ffc`f5855160 user32!SetThreadDpiAwarenessContext
00007ff7`0b46b778  00007ffc`f584af60 user32!ReleaseDC
00007ff7`0b46b780  00007ffc`f5866d60 user32!GetDC
00007ff7`0b46b788  00007ffc`f584dc10 user32!GetDpiForSystem
00007ff7`0b46b790  00007ffc`f584cf10 user32!GetCurrentThreadDesktopHwnd
00007ff7`0b46b798  00007ffc`f58677b0 user32!GetKeyboardLayout
00007ff7`0b46b7a0  00007ffc`f585bba0 user32!DrawTextExW
00007ff7`0b46b7a8  00007ffc`f5842f60 user32!CreateDialogParamW
00007ff7`0b46b7b0  00007ffc`f5866d90 user32!DrawIconEx
```

```
00007ff7`0b46b7b8  00007ffc`f584dfb0 user32!GetWindowTextW
00007ff7`0b46b7c0  00007ffc`f5855be0 user32!SetWindowTextW
00007ff7`0b46b7c8  00007ffc`f58ccc30 user32!SetDlgItemTextW
00007ff7`0b46b7d0  00007ffc`f5855f80 user32!IsDialogMessageW
00007ff7`0b46b7d8  00007ffc`f5852e70 user32!GetDpiForWindow
00007ff7`0b46b7e0  00007ffc`f5859d40 user32!PeekMessageW
00007ff7`0b46b7e8  00007ffc`f5853d70 user32!GetWindowTextLengthW
00007ff7`0b46b7f0  00007ffc`f585ce00 user32!GetDlgCtrlID
00007ff7`0b46b7f8  00007ffc`f5872ec0 user32!NtUserSetFocus
00007ff7`0b46b800  00007ffc`f5842920 user32!GetDlgItem
00007ff7`0b46b808  00007ffc`f5898bf0 user32!EndDialog
00007ff7`0b46b810  00007ffc`f5849080 user32!LoadImageW
00007ff7`0b46b818  00007ffc`f588fbd0 user32!DialogBoxParamW
00007ff7`0b46b820  00007ffc`f584b110 user32!LoadCursorW
00007ff7`0b46b828  00007ffc`f5850600 user32!SendMessageW
00007ff7`0b46b830  00007ffc`f586a010 user32!SetCursorStub
00007ff7`0b46b838  00007ffc`f58bfd30 user32!SetScrollPos
00007ff7`0b46b840  00007ffc`f5864820 user32!GetSystemMetrics
00007ff7`0b46b848  00007ffc`f586b6c0 user32!SendMessageA
00007ff7`0b46b850  00007ffc`f58522f0 user32!GetWindowLongPtrW
00007ff7`0b46b858  00007ffc`f5865e90 user32!SetRectEmpty
00007ff7`0b46b860  00007ffc`f584ff10 user32!MapWindowPoints
00007ff7`0b46b868  00007ffc`f5849760 user32!LoadIconW
00007ff7`0b46b870  00007ffc`f58cc180 user32!SetProcessDefaultLayout
00007ff7`0b46b878  00007ffc`f5867130 user32!RegisterWindowMessageW
00007ff7`0b46b880  00007ffc`f58725c0 user32!NtUserGetSystemMenu
00007ff7`0b46b888  00007ffc`f5865f30 user32!CharUpperWStub
00007ff7`0b46b890  00007ffc`f58730c0 user32!NtUserSetWindowPlacement
00007ff7`0b46b898  00007ffc`f5871f30 user32!NtUserCreateAcceleratorTable
00007ff7`0b46b8a0  00007ffc`f5859a20 user32!GetParent
00007ff7`0b46b8a8  00007ffc`f586b1a0 user32!EnableWindow
00007ff7`0b46b8b0  00007ffc`f584d440 user32!SetWindowLongW
00007ff7`0b46b8b8  00007ffc`f5872980 user32!NtUserMoveWindow
00007ff7`0b46b8c0  00007ffc`f5864990 user32!IsIconic
00007ff7`0b46b8c8  00007ffc`f5872810 user32!NtUserInvalidateRect
00007ff7`0b46b8d0  00007ffc`f5869940 user32!UpdateWindow
00007ff7`0b46b8d8  00007ffc`f585d6f0 user32!ClientToScreen
00007ff7`0b46b8e0  00007ffc`f5872320 user32!NtUserGetGUIThreadInfo
00007ff7`0b46b8e8  00007ffc`f584c780 user32!ScreenToClient
00007ff7`0b46b8f0  00007ffc`f5868520 user32!GetCursorPos
00007ff7`0b46b8f8  00007ffc`f586a2b0 user32!GetActiveWindow
00007ff7`0b46b900  00007ffc`f58ccb90 user32!GetDlgItemTextW
00007ff7`0b46b908  00007ffc`f5872d50 user32!NtUserSetActiveWindow
00007ff7`0b46b910  00007ffc`f5850bf0 user32!DispatchMessageW
00007ff7`0b46b918  00007ffc`f58563e0 user32!TranslateMessage
00007ff7`0b46b920  00007ffc`f5864ea0 user32!TranslateAcceleratorW
00007ff7`0b46b928  00007ffc`f5864620 user32!GetMessageW
00007ff7`0b46b930  00007ffc`f5841d10 user32!HideCaretStub
00007ff7`0b46b938  00007ffc`f586b130 user32!CharNextWStub
00007ff7`0b46b940  00007ffc`f5855670 user32!PtInRect
00007ff7`0b46b948  00007ffc`f6343420 ntdll!NtdllDefWindowProc_W
00007ff7`0b46b950  00007ffc`f5859480 user32!GetFocus
00007ff7`0b46b958  00007ffc`f09d3160 apphelp!MbHook_IsWindowVisible
00007ff7`0b46b960  00007ffc`f5872310 user32!NtUserGetForegroundWindow
00007ff7`0b46b968  00007ffc`f58730d0 user32!NtUserSetWindowPos
00007ff7`0b46b970  00007ffc`f586bc80 user32!SetForegroundWindow
00007ff7`0b46b978  00007ffc`f586b7b0 user32!PostQuitMessage
00007ff7`0b46b980  00007ffc`f5872180 user32!NtUserEndPaint
00007ff7`0b46b988  00007ffc`f5871dd0 user32!NtUserBeginPaint
00007ff7`0b46b990  00007ffc`f58728d0 user32!NtUserKillTimer
00007ff7`0b46b998  00007ffc`f5865b20 user32!FillRect
00007ff7`0b46b9a0  00007ffc`f5868410 user32!GetSysColorBrush
00007ff7`0b46b9a8  00007ffc`f5871fe0 user32!NtUserDestroyWindow
00007ff7`0b46b9b0  00007ffc`f5866bf0 user32!GetSysColor
00007ff7`0b46b9b8  00007ffc`f58525f0 user32!GetWindowLongW
00007ff7`0b46b9c0  00007ffc`f5855a40 user32!GetScrollInfo
00007ff7`0b46b9c8  00007ffc`f5865860 user32!SetTimer
00007ff7`0b46b9d0  00007ffc`f585ace0 user32!GetKeyState
00007ff7`0b46b9d8  00007ffc`f584c710 user32!IsChild
00007ff7`0b46b9e0  00007ffc`f586c540 user32!CloseClipboardStub
00007ff7`0b46b9e8  00007ffc`f58650a0 user32!IsClipboardFormatAvailableStub
00007ff7`0b46b9f0  00007ffc`f586bb80 user32!OpenClipboard
00007ff7`0b46b9f8  00007ffc`f5869ef0 user32!GetSubMenu
00007ff7`0b46ba00  00007ffc`f58690c0 user32!EnableMenuItem
00007ff7`0b46ba08  00007ffc`f586b350 user32!GetMenu
00007ff7`0b46ba10  00007ffc`f09d5da0 apphelp!SrHook_ShowWindow
00007ff7`0b46ba18  00007ffc`f5872be0 user32!NtUserRedrawWindow
00007ff7`0b46ba20  00007ffc`f5854ac0 user32!SetWindowRgn
00007ff7`0b46ba28  00007ffc`f584fd40 user32!GetSystemMetricsForDpi
00007ff7`0b46ba30  00007ffc`f58646a0 user32!MonitorFromWindow
00007ff7`0b46ba38  00007ffc`f5856b90 user32!GetClientRect
00007ff7`0b46ba40  00007ffc`f5847bb0 user32!RegisterClassExW
00007ff7`0b46ba48  00007ffc`f5857070 user32!PostMessageW
00007ff7`0b46ba50  00007ffc`f58503b0 user32!GetWindowRect
00007ff7`0b46ba58  00007ffc`f58726a0 user32!NtUserGetWindowPlacement
00007ff7`0b46ba60  00000000`00000000
00007ff7`0b46ba68  00007ffc`f0a5cd70 uxtheme!CloseThemeData
00007ff7`0b46ba70  00007ffc`f0a6c340 uxtheme!DrawThemeTextEx
00007ff7`0b46ba78  00007ffc`f0a611c0 uxtheme!OpenThemeData
00007ff7`0b46ba80  00007ffc`f0aac230 uxtheme!GetThemeSysFont
00007ff7`0b46ba88  00000000`00000000
00007ff7`0b46ba90  00007ffc`b1bf24e0 VCRUNTIME140!__current_exception [d:\a01\_work\3\s\src\vctools\crt\vcruntime\src\eh\ehhelpers.cpp @ 114]
00007ff7`0b46ba98  00007ffc`b1bf1fd0 VCRUNTIME140!wcschr [d:\a01\_work\3\s\src\vctools\crt\vcruntime\src\string\amd64\wcschr.c @ 48]
00007ff7`0b46baa0  00007ffc`b1bf2540 VCRUNTIME140!__std_terminate [d:\a01\_work\3\s\src\vctools\crt\vcruntime\src\eh\ehhelpers.cpp @ 191]
00007ff7`0b46baa8  00007ffc`b1bf6430 VCRUNTIME140!_CxxThrowException [d:\a01\_work\3\s\src\vctools\crt\vcruntime\src\eh\throw.cpp @ 30]
00007ff7`0b46bab0  00007ffc`b1bf2500 VCRUNTIME140!__current_exception_context [d:\a01\_work\3\s\src\vctools\crt\vcruntime\src\eh\ehhelpers.cpp @ 119]
00007ff7`0b46bab8  00007ffc`b1bf6c30 VCRUNTIME140!_purecall [d:\a01\_work\3\s\src\vctools\crt\vcruntime\src\misc\purevirt.cpp @ 19]
00007ff7`0b46bac0  00007ffc`b1bf6220 VCRUNTIME140!__std_exception_destroy [d:\a01\_work\3\s\src\vctools\crt\vcruntime\src\eh\std_exception.cpp @ 43]
```

```
00007ff7`0b46bac8  00007ffc`b1bf6190 VCRUNTIME140!__std_exception_copy [d:\a01\_work\3\s\src\vctools\crt\vcruntime\src\eh\std_exception.cpp @ 17]
00007ff7`0b46bad0  00007ffc`b1bf19a0 VCRUNTIME140!memset [d:\a01\_work\3\s\src\vctools\crt\vcruntime\src\string\amd64\memset.asm @ 79]
00007ff7`0b46bad8  00007ffc`b1bf11f0 VCRUNTIME140!memcmp [d:\a01\_work\3\s\src\vctools\crt\vcruntime\src\string\amd64\memcmp.asm @ 56]
00007ff7`0b46bae0  00007ffc`b1bf12f0 VCRUNTIME140!memcpy [d:\a01\_work\3\s\src\vctools\crt\vcruntime\src\string\amd64\memcpy.asm @ 68]
00007ff7`0b46bae8  00007ffc`b1bf12f0 VCRUNTIME140!memcpy [d:\a01\_work\3\s\src\vctools\crt\vcruntime\src\string\amd64\memcpy.asm @ 68]
00007ff7`0b46baf0  00000000`00000000
00007ff7`0b46baf8  00007ffc`cc544070 VCRUNTIME140_1!__CxxFrameHandler4 [d:\a01\_work\3\s\src\vctools\crt\vcruntime\src\eh\risctrnsctrl.cpp @ 291]
00007ff7`0b46bb00  00000000`00000000
00007ff7`0b46bb08  00007ffc`d85a5bc0 winspool!ClosePrinter
00007ff7`0b46bb10  00007ffc`d85a67d0 winspool!GetPrinterDriverW
00007ff7`0b46bb18  00007ffc`d85b1420 winspool!OpenPrinterW
00007ff7`0b46bb20  00000000`00000000
00007ff7`0b46bb28  00007ffc`f5da6520 combase!RoGetActivationFactory [onecore\com\combase\winrtbase\winrtbase.cpp @ 1060]
00007ff7`0b46bb30  00000000`00000000
00007ff7`0b46bb38  00007ffc`f5d70ac0 combase!WindowsCreateStringReference [onecore\com\combase\winrt\string\string.cpp @ 70]
00007ff7`0b46bb40  00007ffc`f5da00b0 combase!WindowsDeleteString [onecore\com\combase\winrt\string\string.cpp @ 146]
00007ff7`0b46bb48  00007ffc`f5d77280 combase!WindowsGetStringRawBuffer [onecore\com\combase\winrt\string\string.cpp @ 226]
00007ff7`0b46bb50  00007ffc`f5d73870 combase!WindowsCreateString [onecore\com\combase\winrt\string\string.cpp @ 30]
00007ff7`0b46bb58  00000000`00000000
00007ff7`0b46bb60  00007ffc`f395e8a0 ucrtbase!wcstol
00007ff7`0b46bb68  00000000`00000000
00007ff7`0b46bb70  00007ffc`f39c8870 ucrtbase!callnewh
00007ff7`0b46bb78  00007ffc`f3962150 ucrtbase!free
00007ff7`0b46bb80  00007ffc`f3976ae0 ucrtbase!_set_new_mode
00007ff7`0b46bb88  00007ffc`f3960060 ucrtbase!malloc
00007ff7`0b46bb90  00000000`00000000
00007ff7`0b46bb98  00007ffc`f3976900 ucrtbase!_configthreadlocale
00007ff7`0b46bba0  00000000`00000000
00007ff7`0b46bba8  00007ffc`f39f0d20 ucrtbase!_setusermatherr
00007ff7`0b46bbb0  00007ffc`f39873d0 ucrtbase!ceilf
00007ff7`0b46bbb8  00000000`00000000
00007ff7`0b46bbc0  00007ffc`f39cfb30 ucrtbase!c_exit
00007ff7`0b46bbc8  00007ffc`f39cfb70 ucrtbase!register_thread_local_exe_atexit_callback
00007ff7`0b46bbd0  00007ffc`f39cfb10 ucrtbase!Exit
00007ff7`0b46bbd8  00007ffc`f3969f40 ucrtbase!exit
00007ff7`0b46bbe0  00007ffc`f3972fc0 ucrtbase!configure_narrow_argv
00007ff7`0b46bbe8  00007ffc`f3974eb0 ucrtbase!initialize_narrow_environment
00007ff7`0b46bbf0  00007ffc`f3972d00 ucrtbase!initialize_onexit_table
00007ff7`0b46bbf8  00007ffc`f3972d80 ucrtbase!initterm_e
00007ff7`0b46bc00  00007ffc`f3972d30 ucrtbase!initterm
00007ff7`0b46bc08  00007ffc`f39cdd30 ucrtbase!abort
00007ff7`0b46bc10  00007ffc`f3972fb0 ucrtbase!get_narrow_winmain_command_line
00007ff7`0b46bc18  00007ffc`f39774e0 ucrtbase!set_app_type
00007ff7`0b46bc20  00007ffc`f3978d10 ucrtbase!_seh_filter_exe
00007ff7`0b46bc28  00007ffc`f39cfb50 ucrtbase!cexit
00007ff7`0b46bc30  00007ffc`f395f170 ucrtbase!register_onexit_function
00007ff7`0b46bc38  00007ffc`f3975c00 ucrtbase!crt_atexit
00007ff7`0b46bc40  00007ffc`f39cd470 ucrtbase!terminate
00007ff7`0b46bc48  00007ffc`f3978670 ucrtbase!invalid_parameter_noinfo
00007ff7`0b46bc50  00007ffc`f3968770 ucrtbase!_errno
00007ff7`0b46bc58  00007ffc`f39cc600 ucrtbase!invalid_parameter_noinfo_noreturn
00007ff7`0b46bc60  00000000`00000000
00007ff7`0b46bc68  00007ffc`f3976b10 ucrtbase!_set_fmode
00007ff7`0b46bc70  00007ffc`f3961f10 ucrtbase!__stdio_common_vswprintf
00007ff7`0b46bc78  00007ffc`f39774d0 ucrtbase!_p__commode
00007ff7`0b46bc80  00007ffc`f395dfc0 ucrtbase!_stdio_common_vsnprintf_s
00007ff7`0b46bc88  00000000`00000000
00007ff7`0b46bc90  00007ffc`f397c920 ucrtbase!wcscpy_s
00007ff7`0b46bc98  00007ffc`f397b430 ucrtbase!wcsnlen
00007ff7`0b46bca0  00007ffc`f397cab0 ucrtbase!wcsicmp
00007ff7`0b46bca8  00007ffc`f3959320 ucrtbase!iswspace
00007ff7`0b46bcb0  00007ffc`f3958fb0 ucrtbase!iswdigit
00007ff7`0b46bcb8  00000000`00000000
00007ff7`0b46bcc0  00007ffc`f50ae5d0 SHCore!GetDpiForMonitor
00007ff7`0b46bcc8  00000000`00000000
00007ff7`0b46bcd0  00007ffc`f0cf3f90 dwmapi!DwmExtendFrameIntoClientArea
00007ff7`0b46bcd8  00007ffc`f0cf3280 dwmapi!DwmGetWindowAttribute
00007ff7`0b46bce0  00007ffc`f0cf3a30 dwmapi!DwmSetWindowAttribute
00007ff7`0b46bce8  00007ffc`f0cf1570 dwmapi!DwmDefWindowProc
00007ff7`0b46bcf0  00000000`00000000
00007ff7`0b46bcf8  00007ffc`f5d8ef50 combase!CoCreateGuid [onecore\com\combase\class\cocrguid.cxx @ 49]
00007ff7`0b46bd00  00007ffc`f5de54d0 combase!CoTaskMemFree [onecore\com\combase\class\memapi.cxx @ 453]
00007ff7`0b46bd08  00007ffc`f5da3f70 combase!CoCreateInstance [onecore\com\combase\objact\actapi.cxx @ 252]
00007ff7`0b46bd10  00007ffc`f5de6640 combase!CoTaskMemAlloc [onecore\com\combase\class\memapi.cxx @ 437]
00007ff7`0b46bd18  00007ffc`f5da15f0 combase!CoUninitialize [onecore\com\combase\class\compobj.cxx @ 3793]
00007ff7`0b46bd20  00007ffc`f5da0f00 combase!CoInitializeEx [onecore\com\combase\class\compobj.cxx @ 3734]
00007ff7`0b46bd28  00007ffc`f5de4340 combase!PropVariantClear [onecore\com\combase\util\propvar.cxx @ 278]
00007ff7`0b46bd30  00007ffc`f5d9d620 combase!CoWaitForMultipleHandles [onecore\com\combase\dcomrem\sync.cxx @ 86]
00007ff7`0b46bd38  00007ffc`f5de3f50 combase!CoGetApartmentType [onecore\com\combase\dcomrem\coapi.cxx @ 2754]
00007ff7`0b46bd40  00007ffc`f5e02ac0 combase!CoGetObjectContext [onecore\com\combase\dcomrem\pstable.cxx @ 2959]
00007ff7`0b46bd48  00007ffc`f5db3d20 combase!CoCreateFreeThreadedMarshaler [onecore\com\combase\dcomrem\ipmrshl.cxx @ 201]
00007ff7`0b46bd50  00000000`00000000
00007ff7`0b46bd58  00007ffc`ecb09dc0 urlmon!FindMimeFromData
00007ff7`0b46bd60  00000000`00000000
```

Note: We see it uses registry-related functions from the *advapi32* module. Let's see the latter's import dependencies:

```
0:000> !dh advapi32

File Type: DLL
FILE HEADER VALUES
```

```
    8664 machine (X64)
       7 number of sections
CE622C7B time date stamp Thu Sep 21 17:46:51 2079

       0 file pointer to symbol table
       0 number of symbols
      F0 size of optional header
    2022 characteristics
            Executable
            App can handle >2gb addresses
            DLL

OPTIONAL HEADER VALUES
     20B magic #
   14.28 linker version
   68000 size of code
   45000 size of initialized data
       0 size of uninitialized data
    4C70 address of entry point
    1000 base of code
         ----- new -----
00007ffcf5240000 image base
    1000 section alignment
    1000 file alignment
       3 subsystem (Windows CUI)
   10.00 operating system version
   10.00 image version
   10.00 subsystem version
   AE000 size of image
    1000 size of headers
   B3924 checksum
0000000000040000 size of stack reserve
0000000000001000 size of stack commit
0000000000100000 size of heap reserve
0000000000001000 size of heap commit
    4160  DLL characteristics
            High entropy VA supported
            Dynamic base
            NX compatible
            Guard
   92F70 [    76D8] address [size] of Export Directory
   9A648 [     2BC] address [size] of Import Directory
   AB000 [     5C8] address [size] of Resource Directory
   A5000 [    45FC] address [size] of Exception Directory
   AC000 [    3E00] address [size] of Security Directory
   AC000 [    1934] address [size] of Base Relocation Directory
   7E42C [      70] address [size] of Debug Directory
       0 [       0] address [size] of Description Directory
       0 [       0] address [size] of Special Directory
       0 [       0] address [size] of Thread Storage Directory
   69590 [     138] address [size] of Load Configuration Directory
       0 [       0] address [size] of Bound Import Directory
   78F38 [    14D8] address [size] of Import Address Table Directory
   91910 [     1C0] address [size] of Delay Import Directory
       0 [       0] address [size] of COR20 Header Directory
       0 [       0] address [size] of Reserved Directory
```

SECTION HEADER #1

```
   .text name
   67E67 virtual size
    1000 virtual address
   68000 size of raw data
    1000 file pointer to raw data
       0 file pointer to relocation table
       0 file pointer to line numbers
       0 number of relocations
       0 number of line numbers
60000020 flags
         Code
         (no align specified)
         Execute Read

SECTION HEADER #2
  .rdata name
   3694A virtual size
   69000 virtual address
   37000 size of raw data
   69000 file pointer to raw data
       0 file pointer to relocation table
       0 file pointer to line numbers
       0 number of relocations
       0 number of line numbers
40000040 flags
         Initialized Data
         (no align specified)
         Read Only


Debug Directories(4)
      Type       Size      Address  Pointer
      cv           25        8b340    8b340    Format: RSDS, guid, 1, advapi32.pdb
      (   13)     5c4        8b368    8b368
      (   16)      24        8b92c    8b92c
      dllchar       4        8b950    8b950

00000001 extended DLL characteristics
           CET compatible


SECTION HEADER #3
   .data name
    455A virtual size
   A0000 virtual address
    3000 size of raw data
   A0000 file pointer to raw data
       0 file pointer to relocation table
       0 file pointer to line numbers
       0 number of relocations
       0 number of line numbers
C0000040 flags
         Initialized Data
         (no align specified)
         Read Write

SECTION HEADER #4
  .pdata name
    45FC virtual size
```

```
   A5000 virtual address
    5000 size of raw data
   A3000 file pointer to raw data
       0 file pointer to relocation table
       0 file pointer to line numbers
       0 number of relocations
       0 number of line numbers
40000040 flags
         Initialized Data
         (no align specified)
         Read Only

SECTION HEADER #5
  .didat name
     478 virtual size
   AA000 virtual address
    1000 size of raw data
   A8000 file pointer to raw data
       0 file pointer to relocation table
       0 file pointer to line numbers
       0 number of relocations
       0 number of line numbers
C0000040 flags
         Initialized Data
         (no align specified)
         Read Write

SECTION HEADER #6
   .rsrc name
     5C8 virtual size
   AB000 virtual address
    1000 size of raw data
   A9000 file pointer to raw data
       0 file pointer to relocation table
       0 file pointer to line numbers
       0 number of relocations
       0 number of line numbers
40000040 flags
         Initialized Data
         (no align specified)
         Read Only

SECTION HEADER #7
  .reloc name
    1934 virtual size
   AC000 virtual address
    2000 size of raw data
   AA000 file pointer to raw data
       0 file pointer to relocation table
       0 file pointer to line numbers
       0 number of relocations
       0 number of line numbers
42000040 flags
         Initialized Data
         Discardable
         (no align specified)
         Read Only
```

```
0:000> dps 00007ffcf5240000 + 78F38 L14D8/8
0007ffc`f52b8f38  00007ffc`f4fdbfa0 kernel32!GetModuleFileNameWStub
00007ffc`f52b8f40  00007ffc`f4fd5fc0 kernel32!MultiByteToWideCharStub
00007ffc`f52b8f48  00007ffc`f4fd82d0 kernel32!LocalFreeStub
00007ffc`f52b8f50  00007ffc`f4fd6010 kernel32!WideCharToMultiByteStub
00007ffc`f52b8f58  00007ffc`f62da4e0 ntdll!RtlEnterCriticalSection
00007ffc`f52b8f60  00007ffc`f62db4d0 ntdll!RtlLeaveCriticalSection
00007ffc`f52b8f68  00007ffc`f4fd62e0 kernel32!GetLastErrorStub
00007ffc`f52b8f70  00007ffc`f4fd93b0 kernel32!GetProcAddressStub
00007ffc`f52b8f78  00007ffc`f4ff8650 kernel32!DelayLoadFailureHookStub
00007ffc`f52b8f80  00007ffc`f62d5eb0 ntdll!LdrResolveDelayLoadedAPI
00007ffc`f52b8f88  00007ffc`f4fda650 kernel32!FreeLibraryStub
00007ffc`f52b8f90  00007ffc`f4fd93f0 kernel32!LoadLibraryExWStub
00007ffc`f52b8f98  00007ffc`f4ffa370 kernel32!UnhandledExceptionFilterStub
00007ffc`f52b8fa0  00007ffc`f4fde6d0 kernel32!SetUnhandledExceptionFilterStub
00007ffc`f52b8fa8  00007ffc`f4fd6670 kernel32!QueryPerformanceCounterStub
00007ffc`f52b8fb0  00007ffc`f4fe2cd0 kernel32!CreateEventW
00007ffc`f52b8fb8  00007ffc`f4fe2c50 kernel32!CloseHandle
00007ffc`f52b8fc0  00007ffc`f4ff8e70 kernel32!GetThreadUILanguageStub
00007ffc`f52b8fc8  00007ffc`f4fdd660 kernel32!GetCommandLineWStub
00007ffc`f52b8fd0  00007ffc`f4fde090 kernel32!GetModuleHandleExWStub
00007ffc`f52b8fd8  00007ffc`f4fe3360 kernel32!WriteFile
00007ffc`f52b8fe0  00007ffc`f4fda0e0 kernel32!ExpandEnvironmentStringsWStub
00007ffc`f52b8fe8  00007ffc`f4fe3300 kernel32!SetFilePointer
00007ffc`f52b8ff0  00007ffc`f4fe2ed0 kernel32!CreateFileW
00007ffc`f52b8ff8  00007ffc`f4fdbf80 kernel32!FormatMessageWStub
00007ffc`f52b9000  00007ffc`f4fe30c0 kernel32!GetFileAttributesExW
00007ffc`f52b9008  00007ffc`f4fdaf10 kernel32!OutputDebugStringWStub
00007ffc`f52b9010  00007ffc`f4fe2f00 kernel32!DeleteFileW
00007ffc`f52b9018  00007ffc`f4fe0e60 kernel32!MoveFileW
00007ffc`f52b9020  00007ffc`f4fdb790 kernel32!GetModuleHandleWStub
00007ffc`f52b9028  00007ffc`f4fe3100 kernel32!GetFileSizeEx
00007ffc`f52b9030  00007ffc`f4fda7c0 kernel32!CreateFileMappingWStub
00007ffc`f52b9038  00007ffc`f62c8ac0 ntdll!RtlAllocateHeap
00007ffc`f52b9040  00007ffc`f4fd6340 kernel32!GetProcessHeapStub
00007ffc`f52b9048  00007ffc`f4fdb9e0 kernel32!MapViewOfFileStub
00007ffc`f52b9050  00007ffc`f4fdc070 kernel32!UnmapViewOfFileStub
00007ffc`f52b9058  00007ffc`f4fd5ef0 kernel32!HeapFreeStub
00007ffc`f52b9060  00007ffc`f4fc6d40 kernel32!GetLongPathNameW
00007ffc`f52b9068  00007ffc`f4fe2e80 kernel32!CompareFileTime
00007ffc`f52b9070  00007ffc`f4fd9ac0 kernel32!FindResourceExWStub
00007ffc`f52b9078  00007ffc`f4fd9570 kernel32!LoadResourceStub
00007ffc`f52b9080  00007ffc`f4fe3210 kernel32!GetVolumePathNameW
00007ffc`f52b9088  00007ffc`f62be080 ntdll!RtlDeleteCriticalSection
00007ffc`f52b9090  00007ffc`f4fe2e40 kernel32!WaitForSingleObject
00007ffc`f52b9098  00007ffc`f3b450c0 KERNELBASE!InitOnceBeginInitialize
00007ffc`f52b90a0  00007ffc`f4fd7b20 kernel32!CompareStringOrdinalStub
00007ffc`f52b90a8  00007ffc`f4fe2dc0 kernel32!ReleaseMutex
00007ffc`f52b90b0  00007ffc`f3b99220 KERNELBASE!InitOnceComplete
00007ffc`f52b90b8  00007ffc`f4fcfea0 kernel32!GetComputerNameW
00007ffc`f52b90c0  00007ffc`f4fe5820 kernel32!ExpandEnvironmentStringsAStub
00007ffc`f52b90c8  00007ffc`f4fdf7c0 kernel32!AreFileApisANSIStub
00007ffc`f52b90d0  00007ffc`f4ff9fa0 kernel32!SearchPathWStub
00007ffc`f52b90d8  00007ffc`f4fe3160 kernel32!GetFullPathNameW
00007ffc`f52b90e0  00007ffc`f4fe30d0 kernel32!GetFileAttributesW
00007ffc`f52b90e8  00007ffc`f4fe2e10 kernel32!SleepEx
00007ffc`f52b90f0  00007ffc`f4fdd680 kernel32!LoadLibraryExAStub
00007ffc`f52b90f8  00007ffc`f4fdf500 kernel32!LoadLibraryAStub
00007ffc`f52b9100  00007ffc`f4fe2d10 kernel32!CreateMutexW
00007ffc`f52b9108  00007ffc`f6308fe0 ntdll!RtlInitializeCriticalSection
00007ffc`f52b9110  00007ffc`f4fd9aa0 kernel32!SizeofResourceStub
00007ffc`f52b9118  00007ffc`f4fd9a80 kernel32!LockResourceStub
00007ffc`f52b9120  00007ffc`f4fe2df0 kernel32!SetEvent
00007ffc`f52b9128  00007ffc`f4fe2de0 kernel32!ResetEvent
00007ffc`f52b9130  00007ffc`f4fe3110 kernel32!GetFileTime
00007ffc`f52b9138  00007ffc`f4fc86e0 kernel32!FileTimeToDosDateTime
00007ffc`f52b9140  00007ffc`f4fc82a0 kernel32!DosDateTimeToFileTime
00007ffc`f52b9148  00007ffc`f4fe30f0 kernel32!GetFileSize
00007ffc`f52b9150  00007ffc`f4fe2f90 kernel32!FindFirstFileExW
00007ffc`f52b9158  00007ffc`f4fe3000 kernel32!FindNextFileW
00007ffc`f52b9160  00007ffc`f4fe2f30 kernel32!FindClose
00007ffc`f52b9168  00007ffc`f4fdb8c0 kernel32!TermsrvDeleteKey
00007ffc`f52b9170  00007ffc`f4fda8d0 kernel32!TermsrvOpenUserClasses
00007ffc`f52b9178  00007ffc`f4fda6e0 kernel32!ReadProcessMemoryStub
00007ffc`f52b9180  00007ffc`f63139d0 ntdll!RtlDecodePointer
00007ffc`f52b9188  00007ffc`f4fde880 kernel32!LoadLibraryWStub
00007ffc`f52b9190  00007ffc`f4fe2c60 kernel32!DuplicateHandle
00007ffc`f52b9198  00007ffc`f4fdfc80 kernel32!FreeLibraryAndExitThreadStub
00007ffc`f52b91a0  00007ffc`f631a2b0 ntdll!RtlEncodePointer
00007ffc`f52b91a8  00007ffc`f63244a0 ntdll!TpCallbackUnloadDllOnCompletion
00007ffc`f52b91b0  00007ffc`f6324a90 ntdll!TpReleaseIoCompletion
00007ffc`f52b91b8  00007ffc`f4fde9a0 kernel32!CancelIoExStub
00007ffc`f52b91c0  00007ffc`f6321e60 ntdll!TpCancelAsyncIoOperation
00007ffc`f52b91c8  00007ffc`f4fe24a0 kernel32!CreateThreadpoolIoStub
00007ffc`f52b91d0  00007ffc`f4fd5f10 kernel32!DeviceIoControlImplementation
00007ffc`f52b91d8  00007ffc`f62bb7b0 ntdll!TpStartAsyncIoOperation
00007ffc`f52b91e0  00007ffc`f4fde030 kernel32!GetFileMUIPathStub
00007ffc`f52b91e8  00007ffc`f4ff8930 kernel32!EnumUILanguagesWStub
00007ffc`f52b91f0  00007ffc`f4fda6c0 kernel32!SetErrorModeStub
00007ffc`f52b91f8  00007ffc`f4fdbe40 kernel32!RaiseExceptionStub
00007ffc`f52b9200  00007ffc`f4fe32f0 kernel32!SetFileInformationByHandle
00007ffc`f52b9208  00007ffc`f4fdf440 kernel32!CopyFileExWStub
00007ffc`f52b9210  00007ffc`f4fd6360 kernel32!SetLastErrorStub
00007ffc`f52b9218  00000000`00000000
00007ffc`f52b9220  00007ffc`f3b6cb30 KERNELBASE!lstrcmpiW
00007ffc`f52b9228  00007ffc`f3c396f0 KERNELBASE!RegKrnGetHKEY_ClassesRootAddress
00007ffc`f52b9230  00007ffc`f3b9b380 KERNELBASE!RegKrnGetClassesEnumTableAddressInternal
00007ffc`f52b9238  00007ffc`f3ba4210 KERNELBASE!RegKrnGetTermsrvRegistryExtensionFlags
```

```
00007ffc`f52b9240  00007ffc`f3b61700 KERNELBASE!lstrlenW
00007ffc`f52b9248  00007ffc`f3b60d60 KERNELBASE!LocalAlloc
00007ffc`f52b9250  00007ffc`f3b88f80 KERNELBASE!LocalReAlloc
00007ffc`f52b9258  00007ffc`f3b9e970 KERNELBASE!CreateProcessAsUserW
00007ffc`f52b9260  00007ffc`f3c6f020 KERNELBASE!CreateProcessAsUserA
00007ffc`f52b9268  00007ffc`f3ba02e0 KERNELBASE!GetSystemDefaultUILanguage
00007ffc`f52b9270  00007ffc`f3b85280 KERNELBASE!RegDeleteKeyExInternalW
00007ffc`f52b9278  00007ffc`f3b64650 KERNELBASE!RegCreateKeyExInternalW
00007ffc`f52b9280  00007ffc`f3b67bf0 KERNELBASE!RegOpenKeyExInternalW
00007ffc`f52b9288  00007ffc`f3b6ad20 KERNELBASE!CLOSE_LOCAL_HANDLE_INTERNAL
00007ffc`f52b9290  00007ffc`f3b67ef0 KERNELBASE!MapPredefinedHandleInternal
00007ffc`f52b9298  00007ffc`f3c39770 KERNELBASE!RegDeleteKeyExInternalA
00007ffc`f52b92a0  00007ffc`f3c39300 KERNELBASE!RemapPredefinedHandleInternal
00007ffc`f52b92a8  00007ffc`f3b83440 KERNELBASE!RegCreateKeyExInternalA
00007ffc`f52b92b0  00007ffc`f3b934d0 KERNELBASE!DisablePredefinedHandleTableInternal
00007ffc`f52b92b8  00007ffc`f3b64a30 KERNELBASE!RegOpenKeyExInternalA
00007ffc`f52b92c0  00007ffc`f3b4aac0 KERNELBASE!GetStagedPackagePathByFullName
00007ffc`f52b92c8  00007ffc`f3baabc0 KERNELBASE!GetUserDefaultUILanguage
00007ffc`f52b92d0  00007ffc`f3b6c240 KERNELBASE!lstrcmpW
00007ffc`f52b92d8  00007ffc`f3b7b320 KERNELBASE!Sleep
00007ffc`f52b92e0  00007ffc`f3b94370 KERNELBASE!PackageIdFromFullName
00007ffc`f52b92e8  00000000`00000000
00007ffc`f52b92f0  00007ffc`f4368ec0 rpcrt4!RpcBindingBind
00007ffc`f52b92f8  00007ffc`f43751f0 rpcrt4!RpcBindingCreateW
00007ffc`f52b9300  00007ffc`f43552c0 rpcrt4!RpcSsDestroyClientContext
00007ffc`f52b9308  00007ffc`f4375d60 rpcrt4!RpcBindingSetAuthInfoExW
00007ffc`f52b9310  00007ffc`f43580e0 rpcrt4!RpcStringFreeW
00007ffc`f52b9318  00007ffc`f43576d0 rpcrt4!RpcBindingFree
00007ffc`f52b9320  00007ffc`f442d480 rpcrt4!NdrClientCall3
00007ffc`f52b9328  00007ffc`f43a60f0 rpcrt4!I_RpcExceptionFilter
00007ffc`f52b9330  00007ffc`f4369920 rpcrt4!UuidToStringW
00007ffc`f52b9338  00007ffc`f4369350 rpcrt4!UuidFromStringW
00007ffc`f52b9340  00007ffc`f4351d20 rpcrt4!NdrClientCall2
00007ffc`f52b9348  00007ffc`f43765e0 rpcrt4!RpcBindingFromStringBindingW
00007ffc`f52b9350  00007ffc`f43ba9a0 rpcrt4!RpcMgmtInqServerPrincNameW
00007ffc`f52b9358  00007ffc`f4358210 rpcrt4!RpcStringBindingComposeW
00007ffc`f52b9360  00007ffc`f43b8800 rpcrt4!I_RpcSNCHOption
00007ffc`f52b9368  00007ffc`f4403b20 rpcrt4!RpcBindingSetAuthInfoA
00007ffc`f52b9370  00007ffc`f43b47d0 rpcrt4!RpcEpResolveBinding
00007ffc`f52b9378  00007ffc`f43b1510 rpcrt4!RpcBindingSetAuthInfoW
00007ffc`f52b9380  00007ffc`f43a3310 rpcrt4!I_RpcMapWin32Status
00007ffc`f52b9388  00007ffc`f43a60f0 rpcrt4!I_RpcExceptionFilter
00007ffc`f52b9390  00000000`00000000
00007ffc`f52b9398  00007ffc`f533e660 sechost!QueryAllTracesA
00007ffc`f52b93a0  00007ffc`f52fd350 sechost!ControlTraceA
00007ffc`f52b93a8  00007ffc`f52fce50 sechost!StartTraceA
00007ffc`f52b93b0  00000000`00000000
00007ffc`f52b93b8  00007ffc`f631d920 ntdll!ApiSetQueryApiSetPresence
00007ffc`f52b93c0  00000000`00000000
00007ffc`f52b93c8  00007ffc`f3b8fba0 KERNELBASE!ImpersonateNamedPipeClient
00007ffc`f52b93d0  00000000`00000000
00007ffc`f52b93d8  00007ffc`f3b9b5a0 KERNELBASE!PcwCollectData
00007ffc`f52b93e0  00007ffc`f3b9b4b0 KERNELBASE!PcwSetQueryItemUserData
00007ffc`f52b93e8  00007ffc`f3b9b500 KERNELBASE!PcwAddQueryItem
00007ffc`f52b93f0  00007ffc`f3c5b510 KERNELBASE!PcwRemoveQueryItem
00007ffc`f52b93f8  00007ffc`f3c5b560 KERNELBASE!PcwSendNotification
00007ffc`f52b9400  00007ffc`f3befed0 KERNELBASE!PcwSendStatelessNotification
00007ffc`f52b9408  00007ffc`f3c5b3c0 KERNELBASE!PcwCreateNotifier
00007ffc`f52b9410  00007ffc`f3b9b440 KERNELBASE!PcwEnumerateInstances
00007ffc`f52b9418  00007ffc`f3b9b3f0 KERNELBASE!PcwCreateQuery
00007ffc`f52b9420  00000000`00000000
00007ffc`f52b9428  00007ffc`f4fdb840 kernel32!OpenThreadStub
00007ffc`f52b9430  00007ffc`f3b645d0 KERNELBASE!OpenThreadToken
00007ffc`f52b9438  00007ffc`f4fdb370 kernel32!GetProcessIdStub
00007ffc`f52b9440  00007ffc`f4fd9e40 kernel32!CreateThreadStub
00007ffc`f52b9448  00007ffc`f4fdfdc0 kernel32!GetPriorityClassStub
00007ffc`f52b9450  00007ffc`f4fd62a0 kernel32!GetCurrentThread
00007ffc`f52b9458  00007ffc`f4fc6170 kernel32!GetCurrentThreadId
00007ffc`f52b9460  00007ffc`f4fe2be0 kernel32!GetCurrentProcessId
00007ffc`f52b9468  00007ffc`f4fdf800 kernel32!TerminateProcessStub
00007ffc`f52b9470  00007ffc`f4fe2bd0 kernel32!GetCurrentProcess
00007ffc`f52b9478  00007ffc`f3b87d90 KERNELBASE!SetThreadToken
00007ffc`f52b9480  00007ffc`f3b41700 KERNELBASE!OpenProcessToken
00007ffc`f52b9488  00000000`00000000
00007ffc`f52b9490  00007ffc`f4fd9cf0 kernel32!OpenProcessStub
00007ffc`f52b9498  00000000`00000000
00007ffc`f52b94a0  00007ffc`f3b84ee0 KERNELBASE!RegDeleteTreeA
00007ffc`f52b94a8  00007ffc`f3b934a0 KERNELBASE!RegDisablePredefinedCacheEx
00007ffc`f52b94b0  00007ffc`f3b83730 KERNELBASE!RegNotifyChangeKeyValue
00007ffc`f52b94b8  00007ffc`f3b83df0 KERNELBASE!RegGetKeySecurity
00007ffc`f52b94c0  00007ffc`f3b979f0 KERNELBASE!RegLoadAppKeyW
00007ffc`f52b94c8  00007ffc`f3b84f50 KERNELBASE!RegDeleteKeyExW
00007ffc`f52b94d0  00007ffc`f3b95740 KERNELBASE!RegOpenCurrentUser
00007ffc`f52b94d8  00007ffc`f3b65480 KERNELBASE!RegQueryInfoKeyW
00007ffc`f52b94e0  00007ffc`f3b823c0 KERNELBASE!RegGetValueA
00007ffc`f52b94e8  00007ffc`f3c3a490 KERNELBASE!RegSaveKeyExA
00007ffc`f52b94f0  00007ffc`f3c39af0 KERNELBASE!RegLoadMUIStringA
00007ffc`f52b94f8  00007ffc`f3b657c0 KERNELBASE!RegQueryValueExA
00007ffc`f52b9500  00007ffc`f3b833e0 KERNELBASE!RegCreateKeyExA
00007ffc`f52b9508  00007ffc`f3ba2220 KERNELBASE!RegFlushKey
00007ffc`f52b9510  00007ffc`f3b63c20 KERNELBASE!RegCreateKeyExW
00007ffc`f52b9518  00007ffc`f3c399e0 KERNELBASE!RegUnLoadKeyA
00007ffc`f52b9520  00007ffc`f3b8e010 KERNELBASE!RegOpenUserClassesRoot
00007ffc`f52b9528  00007ffc`f3c39750 KERNELBASE!RegDeleteKeyExA
00007ffc`f52b9530  00007ffc`f3b66b20 KERNELBASE!RegEnumKeyExW
00007ffc`f52b9538  00007ffc`f3b83b80 KERNELBASE!RegSetKeySecurity
00007ffc`f52b9540  00007ffc`f3c3a660 KERNELBASE!RegSaveKeyExW
00007ffc`f52b9548  00007ffc`f3b84f70 KERNELBASE!RegDeleteTreeW
```

```
00007ffc`f52b9550  00007ffc`f3b8b8d0 KERNELBASE!RegLoadMUIStringW
00007ffc`f52b9558  00007ffc`f3b6aad0 KERNELBASE!RegSetValueExW
00007ffc`f52b9560  00007ffc`f3b97690 KERNELBASE!RegLoadAppKeyA
00007ffc`f52b9568  00007ffc`f3b82d90 KERNELBASE!RegSetValueExA
00007ffc`f52b9570  00007ffc`f3bdd6f0 KERNELBASE!RegCopyTreeW
00007ffc`f52b9578  00007ffc`f3c39810 KERNELBASE!RegLoadKeyA
00007ffc`f52b9580  00007ffc`f3bdba10 KERNELBASE!RegUnLoadKeyW
00007ffc`f52b9588  00007ffc`f3b972c0 KERNELBASE!RegQueryInfoKeyA
00007ffc`f52b9590  00007ffc`f3bdb780 KERNELBASE!RegLoadKeyW
00007ffc`f52b9598  00007ffc`f3b64a00 KERNELBASE!RegOpenKeyExA
00007ffc`f52b95a0  00007ffc`f3b66d40 KERNELBASE!RegGetValueW
00007ffc`f52b95a8  00007ffc`f3c3a380 KERNELBASE!RegRestoreKeyW
00007ffc`f52b95b0  00007ffc`f3b31580 KERNELBASE!RegEnumValueA
00007ffc`f52b95b8  00007ffc`f3b86340 KERNELBASE!RegDeleteValueW
00007ffc`f52b95c0  00007ffc`f3c3a230 KERNELBASE!RegRestoreKeyA
00007ffc`f52b95c8  00007ffc`f3b85f20 KERNELBASE!RegDeleteValueA
00007ffc`f52b95d0  00007ffc`f3b65c40 KERNELBASE!RegEnumValueW
00007ffc`f52b95d8  00007ffc`f3b88ab0 KERNELBASE!RegEnumKeyExA
00007ffc`f52b95e0  00007ffc`f3b69d30 KERNELBASE!RegCloseKey
00007ffc`f52b95e8  00007ffc`f3b679c0 KERNELBASE!RegOpenKeyExW
00007ffc`f52b95f0  00007ffc`f3b679f0 KERNELBASE!RegQueryValueExW
00007ffc`f52b95f8  00000000`00000000
00007ffc`f52b9600  00007ffc`f3b94b30 KERNELBASE!RegSetKeyValueW
00007ffc`f52b9608  00007ffc`f3b85ec0 KERNELBASE!RegDeleteKeyValueA
00007ffc`f52b9610  00007ffc`f3b85fe0 KERNELBASE!RegDeleteKeyValueW
00007ffc`f52b9618  00007ffc`f3b82cc0 KERNELBASE!RegSetKeyValueA
00007ffc`f52b9620  00000000`00000000
00007ffc`f52b9628  00007ffc`f3c39b00 KERNELBASE!RegQueryMultipleValuesA
00007ffc`f52b9630  00007ffc`f3b90800 KERNELBASE!RegQueryMultipleValuesW
00007ffc`f52b9638  00000000`00000000
00007ffc`f52b9640  00007ffc`f3bde120 KERNELBASE!GetComputerNameExA
00007ffc`f52b9648  00007ffc`f3b89920 KERNELBASE!GetSystemTime
00007ffc`f52b9650  00007ffc`f3bac370 KERNELBASE!GetSystemWindowsDirectoryW
00007ffc`f52b9658  00007ffc`f3b86f00 KERNELBASE!GetSystemTimeAsFileTime
00007ffc`f52b9660  00007ffc`f3bb4ce0 KERNELBASE!GetTickCount
00007ffc`f52b9668  00007ffc`f3baa870 KERNELBASE!GetLocalTime
00007ffc`f52b9670  00007ffc`f3bbc200 KERNELBASE!GetComputerNameExW
00007ffc`f52b9678  00007ffc`f3b8daf0 KERNELBASE!GetSystemDirectoryW
00007ffc`f52b9680  00000000`00000000
00007ffc`f52b9688  00007ffc`f3b31ee0 KERNELBASE!EnumDynamicTimeZoneInformation
00007ffc`f52b9690  00007ffc`f3ba6db0 KERNELBASE!GetDynamicTimeZoneInformationEffectiveYears
00007ffc`f52b9698  00000000`00000000
00007ffc`f52b96a0  00007ffc`f52fb9c0 sechost!CloseTrace
00007ffc`f52b96a8  00007ffc`f52fba80 sechost!ProcessTrace
00007ffc`f52b96b0  00007ffc`f52fbda0 sechost!OpenTraceW
00007ffc`f52b96b8  00000000`00000000
00007ffc`f52b96c0  00007ffc`f533dcf0 sechost!QueryTraceProcessingHandle
00007ffc`f52b96c8  00000000`00000000
00007ffc`f52b96d0  00007ffc`f530a8a0 sechost!StopTraceW
00007ffc`f52b96d8  00007ffc`f530cc50 sechost!EnumerateTraceGuidsEx
00007ffc`f52b96e0  00007ffc`f52f68d0 sechost!ControlTraceW
00007ffc`f52b96e8  00007ffc`f533e670 sechost!TraceSetInformation
00007ffc`f52b96f0  00007ffc`f533e440 sechost!EventAccessRemove
00007ffc`f52b96f8  00007ffc`f533e1f0 sechost!EventAccessQuery
00007ffc`f52b9700  00007ffc`f52fa710 sechost!StartTraceW
00007ffc`f52b9708  00007ffc`f52f6ea0 sechost!EnableTraceEx2
00007ffc`f52b9710  00007ffc`f530a8c0 sechost!QueryAllTracesW
00007ffc`f52b9718  00007ffc`f533e1a0 sechost!EventAccessControl
00007ffc`f52b9720  00000000`00000000
00007ffc`f52b9728  00007ffc`f62b59f0 ntdll!EtwEventRegister
00007ffc`f52b9730  00007ffc`f62b5520 ntdll!EtwEventSetInformation
00007ffc`f52b9738  00007ffc`f62a65e0 ntdll!EtwEventUnregister
00007ffc`f52b9740  00007ffc`f62a4f40 ntdll!EtwEventWriteTransfer
00007ffc`f52b9748  00000000`00000000
00007ffc`f52b9750  00007ffc`f530ce10 sechost!AuditComputeEffectivePolicyBySid
00007ffc`f52b9758  00007ffc`f530a880 sechost!AuditFree
00007ffc`f52b9760  00007ffc`f530bc60 sechost!AuditQuerySystemPolicy
00007ffc`f52b9768  00007ffc`f53421a0 sechost!AuditSetSystemPolicy
00007ffc`f52b9770  00000000`00000000
00007ffc`f52b9778  00007ffc`f5341c60 sechost!AuditLookupSubCategoryNameW
00007ffc`f52b9780  00007ffc`f5341e50 sechost!AuditQuerySecurity
00007ffc`f52b9788  00007ffc`f53418e0 sechost!AuditEnumeratePerUserPolicy
00007ffc`f52b9790  00007ffc`f5341f90 sechost!AuditSetPerUserPolicy
00007ffc`f52b9798  00007ffc`f5341970 sechost!AuditEnumerateSubCategories
00007ffc`f52b97a0  00007ffc`f5341de0 sechost!AuditQueryGlobalSaclW
00007ffc`f52b97a8  00007ffc`f5341790 sechost!AuditEnumerateCategories
00007ffc`f52b97b0  00007ffc`f530bbc0 sechost!AuditQueryPerUserPolicy
00007ffc`f52b97b8  00007ffc`f5341ae0 sechost!AuditLookupCategoryNameW
00007ffc`f52b97c0  00007ffc`f5341f20 sechost!AuditSetGlobalSaclW
00007ffc`f52b97c8  00007ffc`f5342040 sechost!AuditSetSecurity
00007ffc`f52b97d0  00000000`00000000
00007ffc`f52b97d8  00007ffc`f3c65db0 KERNELBASE!AccessCheckByTypeResultListAndAuditAlarmByHandleW
00007ffc`f52b97e0  00007ffc`f3c66700 KERNELBASE!ObjectDeleteAuditAlarmW
00007ffc`f52b97e8  00007ffc`f3b8e660 KERNELBASE!GetSidSubAuthorityCount
00007ffc`f52b97f0  00007ffc`f3ba2900 KERNELBASE!SetTokenInformation
00007ffc`f52b97f8  00007ffc`f3c66410 KERNELBASE!AreAnyAccessesGranted
00007ffc`f52b9800  00007ffc`f3ba53b0 KERNELBASE!SetSecurityDescriptorRMControl
00007ffc`f52b9808  00007ffc`f3b391a0 KERNELBASE!EqualSid
00007ffc`f52b9810  00007ffc`f3b8d6d0 KERNELBASE!GetAce
00007ffc`f52b9818  00007ffc`f3ba4db0 KERNELBASE!GetSecurityDescriptorRMControl
00007ffc`f52b9820  00007ffc`f3b994f0 KERNELBASE!SetSecurityDescriptorOwner
00007ffc`f52b9828  00007ffc`f3b60570 KERNELBASE!IsTokenRestricted
00007ffc`f52b9830  00007ffc`f3ba0f90 KERNELBASE!DeleteAce
00007ffc`f52b9838  00007ffc`f3c66800 KERNELBASE!PrivilegedServiceAuditAlarmW
00007ffc`f52b9840  00007ffc`f3b991b0 KERNELBASE!InitializeSid
00007ffc`f52b9848  00007ffc`f3b9d830 KERNELBASE!GetSecurityDescriptorOwner
00007ffc`f52b9850  00007ffc`f3ba1f30 KERNELBASE!ImpersonateAnonymousToken
00007ffc`f52b9858  00007ffc`f3b9a9b0 KERNELBASE!ImpersonateSelf
```

```
00007ffc`f52b9860  00007ffc`f3b37710 KERNELBASE!EqualDomainSid
00007ffc`f52b9868  00007ffc`f3b9fe90 KERNELBASE!SetSecurityDescriptorSacl
00007ffc`f52b9870  00007ffc`f3b8ac90 KERNELBASE!IsValidSid
00007ffc`f52b9878  00007ffc`f3b8f8b0 KERNELBASE!AddAce
00007ffc`f52b9880  00007ffc`f3c66220 KERNELBASE!AddAuditAccessAce
00007ffc`f52b9888  00007ffc`f3c65f50 KERNELBASE!AccessCheckByTypeResultListAndAuditAlarmW
00007ffc`f52b9890  00007ffc`f3c668a0 KERNELBASE!SetAclInformation
00007ffc`f52b9898  00007ffc`f3b91ec0 KERNELBASE!IsValidSecurityDescriptor
00007ffc`f52b98a0  00007ffc`f3bde540 KERNELBASE!AddAccessDeniedAce
00007ffc`f52b98a8  00007ffc`f3ba7620 KERNELBASE!CreateRestrictedToken
00007ffc`f52b98b0  00007ffc`f3b8e6a0 KERNELBASE!FreeSid
00007ffc`f52b98b8  00007ffc`f3b358c0 KERNELBASE!EqualPrefixSid
00007ffc`f52b98c0  00007ffc`f3b8c810 KERNELBASE!GetFileSecurityW
00007ffc`f52b98c8  00007ffc`f3b9e9f0 KERNELBASE!GetSecurityDescriptorGroup
00007ffc`f52b98d0  00007ffc`f3b8a360 KERNELBASE!CheckTokenMembership
00007ffc`f52b98d8  00007ffc`f3b94d90 KERNELBASE!InitializeSecurityDescriptor
00007ffc`f52b98e0  00007ffc`f3b91bf0 KERNELBASE!InitializeAcl
00007ffc`f52b98e8  00007ffc`f3b8be90 KERNELBASE!DuplicateToken
00007ffc`f52b98f0  00007ffc`f3c66970 KERNELBASE!SetPrivateObjectSecurityEx
00007ffc`f52b98f8  00007ffc`f3c660e0 KERNELBASE!AddAccessAllowedObjectAce
00007ffc`f52b9900  00007ffc`f3b62fd0 KERNELBASE!GetKernelObjectSecurity
00007ffc`f52b9908  00007ffc`f3b9acb0 KERNELBASE!MapGenericMask
00007ffc`f52b9910  00007ffc`f3b9fb90 KERNELBASE!SetKernelObjectSecurity
00007ffc`f52b9918  00007ffc`f3b992e0 KERNELBASE!AddAccessAllowedAceEx
00007ffc`f52b9920  00007ffc`f3b882b0 KERNELBASE!GetLengthSid
00007ffc`f52b9928  00007ffc`f3ba3dc0 KERNELBASE!SetSecurityDescriptorControl
00007ffc`f52b9930  00007ffc`f3b8bec0 KERNELBASE!DuplicateTokenEx
00007ffc`f52b9938  00007ffc`f3b8d710 KERNELBASE!IsValidAcl
00007ffc`f52b9940  00007ffc`f3b9ad00 KERNELBASE!GetSecurityDescriptorLength
00007ffc`f52b9948  00007ffc`f3b94e90 KERNELBASE!AddAccessAllowedAce
00007ffc`f52b9950  00007ffc`f3c661b0 KERNELBASE!AddAccessDeniedObjectAce
00007ffc`f52b9958  00007ffc`f3b92f70 KERNELBASE!MakeSelfRelativeSD
00007ffc`f52b9960  00007ffc`f3b8a890 KERNELBASE!AccessCheckByType
00007ffc`f52b9968  00007ffc`f3c66270 KERNELBASE!AddAuditAccessAceEx
00007ffc`f52b9970  00007ffc`f3b90430 KERNELBASE!MakeAbsoluteSD
00007ffc`f52b9978  00007ffc`f3b99c00 KERNELBASE!SetSecurityDescriptorGroup
00007ffc`f52b9980  00007ffc`f3b94e50 KERNELBASE!GetSidIdentifierAuthority
00007ffc`f52b9988  00007ffc`f3b63c80 KERNELBASE!GetTokenInformation
00007ffc`f52b9990  00007ffc`f3c65cf0 KERNELBASE!AccessCheckByTypeResultList
00007ffc`f52b9998  00007ffc`f3b94390 KERNELBASE!PrivilegeCheck
00007ffc`f52b99a0  00007ffc`f3ba3ec0 KERNELBASE!ObjectCloseAuditAlarmW
00007ffc`f52b99a8  00007ffc`f3b897e0 KERNELBASE!AccessCheck
00007ffc`f52b99b0  00007ffc`f3b38750 KERNELBASE!GetSidSubAuthority
00007ffc`f52b99b8  00007ffc`f3b896c0 KERNELBASE!AllocateAndInitializeSid
00007ffc`f52b99c0  00007ffc`f3b37d90 KERNELBASE!IsWellKnownSid
00007ffc`f52b99c8  00007ffc`f3c66770 KERNELBASE!ObjectPrivilegeAuditAlarmW
00007ffc`f52b99d0  00007ffc`f3b41890 KERNELBASE!CopySid
00007ffc`f52b99d8  00007ffc`f3b9f520 KERNELBASE!DestroyPrivateObjectSecurity
00007ffc`f52b99e0  00007ffc`f3b971f0 KERNELBASE!AdjustTokenPrivileges
00007ffc`f52b99e8  00007ffc`f3bb1060 KERNELBASE!GetSecurityDescriptorControl
00007ffc`f52b99f0  00007ffc`f3ba2830 KERNELBASE!SetSecurityAccessMask
00007ffc`f52b99f8  00007ffc`f3c66150 KERNELBASE!AddAccessDeniedAceEx
00007ffc`f52b9a00  00007ffc`f3b37120 KERNELBASE!CreatePrivateObjectSecurity
00007ffc`f52b9a08  00007ffc`f3b37ad0 KERNELBASE!CreateWellKnownSid
00007ffc`f52b9a10  00007ffc`f3c66640 KERNELBASE!GetPrivateObjectSecurity
00007ffc`f52b9a18  00007ffc`f3b34c90 KERNELBASE!AreAllAccessesGranted
00007ffc`f52b9a20  00007ffc`f3b36fe0 KERNELBASE!CreatePrivateObjectSecurityWithMultipleInheritance
00007ffc`f52b9a28  00007ffc`f3b99320 KERNELBASE!GetSidLengthRequired
00007ffc`f52b9a30  00007ffc`f3c66440 KERNELBASE!ConvertToAutoInheritPrivateObjectSecurity
00007ffc`f52b9a38  00007ffc`f3b91070 KERNELBASE!AccessCheckAndAuditAlarmW
00007ffc`f52b9a40  00007ffc`f3ba2930 KERNELBASE!AdjustTokenGroups
00007ffc`f52b9a48  00007ffc`f3b9d200 KERNELBASE!CreatePrivateObjectSecurityEx
00007ffc`f52b9a50  00007ffc`f3b945b0 KERNELBASE!GetAclInformation
00007ffc`f52b9a58  00007ffc`f3ba26d0 KERNELBASE!SetFileSecurityW
00007ffc`f52b9a60  00007ffc`f3c662e0 KERNELBASE!AddAuditAccessObjectAce
00007ffc`f52b9a68  00007ffc`f3ba4650 KERNELBASE!AllocateLocallyUniqueId
00007ffc`f52b9a70  00007ffc`f3b8bf90 KERNELBASE!GetSecurityDescriptorDacl
00007ffc`f52b9a78  00007ffc`f3b895b0 KERNELBASE!RevertToSelf
00007ffc`f52b9a80  00007ffc`f3b8c990 KERNELBASE!QuerySecurityAccessMask
00007ffc`f52b9a88  00007ffc`f3c66930 KERNELBASE!SetPrivateObjectSecurity
00007ffc`f52b9a90  00007ffc`f3b37650 KERNELBASE!GetWindowsAccountDomainSid
00007ffc`f52b9a98  00007ffc`f3ba1290 KERNELBASE!ObjectOpenAuditAlarmW
00007ffc`f52b9aa0  00007ffc`f3b94df0 KERNELBASE!GetSecurityDescriptorSacl
00007ffc`f52b9aa8  00007ffc`f3c665a0 KERNELBASE!FindFirstFreeAce
00007ffc`f52b9ab0  00007ffc`f3c65bb0 KERNELBASE!AccessCheckByTypeAndAuditAlarmW
00007ffc`f52b9ab8  00007ffc`f3b93940 KERNELBASE!SetSecurityDescriptorDacl
00007ffc`f52b9ac0  00007ffc`f3b8d9a0 KERNELBASE!ImpersonateLoggedOnUser
00007ffc`f52b9ac8  00000000`00000000
00007ffc`f52b9ad0  00007ffc`f3c666c0 KERNELBASE!MakeAbsoluteSD2
00007ffc`f52b9ad8  00000000`00000000
00007ffc`f52b9ae0  00007ffc`f52f36f0 sechost!StartServiceCtrlDispatcherW
00007ffc`f52b9ae8  00007ffc`f52f5070 sechost!SetServiceStatus
00007ffc`f52b9af0  00007ffc`f52f3180 sechost!RegisterServiceCtrlHandlerExW
00007ffc`f52b9af8  00000000`00000000
00007ffc`f52b9b00  00007ffc`f533b560 sechost!QueryServiceDynamicInformation
00007ffc`f52b9b08  00007ffc`f52f54d0 sechost!EnumServicesStatusExW
00007ffc`f52b9b10  00007ffc`f530ae70 sechost!EnumDependentServicesW
00007ffc`f52b9b18  00000000`00000000
00007ffc`f52b9b20  00007ffc`f530acb0 sechost!GetServiceKeyNameW
00007ffc`f52b9b28  00007ffc`f530abf0 sechost!GetServiceDisplayNameW
00007ffc`f52b9b30  00000000`00000000
00007ffc`f52b9b38  00007ffc`f5339fb0 sechost!CreateServiceW
00007ffc`f52b9b40  00007ffc`f52f58f0 sechost!OpenServiceW
00007ffc`f52b9b48  00007ffc`f52f2cc0 sechost!StartServiceW
00007ffc`f52b9b50  00007ffc`f52f25b0 sechost!ControlServiceExW
00007ffc`f52b9b58  00007ffc`f533a470 sechost!DeleteService
00007ffc`f52b9b60  00007ffc`f52f5970 sechost!OpenSCManagerW
00007ffc`f52b9b68  00007ffc`f52f5a70 sechost!CloseServiceHandle
```

```
00007ffc`f52b9b70  00000000`00000000
00007ffc`f52b9b78  00007ffc`f52f57d0 sechost!QueryServiceConfigW
00007ffc`f52b9b80  00007ffc`f530a330 sechost!SetServiceObjectSecurity
00007ffc`f52b9b88  00007ffc`f530a720 sechost!ChangeServiceConfig2W
00007ffc`f52b9b90  00007ffc`f52f2590 sechost!NotifyServiceStatusChangeW
00007ffc`f52b9b98  00007ffc`f533b050 sechost!QueryServiceObjectSecurity
00007ffc`f52b9ba0  00007ffc`f52f47e0 sechost!QueryServiceConfig2W
00007ffc`f52b9ba8  00007ffc`f530a5a0 sechost!ChangeServiceConfigW
00007ffc`f52b9bb0  00007ffc`f52f4f70 sechost!QueryServiceStatusEx
00007ffc`f52b9bb8  00000000`00000000
00007ffc`f52b9bc0  00007ffc`f530d510 sechost!I_ScSetServiceBitsW
00007ffc`f52b9bc8  00007ffc`f5338c20 sechost!I_ScSetServiceBitsA
00007ffc`f52b9bd0  00007ffc`f52f21d0 sechost!WaitServiceState
00007ffc`f52b9bd8  00007ffc`f533c180 sechost!I_ScRpcBindA
00007ffc`f52b9be0  00007ffc`f530cda0 sechost!I_ScRpcBindW
00007ffc`f52b9be8  00000000`00000000
00007ffc`f52b9bf0  00007ffc`f533a850 sechost!QueryLocalUserServiceName
00007ffc`f52b9bf8  00007ffc`f52f4ac0 sechost!QueryUserServiceName
00007ffc`f52b9c00  00007ffc`f533a790 sechost!I_ScReparseServiceDatabase
00007ffc`f52b9c08  00000000`00000000
00007ffc`f52b9c10  00007ffc`f533b130 sechost!QueryUserServiceNameForContext
00007ffc`f52b9c18  00000000`00000000
00007ffc`f52b9c20  00007ffc`f5339ab0 sechost!CreateServiceEx
00007ffc`f52b9c28  00000000`00000000
00007ffc`f52b9c30  00007ffc`f53391f0 sechost!ControlServiceExA
00007ffc`f52b9c38  00007ffc`f52f1e50 sechost!RegisterServiceCtrlHandlerW
00007ffc`f52b9c40  00007ffc`f5338f10 sechost!ChangeServiceConfigA
00007ffc`f52b9c48  00007ffc`f52f1b60 sechost!RegisterServiceCtrlHandlerExA
00007ffc`f52b9c50  00007ffc`f52f1ca0 sechost!StartServiceA
00007ffc`f52b9c58  00007ffc`f530a220 sechost!QueryServiceConfigA
00007ffc`f52b9c60  00007ffc`f52f1e70 sechost!NotifyServiceStatusChangeA
00007ffc`f52b9c68  00007ffc`f52f5010 sechost!QueryServiceStatus
00007ffc`f52b9c70  00007ffc`f533b670 sechost!StartServiceCtrlDispatcherA
00007ffc`f52b9c78  00007ffc`f52f1ed0 sechost!OpenServiceA
00007ffc`f52b9c80  00007ffc`f5339350 sechost!CreateServiceA
00007ffc`f52b9c88  00007ffc`f533abe0 sechost!QueryServiceConfig2A
00007ffc`f52b9c90  00007ffc`f5338d50 sechost!ChangeServiceConfig2A
00007ffc`f52b9c98  00007ffc`f533b600 sechost!RegisterServiceCtrlHandlerA
00007ffc`f52b9ca0  00007ffc`f530a820 sechost!ControlService
00007ffc`f52b9ca8  00007ffc`f52f5740 sechost!OpenSCManagerA
00007ffc`f52b9cb0  00000000`00000000
00007ffc`f52b9cb8  00007ffc`f41a16e0 msvcrt!wcscpy_s
00007ffc`f52b9cc0  00007ffc`f41a15c0 msvcrt!wcscat_s
00007ffc`f52b9cc8  00007ffc`f4179a70 msvcrt!swprintf_s
00007ffc`f52b9cd0  00007ffc`f419f0f0 msvcrt!wcsicmp
00007ffc`f52b9cd8  00007ffc`f419f980 msvcrt!wcsnicmp
00007ffc`f52b9ce0  00007ffc`f4144bc0 msvcrt!tolower
00007ffc`f52b9ce8  00007ffc`f41a1240 msvcrt!strstr
00007ffc`f52b9cf0  00007ffc`f41a07a0 msvcrt!strchr
00007ffc`f52b9cf8  00007ffc`f4143110 msvcrt!ultow_s
00007ffc`f52b9d00  00007ffc`f4143810 msvcrt!iswctype
00007ffc`f52b9d08  00007ffc`f41457e0 msvcrt!wcstoul
00007ffc`f52b9d10  00007ffc`f4145410 msvcrt!wcstoui64
00007ffc`f52b9d18  00007ffc`f41a1bf0 msvcrt!wcsstr
00007ffc`f52b9d20  00007ffc`f4142f10 msvcrt!ultow
00007ffc`f52b9d28  00007ffc`f41a1d20 msvcrt!wcstok_s
00007ffc`f52b9d30  00007ffc`f4148010 msvcrt!errno
00007ffc`f52b9d38  00007ffc`f41430f0 msvcrt!ui64tow_s
00007ffc`f52b9d40  00007ffc`f4143060 msvcrt!i64tow_s
00007ffc`f52b9d48  00007ffc`f419dd90 msvcrt!stricmp
00007ffc`f52b9d50  00007ffc`f41a1af0 msvcrt!wcsnlen
00007ffc`f52b9d58  00007ffc`f415c500 msvcrt!resetstkoflw
00007ffc`f52b9d60  00007ffc`f41435e0 msvcrt!iswalpha
00007ffc`f52b9d68  00007ffc`f41a1960 msvcrt!wcsncmp
00007ffc`f52b9d70  00007ffc`f418dd40 msvcrt!vsnprintf
00007ffc`f52b9d78  00007ffc`f414d500 msvcrt!_CxxFrameHandler3
00007ffc`f52b9d80  00007ffc`f41453a0 msvcrt!wcstoi64
00007ffc`f52b9d88  00007ffc`f41a0500 msvcrt!memcmp
00007ffc`f52b9d90  00007ffc`f41b7d40 msvcrt!memcpy
00007ffc`f52b9d98  00007ffc`f41b7d40 msvcrt!memcpy
00007ffc`f52b9da0  00007ffc`f418e200 msvcrt!vsnwprintf
00007ffc`f52b9da8  00007ffc`f41a1b60 msvcrt!wcsrchr
00007ffc`f52b9db0  00007ffc`f41a1670 msvcrt!wcschr
00007ffc`f52b9db8  00007ffc`f4179b20 msvcrt!swscanf_s
00007ffc`f52b9dc0  00007ffc`f416acc0 msvcrt!_C_specific_handler
00007ffc`f52b9dc8  00007ffc`f41a19f0 msvcrt!wcsncpy_s
00007ffc`f52b9dd0  00007ffc`f41b8000 msvcrt!memset
00007ffc`f52b9dd8  00000000`00000000
00007ffc`f52b9de0  00007ffc`f6302880 ntdll!RtlIsValidIndexHandle
00007ffc`f52b9de8  00007ffc`f6302200 ntdll!RtlFreeHandle
00007ffc`f52b9df0  00007ffc`f6343980 ntdll!NtOpenKey
00007ffc`f52b9df8  00007ffc`f6343a20 ntdll!NtQueryValueKey
00007ffc`f52b9e00  00007ffc`f6343920 ntdll!NtClose
00007ffc`f52b9e08  00007ffc`f6343bc0 ntdll!NtOpenThreadToken
00007ffc`f52b9e10  00007ffc`f6345cf0 ntdll!NtOpenProcessToken
00007ffc`f52b9e18  00007ffc`f62b2470 ntdll!RtlEqualSid
00007ffc`f52b9e20  00007ffc`f6314980 ntdll!RtlLengthSid
00007ffc`f52b9e28  00007ffc`f63211a0 ntdll!RtlAddAccessAllowedAceEx
00007ffc`f52b9e30  00007ffc`f6346bd0 ntdll!NtSetInformationToken
00007ffc`f52b9e38  00007ffc`f62ed450 ntdll!RtlCreateSecurityDescriptor
00007ffc`f52b9e40  00007ffc`f62e9890 ntdll!RtlSetOwnerSecurityDescriptor
00007ffc`f52b9e48  00007ffc`f6343f80 ntdll!NtDuplicateToken
00007ffc`f52b9e50  00007ffc`f6344b10 ntdll!NtCompareTokens
00007ffc`f52b9e58  00007ffc`f63146f0 ntdll!RtlAllocateAndInitializeSid
00007ffc`f52b9e60  00007ffc`f631bfd0 ntdll!RtlFreeSid
00007ffc`f52b9e68  00007ffc`f6321ab0 ntdll!RtlIsGenericTableEmpty
00007ffc`f52b9e70  00007ffc`f630a120 ntdll!RtlEnumerateGenericTableWithoutSplaying
00007ffc`f52b9e78  00007ffc`f62d4620 ntdll!RtlCopyUnicodeString
```

```
00007ffc`f52b9e80  00007ffc`f62ea260 ntdll!RtlDuplicateUnicodeString
00007ffc`f52b9e88  00007ffc`f62d2f70 ntdll!RtlExpandEnvironmentStrings_U
00007ffc`f52b9e90  00007ffc`f6343da0 ntdll!NtOpenFile
00007ffc`f52b9e98  00007ffc`f62ea6d0 ntdll!RtlCreateUnicodeString
00007ffc`f52b9ea0  00007ffc`f6343a60 ntdll!NtQueryInformationProcess
00007ffc`f52b9ea8  00007ffc`f6323bc0 ntdll!RtlGetLastNtStatus
00007ffc`f52b9eb0  00007ffc`f6343a00 ntdll!NtQueryKey
00007ffc`f52b9eb8  00007ffc`f62e8420 ntdll!RtlValidSid
00007ffc`f52b9ec0  00007ffc`f62dad00 ntdll!LdrLoadDll
00007ffc`f52b9ec8  00007ffc`f62eeea0 ntdll!RtlImageNtHeader
00007ffc`f52b9ed0  00007ffc`f62d8130 ntdll!LdrUnloadDll
00007ffc`f52b9ed8  00007ffc`f6343820 ntdll!NtDeviceIoControlFile
00007ffc`f52b9ee0  00007ffc`f6343e00 ntdll!NtQuerySystemInformation
00007ffc`f52b9ee8  00007ffc`f62b59f0 ntdll!EtwEventRegister
00007ffc`f52b9ef0  00007ffc`f62a61f0 ntdll!EtwEventWrite
00007ffc`f52b9ef8  00007ffc`f6343ae0 ntdll!NtCreateKey
00007ffc`f52b9f00  00007ffc`f6344330 ntdll!NtSetValueKey
00007ffc`f52b9f08  00007ffc`f6309930 ntdll!RtlDeleteElementGenericTable
00007ffc`f52b9f10  00007ffc`f62ef1e0 ntdll!RtlAppendUnicodeToString
00007ffc`f52b9f18  00007ffc`f6345230 ntdll!NtDeleteKey
00007ffc`f52b9f20  00007ffc`f63099d0 ntdll!RtlInsertElementGenericTable
00007ffc`f52b9f28  00007ffc`f62af070 ntdll!RtlCopySid
00007ffc`f52b9f30  00007ffc`f63262a0 ntdll!RtlInitializeHandleTable
00007ffc`f52b9f38  00007ffc`f632c860 ntdll!RtlDestroyHandleTable
00007ffc`f52b9f40  00007ffc`f62a65e0 ntdll!EtwEventUnregister
00007ffc`f52b9f48  00007ffc`f6343d80 ntdll!NtEnumerateKey
00007ffc`f52b9f50  00007ffc`f6313710 ntdll!RtlIntegerToUnicodeString
00007ffc`f52b9f58  00007ffc`f6314c40 ntdll!RtlStringFromGUID
00007ffc`f52b9f60  00007ffc`f62e9110 ntdll!RtlAppendUnicodeStringToString
00007ffc`f52b9f68  00007ffc`f62e7ac0 ntdll!RtlFormatCurrentUserKeyPath
00007ffc`f52b9f70  00007ffc`f6325d20 ntdll!RtlInitializeGenericTable
00007ffc`f52b9f78  00007ffc`f62a9b10 ntdll!RtlQueryRegistryValuesEx
00007ffc`f52b9f80  00007ffc`f6309b50 ntdll!RtlLookupElementGenericTable
00007ffc`f52b9f88  00007ffc`f6329ee0 ntdll!RtlNumberGenericTableElements
00007ffc`f52b9f90  00007ffc`f6314b60 ntdll!RtlGUIDFromString
00007ffc`f52b9f98  00007ffc`f62fbaa0 ntdll!RtlUpcaseUnicodeChar
00007ffc`f52b9fa0  00007ffc`f6344060 ntdll!NtQueryVolumeInformationFile
00007ffc`f52b9fa8  00007ffc`f6345d90 ntdll!NtOpenSymbolicLinkObject
00007ffc`f52b9fb0  00007ffc`f63463d0 ntdll!NtQuerySymbolicLinkObject
00007ffc`f52b9fb8  00007ffc`f62e1230 ntdll!RtlPrefixUnicodeString
00007ffc`f52b9fc0  00007ffc`f6300340 ntdll!RtlDetermineDosPathNameType_U
00007ffc`f52b9fc8  00007ffc`f6343960 ntdll!NtQueryInformationFile
00007ffc`f52b9fd0  00007ffc`f6326be0 ntdll!RtlGetFullPathName_U
00007ffc`f52b9fd8  00007ffc`f62add60 ntdll!RtlUnicodeToMultiByteN
00007ffc`f52b9fe0  00007ffc`f631a6a0 ntdll!RtlNtStatusToDosErrorNoTeb
00007ffc`f52b9fe8  00007ffc`f62ae020 ntdll!RtlUnicodeToMultiByteSize
00007ffc`f52b9ff0  00007ffc`f62ff5a0 ntdll!RtlAnsiCharToUnicodeChar
00007ffc`f52b9ff8  00007ffc`f62add10 ntdll!RtlMultiByteToUnicodeN
00007ffc`f52ba000  00007ffc`f63470f0 ntdll!NtTraceControl
00007ffc`f52ba008  00007ffc`f62a67c0 ntdll!RtlSetLastWin32Error
00007ffc`f52ba010  00007ffc`f63138f0 ntdll!RtlInitAnsiStringEx
00007ffc`f52ba018  00007ffc`f62e7c30 ntdll!RtlInitUnicodeStringEx
00007ffc`f52ba020  00007ffc`f63138b0 ntdll!RtlCreateUnicodeStringFromAsciiz
00007ffc`f52ba028  00007ffc`f6346690 ntdll!NtRenameKey
00007ffc`f52ba030  00007ffc`f63131b0 ntdll!RtlQueryPackageIdentity
00007ffc`f52ba038  00007ffc`f62ac250 ntdll!RtlOemStringToUnicodeString
00007ffc`f52ba040  00007ffc`f6308a40 ntdll!RtlIsTextUnicode
00007ffc`f52ba048  00007ffc`f63438e0 ntdll!NtSetInformationThread
00007ffc`f52ba050  00007ffc`f62abab0 ntdll!RtlAddAce
00007ffc`f52ba058  00007ffc`f62e7e80 ntdll!RtlValidAcl
00007ffc`f52ba060  00007ffc`f62ab690 ntdll!RtlSetSaclSecurityDescriptor
00007ffc`f52ba068  00007ffc`f62b2800 ntdll!RtlInitializeSid
00007ffc`f52ba070  00007ffc`f6321c50 ntdll!RtlGetControlSecurityDescriptor
00007ffc`f52ba078  00007ffc`f6394300 ntdll!RtlAddAuditAccessObjectAce
00007ffc`f52ba080  00007ffc`f62ed3e0 ntdll!RtlSetDaclSecurityDescriptor
00007ffc`f52ba088  00007ffc`f631e6d0 ntdll!RtlGetSaclSecurityDescriptor
00007ffc`f52ba090  00007ffc`f63113f0 ntdll!RtlGetAce
00007ffc`f52ba098  00007ffc`f6330580 ntdll!RtlAddAuditAccessAceEx
00007ffc`f52ba0a0  00007ffc`f62ade90 ntdll!RtlxAnsiStringToUnicodeSize
00007ffc`f52ba0a8  00007ffc`f6311300 ntdll!RtlGetOwnerSecurityDescriptor
00007ffc`f52ba0b0  00007ffc`f6322420 ntdll!RtlGetGroupSecurityDescriptor
00007ffc`f52ba0b8  00007ffc`f63116d0 ntdll!RtlAbsoluteToSelfRelativeSD
00007ffc`f52ba0c0  00007ffc`f632ca70 ntdll!RtlAddAccessDeniedAceEx
00007ffc`f52ba0c8  00007ffc`f6302980 ntdll!RtlAllocateHandle
00007ffc`f52ba0d0  00007ffc`f6394080 ntdll!RtlAddAccessDeniedObjectAce
00007ffc`f52ba0d8  00007ffc`f62e7e20 ntdll!RtlFirstFreeAce
00007ffc`f52ba0e0  00007ffc`f62e9830 ntdll!RtlSetGroupSecurityDescriptor
00007ffc`f52ba0e8  00007ffc`f6311340 ntdll!RtlGetDaclSecurityDescriptor
00007ffc`f52ba0f0  00007ffc`f631d8f0 ntdll!RtlDosPathNameToNtPathName_U
00007ffc`f52ba0f8  00007ffc`f62a4f40 ntdll!EtwEventWriteTransfer
00007ffc`f52ba100  00007ffc`f62b5520 ntdll!EtwEventSetInformation
00007ffc`f52ba108  00007ffc`f631f180 ntdll!RtlImpersonateSelf
00007ffc`f52ba110  00007ffc`f6321b40 ntdll!RtlAdjustPrivilege
00007ffc`f52ba118  00007ffc`f632c710 ntdll!RtlCopyString
00007ffc`f52ba120  00007ffc`f6344280 ntdll!NtQuerySystemTime
00007ffc`f52ba128  00007ffc`f6321860 ntdll!RtlTimeToSecondsSince1970
00007ffc`f52ba130  00007ffc`f62a6490 ntdll!EtwTraceMessage
00007ffc`f52ba138  00007ffc`f63271c0 ntdll!EtwGetTraceLoggerHandle
00007ffc`f52ba140  00007ffc`f6327200 ntdll!EtwGetTraceEnableLevel
00007ffc`f52ba148  00007ffc`f6327240 ntdll!EtwGetTraceEnableFlags
00007ffc`f52ba150  00007ffc`f62b5360 ntdll!EtwRegisterTraceGuidsW
00007ffc`f52ba158  00007ffc`f62a6590 ntdll!EtwUnregisterTraceGuids
00007ffc`f52ba160  00007ffc`f63437c0 ntdll!NtWaitForSingleObject
00007ffc`f52ba168  00007ffc`f62eea20 ntdll!RtlGetVersion
00007ffc`f52ba170  00007ffc`f6343be0 ntdll!NtQueryInformationThread
00007ffc`f52ba178  00007ffc`f6346370 ntdll!NtQuerySecurityObject
00007ffc`f52ba180  00007ffc`f62b62f0 ntdll!RtlRunOnceExecuteOnce
00007ffc`f52ba188  00007ffc`f62f9720 ntdll!RtlRunOnceBeginInitialize
```

```
00007ffc`f52ba190  00007ffc`f62fafd0 ntdll!RtlDllShutdownInProgress
00007ffc`f52ba198  00007ffc`f6327e10 ntdll!RtlRunOnceInitialize
00007ffc`f52ba1a0  00007ffc`f6343d60 ntdll!NtQueryPerformanceCounter
00007ffc`f52ba1a8  00007ffc`f62c5b40 ntdll!NtdllpFreeStringRoutine
00007ffc`f52ba1b0  00007ffc`f6344df0 ntdll!NtCreateMutant
00007ffc`f52ba1b8  00007ffc`f6345cd0 ntdll!NtOpenPrivateNamespace
00007ffc`f52ba1c0  00007ffc`f6344e90 ntdll!NtCreatePrivateNamespace
00007ffc`f52ba1c8  00007ffc`f63278d0 ntdll!RtlAddSIDToBoundaryDescriptor
00007ffc`f52ba1d0  00007ffc`f6329080 ntdll!RtlCreateBoundaryDescriptor
00007ffc`f52ba1d8  00007ffc`f6344290 ntdll!NtWaitForMultipleObjects
00007ffc`f52ba1e0  00007ffc`f62b0de0 ntdll!RtlCreateAcl
00007ffc`f52ba1e8  00007ffc`f631e540 ntdll!RtlValidRelativeSecurityDescriptor
00007ffc`f52ba1f0  00007ffc`f63441e0 ntdll!NtCreateFile
00007ffc`f52ba1f8  00007ffc`f6343840 ntdll!NtWriteFile
00007ffc`f52ba200  00007ffc`f6343800 ntdll!NtReadFile
00007ffc`f52ba208  00007ffc`f62faff0 ntdll!RtlWaitOnAddress
00007ffc`f52ba210  00007ffc`f62fad30 ntdll!RtlWakeAddressAll
00007ffc`f52ba218  00007ffc`f62b28f0 ntdll!RtlQueryPerformanceCounter
00007ffc`f52ba220  00007ffc`f62b9860 ntdll!RtlAcquireSRWLockExclusive
00007ffc`f52ba228  00007ffc`f62ff960 ntdll!RtlInsertElementGenericTableAvl
00007ffc`f52ba230  00007ffc`f62bb270 ntdll!RtlReleaseSRWLockExclusive
00007ffc`f52ba238  00007ffc`f62da8d0 ntdll!RtlAcquireSRWLockShared
00007ffc`f52ba240  00007ffc`f62ffb10 ntdll!RtlLookupElementGenericTableAvl
00007ffc`f52ba248  00007ffc`f62daa90 ntdll!RtlReleaseSRWLockShared
00007ffc`f52ba250  00007ffc`f6309750 ntdll!RtlEnumerateGenericTableAvl
00007ffc`f52ba258  00007ffc`f62ff720 ntdll!RtlDeleteElementGenericTableAvl
00007ffc`f52ba260  00007ffc`f6327d40 ntdll!RtlInitializeGenericTableAvl
00007ffc`f52ba268  00007ffc`f62fad10 ntdll!RtlWakeAddressSingle
00007ffc`f52ba270  00007ffc`f6306e90 ntdll!RtlDosPathNameToRelativeNtPathName_U
00007ffc`f52ba278  00007ffc`f62f0560 ntdll!RtlReleaseRelativeName
00007ffc`f52ba280  00007ffc`f6311d40 ntdll!RtlInitializeSRWLock
00007ffc`f52ba288  00007ffc`f62ea060 ntdll!RtlEqualUnicodeString
00007ffc`f52ba290  00007ffc`f62a22d0 ntdll!RtlDestroyQueryDebugBuffer
00007ffc`f52ba298  00007ffc`f62a13c0 ntdll!RtlQueryProcessDebugInformation
00007ffc`f52ba2a0  00007ffc`f6344850 ntdll!NtAlpcQueryInformation
00007ffc`f52ba2a8  00007ffc`f62a2080 ntdll!RtlCreateQueryDebugBuffer
00007ffc`f52ba2b0  00007ffc`f6343940 ntdll!NtQueryObject
00007ffc`f52ba2b8  00007ffc`f63462b0 ntdll!NtQueryMutant
00007ffc`f52ba2c0  00007ffc`f62d5200 ntdll!RtlInitAnsiString
00007ffc`f52ba2c8  00007ffc`f62ea6a0 ntdll!RtlAddAccessAllowedAce
00007ffc`f52ba2d0  00007ffc`f630e960 ntdll!RtlOpenCurrentUser
00007ffc`f52ba2d8  00007ffc`f63466d0 ntdll!NtReplaceKey
00007ffc`f52ba2e0  00007ffc`f6346890 ntdll!NtSaveKey
00007ffc`f52ba2e8  00007ffc`f63468d0 ntdll!NtSaveMergedKeys
00007ffc`f52ba2f0  00007ffc`f6316bd0 ntdll!RtlLengthSecurityDescriptor
00007ffc`f52ba2f8  00007ffc`f62ea160 ntdll!RtlValidSecurityDescriptor
00007ffc`f52ba300  00007ffc`f62f0220 ntdll!RtlGetNtProductType
00007ffc`f52ba308  00007ffc`f62e8070 ntdll!RtlConvertSidToUnicodeString
00007ffc`f52ba310  00007ffc`f631d710 ntdll!RtlSubAuthorityCountSid
00007ffc`f52ba318  00007ffc`f631a290 ntdll!RtlSubAuthoritySid
00007ffc`f52ba320  00007ffc`f62e91b0 ntdll!RtlGetThreadPreferredUILanguages
00007ffc`f52ba328  00007ffc`f63116f0 ntdll!RtlMakeSelfRelativeSD
00007ffc`f52ba330  00007ffc`f62c75e0 ntdll!RtlFreeHeap
00007ffc`f52ba338  00007ffc`f62ac570 ntdll!RtlxUnicodeStringToAnsiSize
00007ffc`f52ba340  00007ffc`f6343b60 ntdll!NtQueryInformationToken
00007ffc`f52ba348  00007ffc`f62e8510 ntdll!RtlFreeUnicodeString
00007ffc`f52ba350  00007ffc`f62c8ac0 ntdll!RtlAllocateHeap
00007ffc`f52ba358  00007ffc`f6394010 ntdll!RtlAddAccessAllowedObjectAce
00007ffc`f52ba360  00007ffc`f62d8f10 ntdll!RtlVirtualUnwind
00007ffc`f52ba368  00007ffc`f62d9ca0 ntdll!RtlLookupFunctionEntry
00007ffc`f52ba370  00007ffc`f6347970 ntdll!RtlCaptureContext
00007ffc`f52ba378  00007ffc`f6308fe0 ntdll!RtlInitializeCriticalSection
00007ffc`f52ba380  00007ffc`f62be080 ntdll!RtlDeleteCriticalSection
00007ffc`f52ba388  00007ffc`f6346db0 ntdll!NtSetSystemInformation
00007ffc`f52ba390  00007ffc`f62db4d0 ntdll!RtlLeaveCriticalSection
00007ffc`f52ba398  00007ffc`f62da4e0 ntdll!RtlEnterCriticalSection
00007ffc`f52ba3a0  00007ffc`f62a69c0 ntdll!DbgPrint
00007ffc`f52ba3a8  00007ffc`f62a6840 ntdll!RtlNtStatusToDosError
00007ffc`f52ba3b0  00007ffc`f62d5200 ntdll!RtlInitAnsiString
00007ffc`f52ba3b8  00007ffc`f62adf10 ntdll!RtlUnicodeStringToAnsiString
00007ffc`f52ba3c0  00007ffc`f6315830 ntdll!RtlGetCurrentTransaction
00007ffc`f52ba3c8  00007ffc`f62ebd40 ntdll!RtlInitUnicodeString
00007ffc`f52ba3d0  00007ffc`f6345bf0 ntdll!NtOpenKeyEx
00007ffc`f52ba3d8  00007ffc`f6346b70 ntdll!NtSetInformationKey
00007ffc`f52ba3e0  00007ffc`f6327b30 ntdll!LdrGetProcedureAddress
00007ffc`f52ba3e8  00007ffc`f631e400 ntdll!LdrGetDllHandle
00007ffc`f52ba3f0  00007ffc`f62dd400 ntdll!RtlAnsiStringToUnicodeString
00007ffc`f52ba3f8  00007ffc`f6322180 ntdll!RtlFreeAnsiString
00007ffc`f52ba400  00007ffc`f631b940 ntdll!RtlUnicodeStringToInteger
00007ffc`f52ba408  00000000`00000000
```

Note: We see it depends on *rpcrt4* and *sechost* modules and their API.

8. We close logging before exiting WinDbg:

```
0:000> .logclose
Closing open log file C:\AWAPI-Dumps\W4.log
```

Delay-loaded API

- Documentation
- Example:

```
pub func   00007ffc`e85b6d30    0 winmm!_imp_load_waveInOpen (__imp_load_waveInOpen)
pub global 00007ffc`e85db3c0    0 winmm!_imp_waveInOpen = <no type information>
```


If certain Windows API function categories may not be used (for example, lazy evaluation) at all during the program execution, there is no need to load the corresponding module and resolve import address table references to the loaded module. Such Windows API usage is called delay-loaded API, and you may have noticed it in the previous exercises with some global pub functions having a load name after the *_imp_* prefix. Such entries exist in **Delay Import Directory** (DID) entries that point to these "load" functions that load the corresponding module upon the first usage and replace the DID entry value to point to the loaded module. The details are in the documentation link, and we also see that in the next live debugging exercise.

Documentation
https://learn.microsoft.com/en-us/cpp/build/reference/linker-support-for-delay-loaded-dlls?view=msvc-170

API Sets

- Documentation

```
contract_name → module.dll
```

- Example of API contract:

```
api_ms_win_mm_mme_l1_1_0 → winmmbase.dll
```


Another feature added to Windows API is the so-called API sets or API contracts that provide mappings to module names on different platforms. This feature is also used in conjunction with delay-loaded API, as we see in the next exercise. We also mention API sets when we look at API namespaces.

Documentation
https://learn.microsoft.com/en-us/windows/win32/apiindex/windows-apisets

Exercise W5

- **Goal:** Explore the delay-loaded API and API sets
- **Debugging Implementation Patterns:** Code Breakpoint
- **ADDR Patterns:** Call Path
- \AWAPI-Dumps\Exercise-W5.pdf

Exercise W5

Goal: Explore the delay-loaded API and API sets.

Debugging Implementation Patterns: Code Breakpoint.

ADDR Patterns: Call Path.

1. Download and install Visual Studio 2022 Community Edition. Choose Desktop development with C++ option.

2. Copy the example from https://learn.microsoft.com/en-us/cpp/build/reference/linker-support-for-delay-loaded-dlls?view=msvc-170 to *\AWAPI-Dumps\t.cpp*. Add the following code and compiler options, highlighted in blue:

```
// cl t.cpp user32.lib delayimp.lib /Zi /link /DELAYLOAD:user32.dll
#include <windows.h>
// uncomment these lines to remove .libs from command line
// #pragma comment(lib, "delayimp")
// #pragma comment(lib, "user32")

#pragma comment(lib, "winmm")

int main() {
   waveInReset(nullptr);

   // user32.dll will load at this point
   MessageBox(NULL, "Hello", "Hello", MB_OK);
}
```

3. Launch x64 Native Tools Command Prompt for VS 2022:

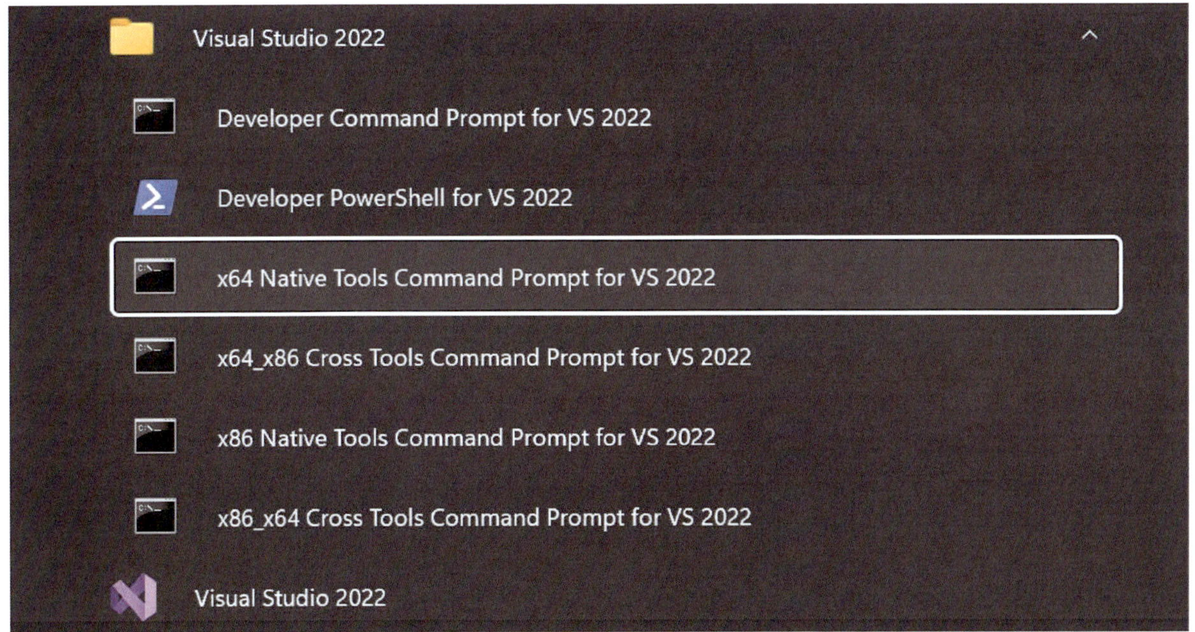

4. Compile and link the program:

```
c:\AWAPI-Dumps>cl t.cpp user32.lib delayimp.lib /Zi /link /DELAYLOAD:user32.dll
Microsoft (R) C/C++ Optimizing Compiler Version 19.32.31332 for x64
Copyright (C) Microsoft Corporation. All rights reserved.

t.cpp
Microsoft (R) Incremental Linker Version 14.32.31332.0
Copyright (C) Microsoft Corporation. All rights reserved.

/out:t.exe
/debug
/DELAYLOAD:user32.dll
t.obj
user32.lib
delayimp.lib

c:\AWAPI-Dumps>
```

5. Launch WinDbg.

6. Choose Launch Executable and *\AWAPI-Dumps\t.exe*.

7. We get the process launched with an initial breakpoint:

```
Microsoft (R) Windows Debugger Version 10.0.27725.1000 AMD64
Copyright (c) Microsoft Corporation. All rights reserved.

CommandLine: C:\AWAPI-Dumps\t.exe

************* Path validation summary **************
Response                         Time (ms)     Location
Deferred                                       srv*
Symbol search path is: srv*
Executable search path is:
ModLoad: 00007ff7`03000000 00007ff7`030a5000   t.exe
ModLoad: 00007ffd`50350000 00007ffd`50567000   ntdll.dll
ModLoad: 00007ffd`4e900000 00007ffd`4e9c4000   C:\WINDOWS\System32\KERNEL32.DLL
ModLoad: 00007ffd`4dbe0000 00007ffd`4df9a000   C:\WINDOWS\System32\KERNELBASE.dll
ModLoad: 00007ffd`46c70000 00007ffd`46d07000   C:\WINDOWS\SYSTEM32\apphelp.dll
ModLoad: 00007ffd`42aa0000 00007ffd`42ad4000   C:\WINDOWS\SYSTEM32\WINMM.dll
ModLoad: 00007ffd`4da00000 00007ffd`4db11000   C:\WINDOWS\System32\ucrtbase.dll
(8e4c.6fc0): Break instruction exception - code 80000003 (first chance)
ntdll!LdrpDoDebuggerBreak+0x30:
00007ffd`5042c134 int     3
```

8. Open a log file using **.logopen**:

```
0:000> .logopen C:\AWAPI-Dumps\W5.log
Opened log file 'C:\AWAPI-Dumps\W5.log'
```

9. Put a breakpoint on the *main* function and resume:

```
0:000> bp main
```

```
0:000> g
Breakpoint 0 hit
t!main:
00007ff6`eef271d0 4883ec28        sub     rsp,28h
```

Note: You may also see the source code window:

t.cpp

```
// #pragma comment(lib, "user32")

#pragma comment(lib, "winmm")

int main() {
    waveInReset(nullptr);

    // user32.dll will load at this point
    MessageBox(NULL, "Hello", "Hello", MB_OK);
```

10. Let's examine IAT at the start of the *main* function execution:

```
0:000> !dh t

File Type: EXECUTABLE IMAGE
FILE HEADER VALUES
    8664 machine (X64)
       9 number of sections
6369908E time date stamp Mon Nov  7 23:11:10 2022

       0 file pointer to symbol table
       0 number of symbols
      F0 size of optional header
      22 characteristics
            Executable
            App can handle >2gb addresses

OPTIONAL HEADER VALUES
     20B magic #
   14.32 linker version
   7D000 size of code
   20A00 size of initialized data
       0 size of uninitialized data
    2DFB address of entry point
    1000 base of code
         ----- new -----
00007ff6eef20000 image base
    1000 section alignment
     200 file alignment
       3 subsystem (Windows CUI)
    6.00 operating system version
    0.00 image version
    6.00 subsystem version
   A3000 size of image
     400 size of headers
       0 checksum
0000000000100000 size of stack reserve
```

```
0000000000001000 size of stack commit
0000000000100000 size of heap reserve
0000000000001000 size of heap commit
    8160 DLL characteristics
           High entropy VA supported
           Dynamic base
           NX compatible
           Terminal server aware
       0 [        0] address [size] of Export Directory
   9D430 [       3C] address [size] of Import Directory
       0 [        0] address [size] of Resource Directory
   98000 [     477C] address [size] of Exception Directory
       0 [        0] address [size] of Security Directory
   A2000 [      828] address [size] of Base Relocation Directory
   8B414 [       38] address [size] of Debug Directory
       0 [        0] address [size] of Description Directory
       0 [        0] address [size] of Special Directory
       0 [        0] address [size] of Thread Storage Directory
   7EFC0 [      140] address [size] of Load Configuration Directory
       0 [        0] address [size] of Bound Import Directory
   9D000 [      430] address [size] of Import Address Table Directory
   9F000 [       40] address [size] of Delay Import Directory
       0 [        0] address [size] of COR20 Header Directory
       0 [        0] address [size] of Reserved Directory

SECTION HEADER #1
   .text name
   7CE43 virtual size
    1000 virtual address
   7D000 size of raw data
     400 file pointer to raw data
       0 file pointer to relocation table
       0 file pointer to line numbers
       0 number of relocations
       0 number of line numbers
60000020 flags
         Code
         (no align specified)
         Execute Read

SECTION HEADER #2
  .rdata name
   15A1E virtual size
   7E000 virtual address
   15C00 size of raw data
   7D400 file pointer to raw data
       0 file pointer to relocation table
       0 file pointer to line numbers
       0 number of relocations
       0 number of line numbers
40000040 flags
         Initialized Data
         (no align specified)
         Read Only


Debug Directories(2)
      Type       Size     Address  Pointer
      cv           2d       8c32c    8b72c    Format: RSDS, guid, 4, c:\AWAPI-Dumps\t.pdb
```

```
      (   12)      14        8c35c    8b75c

SECTION HEADER #3
   .data name
    31E1 virtual size
   94000 virtual address
    1200 size of raw data
   93000 file pointer to raw data
       0 file pointer to relocation table
       0 file pointer to line numbers
       0 number of relocations
       0 number of line numbers
C0000040 flags
         Initialized Data
         (no align specified)
         Read Write

SECTION HEADER #4
  .pdata name
    4FC8 virtual size
   98000 virtual address
    5000 size of raw data
   94200 file pointer to raw data
       0 file pointer to relocation table
       0 file pointer to line numbers
       0 number of relocations
       0 number of line numbers
40000040 flags
         Initialized Data
         (no align specified)
         Read Only

SECTION HEADER #5
  .idata name
    112D virtual size
   9D000 virtual address
    1200 size of raw data
   99200 file pointer to raw data
       0 file pointer to relocation table
       0 file pointer to line numbers
       0 number of relocations
       0 number of line numbers
40000040 flags
         Initialized Data
         (no align specified)
         Read Only

SECTION HEADER #6
  .didat name
     321 virtual size
   9F000 virtual address
     400 size of raw data
   9A400 file pointer to raw data
       0 file pointer to relocation table
       0 file pointer to line numbers
       0 number of relocations
       0 number of line numbers
C0000040 flags
         Initialized Data
```

```
         (no align specified)
         Read Write

SECTION HEADER #7
   .00cfg name
      175 virtual size
    A0000 virtual address
      200 size of raw data
    9A800 file pointer to raw data
        0 file pointer to relocation table
        0 file pointer to line numbers
        0 number of relocations
        0 number of line numbers
 40000040 flags
         Initialized Data
         (no align specified)
         Read Only

SECTION HEADER #8
   _RDATA name
      29F virtual size
    A1000 virtual address
      400 size of raw data
    9AA00 file pointer to raw data
        0 file pointer to relocation table
        0 file pointer to line numbers
        0 number of relocations
        0 number of line numbers
 40000040 flags
         Initialized Data
         (no align specified)
         Read Only

SECTION HEADER #9
   .reloc name
      FE4 virtual size
    A2000 virtual address
     1000 size of raw data
    9AE00 file pointer to raw data
        0 file pointer to relocation table
        0 file pointer to line numbers
        0 number of relocations
        0 number of line numbers
 42000040 flags
         Initialized Data
         Discardable
         (no align specified)
         Read Only
```

```
0:000> dps 00007ff6eef20000 + 9D000 L430/8
00007ff6`eefbd000  00007ff8`0b037b00 KERNEL32!GetCommandLineWStub
00007ff6`eefbd008  00007ff8`0b040a30 KERNEL32!WriteConsoleW
00007ff6`eefbd010  00007ff8`0b037d50 KERNEL32!RaiseExceptionStub
00007ff6`eefbd018  00007ff8`0b030cc0 KERNEL32!GetLastErrorStub
00007ff6`eefbd020  00007ff8`0b036c90 KERNEL32!GetSystemInfoStub
00007ff6`eefbd028  00007ff8`0b035340 KERNEL32!VirtualProtectStub
00007ff6`eefbd030  00007ff8`0b035360 KERNEL32!VirtualQueryStub
00007ff6`eefbd038  00007ff8`0b0352f0 KERNEL32!FreeLibraryStub
00007ff6`eefbd040  00007ff8`0b036580 KERNEL32!GetModuleHandleWStub
```

```
00007ff6`eefbd048  00007ff8`0b033ae0 KERNEL32!GetProcAddressStub
00007ff6`eefbd050  00007ff8`0b037d70 KERNEL32!LoadLibraryExAStub
00007ff6`eefbd058  00007ff8`0b030d20 KERNEL32!QueryPerformanceCounterStub
00007ff6`eefbd060  00007ff8`0b03fe90 KERNEL32!GetCurrentProcessId
00007ff6`eefbd068  00007ff8`0b022750 KERNEL32!GetCurrentThreadId
00007ff6`eefbd070  00007ff8`0b030fd0 KERNEL32!GetSystemTimeAsFileTimeStub
00007ff6`eefbd078  00007ff8`0c1a0cb0 ntdll!RtlInitializeSListHead
00007ff6`eefbd080  00007ff8`0b03fcb0 KERNEL32!RtlCaptureContext
00007ff6`eefbd088  00007ff8`0b035410 KERNEL32!RtlLookupFunctionEntryStub
00007ff6`eefbd090  00007ff8`0b033a70 KERNEL32!RtlVirtualUnwindStub
00007ff6`eefbd098  00007ff8`0b037c40 KERNEL32!IsDebuggerPresentStub
00007ff6`eefbd0a0  00007ff8`0b05c2b0 KERNEL32!UnhandledExceptionFilterStub
00007ff6`eefbd0a8  00007ff8`0b038850 KERNEL32!SetUnhandledExceptionFilterStub
00007ff6`eefbd0b0  00007ff8`0b037190 KERNEL32!GetStartupInfoWStub
00007ff6`eefbd0b8  00007ff8`0b036e10 KERNEL32!IsProcessorFeaturePresentStub
00007ff6`eefbd0c0  00007ff8`0b03ff00 KERNEL32!CloseHandle
00007ff6`eefbd0c8  00007ff8`0b037e70 KERNEL32!RtlUnwindExStub
00007ff6`eefbd0d0  00007ff8`0c19e990 ntdll!RtlInterlockedPushEntrySList
00007ff6`eefbd0d8  00007ff8`0c1a2c50 ntdll!RtlInterlockedFlushSList
00007ff6`eefbd0e0  00007ff8`0b030f90 KERNEL32!SetLastErrorStub
00007ff6`eefbd0e8  00007ff8`0c151d10 ntdll!RtlEnterCriticalSection
00007ff6`eefbd0f0  00007ff8`0c1571b0 ntdll!RtlLeaveCriticalSection
00007ff6`eefbd0f8  00007ff8`0c18dc90 ntdll!RtlDeleteCriticalSection
00007ff6`eefbd100  00007ff8`0b040000 KERNEL32!InitializeCriticalSectionAndSpinCount
00007ff6`eefbd108  00007ff8`0b036810 KERNEL32!TlsAllocStub
00007ff6`eefbd110  00007ff8`0b030c00 KERNEL32!TlsGetValueStub
00007ff6`eefbd118  00007ff8`0b030d00 KERNEL32!TlsSetValueStub
00007ff6`eefbd120  00007ff8`0b037950 KERNEL32!TlsFreeStub
00007ff6`eefbd128  00007ff8`0b033c60 KERNEL32!LoadLibraryExWStub
00007ff6`eefbd130  00007ff8`0c1a3770 ntdll!RtlEncodePointer
00007ff6`eefbd138  00007ff8`0b038db0 KERNEL32!RtlPcToFileHeaderStub
00007ff6`eefbd140  00007ff8`0b037a20 KERNEL32!GetStdHandleStub
00007ff6`eefbd148  00007ff8`0b040610 KERNEL32!WriteFile
00007ff6`eefbd150  00007ff8`0b036f00 KERNEL32!GetModuleFileNameWStub
00007ff6`eefbd158  00007ff8`0b03fe80 KERNEL32!GetCurrentProcess
00007ff6`eefbd160  00007ff8`0b037cc0 KERNEL32!ExitProcessImplementation
00007ff6`eefbd168  00007ff8`0b0394e0 KERNEL32!TerminateProcessStub
00007ff6`eefbd170  00007ff8`0b0381c0 KERNEL32!GetModuleHandleExWStub
00007ff6`eefbd178  00007ff8`0b0387f0 KERNEL32!GetCommandLineAStub
00007ff6`eefbd180  00007ff8`0b046060 KERNEL32!RtlUnwindStub
00007ff6`eefbd188  00007ff8`0b030c60 KERNEL32!GetCurrentThread
00007ff6`eefbd190  00007ff8`0b039360 KERNEL32!OutputDebugStringWStub
00007ff6`eefbd198  00007ff8`0c16cca0 ntdll!RtlAllocateHeap
00007ff6`eefbd1a0  00007ff8`0b030be0 KERNEL32!HeapFreeStub
00007ff6`eefbd1a8  00007ff8`0b0401e0 KERNEL32!FindClose
00007ff6`eefbd1b0  00007ff8`0b040240 KERNEL32!FindFirstFileExW
00007ff6`eefbd1b8  00007ff8`0b0402b0 KERNEL32!FindNextFileW
00007ff6`eefbd1c0  00007ff8`0b0386d0 KERNEL32!IsValidCodePageStub
00007ff6`eefbd1c8  00007ff8`0b037e50 KERNEL32!GetACPStub
00007ff6`eefbd1d0  00007ff8`0b039420 KERNEL32!GetOEMCPStub
00007ff6`eefbd1d8  00007ff8`0b037b20 KERNEL32!GetCPInfoStub
00007ff6`eefbd1e0  00007ff8`0b030c10 KERNEL32!MultiByteToWideCharStub
00007ff6`eefbd1e8  00007ff8`0b030ce0 KERNEL32!WideCharToMultiByteStub
00007ff6`eefbd1f0  00007ff8`0b038580 KERNEL32!GetEnvironmentStringsWStub
00007ff6`eefbd1f8  00007ff8`0b0385a0 KERNEL32!FreeEnvironmentStringsWStub
00007ff6`eefbd200  00007ff8`0b037ae0 KERNEL32!SetEnvironmentVariableWStub
00007ff6`eefbd208  00007ff8`0b05c1c0 KERNEL32!SetStdHandleStub
00007ff6`eefbd210  00007ff8`0b0403d0 KERNEL32!GetFileType
00007ff6`eefbd218  00007ff8`0b037dc0 KERNEL32!GetStringTypeWStub
```

```
00007ff6`eefbd220  00007ff8`0b038d90 KERNEL32!GetLocaleInfoWStub
00007ff6`eefbd228  00007ff8`0b038ec0 KERNEL32!IsValidLocaleStub
00007ff6`eefbd230  00007ff8`0b0390e0 KERNEL32!GetUserDefaultLCIDStub
00007ff6`eefbd238  00007ff8`0b05ac00 KERNEL32!EnumSystemLocalesWStub
00007ff6`eefbd240  00007ff8`0b040480 KERNEL32!GetTempPathW
00007ff6`eefbd248  00007ff8`0b038830 KERNEL32!FlsAllocStub
00007ff6`eefbd250  00007ff8`0b0331e0 KERNEL32!FlsGetValueStub
00007ff6`eefbd258  00007ff8`0b034fc0 KERNEL32!FlsSetValueStub
00007ff6`eefbd260  00007ff8`0b038e20 KERNEL32!FlsFreeStub
00007ff6`eefbd268  00007ff8`0b038e60 KERNEL32!GetDateFormatWStub
00007ff6`eefbd270  00007ff8`0b0395e0 KERNEL32!GetTimeFormatWStub
00007ff6`eefbd278  00007ff8`0b035b70 KERNEL32!CompareStringWStub
00007ff6`eefbd280  00007ff8`0b033160 KERNEL32!LCMapStringWStub
00007ff6`eefbd288  00007ff8`0b030ca0 KERNEL32!GetProcessHeapStub
00007ff6`eefbd290  00007ff8`0b040a00 KERNEL32!SetConsoleCtrlHandler
00007ff6`eefbd298  00007ff8`0c16ab70 ntdll!RtlSizeHeap
00007ff6`eefbd2a0  00007ff8`0c1722e0 ntdll!RtlReAllocateHeap
00007ff6`eefbd2a8  00007ff8`0b0402e0 KERNEL32!FlushFileBuffers
00007ff6`eefbd2b0  00007ff8`0b040970 KERNEL32!GetConsoleOutputCP
00007ff6`eefbd2b8  00007ff8`0b040960 KERNEL32!GetConsoleMode
00007ff6`eefbd2c0  00007ff8`0b0403b0 KERNEL32!GetFileSizeEx
00007ff6`eefbd2c8  00007ff8`0b0405c0 KERNEL32!SetFilePointerEx
00007ff6`eefbd2d0  00007ff8`0b040520 KERNEL32!ReadFile
00007ff6`eefbd2d8  00007ff8`0b0409e0 KERNEL32!ReadConsoleW
00007ff6`eefbd2e0  00007ff8`0b040180 KERNEL32!CreateFileW
00007ff6`eefbd2e8  00000000`00000000
00007ff6`eefbd2f0  00000000`00000000
00007ff6`eefbd2f8  00000000`00000000
00007ff6`eefbd300  00000000`00000000
00007ff6`eefbd308  00000000`00000000
00007ff6`eefbd310  00000000`00000000
00007ff6`eefbd318  00000000`00000000
00007ff6`eefbd320  00000000`00000000
00007ff6`eefbd328  00000000`00000000
00007ff6`eefbd330  00000000`00000000
00007ff6`eefbd338  00000000`00000000
00007ff6`eefbd340  00000000`00000000
00007ff6`eefbd348  00000000`00000000
00007ff6`eefbd350  00000000`00000000
00007ff6`eefbd358  00000000`00000000
00007ff6`eefbd360  00000000`00000000
00007ff6`eefbd368  00000000`00000000
00007ff6`eefbd370  00000000`00000000
00007ff6`eefbd378  00000000`00000000
00007ff6`eefbd380  00000000`00000000
00007ff6`eefbd388  00000000`00000000
00007ff6`eefbd390  00000000`00000000
00007ff6`eefbd398  00000000`00000000
00007ff6`eefbd3a0  00000000`00000000
00007ff6`eefbd3a8  00000000`00000000
00007ff6`eefbd3b0  00000000`00000000
00007ff6`eefbd3b8  00000000`00000000
00007ff6`eefbd3c0  00000000`00000000
00007ff6`eefbd3c8  00000000`00000000
00007ff6`eefbd3d0  00007fff`ff0ef4a0 WINMM!waveInResetStub
00007ff6`eefbd3d8  00000000`00000000
00007ff6`eefbd3e0  00000000`00000000
00007ff6`eefbd3e8  00000000`00000000
00007ff6`eefbd3f0  00000000`00000000
```

```
00007ff6`eefbd3f8  00000000`00000000
00007ff6`eefbd400  00000000`00000000
00007ff6`eefbd408  00000000`00000000
00007ff6`eefbd410  00000000`00000000
00007ff6`eefbd418  00000000`00000000
00007ff6`eefbd420  00000000`00000000
00007ff6`eefbd428  00000000`00000000
```

Note: We don't see any *user32* entry because it is loaded on demand. However, we see something related to the *waveInReset* function we call in our source code.

11. Let's look at the *main* function disassembly:

```
0:000> uf main
t!main [c:\AWAPI-Dumps\t.cpp @ 9]:
    9 00007ff6`eef271d0 4883ec28        sub     rsp,28h
   10 00007ff6`eef271d4 33c9            xor     ecx,ecx
   10 00007ff6`eef271d6 ff15f4610900    call    qword ptr [t!_imp_waveInReset (00007ff6`eefbd3d0)]
   13 00007ff6`eef271dc 4533c9          xor     r9d,r9d
   13 00007ff6`eef271df 4c8d053a7d0700  lea     r8,[t!__xt_z+0x110 (00007ff6`eef9ef20)]
   13 00007ff6`eef271e6 488d153b7d0700  lea     rdx,[t!__xt_z+0x118 (00007ff6`eef9ef28)]
   13 00007ff6`eef271ed 33c9            xor     ecx,ecx
   13 00007ff6`eef271ef ff15ab7e0900    call    qword ptr [t!_imp_MessageBoxA (00007ff6`eefbf0a0)]
   14 00007ff6`eef271f5 33c0            xor     eax,eax
   14 00007ff6`eef271f7 4883c428        add     rsp,28h
   14 00007ff6`eef271fb c3              ret
```

Let's look at *waveInReset* **Call Path**:

```
0:000> dps 00007ff6`eefbd3d0 L1
00007ff6`eefbd3d0  00007fff`ff0ef4a0 WINMM!waveInResetStub

0:000> u WINMM!waveInResetStub
WINMM!waveInResetStub:
00007fff`ff0ef4a0 48ff2589cd0100  jmp     qword ptr [WINMM!_imp_waveInReset (00007fff`ff10c230)]
00007fff`ff0ef4a7 cc              int     3
00007fff`ff0ef4a8 cc              int     3
00007fff`ff0ef4a9 cc              int     3
00007fff`ff0ef4aa cc              int     3
00007fff`ff0ef4ab cc              int     3
00007fff`ff0ef4ac cc              int     3
00007fff`ff0ef4ad cc              int     3
```

Note: We see that instead of being implemented inside the WINMM module, the call is redirected to an import entry in WINMM itself. Let's look where it goes:

```
0:000> dps WINMM!_imp_waveInReset L1
00007fff`ff10c230  00007fff`ff0e7106 WINMM!_imp_load_waveInReset
```

Note: We see it goes to a delay-loaded module specified by *api_ms_win_mm_mme_l1_1_0* contract:

```
0:000> u WINMM!_imp_load_waveInReset
WINMM!_imp_load_waveInReset:
00007fff`ff0e7106 488d0523510200  lea     rax,[WINMM!_imp_waveInReset (00007fff`ff10c230)]
00007fff`ff0e710d e997fbffff      jmp     WINMM!_tailMerge_api_ms_win_mm_mme_l1_1_0_dll (00007fff`ff0e6ca9)
00007fff`ff0e7112 cc              int     3
00007fff`ff0e7113 cc              int     3
00007fff`ff0e7114 cc              int     3
00007fff`ff0e7115 cc              int     3
00007fff`ff0e7116 cc              int     3
```

```
00007fff`ff0e7117 cc              int     3
```

12. Let's now execute the *waveInReset* call in the *main* function:

```
0:000> p
t!main+0x4:
00007ff6`eef271d4 33c9            xor     ecx,ecx

0:000> p
ModLoad: 00007fff`ee1e0000 00007fff`ee209000   C:\WINDOWS\SYSTEM32\winmmbase.dll
ModLoad: 00007ff8`0bd70000 00007ff8`0be14000   C:\WINDOWS\System32\sechost.dll
ModLoad: 00007ff8`0b710000 00007ff8`0b825000   C:\WINDOWS\System32\RPCRT4.dll
ModLoad: 00007fff`f6b10000 00007fff`f6bad000   C:\WINDOWS\SYSTEM32\MMDevAPI.DLL
ModLoad: 00007ff8`095e0000 00007ff8`0967a000   C:\WINDOWS\System32\msvcp_win.dll
ModLoad: 00007fff`7b990000 00007fff`7b9d6000   C:\WINDOWS\SYSTEM32\wdmaud.drv
ModLoad: 00007ff8`09f30000 00007ff8`0a2b9000   C:\WINDOWS\System32\combase.dll
ModLoad: 00007fff`b2f30000 00007fff`b2f39000   C:\WINDOWS\SYSTEM32\ksuser.dll
ModLoad: 00007ff8`07690000 00007ff8`076a8000   C:\WINDOWS\SYSTEM32\kernel.appcore.dll
ModLoad: 00007ff8`07e60000 00007ff8`07e6b000   C:\WINDOWS\SYSTEM32\AVRT.dll
ModLoad: 00007ff8`0a540000 00007ff8`0a5e7000   C:\WINDOWS\System32\msvcrt.dll
ModLoad: 00007ff8`091f0000 00007ff8`0921c000   C:\WINDOWS\SYSTEM32\DEVOBJ.dll
ModLoad: 00007ff8`091a0000 00007ff8`091ef000   C:\WINDOWS\SYSTEM32\cfgmgr32.dll
ModLoad: 0000022a`485f0000 0000022a`4863f000   C:\WINDOWS\SYSTEM32\cfgmgr32.dll
ModLoad: 0000022a`48640000 0000022a`4868f000   C:\WINDOWS\SYSTEM32\CFGMGR32.dll
ModLoad: 00007fff`f2730000 00007fff`f2914000   C:\WINDOWS\SYSTEM32\AUDIOSES.DLL
ModLoad: 00007ff8`0bc70000 00007ff8`0bd61000   C:\WINDOWS\System32\shcore.dll
ModLoad: 00007ff8`09560000 00007ff8`095db000   C:\WINDOWS\System32\bcryptPrimitives.dll
ModLoad: 00007fff`b2af0000 00007fff`b2afe000   C:\WINDOWS\SYSTEM32\msacm32.drv
ModLoad: 00007fff`901b0000 00007fff`901ce000   C:\WINDOWS\SYSTEM32\MSACM32.dll
ModLoad: 00007fff`b1e50000 00007fff`b1e5b000   C:\WINDOWS\SYSTEM32\midimap.dll
t!main+0xc:
00007ff6`eef271dc 4533c9          xor     r9d,r9d
```

t.cpp

```
#pragma comment(lib, "winmm")

int main() {
    waveInReset(nullptr);

    // user32.dll will load at this point
    MessageBox(NULL, "Hello", "Hello", MB_OK);
}
```

Note: We see that the import table entry in the WINMM module is now replaced with the loaded *winmmbase* module entry:

```
0:000> dps WINMM!_imp_waveInReset L1
00007fff`ff10c230  00007fff`ee1f2350 winmmbase!waveInReset
```

13. If we look at the *main* function disassembly again and check *t!_imp_MessageBoxA* address, we see it points to the Delay Import Directory entry that requires loading the *user32* module:

```
0:000> dps t!_imp_MessageBoxA L1
00007ff6`eefbf0a0  00007ff6`eef27207 t!_imp_load_MessageBoxA

0:000> dps 00007ff6eef20000 + 9F000 L40/8
00007ff6`eefbf000  0007ef30`00000001
00007ff6`eefbf008  0009f0a0`00095170
00007ff6`eefbf010  0009f210`0009f040
00007ff6`eefbf018  00000000`00000000
00007ff6`eefbf020  00000000`00000000
00007ff6`eefbf028  00000000`00000000
00007ff6`eefbf030  00000000`00000000
00007ff6`eefbf038  00000000`00000000

0:000> ? 00007ff6eef20000 + 0009f0a0
Evaluate expression: 140698548170912 = 00007ff6`eefbf0a0

0:000> u t!_imp_load_MessageBoxA
t!_imp_load_MessageBoxA:
00007ff6`eef27207 488d05927e0900  lea     rax,[t!_imp_MessageBoxA (00007ff6`eefbf0a0)]
00007ff6`eef2720e e906000000      jmp     t!_tailMerge_user32_dll (00007ff6`eef27219)
t!MessageBoxA:
00007ff6`eef27213 ff25877e0900    jmp     qword ptr [t!_imp_MessageBoxA (00007ff6`eefbf0a0)]
t!_tailMerge_user32_dll:
00007ff6`eef27219 48894c2408      mov     qword ptr [rsp+8],rcx
00007ff6`eef2721e 4889542410      mov     qword ptr [rsp+10h],rdx
00007ff6`eef27223 4c89442418      mov     qword ptr [rsp+18h],r8
00007ff6`eef27228 4c894c2420      mov     qword ptr [rsp+20h],r9
00007ff6`eef2722d 4883ec68        sub     rsp,68h
```

14. Let's now execute the *MessageBox* function in our source code:

```
0:000> p
ModLoad: 00007ff8`0be20000 00007ff8`0bfca000   C:\WINDOWS\System32\USER32.dll
ModLoad: 00007ff8`097f0000 00007ff8`09816000   C:\WINDOWS\System32\win32u.dll
ModLoad: 00007ff8`0a5f0000 00007ff8`0a619000   C:\WINDOWS\System32\GDI32.dll
ModLoad: 00007ff8`09820000 00007ff8`09932000   C:\WINDOWS\System32\gdi32full.dll
ModLoad: 00007ff8`0a6d0000 00007ff8`0a701000   C:\WINDOWS\System32\IMM32.DLL
ModLoad: 00007fff`f2ca0000 00007fff`f2d50000   C:\WINDOWS\SYSTEM32\TextShaping.dll
ModLoad: 00007ff8`05b40000 00007ff8`05beb000   C:\WINDOWS\system32\uxtheme.dll
ModLoad: 00007ff8`0bfd0000 00007ff8`0c0ee000   C:\WINDOWS\System32\MSCTF.dll
ModLoad: 00007fff`f9fd0000 00007fff`fa0fd000   C:\WINDOWS\SYSTEM32\textinputframework.dll
ModLoad: 00007ff8`0bb90000 00007ff8`0bc67000   C:\WINDOWS\System32\OLEAUT32.dll
ModLoad: 00007fff`e1400000 00007fff`e1469000   C:\WINDOWS\system32\Oleacc.dll
```

We get this result:

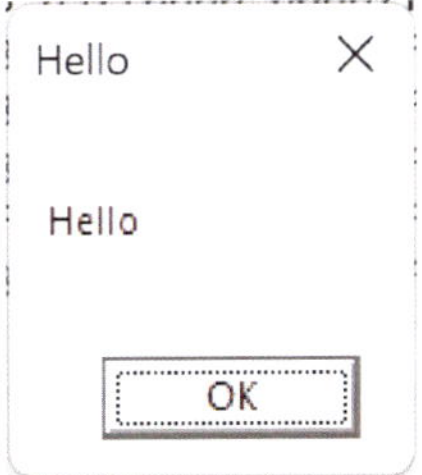

If we dismiss it, we should get this source code window:

t.cpp

```
#pragma comment(lib, "winmm")

int main() {
    waveInReset(nullptr);

    // user32.dll will load at this point
    MessageBox(NULL, "Hello", "Hello", MB_OK);
}
```

Note: The Delay Import Directory entry should point to the loaded *USER32* module function:

```
0:000> dps t!_imp_MessageBoxA L1
00007ff6`eefbf0a0  00007ff8`0be96e00 USER32!MessageBoxA
```

15. We close logging before exiting WinDbg:

```
0:000> .logclose
Closing open log file C:\AWAPI-Dumps\W5.log
```

Exports and Imports

- WinDbg (manual/scripts)
- 3rd-party WinDbg extensions (SwishDbgEx)
- DUMPBIN

It is very easy to dump module imports manually after all referenced modules are loaded. However, it is not so easy to do manually for module exports in one simple **dps** command. We can also guess exports from the output of the **x** command (*pub* functions without the *_imp_* prefix). Fortunately, there are 3rd-party WinDbg extensions that have such commands. One is referenced on the slide. Microsoft also has the DUMPBIN tool for the same purpose, and we play with it during the next exercise. There are also some GUI-based tools that allow browsing import and export tables and much more, but these tools are outside the scope of this training.

SwishDbgEx
https://github.com/comaeio/SwishDbgExt

API and System Calls

- API that do not require kernel services
  - `GetCurrentThreadId`
- API that require kernel services
  - `user32!CreateWindowExW → win32u!NtUserCreateWindowEx`
  - `kernel32!ReadFile → ntdll!NtReadFile`

Sometimes, we want to investigate the code of an API function to know what other API functions it uses and whether it is translated to a system call. For example, some API functions do not require kernel services, such as the current process and thread ids. Other functions may cache the result from the kernel or the different execution paths without kernel services.

Exercise W6

- **Goal:** Explore exports and imports using dumpbin. Check whether the selected API functions use a system call
- **ADDR Patterns:** Call Path
- \AWAPI-Dumps\Exercise-W6.pdf

Exercise W6

Goal: Explore exports and imports using dumpbin. Check whether the selected API functions use a system call.

ADDR Patterns: Call Path.

1. This exercise uses the *t.exe* executable we built in the previous Exercise W5.

2. Launch x64 Native Tools Command Prompt for VS 2022:

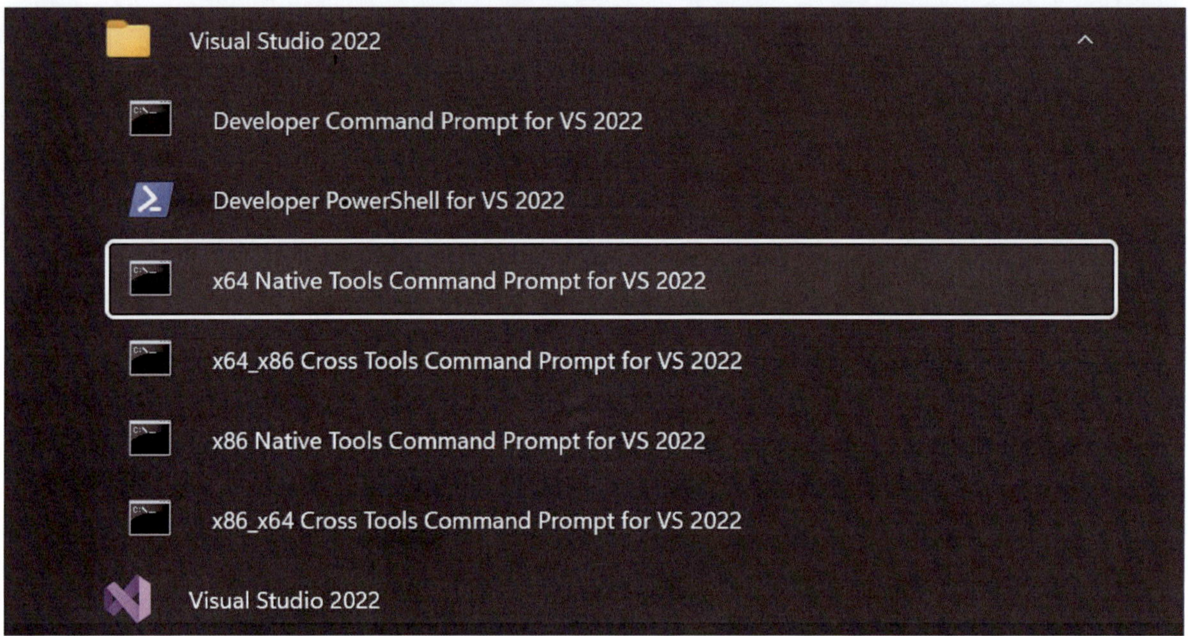

3. Discover exports and imports of *t.exe*.

```
c:\AWAPI-Dumps>dumpbin
Microsoft (R) COFF/PE Dumper Version 14.32.31332.0
Copyright (C) Microsoft Corporation.  All rights reserved.

usage: DUMPBIN [options] [files]

   options:

      /ALL
      /ARCHIVEMEMBERS
      /CLRHEADER
      /DEPENDENTS
      /DIRECTIVES
      /DISASM[:{BYTES|NOBYTES|NOWIDE|WIDE}]
      /ERRORREPORT:{NONE|PROMPT|QUEUE|SEND}
      /EXPORTS
      /FPO
      /HEADERS
      /IMPORTS[:filename]
      /LINENUMBERS
      /LINKERMEMBER[:{1|2|4|8|16|32}]
      /LOADCONFIG
```

```
      /NOLOGO
      /NOPDB
      /NOSECTION:name
      /OUT:filename
      /PDATA
      /PDBPATH[:VERBOSE]
      /RANGE:vaMin[,vaMax]
      /RAWDATA[:{NONE|1|2|4|8}[,#]]
      /RELOCATIONS
      /SECTION:name
      /SUMMARY
      /SYMBOLS
      /TLS
      /UNWINDINFO
```

```
c:\AWAPI-Dumps>dumpbin /EXPORTS t.exe
Microsoft (R) COFF/PE Dumper Version 14.32.31332.0
Copyright (C) Microsoft Corporation.  All rights reserved.


Dump of file t.exe

File Type: EXECUTABLE IMAGE

  Summary

        1000 .00cfg
        4000 .data
        1000 .didat
        2000 .idata
        5000 .pdata
       16000 .rdata
        1000 .reloc
       7D000 .text
        1000 _RDATA

c:\AWAPI-Dumps>dumpbin /IMPORTS t.exe
Microsoft (R) COFF/PE Dumper Version 14.32.31332.0
Copyright (C) Microsoft Corporation.  All rights reserved.


Dump of file t.exe

File Type: EXECUTABLE IMAGE

  Section contains the following imports:

    WINMM.dll
             14009D3D0 Import Address Table
             14009D840 Import Name Table
                     0 time date stamp
                     0 Index of first forwarder reference

                         99 waveInReset

    KERNEL32.dll
             14009D000 Import Address Table
             14009D470 Import Name Table
                     0 time date stamp
```

```
0 Index of first forwarder reference

  1F1 GetCommandLineW
  64A WriteConsoleW
  487 RaiseException
  27D GetLastError
  304 GetSystemInfo
  605 VirtualProtect
  607 VirtualQuery
  1C5 FreeLibrary
  295 GetModuleHandleW
  2CD GetProcAddress
  3E5 LoadLibraryExA
  470 QueryPerformanceCounter
  233 GetCurrentProcessId
  237 GetCurrentThreadId
  30A GetSystemTimeAsFileTime
  38A InitializeSListHead
  4F5 RtlCaptureContext
  4FD RtlLookupFunctionEntry
  504 RtlVirtualUnwind
  3A0 IsDebuggerPresent
  5E6 UnhandledExceptionFilter
  5A4 SetUnhandledExceptionFilter
  2F1 GetStartupInfoW
  3A8 IsProcessorFeaturePresent
   94 CloseHandle
  503 RtlUnwindEx
  390 InterlockedPushEntrySList
  38E InterlockedFlushSList
  564 SetLastError
  149 EnterCriticalSection
  3E0 LeaveCriticalSection
  123 DeleteCriticalSection
  386 InitializeCriticalSectionAndSpinCount
  5D6 TlsAlloc
  5D8 TlsGetValue
  5D9 TlsSetValue
  5D7 TlsFree
  3E6 LoadLibraryExW
  145 EncodePointer
  4FF RtlPcToFileHeader
  2F3 GetStdHandle
  64B WriteFile
  291 GetModuleFileNameW
  232 GetCurrentProcess
  178 ExitProcess
  5C4 TerminateProcess
  294 GetModuleHandleExW
  1F0 GetCommandLineA
  502 RtlUnwind
  236 GetCurrentThread
  43A OutputDebugStringW
  36C HeapAlloc
  370 HeapFree
  18F FindClose
  195 FindFirstFileExW
  1A6 FindNextFileW
  3AE IsValidCodePage
```

```
                  1CC GetACP
                  2B6 GetOEMCP
                  1DB GetCPInfo
                  412 MultiByteToWideChar
                  637 WideCharToMultiByte
                  253 GetEnvironmentStringsW
                  1C4 FreeEnvironmentStringsW
                  546 SetEnvironmentVariableW
                  57F SetStdHandle
                  26A GetFileType
                  2F8 GetStringTypeW
                  281 GetLocaleInfoW
                  3B0 IsValidLocale
                  339 GetUserDefaultLCID
                  16D EnumSystemLocalesW
                  319 GetTempPathW
                  1B4 FlsAlloc
                  1B6 FlsGetValue
                  1B7 FlsSetValue
                  1B5 FlsFree
                  23D GetDateFormatW
                  331 GetTimeFormatW
                   AA CompareStringW
                  3D4 LCMapStringW
                  2D4 GetProcessHeap
                  51B SetConsoleCtrlHandler
                  375 HeapSize
                  373 HeapReAlloc
                  1B9 FlushFileBuffers
                  21A GetConsoleOutputCP
                  216 GetConsoleMode
                  268 GetFileSizeEx
                  555 SetFilePointerEx
                  498 ReadFile
                  495 ReadConsoleW
                   DA CreateFileW

Section contains the following delay load imports:

  USER32.dll
          00000001 Characteristics
  0000000140095170 Address of HMODULE
  000000014009F0A0 Import Address Table
  000000014009F040 Import Name Table
  000000014009F210 Bound Import Name Table
  0000000000000000 Unload Import Name Table
                 0 time date stamp

                        0000000140007207    284 MessageBoxA

Summary

    1000 .00cfg
    4000 .data
    1000 .didat
    2000 .idata
    5000 .pdata
   16000 .rdata
    1000 .reloc
```

```
      7D000 .text
       1000 _RDATA
```

Note: We don't have exports for *t.exe*.

4. Discover exports and imports of *winmm*.

```
c:\AWAPI-Dumps>dumpbin /EXPORTS c:\Windows\System32\winmm.dll
Microsoft (R) COFF/PE Dumper Version 14.32.31332.0
Copyright (C) Microsoft Corporation.  All rights reserved.


Dump of file c:\Windows\System32\winmm.dll

File Type: DLL

  Section contains the following exports for WINMM.dll

    00000000 characteristics
    D59AB6A2 time date stamp
        0.00 version
           2 ordinal base
         181 number of functions
         180 number of names

    ordinal hint RVA      name

          4    0 0000E660 CloseDriver
          5    1 0000E680 DefDriverProc
          6    2 0000E6A0 DriverCallback
          7    3 0000E6B0 DrvGetModuleHandle
          8    4 0000E6D0 GetDriverModuleHandle
          9    5 0000E700 OpenDriver
         10    6 0000E6F0 PlaySound
         11    7 0001C2E0 PlaySoundA
         12    8 00001270 PlaySoundW
         13    9 0000E720 SendDriverMessage
         14    A 0001CBF0 WOWAppExit
         15    B 0000E740 auxGetDevCapsA
         16    C 0000E760 auxGetDevCapsW
         17    D 0000E780 auxGetNumDevs
         18    E 0000E7A0 auxGetVolume
         19    F 0000E7C0 auxOutMessage
         20   10 0000E7E0 auxSetVolume
         21   11 0000C890 joyConfigChanged
         22   12 0000C900 joyGetDevCapsA
         23   13 0000CBC0 joyGetDevCapsW
         24   14 0000CCD0 joyGetNumDevs
         25   15 0000CCE0 joyGetPos
         26   16 0000CD40 joyGetPosEx
         27   17 0000D300 joyGetThreshold
         28   18 0000E1E0 joyReleaseCapture
         29   19 0000E2A0 joySetCapture
         30   1A 0000E510 joySetThreshold
         31   1B 00015710 mciDriverNotify
         32   1C 00015730 mciDriverYield
          3   1D 00010140 mciExecute
         33   1E 00013550 mciFreeCommandResource
```

```
34    1F 00015BE0 mciGetCreatorTask
35    20 00015C60 mciGetDeviceIDA
36    21 00015CC0 mciGetDeviceIDFromElementIDA
37    22 00015D20 mciGetDeviceIDFromElementIDW
38    23 00015E40 mciGetDeviceIDW
39    24 00015E70 mciGetDriverData
40    25 00011720 mciGetErrorStringA
41    26 00011850 mciGetErrorStringW
42    27 00015F20 mciGetYieldProc
43    28 00013770 mciLoadCommandResource
44    29 00011B80 mciSendCommandA
45    2A 00012410 mciSendCommandW
46    2B 00012550 mciSendStringA
47    2C 00012760 mciSendStringW
48    2D 00016570 mciSetDriverData
49    2E 00016630 mciSetYieldProc
50    2F 0000E800 midiConnect
51    30 0000E820 midiDisconnect
52    31 0000E840 midiInAddBuffer
53    32 0000E860 midiInClose
54    33 0000E880 midiInGetDevCapsA
55    34 0000E8A0 midiInGetDevCapsW
56    35 0000E8C0 midiInGetErrorTextA
57    36 0000E8E0 midiInGetErrorTextW
58    37 0000E900 midiInGetID
59    38 0000E920 midiInGetNumDevs
60    39 0000E940 midiInMessage
61    3A 0000E960 midiInOpen
62    3B 0000E980 midiInPrepareHeader
63    3C 0000E9A0 midiInReset
64    3D 0000E9C0 midiInStart
65    3E 0000E9E0 midiInStop
66    3F 0000EA00 midiInUnprepareHeader
67    40 0000EA20 midiOutCacheDrumPatches
68    41 0000EA40 midiOutCachePatches
69    42 0000EA60 midiOutClose
70    43 0000EA80 midiOutGetDevCapsA
71    44 0000EAA0 midiOutGetDevCapsW
72    45 0000EAC0 midiOutGetErrorTextA
73    46 0000EAE0 midiOutGetErrorTextW
74    47 0000EB00 midiOutGetID
75    48 0000EB20 midiOutGetNumDevs
76    49 0000EB40 midiOutGetVolume
77    4A 0000EB60 midiOutLongMsg
78    4B 0000EB80 midiOutMessage
79    4C 0000EBA0 midiOutOpen
80    4D 0000EBC0 midiOutPrepareHeader
81    4E 0000EBE0 midiOutReset
82    4F 0000EC00 midiOutSetVolume
83    50 0000EC20 midiOutShortMsg
84    51 0000EC40 midiOutUnprepareHeader
85    52 0000EC60 midiStreamClose
86    53 0000EC80 midiStreamOpen
87    54 0000ECA0 midiStreamOut
88    55 0000ECC0 midiStreamPause
89    56 0000ECE0 midiStreamPosition
90    57 0000ED00 midiStreamProperty
91    58 0000ED20 midiStreamRestart
92    59 0000ED40 midiStreamStop
```

```
 93   5A 0000ED60 mixerClose
 94   5B 0000ED80 mixerGetControlDetailsA
 95   5C 0000EDA0 mixerGetControlDetailsW
 96   5D 0000EDC0 mixerGetDevCapsA
 97   5E 0000EDE0 mixerGetDevCapsW
 98   5F 0000EE00 mixerGetID
 99   60 0000EE20 mixerGetLineControlsA
100   61 0000EE40 mixerGetLineControlsW
101   62 0000EE60 mixerGetLineInfoA
102   63 0000EE80 mixerGetLineInfoW
103   64 0000EEA0 mixerGetNumDevs
104   65 0000EEC0 mixerMessage
105   66 0000EEE0 mixerOpen
106   67 0000EF00 mixerSetControlDetails
107   68 0000EF20 mmDrvInstall
108   69 0000EF40 mmGetCurrentTask
109   6A 0000EF60 mmTaskBlock
110   6B 0000EF80 mmTaskCreate
111   6C 0000EFA0 mmTaskSignal
112   6D 0000EFC0 mmTaskYield
113   6E 0000EFE0 mmioAdvance
114   6F 0000F000 mmioAscend
115   70 0000F020 mmioClose
116   71 0000F040 mmioCreateChunk
117   72 0000F060 mmioDescend
118   73 0000F080 mmioFlush
119   74 0000F0A0 mmioGetInfo
120   75 0000F0C0 mmioInstallIOProcA
121   76 0000F0E0 mmioInstallIOProcW
122   77 0000F100 mmioOpenA
123   78 0000F120 mmioOpenW
124   79 0000F140 mmioRead
125   7A 0000F160 mmioRenameA
126   7B 0000F180 mmioRenameW
127   7C 0000F1A0 mmioSeek
128   7D 0000F1C0 mmioSendMessage
129   7E 0000F1E0 mmioSetBuffer
130   7F 0000F200 mmioSetInfo
131   80 0000F220 mmioStringToFOURCCA
132   81 0000F240 mmioStringToFOURCCW
133   82 0000F260 mmioWrite
134   83 0001CEA0 mmsystemGetVersion
135   84 0001C770 sndPlaySoundA
136   85 0001C790 sndPlaySoundW
137   86 0000F280 timeBeginPeriod
138   87 0000F2A0 timeEndPeriod
139   88 0000F2C0 timeGetDevCaps
140   89 0000F2E0 timeGetSystemTime
141   8A 0000F300 timeGetTime
142   8B 000039A0 timeKillEvent
143   8C 00002400 timeSetEvent
144   8D 0000F320 waveInAddBuffer
145   8E 0000F340 waveInClose
146   8F 0000F360 waveInGetDevCapsA
147   90 0000F380 waveInGetDevCapsW
148   91 0000F3A0 waveInGetErrorTextA
149   92 0000F3C0 waveInGetErrorTextW
150   93 0000F3E0 waveInGetID
151   94 0000F400 waveInGetNumDevs
```

```
        152   95 0000F420 waveInGetPosition
        153   96 0000F440 waveInMessage
        154   97 0000F460 waveInOpen
        155   98 0000F480 waveInPrepareHeader
        156   99 0000F4A0 waveInReset
        157   9A 0000F4C0 waveInStart
        158   9B 0000F4E0 waveInStop
        159   9C 0000F500 waveInUnprepareHeader
        160   9D 0000F520 waveOutBreakLoop
        161   9E 0000F540 waveOutClose
        162   9F 0000F560 waveOutGetDevCapsA
        163   A0 0000F580 waveOutGetDevCapsW
        164   A1 0000F5A0 waveOutGetErrorTextA
        165   A2 0000F5C0 waveOutGetErrorTextW
        166   A3 0000F5E0 waveOutGetID
        167   A4 0000F600 waveOutGetNumDevs
        168   A5 0000F620 waveOutGetPitch
        169   A6 0000F640 waveOutGetPlaybackRate
        170   A7 0000F660 waveOutGetPosition
        171   A8 0000F680 waveOutGetVolume
        172   A9 0000F6A0 waveOutMessage
        173   AA 0000F6C0 waveOutOpen
        174   AB 0000F6E0 waveOutPause
        175   AC 0000F700 waveOutPrepareHeader
        176   AD 0000F720 waveOutReset
        177   AE 0000F740 waveOutRestart
        178   AF 0000F760 waveOutSetPitch
        179   B0 0000F780 waveOutSetPlaybackRate
        180   B1 0000F7A0 waveOutSetVolume
        181   B2 0000F7C0 waveOutUnprepareHeader
        182   B3 0000F7E0 waveOutWrite
          2      0000E6F0 [NONAME]

  Summary

        3000 .data
        1000 .didat
        1000 .guids
        2000 .pdata
        A000 .rdata
        1000 .reloc
        5000 .rsrc
       1C000 .text
```

```
c:\AWAPI-Dumps>dumpbin /IMPORTS c:\Windows\System32\winmm.dll
Microsoft (R) COFF/PE Dumper Version 14.32.31332.0
Copyright (C) Microsoft Corporation.  All rights reserved.


Dump of file c:\Windows\System32\winmm.dll

File Type: DLL

  Section contains the following imports:

    ntdll.dll
             18001D920 Import Address Table
             180024E28 Import Name Table
                     0 time date stamp
```

```
                    0 Index of first forwarder reference

                       46 EtwGetTraceEnableFlags
                       56 EtwUnregisterTraceGuids
                      970 memmove
                       48 EtwGetTraceLoggerHandle
                       47 EtwGetTraceEnableLevel
                       4F EtwRegisterTraceGuidsW
                      9AF wcsstr
                      935 _vsnprintf_s
                      510 RtlMultiByteToUnicodeN
                      622 RtlUnicodeToMultiByteN
                      41B RtlGetActiveConsoleId
                      9A0 wcschr
                      8FB __C_specific_handler
                      936 _vsnwprintf
                      209 NtQueryTimerResolution
                      13B NtCreateTimer
                      29B NtWaitForMultipleObjects
                      24B NtSetEvent
                      11A NtCreateEvent
                       FB NtCancelTimer
                      101 NtClose
                      26F NtSetTimer
                      8FC __chkstk
                      96D memcmp
                      96E memcpy
                       54 EtwTraceMessage

api-ms-win-crt-string-l1-1-0.dll
         18001D8C8 Import Address Table
         180024DD0 Import Name Table
                 0 time date stamp
                 0 Index of first forwarder reference

                       83 memset

api-ms-win-crt-runtime-l1-1-0.dll
         18001D8B0 Import Address Table
         180024DB8 Import Name Table
                 0 time date stamp
                 0 Index of first forwarder reference

                       36 _initterm
                       37 _initterm_e

api-ms-win-crt-private-l1-1-0.dll
         18001D7D8 Import Address Table
         180024CE0 Import Name Table
                 0 time date stamp
                 0 Index of first forwarder reference

                       C2 _o__execute_onexit_table
                      13C _o__initialize_narrow_environment
                      13D _o__initialize_onexit_table
                      13F _o__invalid_parameter_noinfo
                      243 _o__purecall
                      251 _o__register_onexit_function
                      259 _o__seh_filter_dll
```

```
                        1 _CxxThrowException
                       C0 _o__errno
                      2D3 _o__wcslwr_s
                      38D _o_free
                      3B1 _o_iswdigit
                      3D8 _o_malloc
                      436 _o_terminate
                      43E _o_towlower
                       20 __current_exception
                       21 __current_exception_context
                       A7 _o__crt_atexit
                       A0 _o__configure_narrow_argv
                       93 _o__cexit
                       91 _o__callnewh
                       60 _o___std_type_info_destroy_list
                       5F _o___std_exception_destroy
                       5E _o___std_exception_copy
                       2A __std_terminate
                       13 __CxxFrameHandler4

    api-ms-win-mm-time-l1-1-0.dll
             18001D8F0 Import Address Table
             180024DF8 Import Name Table
                     0 time date stamp
                     0 Index of first forwarder reference

                        0 timeBeginPeriod
                        2 timeGetDevCaps
                        4 timeGetTime
                        3 timeGetSystemTime
                        1 timeEndPeriod

    api-ms-win-core-synch-l1-2-0.dll
             18001D770 Import Address Table
             180024C78 Import Name Table
                     0 time date stamp
                     0 Index of first forwarder reference

                       15 InitOnceExecuteOnce
                       2D Sleep

    api-ms-win-core-registry-l1-1-0.dll
             18001D620 Import Address Table
             180024B28 Import Name Table
                     0 time date stamp
                     0 Index of first forwarder reference

                       2E RegSetValueExW
                       25 RegQueryValueExW
                        0 RegCloseKey
                       1C RegOpenCurrentUser
                       14 RegGetValueW
                       1E RegOpenKeyExW
                        3 RegCreateKeyExW

    api-ms-win-core-synch-l1-1-0.dll
             18001D6B8 Import Address Table
             180024BC0 Import Name Table
                     0 time date stamp
```

```
                    0 Index of first forwarder reference

                       23 ReleaseMutex
                       24 ReleaseSRWLockExclusive
                        0 AcquireSRWLockExclusive
                       21 OpenSemaphoreW
                       25 ReleaseSRWLockShared
                        9 CreateMutexExW
                       30 SleepEx
                        F DeleteCriticalSection
                        6 CreateEventW
                       37 WaitForSingleObjectEx
                       1F OpenEventW
                       18 InitializeCriticalSection
                       36 WaitForSingleObject
                       1D LeaveCriticalSection
                       1A InitializeCriticalSectionEx
                        B CreateSemaphoreExW
                       11 EnterCriticalSection
                        3 CreateEventA
                       26 ReleaseSemaphore
                       29 SetEvent
                        1 AcquireSRWLockShared
                       27 ResetEvent

api-ms-win-core-processthreads-l1-1-0.dll
         18001D5A0 Import Address Table
         180024AA8 Import Name Table
                    0 time date stamp
                    0 Index of first forwarder reference

                       55 TerminateProcess
                        D GetCurrentProcessId
                        C GetCurrentProcess
                       10 GetCurrentThread
                       37 ProcessIdToSessionId
                       57 TlsAlloc
                       58 TlsFree
                        6 CreateThread
                        9 ExitThread
                       4D SetThreadPriority
                       11 GetCurrentThreadId

api-ms-win-core-heap-l2-1-0.dll
         18001D450 Import Address Table
         180024958 Import Name Table
                    0 time date stamp
                    0 Index of first forwarder reference

                        A LocalFree
                        8 LocalAlloc

api-ms-win-core-handle-l1-1-0.dll
         18001D418 Import Address Table
         180024920 Import Name Table
                    0 time date stamp
                    0 Index of first forwarder reference

                        0 CloseHandle
```

```
api-ms-win-core-libraryloader-l1-2-0.dll
          18001D4A8 Import Address Table
          1800249B0 Import Name Table
                  0 time date stamp
                  0 Index of first forwarder reference

                       10 GetModuleFileNameA
                       1E LockResource
                       1B LoadResource
                       1D LoadStringW
                       15 GetModuleHandleW
                        F FreeResource
                       19 LoadLibraryExW
                       14 GetModuleHandleExW
                       11 GetModuleFileNameW
                       21 SizeofResource
                        1 DisableThreadLibraryCalls
                        D FreeLibrary
                       18 LoadLibraryExA
                       16 GetProcAddress
                       1C LoadStringA

api-ms-win-core-string-l1-1-0.dll
          18001D690 Import Address Table
          180024B98 Import Name Table
                  0 time date stamp
                  0 Index of first forwarder reference

                        7 WideCharToMultiByte
                        6 MultiByteToWideChar

api-ms-win-core-kernel32-legacy-l1-1-0.dll
          18001D478 Import Address Table
          180024980 Import Name Table
                  0 time date stamp
                  0 Index of first forwarder reference

                       50 PulseEvent

api-ms-win-core-debug-l1-1-0.dll
          18001D370 Import Address Table
          180024878 Import Name Table
                  0 time date stamp
                  0 Index of first forwarder reference

                        7 OutputDebugStringW
                        5 IsDebuggerPresent
                        4 DebugBreak

api-ms-win-core-heap-l1-1-0.dll
          18001D428 Import Address Table
          180024930 Import Name Table
                  0 time date stamp
                  0 Index of first forwarder reference

                        6 HeapFree
                        2 HeapAlloc
                        0 GetProcessHeap
```

```
                          B HeapSize

    api-ms-win-core-processenvironment-l1-1-0.dll
              18001D580 Import Address Table
              180024A88 Import Name Table
                      0 time date stamp
                      0 Index of first forwarder reference

                         10 SearchPathW
                          7 GetCurrentDirectoryW
                         12 SetCurrentDirectoryW

    api-ms-win-core-string-obsolete-l1-1-0.dll
              18001D6A8 Import Address Table
              180024BB0 Import Name Table
                      0 time date stamp
                      0 Index of first forwarder reference

                          5 lstrcmpiW

    api-ms-win-core-errorhandling-l1-1-0.dll
              18001D3B0 Import Address Table
              1800248B8 Import Name Table
                      0 time date stamp
                      0 Index of first forwarder reference

                          5 GetLastError
                          C SetErrorMode
                          7 RaiseException
                          D SetLastError
                         11 UnhandledExceptionFilter
                          F SetUnhandledExceptionFilter

    api-ms-win-core-libraryloader-l1-2-1.dll
              18001D528 Import Address Table
              180024A30 Import Name Table
                      0 time date stamp
                      0 Index of first forwarder reference

                         1D FindResourceW

    api-ms-win-core-privateprofile-l1-1-0.dll
              18001D568 Import Address Table
              180024A70 Import Name Table
                      0 time date stamp
                      0 Index of first forwarder reference

                          5 GetPrivateProfileStringW
                          B GetProfileStringW

    api-ms-win-core-file-l1-1-0.dll
              18001D3E8 Import Address Table
              1800248F0 Import Name Table
                      0 time date stamp
                      0 Index of first forwarder reference

                          7 CreateFileW
                          2 CompareFileTime
                         31 GetFileTime
```

```
                      2D GetFileAttributesW
                      2F GetFileSize

api-ms-win-core-path-l1-1-0.dll
         18001D550 Import Address Table
         180024A58 Import Name Table
                 0 time date stamp
                 0 Index of first forwarder reference

                      14 PathCchStripToRoot
                       C PathCchIsRoot

api-ms-win-core-threadpool-l1-2-0.dll
         18001D7A0 Import Address Table
         180024CA8 Import Name Table
                 0 time date stamp
                 0 Index of first forwarder reference

                      22 WaitForThreadpoolTimerCallbacks
                       C CreateThreadpoolTimer
                      1A SetThreadpoolTimer
                       6 CloseThreadpoolTimer

api-ms-win-core-localization-l1-2-0.dll
         18001D538 Import Address Table
         180024A40 Import Name Table
                 0 time date stamp
                 0 Index of first forwarder reference

                       A GetACP
                       9 FormatMessageW

api-ms-win-core-shlwapi-legacy-l1-1-0.dll
         18001D680 Import Address Table
         180024B88 Import Name Table
                 0 time date stamp
                 0 Index of first forwarder reference

                      14 PathFileExistsW

api-ms-win-core-util-l1-1-0.dll
         18001D7C8 Import Address Table
         180024CD0 Import Name Table
                 0 time date stamp
                 0 Index of first forwarder reference

                       0 Beep

api-ms-win-core-kernel32-private-l1-1-2.dll
         18001D498 Import Address Table
         1800249A0 Import Name Table
                 0 time date stamp
                 0 Index of first forwarder reference

                       3 CheckForReadOnlyResource

api-ms-win-core-sysinfo-l1-1-0.dll
         18001D788 Import Address Table
         180024C90 Import Name Table
```

```
                0 time date stamp
                0 Index of first forwarder reference

                     11 GetSystemInfo
                     16 GetSystemTimeAsFileTime

api-ms-win-core-kernel32-private-l1-1-0.dll
        18001D488 Import Address Table
        180024990 Import Name Table
                0 time date stamp
                0 Index of first forwarder reference

                     10 _lread

api-ms-win-eventing-provider-l1-1-0.dll
        18001D8D8 Import Address Table
        180024DE0 Import Name Table
                0 time date stamp
                0 Index of first forwarder reference

                      5 EventUnregister
                      3 EventRegister

api-ms-win-core-delayload-l1-1-1.dll
        18001D3A0 Import Address Table
        1800248A8 Import Name Table
                0 time date stamp
                0 Index of first forwarder reference

                      1 ResolveDelayLoadedAPI

api-ms-win-core-delayload-l1-1-0.dll
        18001D390 Import Address Table
        180024898 Import Name Table
                0 time date stamp
                0 Index of first forwarder reference

                      0 DelayLoadFailureHook

api-ms-win-core-rtlsupport-l1-1-0.dll
        18001D660 Import Address Table
        180024B68 Import Name Table
                0 time date stamp
                0 Index of first forwarder reference

                    305 RtlCaptureContext
                    509 RtlLookupFunctionEntry
                    64E RtlVirtualUnwind

api-ms-win-core-processthreads-l1-1-1.dll
        18001D600 Import Address Table
        180024B08 Import Name Table
                0 time date stamp
                0 Index of first forwarder reference

                     32 IsProcessorFeaturePresent

api-ms-win-core-profile-l1-1-0.dll
        18001D610 Import Address Table
```

```
          180024B18 Import Name Table
                  0 time date stamp
                  0 Index of first forwarder reference

                        0 QueryPerformanceCounter

  api-ms-win-core-interlocked-l1-1-0.dll
          18001D468 Import Address Table
          180024970 Import Name Table
                  0 time date stamp
                  0 Index of first forwarder reference

                        0 InitializeSListHead

  api-ms-win-core-apiquery-l1-1-0.dll
          18001D360 Import Address Table
          180024868 Import Name Table
                  0 time date stamp
                  0 Index of first forwarder reference

                        0 ApiSetQueryApiSetPresence

Section contains the following delay load imports:

  WINMMBASE.dll
           00000001 Characteristics
   0000000180027820 Address of HMODULE
   000000018002C040 Import Address Table
   0000000180021BE0 Import Name Table
   0000000180022D18 Bound Import Name Table
   0000000000000000 Unload Import Name Table
                  0 time date stamp

                              0000000180006C43     5 OpenDriver
                              0000000180006C1F     1 DefDriverProc
                              0000000180006C67     6 SendDriverMessage
                              0000000180006C79     3 DrvGetModuleHandle
                              0000000180006C8B    89 winmmbaseFreeMMEHandles
                              0000000180006C0D    47 mmGetCurrentTask
                              0000000180006BFB    4B mmTaskYield
                              0000000180006BE9    4A mmTaskSignal
                              0000000180006C31     0 CloseDriver
                              0000000180006BD7    49 mmTaskCreate
                              0000000180006C55     4 GetDriverModuleHandle
                              0000000180006B4C    48 mmTaskBlock

  api-ms-win-mm-mme-l1-1-0.dll
           00000001 Characteristics
   0000000180027828 Address of HMODULE
   000000018002C178 Import Address Table
   0000000180021D18 Import Name Table
   0000000180022D80 Bound Import Name Table
   0000000000000000 Unload Import Name Table
                  0 time date stamp

                              0000000180006F68    18 midiOutCachePatches
                              0000000180006F7A     B midiInGetDevCapsW
                              0000000180006F8C    34 mixerGetDevCapsA
                              0000000180006F9E    14 midiInStart
```

```
0000000180006FB0     62 waveOutSetPlaybackRate
0000000180006FC2     2A midiStreamOpen
0000000180006FD4     2E midiStreamProperty
0000000180006FE6      9 midiInClose
0000000180006FF8     57 waveOutGetPitch
000000018000700A     5A waveOutGetVolume
000000018000701C     2C midiStreamPause
000000018000702E     5F waveOutReset
0000000180007040      D midiInGetErrorTextW
0000000180007052     60 waveOutRestart
0000000180007064     29 midiStreamClose
0000000180007076     42 waveInGetDevCapsW
0000000180007088     41 waveInGetDevCapsA
000000018000709A     50 waveOutClose
00000001800070AC     64 waveOutUnprepareHeader
00000001800070BE     4E waveInUnprepareHeader
00000001800070D0      3 auxGetVolume
00000001800070E2     48 waveInMessage
00000001800070F4     47 waveInGetPosition
0000000180007106     4B waveInReset
0000000180007118      8 midiInAddBuffer
000000018000712A     3A mixerGetLineInfoW
000000018000713C     52 waveOutGetDevCapsW
0000000180006DA6     31 mixerClose
0000000180007160     22 midiOutMessage
0000000180007172     36 mixerGetID
0000000180007184      C midiInGetErrorTextA
0000000180007196      F midiInGetNumDevs
00000001800071A8     2B midiStreamOut
00000001800071BA     56 waveOutGetNumDevs
0000000180006F44      1 auxGetDevCapsW
00000001800071DE      E midiInGetID
00000001800071F0     63 waveOutSetVolume
0000000180007202     1E midiOutGetID
0000000180007214     26 midiOutSetVolume
0000000180007226     5B waveOutMessage
0000000180007238     32 mixerGetControlDetailsA
000000018000724A     37 mixerGetLineControlsA
000000018000725C     33 mixerGetControlDetailsW
000000018000726E     39 mixerGetLineInfoA
0000000180007280      5 auxSetVolume
0000000180007292     27 midiOutShortMsg
00000001800072A4     2D midiStreamPosition
00000001800072B6     1F midiOutGetNumDevs
00000001800072C8     21 midiOutLongMsg
00000001800072DA     19 midiOutClose
00000001800072EC      4 auxOutMessage
00000001800072FE     55 waveOutGetID
0000000180007310     54 waveOutGetErrorTextW
0000000180007322     65 waveOutWrite
0000000180007334     44 waveInGetErrorTextW
0000000180007346     4A waveInPrepareHeader
0000000180007358     2F midiStreamRestart
000000018000736A     4D waveInStop
000000018000737C     53 waveOutGetErrorTextA
000000018000738E     40 waveInClose
00000001800073A0     58 waveOutGetPlaybackRate
00000001800073B2     1C midiOutGetErrorTextA
00000001800073C4     38 mixerGetLineControlsW
```

```
                    00000001800073D6    15 midiInStop
                    00000001800073E8    25 midiOutReset
                    00000001800073FA     A midiInGetDevCapsA
                    000000018000740C    4F waveOutBreakLoop
                    000000018000741E    1A midiOutGetDevCapsA
                    0000000180007430    3D mixerOpen
                    0000000180006F32    30 midiStreamStop
                    0000000180006F56    5D waveOutPause
                    0000000180006F20    28 midiOutUnprepareHeader
                    0000000180006F0E    12 midiInPrepareHeader
                    0000000180006EFC    49 waveInOpen
                    0000000180006EEA     2 auxGetNumDevs
                    0000000180006ED8    13 midiInReset
                    0000000180006EC6     7 midiDisconnect
                    0000000180006EB4    23 midiOutOpen
                    0000000180006EA2    5C waveOutOpen
                    0000000180006E90    3E mixerSetControlDetails
                    0000000180006E7E    24 midiOutPrepareHeader
                    0000000180006E6C    1B midiOutGetDevCapsW
                    0000000180006E5A    3B mixerGetNumDevs
                    0000000180006E48    5E waveOutPrepareHeader
                    0000000180006E36    4C waveInStart
                    0000000180006E24    11 midiInOpen
                    0000000180006E12    17 midiOutCacheDrumPatches
                    0000000180006E00     6 midiConnect
                    0000000180006DEE    16 midiInUnprepareHeader
                    0000000180006DDC    43 waveInGetErrorTextA
                    0000000180006DCA    45 waveInGetID
                    00000001800071CC    35 mixerGetDevCapsW
                    0000000180006D94    51 waveOutGetDevCapsA
                    0000000180006D82    10 midiInMessage
                    0000000180006D70    1D midiOutGetErrorTextW
                    0000000180006D5E    3F waveInAddBuffer
                    0000000180006D4C     0 auxGetDevCapsA
                    0000000180006D3A    20 midiOutGetVolume
                    0000000180006D28    59 waveOutGetPosition
                    0000000180006C9D    3C mixerMessage
                    0000000180006DB8    61 waveOutSetPitch
                    000000018000714E    46 waveInGetNumDevs

api-ms-win-mm-misc-l1-1-0.dll
          00000001 Characteristics
  0000000180027830 Address of HMODULE
  000000018002C0C0 Import Address Table
  0000000180021C60 Import Name Table
  00000001800230B8 Bound Import Name Table
  0000000000000000 Unload Import Name Table
                 0 time date stamp

                    0000000180007623    1B mmioStringToFOURCCW
                    0000000180007611     8 mmioAdvance
                    00000001800075FF     C mmioDescend
                    00000001800075ED    11 mmioOpenA
                    00000001800075DB     D mmioFlush
                    00000001800075C9     9 mmioAscend
                    00000001800075B7    19 mmioSetInfo
                    00000001800075A5     B mmioCreateChunk
                    0000000180007593    17 mmioSendMessage
                    0000000180007581    18 mmioSetBuffer
```

```
                         000000018000756F      A mmioClose
                         000000018000755D     16 mmioSeek
                         000000018000754B      F mmioInstallIOProcA
                         0000000180007539      E mmioGetInfo
                         0000000180007527     13 mmioRead
                         0000000180007515     12 mmioOpenW
                         0000000180007503     14 mmioRenameA
                         00000001800074F1      7 mmDrvInstall
                         00000001800074DF     1C mmioWrite
                         00000001800074CD     1A mmioStringToFOURCCA
                         0000000180007442     10 mmioInstallIOProcW
                         0000000180007635     15 mmioRenameW

AudioSes.DLL
          00000001 Characteristics
  0000000180027838 Address of HMODULE
  000000018002C000 Import Address Table
  0000000180021BA0 Import Name Table
  0000000180023170 Bound Import Name Table
  0000000000000000 Unload Import Name Table
                 0 time date stamp

                         0000000180007647         Ordinal     6

ext-ms-win-ntuser-dialogbox-l1-1-0.dll
          00000001 Characteristics
  0000000180027E38 Address of HMODULE
  000000018002C4B0 Import Address Table
  0000000180022050 Import Name Table
  0000000180023180 Bound Import Name Table
  0000000000000000 Unload Import Name Table
                 0 time date stamp

                         00000001800088EC     27 MessageBoxW

ext-ms-win-ntuser-gui-l1-1-0.dll
          00000001 Characteristics
  0000000180027E40 Address of HMODULE
  000000018002C4C0 Import Address Table
  0000000180022060 Import Name Table
  0000000180023190 Bound Import Name Table
  0000000000000000 Unload Import Name Table
                 0 time date stamp

                         0000000180008977      E LoadIconA

ext-ms-win-ntuser-keyboard-l1-1-0.dll
          00000001 Characteristics
  0000000180027E48 Address of HMODULE
  000000018002C4D0 Import Address Table
  0000000180022070 Import Name Table
  00000001800231A0 Bound Import Name Table
  0000000000000000 Unload Import Name Table
                 0 time date stamp

                         0000000180008A02      8 GetKeyState

ext-ms-win-ntuser-message-l1-1-0.dll
          00000001 Characteristics
```

```
  0000000180027E50 Address of HMODULE
  000000018002C4E0 Import Address Table
  0000000180022080 Import Name Table
  00000001800231B0 Bound Import Name Table
  0000000000000000 Unload Import Name Table
                 0 time date stamp

                              0000000180008BC6     2 DispatchMessageA
                              0000000180008BEA     D PostMessageA
                              0000000180008BD8     4 GetMessageA
                              0000000180008BB4    10 PostThreadMessageA
                              0000000180008BA2     B PeekMessageA
                              0000000180008B6C     E PostMessageW
                              0000000180008B90     C PeekMessageW
                              0000000180008B7E    15 SendMessageA
                              0000000180008A8D    13 RegisterWindowMessageW

ext-ms-win-ntuser-server-l1-1-0.dll
          00000001 Characteristics
  0000000180027E68 Address of HMODULE
  000000018002C530 Import Address Table
  00000001800220D0 Import Name Table
  0000000180023200 Bound Import Name Table
  0000000000000000 Unload Import Name Table
                 0 time date stamp

                              0000000180008BFC     1 NotifyWinEvent

ext-ms-win-ntuser-windowclass-l1-1-0.dll
          00000001 Characteristics
  0000000180027E70 Address of HMODULE
  000000018002C588 Import Address Table
  0000000180022128 Import Name Table
  0000000180023210 Bound Import Name Table
  0000000000000000 Unload Import Name Table
                 0 time date stamp

                              0000000180008D24    11 RegisterClassA
                              0000000180008C87    1E UnregisterClassA
                              0000000180008D12     0 GetClassInfoA

ext-ms-win-ntuser-window-l1-1-2.dll
          00000001 Characteristics
  0000000180027E98 Address of HMODULE
  000000018002C570 Import Address Table
  0000000180022110 Import Name Table
  0000000180023230 Bound Import Name Table
  0000000000000000 Unload Import Name Table
                 0 time date stamp

                              0000000180008DE0    4E KillTimer
                              0000000180008E6B    66 SetTimer

ext-ms-win-ntuser-window-l1-1-0.dll
          00000001 Characteristics
  0000000180027EA0 Address of HMODULE
  000000018002C540 Import Address Table
  00000001800220E0 Import Name Table
  0000000180023248 Bound Import Name Table
```

```
    0000000000000000 Unload Import Name Table
                   0 time date stamp

                          0000000180008F1A    14 DefWindowProcA
                          0000000180008F2C    17 DestroyWindow
                          0000000180008F3E    71 SoundSentry
                          0000000180008E7D    49 IsWindow
                          0000000180008F08     E CreateWindowExA

  ext-ms-win-rtcore-ntuser-sysparams-l1-1-0.dll
            00000001 Characteristics
    0000000180027EA8 Address of HMODULE
    000000018002C5B8 Import Address Table
    0000000180022158 Import Name Table
    0000000180023278 Bound Import Name Table
    0000000000000000 Unload Import Name Table
                   0 time date stamp

                          0000000180008FDB     9 GetSystemMetrics
                          0000000180008F50    10 SystemParametersInfoW

  api-ms-win-core-com-l1-1-0.dll
            00000001 Characteristics
    0000000180027EB0 Address of HMODULE
    000000018002C0A8 Import Address Table
    0000000180021C48 Import Name Table
    0000000180023290 Bound Import Name Table
    0000000000000000 Unload Import Name Table
                   0 time date stamp

                          0000000180009078    28 CoInitializeEx
                          0000000180008FED    46 CoUninitialize

  OLEAUT32.dll
            00000001 Characteristics
    0000000180027EB8 Address of HMODULE
    000000018002C010 Import Address Table
    0000000180021BB0 Import Name Table
    00000001800232A8 Bound Import Name Table
    0000000000000000 Unload Import Name Table
                   0 time date stamp

                          0000000180009115      Ordinal     2
                          000000018000908A      Ordinal     7
                          000000018000914B      Ordinal   162
                          0000000180009139      Ordinal   161
                          0000000180009127      Ordinal     6

  ext-ms-win-oleacc-l1-1-1.dll
            00000001 Characteristics
    0000000180027EC0 Address of HMODULE
    000000018002C5A8 Import Address Table
    0000000180022148 Import Name Table
    00000001800232D8 Bound Import Name Table
    0000000000000000 Unload Import Name Table
                   0 time date stamp

                          000000018000916C     8 LresultFromObject
```

```
  Summary

        3000 .data
        1000 .didat
        1000 .guids
        2000 .pdata
        A000 .rdata
        1000 .reloc
        5000 .rsrc
       1C000 .text
```

Note: We see that the *winmm* module uses a lot of delay-loaded API sets.

5. Launch WinDbg.

6. Choose Launch Executable and *\AWAPI-Dumps\t.exe.*

7. We get the process launched with an initial breakpoint:

```
Microsoft (R) Windows Debugger Version 10.0.27725.1000 AMD64
Copyright (c) Microsoft Corporation. All rights reserved.

CommandLine: C:\AWAPI-Dumps\t.exe

************* Path validation summary **************
Response                         Time (ms)     Location
Deferred                                       srv*
Symbol search path is: srv*
Executable search path is:
ModLoad: 00007ff7`03000000 00007ff7`030a5000   t.exe
ModLoad: 00007ffd`50350000 00007ffd`50567000   ntdll.dll
ModLoad: 00007ffd`4e900000 00007ffd`4e9c4000   C:\WINDOWS\System32\KERNEL32.DLL
ModLoad: 00007ffd`4dbe0000 00007ffd`4df9a000   C:\WINDOWS\System32\KERNELBASE.dll
ModLoad: 00007ffd`46c70000 00007ffd`46d07000   C:\WINDOWS\SYSTEM32\apphelp.dll
ModLoad: 00007ffd`42aa0000 00007ffd`42ad4000   C:\WINDOWS\SYSTEM32\WINMM.dll
ModLoad: 00007ffd`4da00000 00007ffd`4db11000   C:\WINDOWS\System32\ucrtbase.dll
(8e4c.6fc0): Break instruction exception - code 80000003 (first chance)
ntdll!LdrpDoDebuggerBreak+0x30:
00007ffd`5042c134 int     3
```

8. Open a log file using **.logopen**:

```
0:000> .logopen C:\AWAPI-Dumps\W6.log
Opened log file 'C:\AWAPI-Dumps\W6.log'
```

9. Let's look at the internals of *GetCurrentProcessId* and *GetCurrentThreadId* functions:

```
0:000> uf kernel32!GetCurrentProcessId
Flow analysis was incomplete, some code may be missing
KERNEL32!GetCurrentProcessId:
00007ff8`0b03fe90 ff25a2400600    jmp     qword ptr [KERNEL32!_imp_GetCurrentProcessId (00007ff8`0b0a3f38)]

0:000> dps KERNEL32!_imp_GetCurrentProcessId L1
00007ff8`0b0a3f38  00007ff8`09bf6060 KERNELBASE!GetCurrentProcessId

0:000> uf KERNELBASE!GetCurrentProcessId
```

```
KERNELBASE!GetCurrentProcessId:
00007ff8`09bf6060 65488b042530000000 mov   rax,qword ptr gs:[30h]
00007ff8`09bf6069 8b4040             mov   eax,dword ptr [rax+40h]
00007ff8`09bf606c c3                 ret

0:000> uf KERNELBASE!GetCurrentThreadId
KERNELBASE!GetCurrentThreadId:
00007ff8`09ba1c60 65488b042530000000 mov   rax,qword ptr gs:[30h]
00007ff8`09ba1c69 8b4048             mov   eax,dword ptr [rax+48h]
00007ff8`09ba1c6c c3                 ret

0:000> uf /c KERNELBASE!GetCurrentThreadId
KERNELBASE!GetCurrentThreadId (00007ff8`09ba1c60)
    no calls found

0:000> uf /c KERNELBASE!GetCurrentProcessId
KERNELBASE!GetCurrentProcessId (00007ff8`09bf6060)
    no calls found
```

Note: We don't see any system calls or API that may potentially use system calls.

10. Let's look at the internals of the *ReadFile* function:

```
0:000> uf /c KERNELBASE!ReadFile
KERNELBASE!ReadFile (00007ff8`09bbc0c0)
  KERNELBASE!ReadFile+0x74 (00007ff8`09bbc134):
    call to ntdll!NtReadFile (00007ff8`0c1cee20)
  KERNELBASE!ReadFile+0x101 (00007ff8`09bbc1c1):
    call to ntdll!NtReadFile (00007ff8`0c1cee20)
  KERNELBASE!ReadFile+0x11a (00007ff8`09bbc1da):
    call to KERNELBASE!BaseSetLastNTError (00007ff8`09bbc270)
  KERNELBASE!ReadFile+0x18f (00007ff8`09bbc24f):
    call to KERNELBASE!BaseSetLastNTError (00007ff8`09bbc270)
  KERNELBASE!ReadFile+0x96c98 (00007ff8`09c52d58):
    call to ntdll!NtWaitForSingleObject (00007ff8`0c1cede0)
```

We we 3 calls to 2 different *ntdll* functions:

```
0:000> uf ntdll!NtReadFile
ntdll!NtReadFile:
00007ff8`0c1cee20 4c8bd1             mov     r10,rcx
00007ff8`0c1cee23 b806000000         mov     eax,6
00007ff8`0c1cee28 f604250803fe7f01 test    byte ptr [SharedUserData+0x308 (00000000`7ffe0308)],1
00007ff8`0c1cee30 7503               jne     ntdll!NtReadFile+0x15 (00007ff8`0c1cee35)  Branch

ntdll!NtReadFile+0x12:
00007ff8`0c1cee32 0f05               syscall
00007ff8`0c1cee34 c3                 ret

ntdll!NtReadFile+0x15:
00007ff8`0c1cee35 cd2e               int     2Eh
00007ff8`0c1cee37 c3                 ret

0:000> uf ntdll!NtWaitForSingleObject
ntdll!NtWaitForSingleObject:
00007ff8`0c1cede0 4c8bd1             mov     r10,rcx
00007ff8`0c1cede3 b804000000         mov     eax,4
00007ff8`0c1cede8 f604250803fe7f01 test    byte ptr [SharedUserData+0x308 (00000000`7ffe0308)],1
00007ff8`0c1cedf0 7503               jne     ntdll!NtWaitForSingleObject+0x15 (00007ff8`0c1cedf5)  Branch

ntdll!NtWaitForSingleObject+0x12:
00007ff8`0c1cedf2 0f05               syscall
```

```
00007ff8`0c1cedf4 c3              ret

ntdll!NtWaitForSingleObject+0x15:
00007ff8`0c1cedf5 cd2e            int     2Eh
00007ff8`0c1cedf7 c3              ret
```

Note: We see that each function is translated to a system call mechanism transitioning to kernel space and mode. Let's also check one *win32u* call we need to compare to in the next exercise:

```
0:000> uf win32u!NtUserGetMessage
win32u!NtUserGetMessage:
00007ffc`f3811400 4c8bd1          mov     r10,rcx
00007ffc`f3811403 b804100000      mov     eax,1004h
00007ffc`f3811408 f604250803fe7f01 test    byte ptr [SharedUserData+0x308 (00000000`7ffe0308)],1
00007ffc`f3811410 7503            jne     win32u!NtUserGetMessage+0x15 (00007ffc`f3811415)  Branch

win32u!NtUserGetMessage+0x12:
00007ffc`f3811412 0f05            syscall
00007ffc`f3811414 c3              ret

win32u!NtUserGetMessage+0x15:
00007ffc`f3811415 cd2e            int     2Eh
00007ffc`f3811417 c3              ret
```

11. We close logging before exiting WinDbg:

```
0:000> .logclose
Closing open log file C:\AWAPI-Dumps\W6.log
```

Documented API

- Online documentation
- Present in headers
- Example:

Documentation

```
kernel32!SuspendThread →
        KERNELBASE!SuspendThread →
                ntdll!NtSuspendThread
```


Most of Windows API is documented well. By documented API, we mean either available online documentation or the available function declaration in headers. As a typical example, I show the *SuspendThread* function on this slide.

Documentation
https://learn.microsoft.com/en-us/windows/win32/api/processthreadsapi/nf-processthreadsapi-suspendthread

Undocumented API

ntdll!NtSuspendProcess

```
0:000> x /v ntdll!*
…
pub func   00007ffc`f6346ff0    0 ntdll!NtSuspendProcess (NtSuspendProcess)
…
```


There may be publically available functions such as *NtSuspendProcess* that are not officially documented, although you can find third-party usage examples. These undocumented APIs may sometimes be used programmatically via *LoadLibrary* and *GetProcAddress* API functions. One useful undocumented API is *NtSuspendProcess* which I used in some projects in the past.

API Source Code

- Wine (GitLab) / Wine-Mirror (GitHub)
- Example:

user32!CreateWindowExW

https://gitlab.winehq.org/wine/wine/-/blob/master/dlls/user32/win.c
https://github.com/wine-mirror/wine/blob/master/dlls/user32/win.c

Unless some module source code is made public, or you have access to its source code officially, one of the methods to get approximate source code is to use disassemblers. The other way is to see how API is emulated. One such project is Wine, and I put some links on this slide.

Wine (GitLab)
https://gitlab.winehq.org/wine/wine

Wine-Mirror (GitHub)
https://github.com/wine-mirror/wine/

user32!CreateWindowExW
https://gitlab.winehq.org/wine/wine/-/blob/master/dlls/user32/win.c
https://github.com/wine-mirror/wine/blob/master/dlls/user32/win.c

API Name Patterns

- Create/Open/Delete/Close
- Process/Thread
- Memory
- Read/Write

WinDbg Commands

```
0:000> x /v module!fpattern
```


One way to learn about various available API functions, documented and undocumented, is to search for name patterns, such as all public functions that have *Memory* as their substring in their symbol name.

API Namespaces

- API sets / contracts
  - Example: CreateDialogParamW
- Functions required to accomplish a particular task
  - Example: screen capture / saving image

```
gdi32!CreateCompatibleDC        GdiPlus!GdiplusStartup
gdi32!StretchBlt                GdiPlus!GdipSaveImageToStream
gdi32!CreateDIBSection          GdiPlus!GdipGetImageEncodersSize
gdi32!SelectObject              GdiPlus!GdipDisposeImage
                                GdiPlus!GdipCreateBitmapFromHBITMAP
user32!ReleaseDC                GdiPlus!GdipGetImageEncoders
user32!NtUserGetWindowDC
user32!GetWindowRect            ole32!CreateStreamOnHGlobal
```


API Namespace is a group or several groups of functions required to accomplish a particular task. Some namespaces are defined by API set contracts, for example, functions for creating dialogs windows. A namespace group usually belongs to one particular module. As we mentioned before, malware analysis patterns include the **Namespace** analysis pattern that views groups of imported functions from several modules as potentially serving some particular malicious (but maybe just some utilitarian) need, for example, a screen capture that needs to be saved somewhere.

CreateDialogParamW
https://learn.microsoft.com/en-us/windows/win32/api/winuser/nf-winuser-createdialogparamw

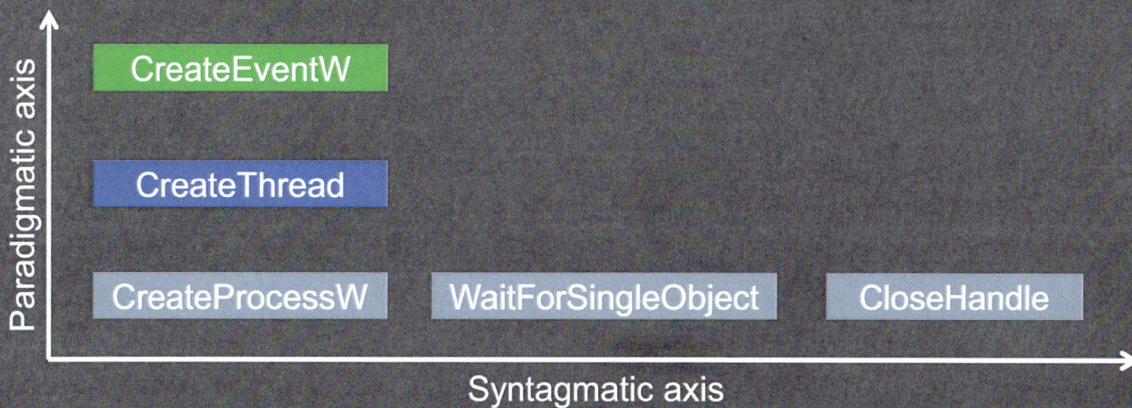

API Syntagms/Paradigms

- Syntagms / syntagmatic analysis
- Paradigms / paradigmatic analysis

Paradigmatic axis

CreateEventW

CreateThread

CreateProcessW

WaitForSingleObject

CloseHandle

Syntagmatic axis

If we look at prescriptive API sequences to which we devoted one of the previous slides, they can belong to a syntagmatic axis. Individual API functions are considered syntactic units (the so-called syntagms) arranged in sequences (sentences). They are encoded by a developer and decoded by a maintainer. The set of API functions of a similar purpose (meaning) from which sequences are formed is called a paradigm. They may be considered equivalent in the sense that some calls may be replaced by other API functions, leaving the overall sequence correct (or at least meaningful). Here, on the slide, we provide a typical simple example of creating various synchronization objects that can be waited for before closing their handles. We encounter additional examples when we look at Windows API categories.

Syntagms
https://en.wikipedia.org/wiki/Syntagma_(linguistics)

Syntagmatic analysis
https://en.wikipedia.org/wiki/Syntagmatic_analysis

Paradigmatic analysis
https://en.wikipedia.org/wiki/Paradigmatic_analysis

Marked API

- Marked Message trace and log analysis pattern
- Points to presence or absence of activity
- Example:
  - CreateThread [-]
  - socket [+]
  - GetMessageW [-]
  - ReadConsoleW [+]

WinDbg Commands

```
0:000> x app!_imp_pattern
```


We call by marked API certain functions imported or present in call paths that point to the presence of activity. If they are not imported and not used, they point to the absence of activity. Such API functions may be compiled into a checklist. This API aspect is borrowed from **Marked Message** trace and log analysis pattern where marked messages may point to some domain of software activity related to functional requirements and help in troubleshooting and debugging, and some unmarked messages may directly say about the absences of activity.

Marked Message
https://www.dumpanalysis.org/blog/index.php/2012/01/02/trace-analysis-patterns-part-45/

ADDR Patterns

- From Accelerated Disassembly Deconstruction Reversing
- List of pattern names
- Pattern descriptions

The name ADDR (sounds like an address) comes from the Accelerated Disassembly Deconstruction Reversing abbreviation from the similar sounding training course. The slide provides the link to their names and the link to their descriptions. We used some ADDR patterns, such as **Call Path** and **Function Skeleton**, in some of our exercises.

List of pattern names
https://www.dumpanalysis.org/addr-patterns

Pattern descriptions
https://www.patterndiagnostics.com/Training/Accelerated-Disassembly-Reconstruction-Reversing-Version2.5-Slides.pdf

DebugWare Patterns

- Patterns for troubleshooting and debugging tools
- API Query

  Periodic or asynchronous query of the same set of API and logging of their input and output data.
- Example: WindowHistory

DebugWare patterns are patterns for designing and implementing troubleshooting and debugging tools. One of the first patterns was named **API Query**, where some Windows API functions are used periodically to query OS data and log the output. In the past, I wrote a tool, **WindowHistory**, that periodically queries the state of GUI windows.

API Query
https://www.dumpanalysis.org/blog/index.php/2008/07/19/debugware-patterns-part-1/

WindowHistory
https://support.citrix.com/article/CTX109235/windowhistory-tool

Patterns vs. Analysis Patterns

Diagnostic Pattern: a common recurrent identifiable problem together with a set of recommendations and possible solutions to apply in a specific context.

Diagnostic Problem: a set of indicators (symptoms, signs) describing a problem.

Diagnostic Analysis Pattern: a common recurrent analysis technique and method of diagnostic pattern identification in a specific context.

Diagnostics Pattern Language: common names of diagnostic and diagnostic analysis patterns. The same language for any operating system: Windows, Mac OS X, Linux, ...

We have now come to associating Windows API with various diagnostic analysis patterns. Here, I'd like to stress the informal difference between patterns and analysis patterns. The distinction is highlighted on this slide.

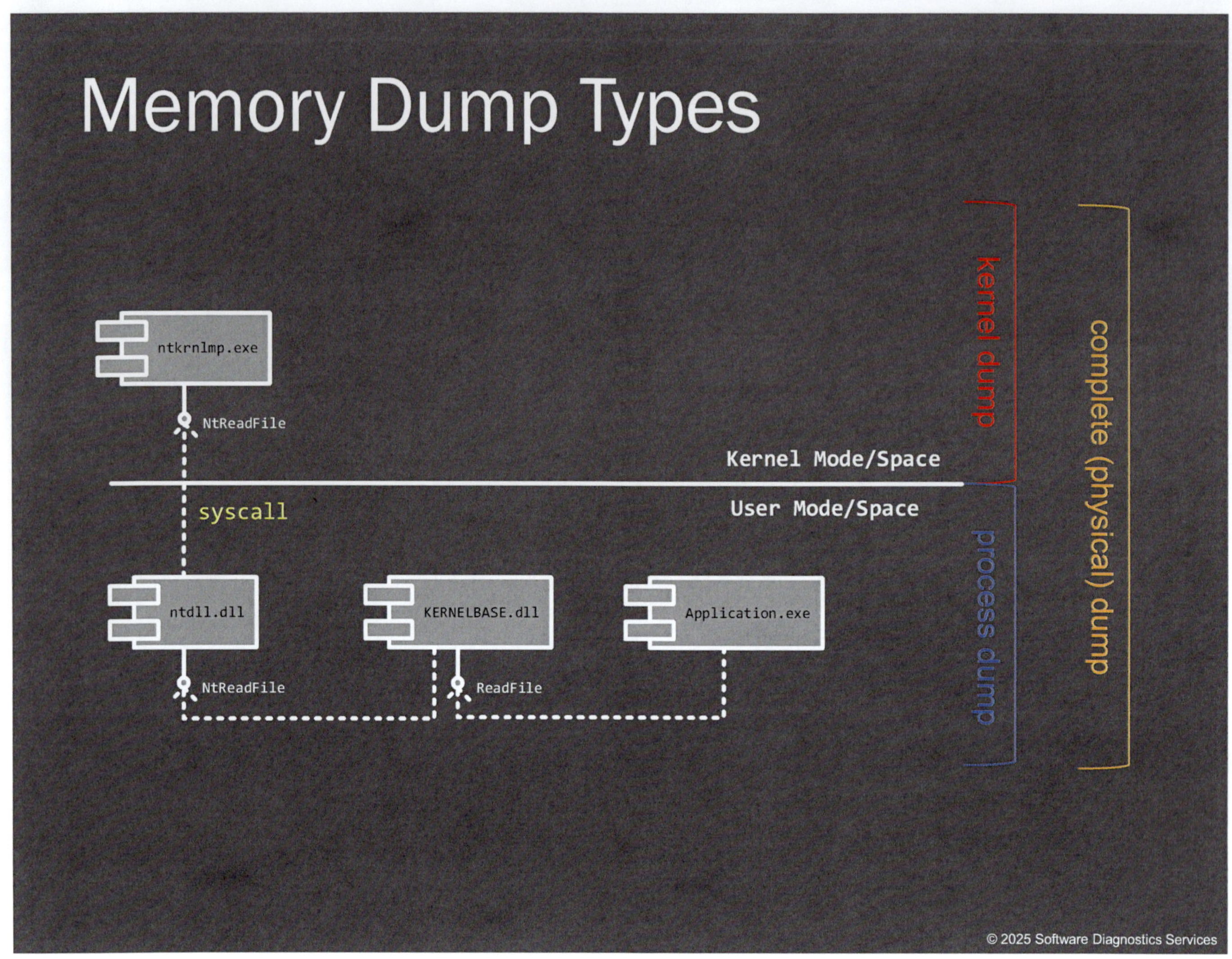

There are several memory dump types, and this picture shows their memory space content relative to the **Call Path**.

Memory Analysis Patterns

- User space
  - Process memory dumps
  - Complete memory dumps

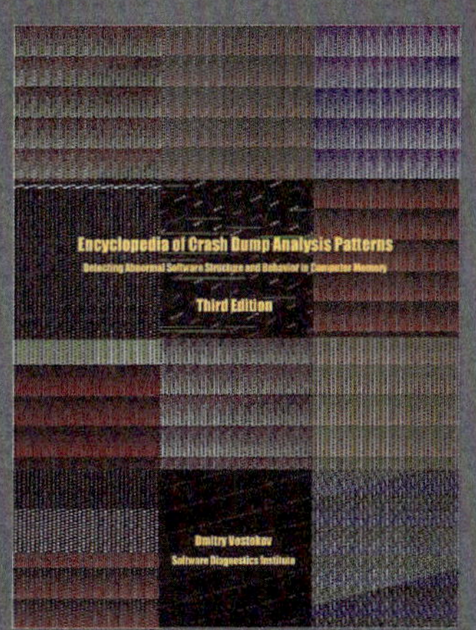

- Function analysis patterns
  - Stack Trace Collection
  - Well-Tested Function
  - False Function Parameters
  - String Parameter
  - Small Value / Design Value
  - Virtualized Process
  - Stack Trace
  - Execution Residue
  - Hidden Parameter
  - Parameter Flow
  - Data Correlation

We pay attention to Windows API function calls when we do diagnostics and postmortem debugging using memory dumps. Since API calls are done in user space, as shown in the previous slide, we are interested only in the process and complete (physical) memory dumps unless we need to follow syscalls into kernel space. On this slide, I put some general analysis patterns related to function parameters. **Execution Residue** allows us to see past execution traces of Windows API, such as their return addresses in the stack regions. We mention specific memory analysis patterns associated with particular Windows API categories when we look at the latter.

Stack Trace Collection
https://www.dumpanalysis.org/blog/index.php/2007/09/14/crash-dump-analysis-patterns-part-27/

Well-Tested Function
https://www.dumpanalysis.org/blog/index.php/2009/10/23/crash-dump-analysis-patterns-part-89/

False Function Parameters
https://www.dumpanalysis.org/blog/index.php/2008/02/15/crash-dump-analysis-patterns-part-50/

String Parameter
https://www.dumpanalysis.org/blog/index.php/2010/12/02/crash-dump-analysis-patterns-part-118/

Small Value
https://www.dumpanalysis.org/blog/index.php/2013/11/09/crash-dump-analysis-patterns-part-202/

Design Value
https://www.dumpanalysis.org/blog/index.php/2014/06/21/crash-dump-analysis-patterns-part-207/

Virtualized Process
https://www.dumpanalysis.org/blog/index.php/2007/09/11/crash-dump-analysis-patterns-part-26/

Stack Trace
https://www.dumpanalysis.org/blog/index.php/2007/09/10/crash-dump-analysis-patterns-part-25/

Execution Residue
https://www.dumpanalysis.org/blog/index.php/2008/04/29/crash-dump-analysis-patterns-part-60/

Hidden Parameter
https://www.dumpanalysis.org/blog/index.php/2011/11/10/crash-dump-analysis-patterns-part-155/

Parameter Flow
https://www.dumpanalysis.org/blog/index.php/2015/11/30/crash-dump-analysis-patterns-part-234/

Data Correlation
https://www.dumpanalysis.org/blog/index.php/2011/03/28/crash-dump-analysis-patterns-part-134a/

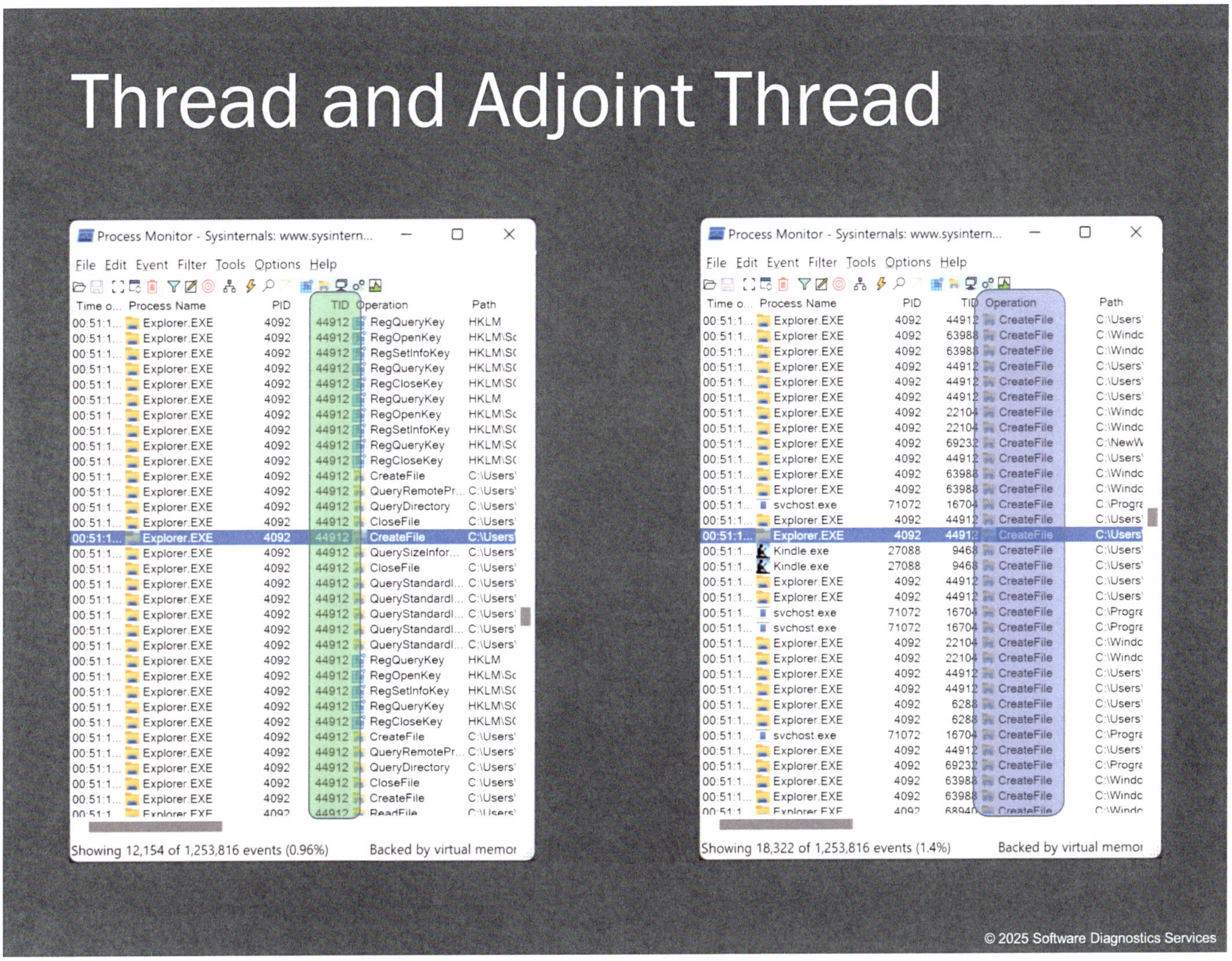

Thread and Adjoint Thread

Before we look at another type of execution history artifacts, trace and log analysis patterns, I illustrate, using the Process Monitor (or Procmon) tool, the notion of **Adjoint Thread** that figures a lot in trace and log analysis pattern descriptions. In **Thread**, we have the same value in the TID column, and other column values vary. In **Adjoint Thread**, we have the same value for some other column, for example, Operation, but the values of other columns, including TID, vary.

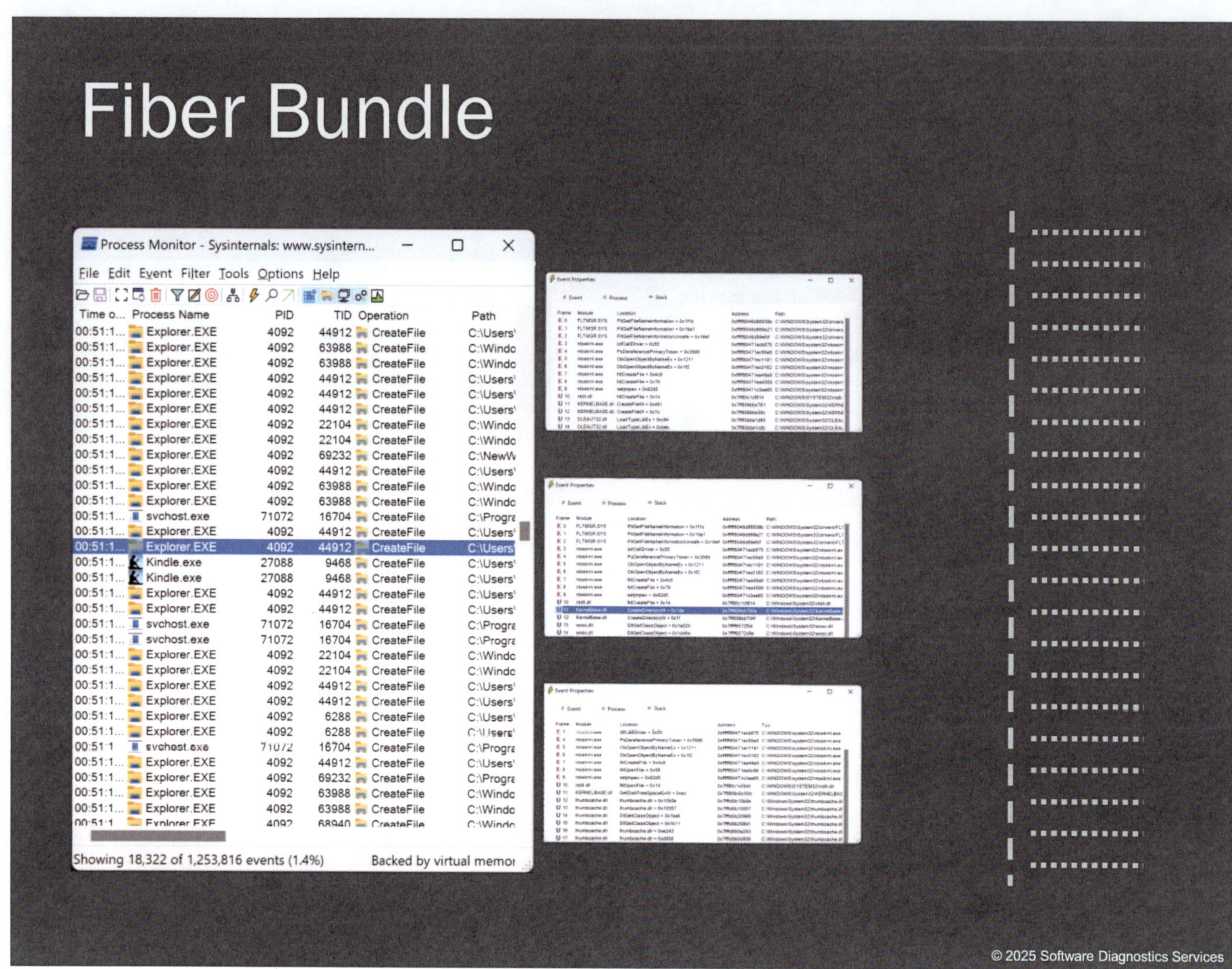

Trace and log messages may not have direct references to Windows API, but stack traces associated with them may have. It is the essence of the **Fiber Bundle** analysis pattern, where other traces, for example, stack traces, are attached to every message.

Trace and Log Analysis Patterns

- Process Monitor

- Function calls:
  - Thread of Activity
  - Fiber of Activity
  - Adjoint Thread of Activity
  - Strand of Activity
  - Discontinuity
  - Fiber Bundle
  - Weave of Activity

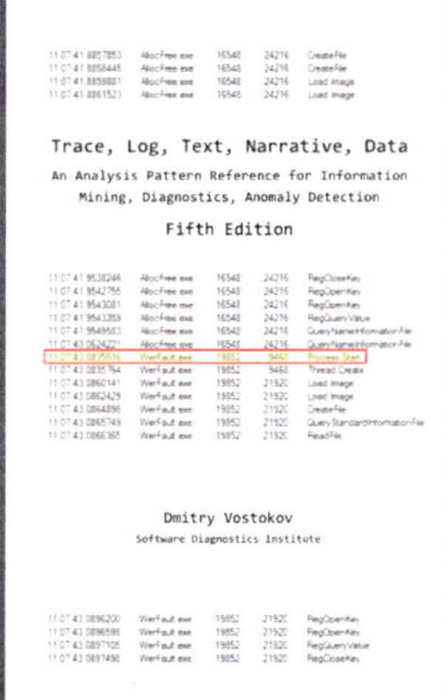

On this slide, I referenced trace and log analysis patterns related to threads, adjoint threads, fiber bundles, and their combinations. For example, discontinuity in messages may show the possible hang when an API function is called.

Process Monitor
https://learn.microsoft.com/en-us/sysinternals/downloads/procmon

Thread of Activity
https://www.dumpanalysis.org/blog/index.php/2009/08/03/trace-analysis-patterns-part-7/

Fiber of Activity
https://www.dumpanalysis.org/blog/index.php/2016/06/29/trace-analysis-patterns-part-126/

Adjoint Thread of Activity
https://www.dumpanalysis.org/blog/index.php/2010/03/04/trace-analysis-patterns-part-17/

Strand of Activity
https://www.dumpanalysis.org/blog/index.php/2020/09/12/trace-analysis-patterns-part-198/

Discontinuity
https://www.dumpanalysis.org/blog/index.php/2009/08/04/trace-analysis-patterns-part-8/

Fiber Bundle
https://www.dumpanalysis.org/blog/index.php/2012/09/26/trace-analysis-patterns-part-52/

Weave of Activity
https://www.dumpanalysis.org/blog/index.php/2020/09/19/trace-analysis-patterns-part-201/

WOW64

- 64-bit process dumps
- 32-bit process dumps
- wow64 (kernel32)
- wow64win (user32, gdi32)
- wow64cpu

Finally, we come to the almost last aspect, **WOW64**. When 32-bit programs are executed, we can save their memory dumps as 32-bit using the 32-bit Task Manager from the *SysWOW64* Windows folder. On the other hand, if we use the 64-bit Task Manager, then the 64-bit process memory dump is saved. In the next exercise, we see that 32-bit Windows API calls imported from 32-bit modules are translated to 64-bit modules, and we also see how to get 32-bit stack traces from 64-bit WOW64 process memory dumps.

Exercise W7

- **Goal:** Explore Windows API calls in the WOW64 context
- **Memory Analysis Patterns:** Stack Trace Collection; Virtualized Process
- \AWAPI-Dumps\Exercise-W7.pdf

Exercise W7

Goal: Explore Windows API calls in the WOW64 context.

Memory Analysis Patterns: Stack Trace Collection; Virtualized Process.

1. Launch WinDbg.

2. Open \AWAPI-Dumps\Process\WOW64\wordpad.DMP

3. We get the dump file loaded:

```
Microsoft (R) Windows Debugger Version 10.0.27725.1000 AMD64
Copyright (c) Microsoft Corporation. All rights reserved.


Loading Dump File [C:\AWAPI-Dumps\Process\WOW64\wordpad.DMP]
User Mini Dump File with Full Memory: Only application data is available


************* Path validation summary **************
Response                         Time (ms)     Location
Deferred                                       srv*
Symbol search path is: srv*
Executable search path is:
Windows 10 Version 22000 MP (2 procs) Free x64
Product: WinNt, suite: SingleUserTS Personal
Edition build lab: 22000.1.amd64fre.co_release.210604-1628
Debug session time: Sat Oct 15 20:22:42.000 2022 (UTC + 0:00)
System Uptime: 0 days 0:35:37.821
Process Uptime: 0 days 0:04:12.000
................................................................
...........
For analysis of this file, run !analyze -v
wow64win!NtUserGetMessage+0x14:
00007ffc`f4eb1424 ret
```

4. Open a log file using **.logopen**:

```
0:000> .logopen C:\AWAPI-Dumps\W7.log
Opened log file 'C:\AWAPI-Dumps\W7.log'
```

5. Let's look at all thread stack traces:

```
0:000> ~*kL

.  0  Id: 2fe0.2fe4 Suspend: 0 Teb: 00000000`00257000 Unfrozen
 # Child-SP          RetAddr               Call Site
00 00000000`000fe208 00007ffc`f4ea510e     wow64win!NtUserGetMessage+0x14
01 00000000`000fe210 00007ffc`f5ae77ca     wow64win!whNtUserGetMessage+0x2e
02 00000000`000fe270 00000000`779b17ba     wow64!Wow64SystemServiceEx+0x15a
03 00000000`000feb30 00000000`779b1d75     wow64cpu!ServiceNoTurbo+0xb
04 00000000`000febe0 00007ffc`f5aee06d     wow64cpu!BTCpuSimulate+0xbb5
05 00000000`000fec20 00007ffc`f5aed8ad     wow64!RunCpuSimulation+0xd
```

```
06 00000000`000fec50 00007ffc`f637f87d     wow64!Wow64LdrpInitialize+0x12d
07 00000000`000fef00 00007ffc`f636d78c     ntdll!LdrpInitializeProcess+0x16d1
08 00000000`000ff2c0 00007ffc`f631a993     ntdll!_LdrpInitialize+0x52dc0
09 00000000`000ff340 00007ffc`f631a8be     ntdll!LdrpInitializeInternal+0x6b
0a 00000000`000ff5c0 00000000`00000000     ntdll!LdrInitializeThunk+0xe

   1  Id: 2fe0.620 Suspend: 0 Teb: 00000000`00267000 Unfrozen
 # Child-SP          RetAddr               Call Site
00 00000000`00f9ed58 00000000`779b184f     wow64cpu!CpupSyscallStub+0x13
01 00000000`00f9ed60 00000000`779b1d75     wow64cpu!WaitForMultipleObjects32+0x1d
02 00000000`00f9ee10 00007ffc`f5aee06d     wow64cpu!BTCpuSimulate+0xbb5
03 00000000`00f9ee50 00007ffc`f5aed8ad     wow64!RunCpuSimulation+0xd
04 00000000`00f9ee80 00007ffc`f631aaa8     wow64!Wow64LdrpInitialize+0x12d
05 00000000`00f9f130 00007ffc`f631a993     ntdll!_LdrpInitialize+0xdc
06 00000000`00f9f1b0 00007ffc`f631a8be     ntdll!LdrpInitializeInternal+0x6b
07 00000000`00f9f430 00000000`00000000     ntdll!LdrInitializeThunk+0xe

   2  Id: 2fe0.748 Suspend: 0 Teb: 00000000`0026b000 Unfrozen
 # Child-SP          RetAddr               Call Site
00 00000000`06f2e138 00007ffc`f5aea76a     ntdll!NtWaitForWorkViaWorkerFactory+0x14
01 00000000`06f2e140 00007ffc`f5ae77ca     wow64!whNtWaitForWorkViaWorkerFactory+0x11a
02 00000000`06f2e1d0 00000000`779b17ba     wow64!Wow64SystemServiceEx+0x15a
03 00000000`06f2ea90 00000000`779b1d75     wow64cpu!ServiceNoTurbo+0xb
04 00000000`06f2eb40 00007ffc`f5aee06d     wow64cpu!BTCpuSimulate+0xbb5
05 00000000`06f2eb80 00007ffc`f5aed8ad     wow64!RunCpuSimulation+0xd
06 00000000`06f2ebb0 00007ffc`f631aaa8     wow64!Wow64LdrpInitialize+0x12d
07 00000000`06f2ee60 00007ffc`f631a993     ntdll!_LdrpInitialize+0xdc
08 00000000`06f2eee0 00007ffc`f631a8be     ntdll!LdrpInitializeInternal+0x6b
09 00000000`06f2f160 00000000`00000000     ntdll!LdrInitializeThunk+0xe

   3  Id: 2fe0.153c Suspend: 0 Teb: 00000000`0026f000 Unfrozen
 # Child-SP          RetAddr               Call Site
00 00000000`06fae428 00007ffc`f5aea76a     ntdll!NtWaitForWorkViaWorkerFactory+0x14
01 00000000`06fae430 00007ffc`f5ae77ca     wow64!whNtWaitForWorkViaWorkerFactory+0x11a
02 00000000`06fae4c0 00000000`779b17ba     wow64!Wow64SystemServiceEx+0x15a
03 00000000`06faed80 00000000`779b1d75     wow64cpu!ServiceNoTurbo+0xb
04 00000000`06faee30 00007ffc`f5aee06d     wow64cpu!BTCpuSimulate+0xbb5
05 00000000`06faee70 00007ffc`f5aed8ad     wow64!RunCpuSimulation+0xd
06 00000000`06faeea0 00007ffc`f631aaa8     wow64!Wow64LdrpInitialize+0x12d
07 00000000`06faf150 00007ffc`f631a993     ntdll!_LdrpInitialize+0xdc
08 00000000`06faf1d0 00007ffc`f631a8be     ntdll!LdrpInitializeInternal+0x6b
09 00000000`06faf450 00000000`00000000     ntdll!LdrInitializeThunk+0xe

   4  Id: 2fe0.2828 Suspend: 0 Teb: 00000000`00273000 Unfrozen
 # Child-SP          RetAddr               Call Site
00 00000000`076be5f8 00007ffc`f4eab95b     wow64win!NtGdiGetCharABCWidthsW+0x14
01 00000000`076be600 00007ffc`f5ae77ca     wow64win!whNtGdiGetCharABCWidthsW+0x2b
02 00000000`076be640 00000000`779b17ba     wow64!Wow64SystemServiceEx+0x15a
03 00000000`076bef00 00000000`779b1d75     wow64cpu!ServiceNoTurbo+0xb
04 00000000`076befb0 00007ffc`f5aee06d     wow64cpu!BTCpuSimulate+0xbb5
05 00000000`076beff0 00007ffc`f5aed8ad     wow64!RunCpuSimulation+0xd
06 00000000`076bf020 00007ffc`f631aaa8     wow64!Wow64LdrpInitialize+0x12d
07 00000000`076bf2d0 00007ffc`f631a993     ntdll!_LdrpInitialize+0xdc
08 00000000`076bf350 00007ffc`f631a8be     ntdll!LdrpInitializeInternal+0x6b
09 00000000`076bf5d0 00000000`00000000     ntdll!LdrInitializeThunk+0xe

   5  Id: 2fe0.2f9c Suspend: 0 Teb: 00000000`00277000 Unfrozen
 # Child-SP          RetAddr               Call Site
00 00000000`08ffe928 00000000`779b1b56     wow64cpu!CpupSyscallStub+0x13
```

```
01 00000000`08ffe930 00000000`779b1d75     wow64cpu!Thunk0ArgReloadState+0x5
02 00000000`08ffe9e0 00007ffc`f5aee06d     wow64cpu!BTCpuSimulate+0xbb5
03 00000000`08ffea20 00007ffc`f5aed8ad     wow64!RunCpuSimulation+0xd
04 00000000`08ffea50 00007ffc`f631aaa8     wow64!Wow64LdrpInitialize+0x12d
05 00000000`08ffed00 00007ffc`f631a993     ntdll!_LdrpInitialize+0xdc
06 00000000`08ffed80 00007ffc`f631a8be     ntdll!LdrpInitializeInternal+0x6b
07 00000000`08fff000 00000000`00000000     ntdll!LdrInitializeThunk+0xe
```

Note: We see familiar API function patterns but from different modules: *wow64*, *wow64win*, and *wow64cpu*.

6. Let's look at API internals.

```
0:000> uf wow64win!NtUserGetMessage
wow64win!NtUserGetMessage:
00007ffc`f4eb1410 4c8bd1           mov     r10,rcx
00007ffc`f4eb1413 b804100000       mov     eax,1004h
00007ffc`f4eb1418 f604250803fe7f01 test    byte ptr [SharedUserData+0x308 (00000000`7ffe0308)],1
00007ffc`f4eb1420 7503             jne     wow64win!NtUserGetMessage+0x15 (00007ffc`f4eb1425)  Branch

wow64win!NtUserGetMessage+0x12:
00007ffc`f4eb1422 0f05             syscall
00007ffc`f4eb1424 c3               ret

wow64win!NtUserGetMessage+0x15:
00007ffc`f4eb1425 cd2e             int     2Eh
00007ffc`f4eb1427 c3               ret
```

We see a syscall that is equivalent to one from the normal 64-bit *win32u* module we saw in the previous Exercise W6:

```
0:000> uf win32u!NtUserGetMessage
win32u!NtUserGetMessage:
00007ffc`f3811400 4c8bd1           mov     r10,rcx
00007ffc`f3811403 b804100000       mov     eax,1004h
00007ffc`f3811408 f604250803fe7f01 test    byte ptr [SharedUserData+0x308 (00000000`7ffe0308)],1
00007ffc`f3811410 7503             jne     win32u!NtUserGetMessage+0x15 (00007ffc`f3811415)  Branch

win32u!NtUserGetMessage+0x12:
00007ffc`f3811412 0f05             syscall
00007ffc`f3811414 c3               ret

win32u!NtUserGetMessage+0x15:
00007ffc`f3811415 cd2e             int     2Eh
00007ffc`f3811417 c3               ret
```

Note: If we try the last command in this memory dump, we see a different *win32u* module, a 32-bit one:

```
0:000> uf win32u!NtUserGetMessage
win32u!NtUserGetMessage:
00000000`758c10c0 b804100000       mov     eax,1004h
00000000`758c10c5 ba606c8c75       mov     edx,offset win32u!Wow64SystemServiceCall (00000000`758c6c60)
00000000`758c10ca ffd2             call    rdx
00000000`758c10cc c21000           ret     10h

0:000> lmv m win32u
Browse full module list
start             end                 module name
00000000`758c0000 00000000`758da000   win32u     (pdb symbols)
C:\WinDbg.Docker.AWAPI\mss\wwin32u.pdb\5119F0FF7995EF5AA74401515B724C001\wwin32u.pdb
    Loaded symbol image file: win32u.dll
    Image path: C:\Windows\SysWOW64\win32u.dll
```

```
    Image name: win32u.dll
    Browse all global symbols  functions  data
    Image was built with /Brepro flag.
    Timestamp:        8A05E54C (This is a reproducible build file hash, not a timestamp)
    CheckSum:         0001A691
    ImageSize:        0001A000
    File version:     10.0.22000.37
    Product version:  10.0.22000.37
    File flags:       0 (Mask 3F)
    File OS:          40004 NT Win32
    File type:        1.0 App
    File date:        00000000.00000000
    Translations:     0409.04b0
    Information from resource tables:
        CompanyName:      Microsoft Corporation
        ProductName:      Microsoft® Windows® Operating System
        InternalName:     Win32u
        OriginalFilename: Win32u.DLL
        ProductVersion:   10.0.22000.37
        FileVersion:      10.0.22000.37 (WinBuild.160101.0800)
        FileDescription:  Win32u
        LegalCopyright:   © Microsoft Corporation. All rights reserved.
```

7. So far, it looks like 32-bit functions and system calls are translated to their 64-bit syscalls (we can see usual *user32* and *kernel32* API patterns if we apply the **x** command to *wow64* and *wow64win* modules). Let's look at 32-bit stack traces (don't forget to reload symbols since we also have different 32-modules instead, loaded at 32-bit addresses, for example, *ntdll_779c0000*):

```
0:000> .effmach x86
Effective machine: x86 compatible (x86)

0:000:x86> .reload
................................................................
...........

************* Symbol Loading Error Summary **************
Module name            Error
SharedUserData         No error - symbol load deferred

You can troubleshoot most symbol related issues by turning on symbol loading diagnostics (!sym
noisy) and repeating the command that caused symbols to be loaded.
You should also verify that your symbol search path (.sympath) is correct.

0:000:x86> ~*kL

.  0  Id: 2fe0.2fe4 Suspend: 0 Teb: 00257000 Unfrozen
 # ChildEBP RetAddr
00 0013f86c 76bb0200     win32u!NtUserGetMessage+0xc
01 0013f8a8 74a18f35     user32!GetMessageW+0x30
02 0013f8c4 74a18fe3     mfc42u!CWinThread::PumpMessage+0x15
03 0013f8e0 749ea242     mfc42u!CWinThread::Run+0x63
04 0013f8f8 01046d0c     mfc42u!AfxWinMain+0xa2
05 0013f988 766e6739     wordpad!__wmainCRTStartup+0x153
06 0013f998 77a28e7f     kernel32!BaseThreadInitThunk+0x19
07 0013f9f0 77a28e4d     ntdll_779c0000!__RtlUserThreadStart+0x2b
08 0013fa00 00000000     ntdll_779c0000!_RtlUserThreadStart+0x1b

   1  Id: 2fe0.620 Suspend: 0 Teb: 00267000 Unfrozen
 # ChildEBP RetAddr
00 06eef9f4 75ea7fa3     ntdll_779c0000!NtWaitForMultipleObjects+0xc
```

```
01 06eefb88 75bc2b88     KERNELBASE!WaitForMultipleObjectsEx+0x133
02 06eefcc4 75bfe431     combase!WaitCoalesced+0xb4
03 06eefcf4 75bc2a6d     combase!CROIDTable::WorkerThreadLoop+0x51
04 06eefd20 75c0029f     combase!CRpcThread::WorkerLoop+0x113
05 06eefd30 766e6739     combase!CRpcThreadCache::RpcWorkerThreadEntry+0x1f
06 06eefd40 77a28e7f     kernel32!BaseThreadInitThunk+0x19
07 06eefd98 77a28e4d     ntdll_779c0000!__RtlUserThreadStart+0x2b
08 06eefda8 00000000     ntdll_779c0000!_RtlUserThreadStart+0x1b

   2  Id: 2fe0.748 Suspend: 0 Teb: 0026b000 Unfrozen
 # ChildEBP RetAddr
00 06f6f7f0 779f1d28     ntdll_779c0000!NtWaitForWorkViaWorkerFactory+0xc
01 06f6f9a4 766e6739     ntdll_779c0000!TppWorkerThread+0x338
02 06f6f9b4 77a28e7f     kernel32!BaseThreadInitThunk+0x19
03 06f6fa0c 77a28e4d     ntdll_779c0000!__RtlUserThreadStart+0x2b
04 06f6fa1c 00000000     ntdll_779c0000!_RtlUserThreadStart+0x1b

   3  Id: 2fe0.153c Suspend: 0 Teb: 0026f000 Unfrozen
 # ChildEBP RetAddr
00 06fef8b8 779f1d28     ntdll_779c0000!NtWaitForWorkViaWorkerFactory+0xc
01 06fefa6c 766e6739     ntdll_779c0000!TppWorkerThread+0x338
02 06fefa7c 77a28e7f     kernel32!BaseThreadInitThunk+0x19
03 06fefad4 77a28e4d     ntdll_779c0000!__RtlUserThreadStart+0x2b
04 06fefae4 00000000     ntdll_779c0000!_RtlUserThreadStart+0x1b

   4  Id: 2fe0.2828 Suspend: 0 Teb: 00273000 Unfrozen
 # ChildEBP RetAddr
00 076ff144 7770320b     win32u!NtGdiGetCharABCWidthsW+0xc
01 076ff480 7770316f     gdi32full!LoadGlyphMetricsWithGetCharABCWidthsI+0x3f
02 076ff4a0 776dd230     gdi32full!LoadGlyphMetrics+0x77
03 076ff4d4 776dd131     gdi32full!GetGlyphAdvanceWidths+0xc0
04 076ff4f8 74cac843     gdi32full!CUspShapingFont::GetGlyphDefaultAdvanceWidths+0x21
05 076ff5dc 776dcd55     TextShaping!ShapingGetGlyphPositions+0x3a3
06 076ff728 776dbb41     gdi32full!ShlPlaceOT+0x315
07 076ff7c8 776db40b     gdi32full!RenderItemNoFallback+0x421
08 076ff808 776db2a8     gdi32full!RenderItemWithFallback+0x12b
09 076ff830 776db02e     gdi32full!RenderItem+0x28
0a 076ff880 776d8f0b     gdi32full!ScriptStringAnalyzeGlyphs+0x1be
0b 076ff984 776d63ab     gdi32full!ScriptStringAnalyse+0x7fb
0c 076ffaf0 776d5c10     gdi32full!LpkCharsetDraw+0x53b
0d 076ffb1c 76bace32     gdi32full!LpkDrawTextEx+0x30
0e 076ffb78 76bbd527     user32!DT_DrawStr+0x61
0f 076ffba8 76bbd45d     user32!DT_DrawJustifiedLine+0x31
10 076ffcf8 76bac744     user32!AddEllipsisAndDrawLine+0xa1
11 076ffdb8 76bac35f     user32!DrawTextExWorker+0x3c4
12 076ffdf4 73d64b8a     user32!DrawTextW+0x3f
13 076ffee0 73d6119c     UIRibbon!CreateFontBitmap+0x16e
14 076fff14 73d61278     UIRibbon!CFontList::_GenerateBitmaps+0x7b
15 076fff2c 73d612cf     UIRibbon!CFontList::_LoadFonts+0x86
16 076fff38 766e6739     UIRibbon!CFontList::s_LoadFontsThreadProc+0x1f
17 076fff48 77a28e7f     kernel32!BaseThreadInitThunk+0x19
18 076fffa0 77a28e4d     ntdll_779c0000!__RtlUserThreadStart+0x2b
19 076fffb0 00000000     ntdll_779c0000!_RtlUserThreadStart+0x1b

   5  Id: 2fe0.2f9c Suspend: 0 Teb: 00277000 Unfrozen
 # ChildEBP RetAddr
00 0934fde0 75e9f649     ntdll_779c0000!NtWaitForSingleObject+0xc
01 0934fe54 75e9f5a2     KERNELBASE!WaitForSingleObjectEx+0x99
02 0934fe68 72d7ab3a     KERNELBASE!WaitForSingleObject+0x12
```

```
03 0934fea0 766e6739      winspool!MonitorRPCServerProcess+0x1a
04 0934feb0 77a28e7f      kernel32!BaseThreadInitThunk+0x19
05 0934ff08 77a28e4d      ntdll_779c0000!__RtlUserThreadStart+0x2b
06 0934ff18 00000000      ntdll_779c0000!_RtlUserThreadStart+0x1b
```

8. To switch back to native 64-bit mode, use this command:

```
0:000:x86> .effmach AMD64
Effective machine: x64 (AMD64)
```

9. We close logging before exiting WinDbg:

```
0:000> .logclose
Closing open log file C:\AWAPI-Dumps\W7.log
```

API and Errors

- Windows protocols
- Windows error codes reference
- Thread Information Block
- Win32 values
- NTSTATUS values
- HRESULT values

Windows API calls may return errors. These are well-documented for individual API functions. However, we may be interested in the overall collection of error values and their meanings since we may find errors in raw stack data (for example, **Execution Residue** memory analysis pattern). Nowadays, Microsoft provides documentation on various protocols, for example, for remote desktops. Since protocols use Windows API, they return their errors, and Microsoft provides their documentation as a part of general protocol documentation. There are 3 classes of errors: **Win32** (like legacy MS-DOS errors), **NTSTATUS** values that are usually returned by kernel system calls and propagated by various drivers, and **HRESULT** values that were introduced by COM if my memory doesn't fail me and allow extension by 3rd-party vendors from their components. Some Windows API functions set Win32 errors only, for example, if higher layers determine invalid parameters, and some set both Win32 and NTSTATUS, for example, from lower layers, after return from syscalls. The last error values are set in the so-called **Thread Information Block** (or Thread Environment Block, **TEB**). Each thread has its own pointer to **TEB**. We see that in our next exercise.

Windows protocols
https://learn.microsoft.com/en-us/openspecs/windows_protocols/MS-WINPROTLP/

Windows error codes reference
https://learn.microsoft.com/en-us/openspecs/windows_protocols/ms-erref/

Thread Information Block
https://en.wikipedia.org/wiki/Win32_Thread_Information_Block

Win32 values
https://learn.microsoft.com/en-us/openspecs/windows_protocols/ms-erref/18d8fbe8-a967-4f1c-ae50-99ca8e491d2d

NTSTATUS values
https://learn.microsoft.com/en-us/openspecs/windows_protocols/ms-erref/87fba13e-bf06-450e-83b1-9241dc81e781

HRESULT values
https://learn.microsoft.com/en-us/openspecs/windows_protocols/ms-erref/0642cb2f-2075-4469-918c-4441e69c548a

Exercise W8

- **Goal:** Explore different Windows API error types
- **Memory Analysis Patterns:** Last Error Collection
- **ADDR Patterns:** Function Skeleton; Call Path; Structure Field
- \AWAPI-Dumps\Exercise-W8.pdf

Exercise W8

Goal: Explore different Windows API error types.

Memory Analysis Patterns: Last Error Collection.

ADDR Patterns: Function Skeleton; Call Path; Structure Field.

1. Launch WinDbg.

2. Open \AWAPI-Dumps\Process\wordpad.DMP

3. We get the dump file loaded:

```
Microsoft (R) Windows Debugger Version 10.0.27725.1000 AMD64
Copyright (c) Microsoft Corporation. All rights reserved.


Loading Dump File [C:\AWAPI-Dumps\Process\wordpad.DMP]
User Mini Dump File with Full Memory: Only application data is available


************* Path validation summary **************
Response                         Time (ms)     Location
Deferred                                       srv*
Symbol search path is: srv*
Executable search path is:
Windows 10 Version 22000 MP (2 procs) Free x64
Product: WinNt, suite: SingleUserTS Personal
Edition build lab: 22000.1.amd64fre.co_release.210604-1628
Debug session time: Sat Oct 15 20:19:05.000 2022 (UTC + 0:00)
System Uptime: 0 days 0:27:57.202
Process Uptime: 0 days 0:00:26.000
................................................................
.....
Loading unloaded module list
.
For analysis of this file, run !analyze -v
win32u!NtUserGetMessage+0x14:
00007ffc`f3811414 ret
```

4. Open a log file using **.logopen**:

```
0:000> .logopen C:\AWAPI-Dumps\W8.log
Opened log file 'C:\AWAPI-Dumps\W8.log'
```

5. Let's look at the *CreateWindowExW* function skeleton and its call path, if necessary, to see what type of error it may set:

```
0:000> uf /c user32!CreateWindowExW
user32!CreateWindowExW (00007ffc`f5848030)
  user32!CreateWindowExW+0x7d (00007ffc`f58480ad):
    call to user32!CreateWindowInternal (00007ffc`f58480c4)
```

```
0:000> uf /c user32!CreateWindowInternal
user32!CreateWindowInternal (00007ffc`f58480c4)
  user32!CreateWindowInternal+0x80 (00007ffc`f5848144):
    call to ntdll!LdrpDispatchUserCallTarget (00007ffc`f63333c0)
  user32!CreateWindowInternal+0x116 (00007ffc`f58481da):
    call to user32!RtlInitLargeUnicodeString (00007ffc`f5848440)
  user32!CreateWindowInternal+0x139 (00007ffc`f58481fd):
    call to user32!RtlInitLargeUnicodeString (00007ffc`f5848440)
  user32!CreateWindowInternal+0x1b8 (00007ffc`f584827c):
    call to user32!VerNtUserCreateWindowEx (00007ffc`f5848480)
  user32!CreateWindowInternal+0x1fe (00007ffc`f58482c2):
    call to KERNELBASE!GetModuleHandleW (00007ffc`f3bab410)
  user32!CreateWindowInternal+0x21f (00007ffc`f58482e3):
    call to user32!RtlCaptureLargeAnsiString (00007ffc`f586b3ec)
  user32!CreateWindowInternal+0x248 (00007ffc`f584830c):
    call to user32!RtlInitLargeAnsiString (00007ffc`f586be04)
  user32!CreateWindowInternal+0x25f (00007ffc`f5848323):
    call to ntdll!RtlFreeHeap (00007ffc`f62c75e0)
  user32!CreateWindowInternal+0x277 (00007ffc`f584833b):
    call to user32!ValidateHwnd (00007ffc`f58505d0)
  user32!CreateWindowInternal+0x30e (00007ffc`f58483d2):
    call to user32!CreateMDIChild (00007ffc`f586f874)
  user32!CreateWindowInternal+0x360 (00007ffc`f5848424):
    call to user32!MDICompleteChildCreation (00007ffc`f5870364)
  user32!CreateWindowInternal+0x322fe (00007ffc`f587a3c2):
    call to ntdll!RtlSetLastWin32Error (00007ffc`f62a67c0)
```

We see that the Win32 error type value is set. Let’s look at the function implementation:

```
0:000> uf ntdll!RtlSetLastWin32Error
ntdll!RtlSetLastWin32Error:
00007ffc`f62a67c0 894c2408         mov     dword ptr [rsp+8],ecx
00007ffc`f62a67c4 4883ec48         sub     rsp,48h
00007ffc`f62a67c8 488b0541bd1800   mov     rax,qword ptr [ntdll!_security_cookie (00007ffc`f6432510)]
00007ffc`f62a67cf 4833c4           xor     rax,rsp
00007ffc`f62a67d2 4889442430       mov     qword ptr [rsp+30h],rax
00007ffc`f62a67d7 65488b042530000000 mov    rax,qword ptr gs:[30h]
00007ffc`f62a67e0 8b15b64e1700     mov     edx,dword ptr [ntdll!g_dwLastErrorToBreakOn (00007ffc`f641b69c)]
00007ffc`f62a67e6 8b4c2450         mov     ecx,dword ptr [rsp+50h]
00007ffc`f62a67ea 85d2             test    edx,edx
00007ffc`f62a67ec 7538             jne     ntdll!RtlSetLastWin32Error+0x66 (00007ffc`f62a6826)  Branch

ntdll!RtlSetLastWin32Error+0x2e:
00007ffc`f62a67ee 394868           cmp     dword ptr [rax+68h],ecx
00007ffc`f62a67f1 7513             jne     ntdll!RtlSetLastWin32Error+0x46 (00007ffc`f62a6806)  Branch

ntdll!RtlSetLastWin32Error+0x33:
00007ffc`f62a67f3 488b4c2430       mov     rcx,qword ptr [rsp+30h]
00007ffc`f62a67f8 4833cc           xor     rcx,rsp
00007ffc`f62a67fb e820c70800       call    ntdll!_security_check_cookie (00007ffc`f6332f20)
00007ffc`f62a6800 4883c448         add     rsp,48h
00007ffc`f62a6804 c3               ret

ntdll!RtlSetLastWin32Error+0x46:
00007ffc`f62a6806 894868           mov     dword ptr [rax+68h],ecx
00007ffc`f62a6809 8b442450         mov     eax,dword ptr [rsp+50h]
00007ffc`f62a680d 85c0             test    eax,eax
00007ffc`f62a680f 74e2             je      ntdll!RtlSetLastWin32Error+0x33 (00007ffc`f62a67f3)  Branch

ntdll!RtlSetLastWin32Error+0x51:
00007ffc`f62a6811 803df447170000   cmp     byte ptr [ntdll!g_isErrorOriginProviderEnabled (00007ffc`f641b00c)],0
00007ffc`f62a6818 74d9             je      ntdll!RtlSetLastWin32Error+0x33 (00007ffc`f62a67f3)  Branch

ntdll!RtlSetLastWin32Error+0x5a:
00007ffc`f62a681a 3de5030000       cmp     eax,3E5h
00007ffc`f62a681f 74d2             je      ntdll!RtlSetLastWin32Error+0x33 (00007ffc`f62a67f3)  Branch
```

```
ntdll!RtlSetLastWin32Error+0x61:
00007ffc`f62a6821 e956bd0a00      jmp     ntdll!RtlSetLastWin32Error+0xabdbc (00007ffc`f635257c)  Branch

ntdll!RtlSetLastWin32Error+0x66:
00007ffc`f62a6826 3bca            cmp     ecx,edx
00007ffc`f62a6828 75c4            jne     ntdll!RtlSetLastWin32Error+0x2e (00007ffc`f62a67ee)  Branch

ntdll!RtlSetLastWin32Error+0x6a:
00007ffc`f62a682a cc              int     3
00007ffc`f62a682b ebc1            jmp     ntdll!RtlSetLastWin32Error+0x2e (00007ffc`f62a67ee)  Branch

ntdll!RtlSetLastWin32Error+0xabdbc:
00007ffc`f635257c 488b0ddd660c00  mov     rcx,qword ptr [ntdll!g_hUserDiagnosticProvider (00007ffc`f6418c60)]
00007ffc`f6352583 488d442450      lea     rax,[rsp+50h]
00007ffc`f6352588 4c8d4c2420      lea     r9,[rsp+20h]
00007ffc`f635258d 4889442420      mov     qword ptr [rsp+20h],rax
00007ffc`f6352592 41b801000000    mov     r8d,1
00007ffc`f6352598 48c74424280400000 mov   qword ptr [rsp+28h],4
00007ffc`f63525a1 488d1528fc0800  lea     rdx,[ntdll!SetLastWin32ErrorEvent (00007ffc`f63e21d0)]
00007ffc`f63525a8 e8433cf5ff      call    ntdll!EtwEventWrite (00007ffc`f62a61f0)
00007ffc`f63525ad 90              nop
00007ffc`f63525ae e94042f5ff      jmp     ntdll!RtlSetLastWin32Error+0x33 (00007ffc`f62a67f3)  Branch
```

We see that the address of _TIB is used to address its field at the 0x68 offset. And it is *LastErrorValue*:

```
0:000> dt ntdll!_TEB LastErrorValue
   +0x068 LastErrorValue : Uint4B
```

We can look at the *CreateWindowExW* function implementation to see what possible error may be saved:

```
0:000> ub user32!CreateWindowInternal+0x322fe
user32!RegisterClassExWOWW+0x32749:
00007ffc`f587a3a1 cc              int     3
00007ffc`f587a3a2 0fbaf30b        btr     ebx,0Bh
00007ffc`f587a3a6 0fbaeb07        bts     ebx,7
00007ffc`f587a3aa e95fddfcff      jmp     user32!CreateWindowInternal+0x4a (00007ffc`f584810e)
00007ffc`f587a3af b900040000      mov     ecx,400h
00007ffc`f587a3b4 663bc1          cmp     ax,cx
00007ffc`f587a3b7 0f82a5ddfcff    jb      user32!CreateWindowInternal+0x9e (00007ffc`f5848162)
00007ffc`f587a3bd b957000000      mov     ecx,57h

0:000> !error 0x57
Error code: (Win32) 0x57 (87) - The parameter is incorrect.
```

6. Let’s look at another API function skeleton, *ReadFile*:

```
0:000> uf /c ReadFile
KERNELBASE!ReadFile (00007ffc`f3b64420)
  KERNELBASE!ReadFile+0x74 (00007ffc`f3b64494):
    call to ntdll!NtReadFile (00007ffc`f6343800)
  KERNELBASE!ReadFile+0x101 (00007ffc`f3b64521):
    call to ntdll!NtReadFile (00007ffc`f6343800)
  KERNELBASE!ReadFile+0x11a (00007ffc`f3b6453a):
    call to KERNELBASE!BaseSetLastNTError (00007ffc`f3b64610)
  KERNELBASE!ReadFile+0x18f (00007ffc`f3b645af):
    call to KERNELBASE!BaseSetLastNTError (00007ffc`f3b64610)
  KERNELBASE!ReadFile+0x9d06c (00007ffc`f3c0148c):
    call to ntdll!NtWaitForSingleObject (00007ffc`f63437c0)
```

It calls a different function to set an error code, NTSTATUS. Let's look at its skeleton and implementation:

```
0:000> uf KERNELBASE!BaseSetLastNTError
KERNELBASE!BaseSetLastNTError:
00007ffc`f3b64610 4053            push    rbx
00007ffc`f3b64612 4883ec20        sub     rsp,20h
00007ffc`f3b64616 48ff1523802000  call    qword ptr [KERNELBASE!_imp_RtlNtStatusToDosError (00007ffc`f3d6c640)]
00007ffc`f3b6461d 0f1f440000      nop     dword ptr [rax+rax]
00007ffc`f3b64622 8bc8            mov     ecx,eax
00007ffc`f3b64624 8bd8            mov     ebx,eax
00007ffc`f3b64626 48ff154b7f2000  call    qword ptr [KERNELBASE!_imp_RtlSetLastWin32Error (00007ffc`f3d6c578)]
00007ffc`f3b6462d 0f1f440000      nop     dword ptr [rax+rax]
00007ffc`f3b64632 8bc3            mov     eax,ebx
00007ffc`f3b64634 4883c420        add     rsp,20h
00007ffc`f3b64638 5b              pop     rbx
00007ffc`f3b64639 c3              ret
```

```
0:000> uf /c KERNELBASE!BaseSetLastNTError
KERNELBASE!BaseSetLastNTError (00007ffc`f3b64610)
  KERNELBASE!BaseSetLastNTError+0x6 (00007ffc`f3b64616):
    call to ntdll!RtlNtStatusToDosError (00007ffc`f62a6840)
  KERNELBASE!BaseSetLastNTError+0x16 (00007ffc`f3b64626):
    call to ntdll!RtlSetLastWin32Error (00007ffc`f62a67c0)
```

We see it converts an NTSTATUS value to a Win32 error (called a legacy DOS error) and then sets the Win32 error value as we saw previously. Let's look at where an NTSTATUS error is saved:

```
0:000> uf ntdll!RtlNtStatusToDosError
ntdll!RtlNtStatusToDosError:
00007ffc`f62a6840 894c2408        mov     dword ptr [rsp+8],ecx
00007ffc`f62a6844 53              push    rbx
00007ffc`f62a6845 4883ec20        sub     rsp,20h
00007ffc`f62a6849 8bd1            mov     edx,ecx
00007ffc`f62a684b 65488b042530000000 mov  rax,qword ptr gs:[30h]
00007ffc`f62a6854 4885c0          test    rax,rax
00007ffc`f62a6857 740c            je      ntdll!RtlNtStatusToDosError+0x25 (00007ffc`f62a6865)  Branch

ntdll!RtlNtStatusToDosError+0x19:
00007ffc`f62a6859 898850120000    mov     dword ptr [rax+1250h],ecx
00007ffc`f62a685f eb04            jmp     ntdll!RtlNtStatusToDosError+0x25 (00007ffc`f62a6865)  Branch

ntdll!RtlNtStatusToDosError+0x25:
00007ffc`f62a6865 85d2            test    edx,edx
00007ffc`f62a6867 0f84f1000000    je      ntdll!RtlNtStatusToDosError+0x11e (00007ffc`f62a695e)  Branch

ntdll!RtlNtStatusToDosError+0x2d:
00007ffc`f62a686d 81fa03010000    cmp     edx,103h
00007ffc`f62a6873 0f84ee000000    je      ntdll!RtlNtStatusToDosError+0x127 (00007ffc`f62a6967)  Branch

ntdll!RtlNtStatusToDosError+0x39:
00007ffc`f62a6879 8bc2            mov     eax,edx
00007ffc`f62a687b 0fbae21d        bt      edx,1Dh
00007ffc`f62a687f 0f82d2000000    jb      ntdll!RtlNtStatusToDosError+0x117 (00007ffc`f62a6957)  Branch

ntdll!RtlNtStatusToDosError+0x45:
00007ffc`f62a6885 250000ff00      and     eax,0FF0000h
00007ffc`f62a688a 3d00000700      cmp     eax,70000h
00007ffc`f62a688f 0f84fc000000    je      ntdll!RtlNtStatusToDosError+0x151 (00007ffc`f62a6991)  Branch

ntdll!RtlNtStatusToDosError+0x55:
00007ffc`f62a6895 8bc2            mov     eax,edx
00007ffc`f62a6897 25000000f0      and     eax,0F0000000h
00007ffc`f62a689c 3d000000d0      cmp     eax,0D0000000h
00007ffc`f62a68a1 0f840dbd0a00    je      ntdll!RtlNtStatusToDosError+0xabd74 (00007ffc`f63525b4)  Branch
```

```
ntdll!RtlNtStatusToDosError+0x67:
00007ffc`f62a68a7 33c0            xor     eax,eax
00007ffc`f62a68a9 41b93a010000    mov     r9d,13Ah
00007ffc`f62a68af 488d1d4a97ffff  lea     rbx,[ntdll!LdrpGetModuleName <PERF> (ntdll+0x0)
(00007ffc`f62a0000)]
00007ffc`f62a68b6 66660f1f840000000000 nop word ptr [rax+rax]

ntdll!RtlNtStatusToDosError+0x80:
00007ffc`f62a68c0 458d0401        lea     r8d,[r9+rax]
00007ffc`f62a68c4 41d1e8          shr     r8d,1
00007ffc`f62a68c7 428b8cc370bd1300 mov     ecx,dword ptr [rbx+r8*8+13BD70h]
00007ffc`f62a68cf 448bd2          mov     r10d,edx
00007ffc`f62a68d2 442bd1          sub     r10d,ecx
00007ffc`f62a68d5 3bd1            cmp     edx,ecx
00007ffc`f62a68d7 734b            jae     ntdll!RtlNtStatusToDosError+0xe4 (00007ffc`f62a6924)  Branch

ntdll!RtlNtStatusToDosError+0x99:
00007ffc`f62a68d9 458d48ff        lea     r9d,[r8-1]

ntdll!RtlNtStatusToDosError+0x9d:
00007ffc`f62a68dd 413bc1          cmp     eax,r9d
00007ffc`f62a68e0 76de            jbe     ntdll!RtlNtStatusToDosError+0x80 (00007ffc`f62a68c0)  Branch

ntdll!RtlNtStatusToDosError+0xa2:
00007ffc`f62a68e2 8bc2            mov     eax,edx
00007ffc`f62a68e4 250000ffff      and     eax,0FFFF0000h
00007ffc`f62a68e9 3d000001c0      cmp     eax,0C0010000h
00007ffc`f62a68ee 0f84b0000000    je      ntdll!RtlNtStatusToDosError+0x164 (00007ffc`f62a69a4)  Branch

ntdll!RtlNtStatusToDosError+0xb4:
00007ffc`f62a68f4 488d0d15a11200  lea     rcx,[ntdll!`string' (00007ffc`f63d0a10)]
00007ffc`f62a68fb e8c0000000      call    ntdll!DbgPrint (00007ffc`f62a69c0)
00007ffc`f62a6900 488d0dd1a01200  lea     rcx,[ntdll!`string' (00007ffc`f63d09d8)]
00007ffc`f62a6907 e8b4000000      call    ntdll!DbgPrint (00007ffc`f62a69c0)
00007ffc`f62a690c 488d0d95a01200  lea     rcx,[ntdll!`string' (00007ffc`f63d09a8)]
00007ffc`f62a6913 e8a8000000      call    ntdll!DbgPrint (00007ffc`f62a69c0)
00007ffc`f62a6918 b83d010000      mov     eax,13Dh
00007ffc`f62a691d 4883c420        add     rsp,20h
00007ffc`f62a6921 5b              pop     rbx
00007ffc`f62a6922 c3              ret

ntdll!RtlNtStatusToDosError+0xe4:
00007ffc`f62a6924 420fb684c374bd1300 movzx eax,byte ptr [rbx+r8*8+13BD74h]
00007ffc`f62a692d 443bd0          cmp     r10d,eax
00007ffc`f62a6930 7206            jb      ntdll!RtlNtStatusToDosError+0xf8 (00007ffc`f62a6938)  Branch

ntdll!RtlNtStatusToDosError+0xf2:
00007ffc`f62a6932 418d4001        lea     eax,[r8+1]
00007ffc`f62a6936 eba5            jmp     ntdll!RtlNtStatusToDosError+0x9d (00007ffc`f62a68dd)  Branch

ntdll!RtlNtStatusToDosError+0xf8:
00007ffc`f62a6938 420fb784c376bd1300 movzx eax,word ptr [rbx+r8*8+13BD76h]
00007ffc`f62a6941 4280bcc375bd130001 cmp   byte ptr [rbx+r8*8+13BD75h],1
00007ffc`f62a694a 7527            jne     ntdll!RtlNtStatusToDosError+0x133 (00007ffc`f62a6973)  Branch

ntdll!RtlNtStatusToDosError+0x10c:
00007ffc`f62a694c 4103c2          add     eax,r10d
00007ffc`f62a694f 0fb78443709e1300 movzx   eax,word ptr [rbx+rax*2+139E70h]

ntdll!RtlNtStatusToDosError+0x117:
00007ffc`f62a6957 4883c420        add     rsp,20h
00007ffc`f62a695b 5b              pop     rbx
00007ffc`f62a695c c3              ret
```

```
ntdll!RtlNtStatusToDosError+0x11e:
00007ffc`f62a695e 33c0            xor     eax,eax
00007ffc`f62a6960 4883c420        add     rsp,20h
00007ffc`f62a6964 5b              pop     rbx
00007ffc`f62a6965 c3              ret

ntdll!RtlNtStatusToDosError+0x127:
00007ffc`f62a6967 b8e5030000      mov     eax,3E5h
00007ffc`f62a696c 4883c420        add     rsp,20h
00007ffc`f62a6970 5b              pop     rbx
00007ffc`f62a6971 c3              ret

ntdll!RtlNtStatusToDosError+0x133:
00007ffc`f62a6973 428d1450        lea     edx,[rax+r10*2]
00007ffc`f62a6977 8d4201          lea     eax,[rdx+1]
00007ffc`f62a697a 0fb78443709e1300 movzx  eax,word ptr [rbx+rax*2+139E70h]
00007ffc`f62a6982 c1e010          shl     eax,10h
00007ffc`f62a6985 0fb78c53709e1300 movzx  ecx,word ptr [rbx+rdx*2+139E70h]
00007ffc`f62a698d 0bc1            or      eax,ecx
00007ffc`f62a698f ebc6            jmp     ntdll!RtlNtStatusToDosError+0x117 (00007ffc`f62a6957)  Branch

ntdll!RtlNtStatusToDosError+0x151:
00007ffc`f62a6991 8bc2            mov     eax,edx
00007ffc`f62a6993 c1e818          shr     eax,18h
00007ffc`f62a6996 83c080          add     eax,0FFFFFF80h
00007ffc`f62a6999 a9bfffffff      test    eax,0FFFFFFBFh
00007ffc`f62a699e 0f85f1feffff    jne     ntdll!RtlNtStatusToDosError+0x55 (00007ffc`f62a6895)  Branch

ntdll!RtlNtStatusToDosError+0x164:
00007ffc`f62a69a4 0fb7c2          movzx   eax,dx
00007ffc`f62a69a7 4883c420        add     rsp,20h
00007ffc`f62a69ab 5b              pop     rbx
00007ffc`f62a69ac c3              ret
```

We see that the address of _TIB is used to address its field at the 0x1250 offset. And it is *LastStatusValue*:

```
0:000> dt ntdll!_TEB LastStatusValue
   +0x1250 LastStatusValue : Uint4B
```

Note: We see that some functions set Win32 errors, and some set both. So, there may be a discrepancy between the last Win32 and status errors when we look at their values across threads:

```
0:000> ~*e !gle
LastErrorValue: (Win32) 0 (0) - The operation completed successfully.
LastStatusValue: (NTSTATUS) 0 - STATUS_SUCCESS
LastErrorValue: (Win32) 0 (0) - The operation completed successfully.
LastStatusValue: (NTSTATUS) 0 - STATUS_SUCCESS
LastErrorValue: (Win32) 0 (0) - The operation completed successfully.
LastStatusValue: (NTSTATUS) 0 - STATUS_SUCCESS
LastErrorValue: (Win32) 0x36b7 (14007) - The requested lookup key was not found in any active activation context.
LastStatusValue: (NTSTATUS) 0xc0150008 - The requested lookup key was not found in any active activation context.
LastErrorValue: (Win32) 0 (0) - The operation completed successfully.
LastStatusValue: (NTSTATUS) 0xc000000d - An invalid parameter was passed to a service or function.
LastErrorValue: (Win32) 0 (0) - The operation completed successfully.
LastStatusValue: (NTSTATUS) 0xc000000d - An invalid parameter was passed to a service or function.
LastErrorValue: (Win32) 0 (0) - The operation completed successfully.
LastStatusValue: (NTSTATUS) 0xc000000d - An invalid parameter was passed to a service or function.
LastErrorValue: (Win32) 0 (0) - The operation completed successfully.
LastStatusValue: (NTSTATUS) 0xc0000034 - Object Name not found.
```

7. We close logging before exiting WinDbg:

```
0:000> .logclose
Closing open log file C:\AWAPI-Dumps\W8.log
```

API and Functional Programming

- Referential transparency
- IsCharLower

```
IsCharLowerA('a'), IsCharLowerA('a') → true, true
```

- Side effects
- ReadFile

```
ReadFile(hFile, buf, len, &num, nullptr),
ReadFile(hFile, buf, len, &num, nullptr) →
    true, false (*buf != *buf, num != num)
```


With the rise of functional programming during the last decade and due to my own experience with FP, I put this new slide that shows the relation of API to referential transparency and side effects. In summary, referential transparency allows the substitution of expressions with their value. This requires that the result of the expression is the same for the same input values and that there are no side effects, modifications of memory, or I/O outside of the expression. For example, the result of the *IsCharLowerA* function is the same for the character, and no memory is modified outside. In contrast, the *ReadFile* function not only may have a different output for the same parameter values but also modifies memory outside with unknown values every time it is called.

Referential transparency
https://en.wikipedia.org/wiki/Referential_transparency

Side effects
https://en.wikipedia.org/wiki/Side_effect_(computer_science)

API and Security

- Maliciousness: What, When, Where
  - SetWindowsHookExW
  - CreateRemoteThread
- Vulnerability: How
  - SAST
  - Static code analysis tools

It is difficult to put all security issues with API on just one slide. So, I attempted to make a view of the surface from a distant star. There are two main aspects: maliciousness of API usage (for example, *SetWindowsHookExW* for keylogging and *CreateRemoteThread* for code injection) and how the API is used that makes the code vulnerable (for example, a wrong buffer size).

SetWindowsHookExW
https://learn.microsoft.com/en-us/windows/win32/api/winuser/nf-winuser-setwindowshookexw

CreateRemoteThread
https://learn.microsoft.com/en-us/windows/win32/api/processthreadsapi/nf-processthreadsapi-createremotethread

SAST
https://en.wikipedia.org/wiki/Static_application_security_testing

Static code analysis tools
https://owasp.org/www-community/Source_Code_Analysis_Tools

API and Versioning

Ex-suffix, longer descriptive function names

CopyFile/CopyFileEx

CreateThread/CreateRemoteThread

A few words about API versioning in case of adding extensions. In Windows, this is often done via the Ex-suffix or by creating longer descriptive names.

API and Unicode

- Character parameters and buffers

```
HICON LoadIconA(
  [in, optional] HINSTANCE hInstance,
  [in]           LPCSTR    lpIconName
);

0:000> uf /c LoadIconA
user32!LoadIconA (00007ffc`f586c2e0)
  user32!LoadIconA+0x2a (00007ffc`f586c30a):
    call to user32!LoadIcoCur (00007ffc`f584cae4)
  user32!LoadIconA+0x1b7f6 (00007ffc`f5887ad6):
    call to user32!MBToWCSEx (00007ffc`f5865d60)
  user32!LoadIconA+0x1b822 (00007ffc`f5887b02):
    call to user32!LoadIcoCur (00007ffc`f584cae4)
  user32!LoadIconA+0x1b838 (00007ffc`f5887b18):
    call to ntdll!RtlFreeHeap (00007ffc`f62c75e0)
```

```
HICON LoadIconW(
  [in, optional] HINSTANCE hInstance,
  [in]           LPCWSTR   lpIconName
);

0:000> uf /c LoadIconW
user32!LoadIconW (00007ffc`f5849760)
  user32!LoadIconW+0x18 (00007ffc`f5849778):
    call to user32!LoadIcoCur (00007ffc`f584cae4)
```

- No character buffers: `ReadFile`

Most Windows API functions that accept characters or character buffers come in two variants: **A**SCII and 16-bit Unicode (**W**ide characters). Internally, Windows works with wide characters, so the **A** Windows API functions usually convert ASCII or multi-byte strings into their wide character format before calling the **W**-variants of Windows API or internal functions and, vice versa, converting output parameters.

Windows API Formalization

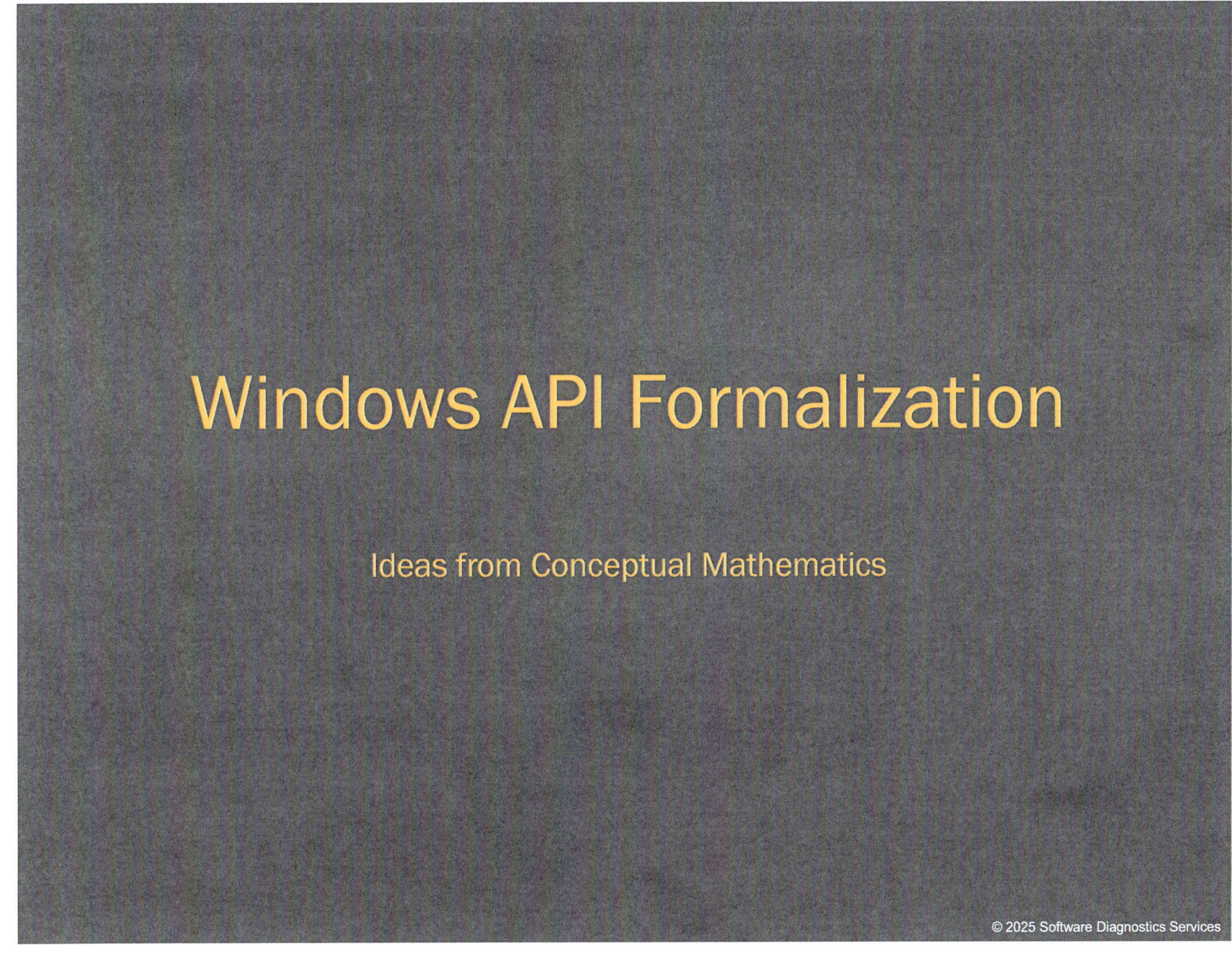
Windows API Formalization
Ideas from Conceptual Mathematics
© 2025 Software Diagnostics Services

Now we have a break. We look at possible API formalization using ideas from conceptual mathematics, such as category theory. The exposition is informal. If you already know category theory, you find another application or even provide objections to the approach if you see it wrong. However, if you don't know category theory, you learn its basic notions and terminology and perhaps apply it to your own areas of interest.

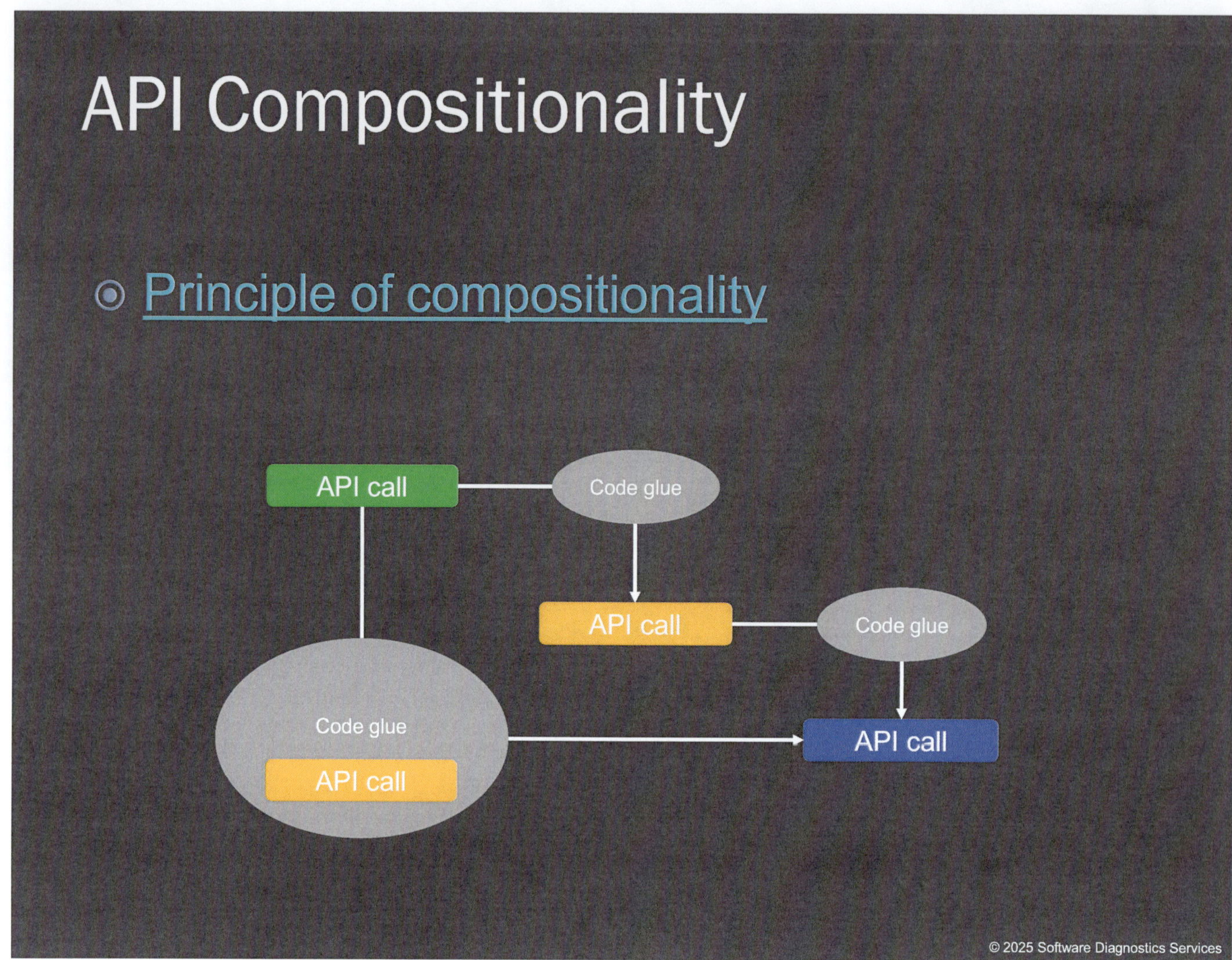

The main principle that allows us to use category theory that I introduce next is the so-called principle of compositionality. We can compose various API calls together via code glue. It's also possible to connect two API calls by some code, including other API calls if necessary. Here, we represent API calls as some objects and code glue as directed arrows.

Principle of compositionality
https://en.wikipedia.org/wiki/Principle_of_compositionality

Category Theory Language

- Category
  - Objects
  - Arrows between objects (must be transitive, if A → B and B → C then A → C)
- Functor
  - Arrow between categories (can be the same category)
  - Maps objects to objects and arrows to arrows
- Natural Transformation
  - Arrows between functors in a category of functors
- Adjunction
  - Relationship between functors, change of perspective, back translation

Category theory allows us to relate different areas analogically via their common structure and behavior. It was introduced in mathematics in the middle of the 20th century. And it has its own language. The three main concepts are **categories** themself, **functors**, and **natural transformations**. To this, I add the so-called **adjunction**. The definitions I give here are all informal but suitable for our application to API. A category consists of objects and arrows between objects. However, arrows, if they exist between objects, must be composable. There are other restrictions and axioms that I omit for our informal purposes. Categories themselves can be considered objects. Arrows between them are called functors. Functors map objects to objects and arrows to arrows between source and target categories. Functors can be considered as objects themselves in a category of functors. Arrows between them are called natural transformations. The 4th concept, adjunction, is a pair of functors, called left and right adjoints, that are in a special relationship to each other that allows changing of perspective when traveling back and forth between categories. We now make all these 4 concepts visual on the next slide.

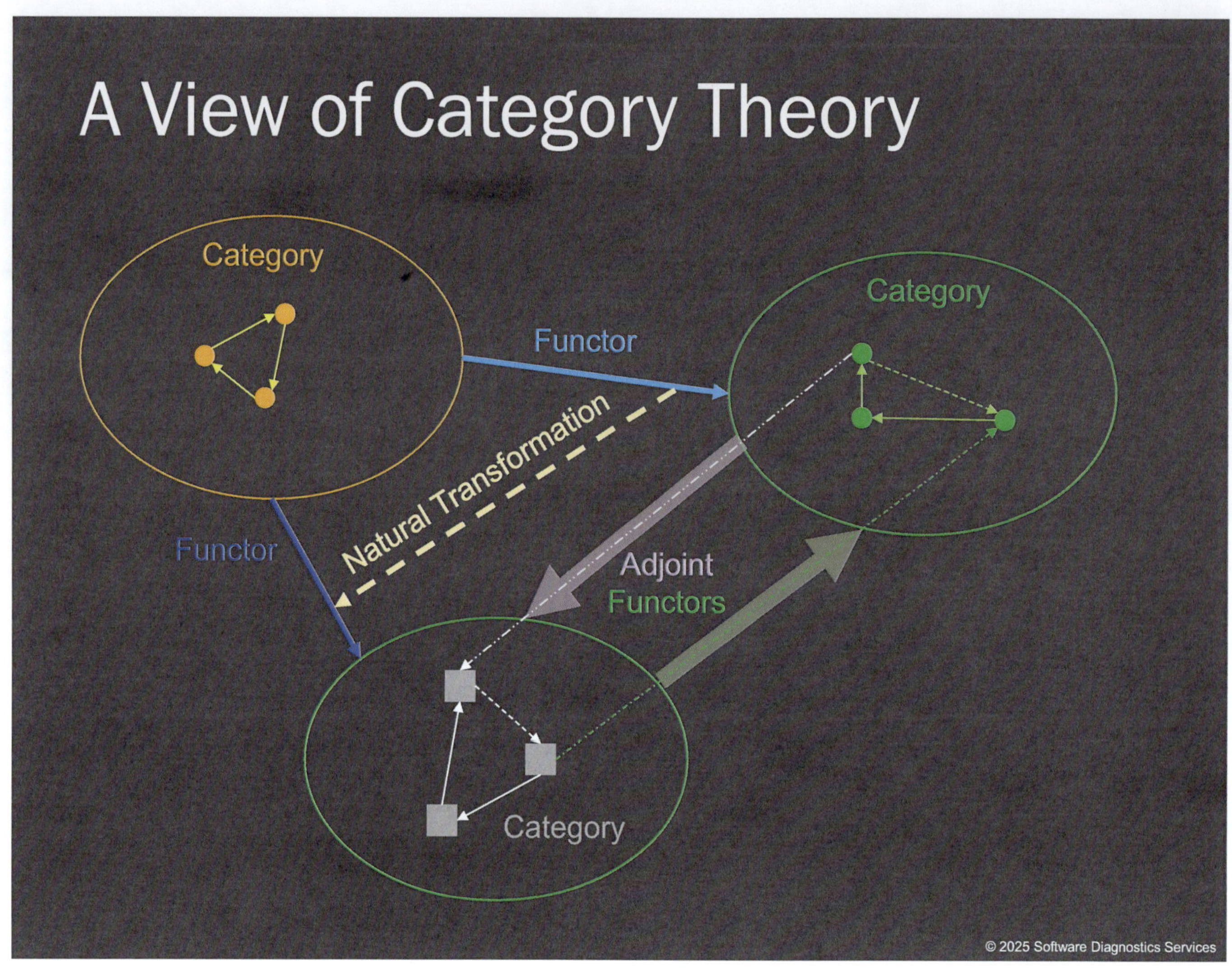

This slide is a visual overview of the four main concepts: category, functor between categories, natural transformation between functors, and adjoint functors. Please notice a kind of tunnel functor arrows that highlight back translation between objects of categories.

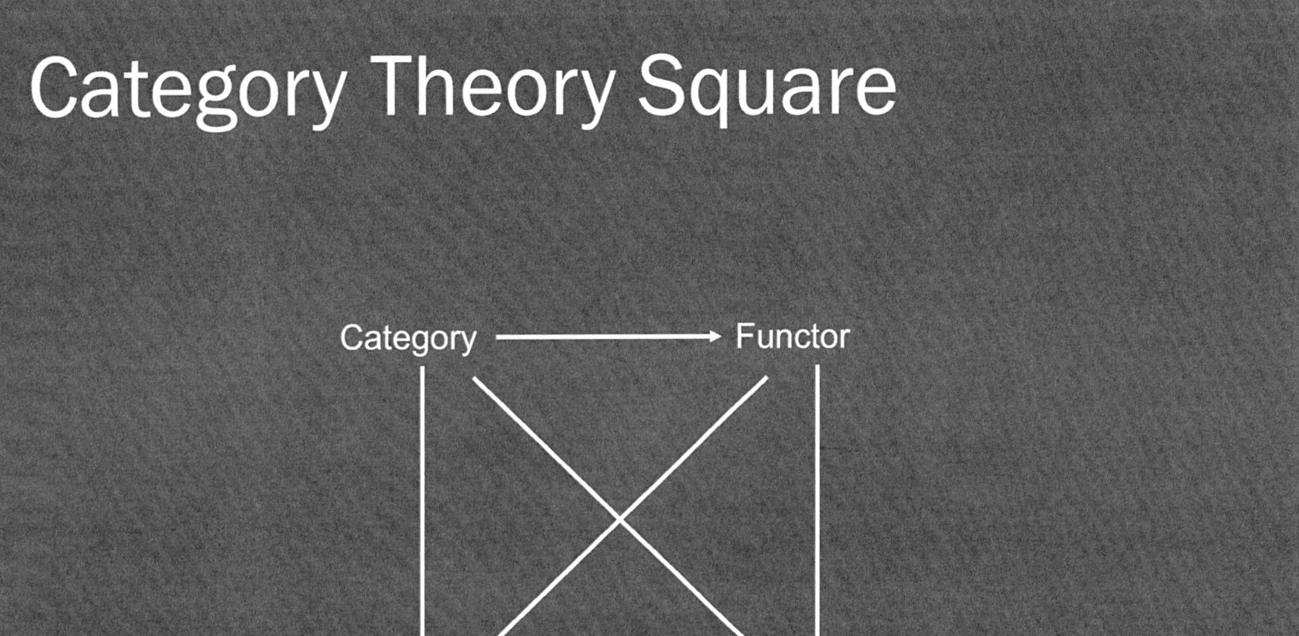

We can arrange all four concepts via the **category theory square** that shows their relationship that is easy to remember.

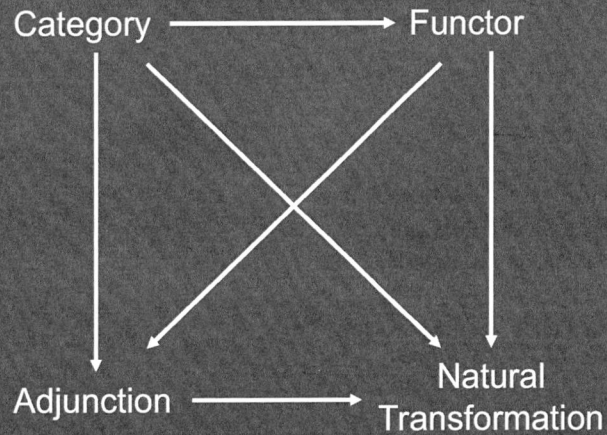

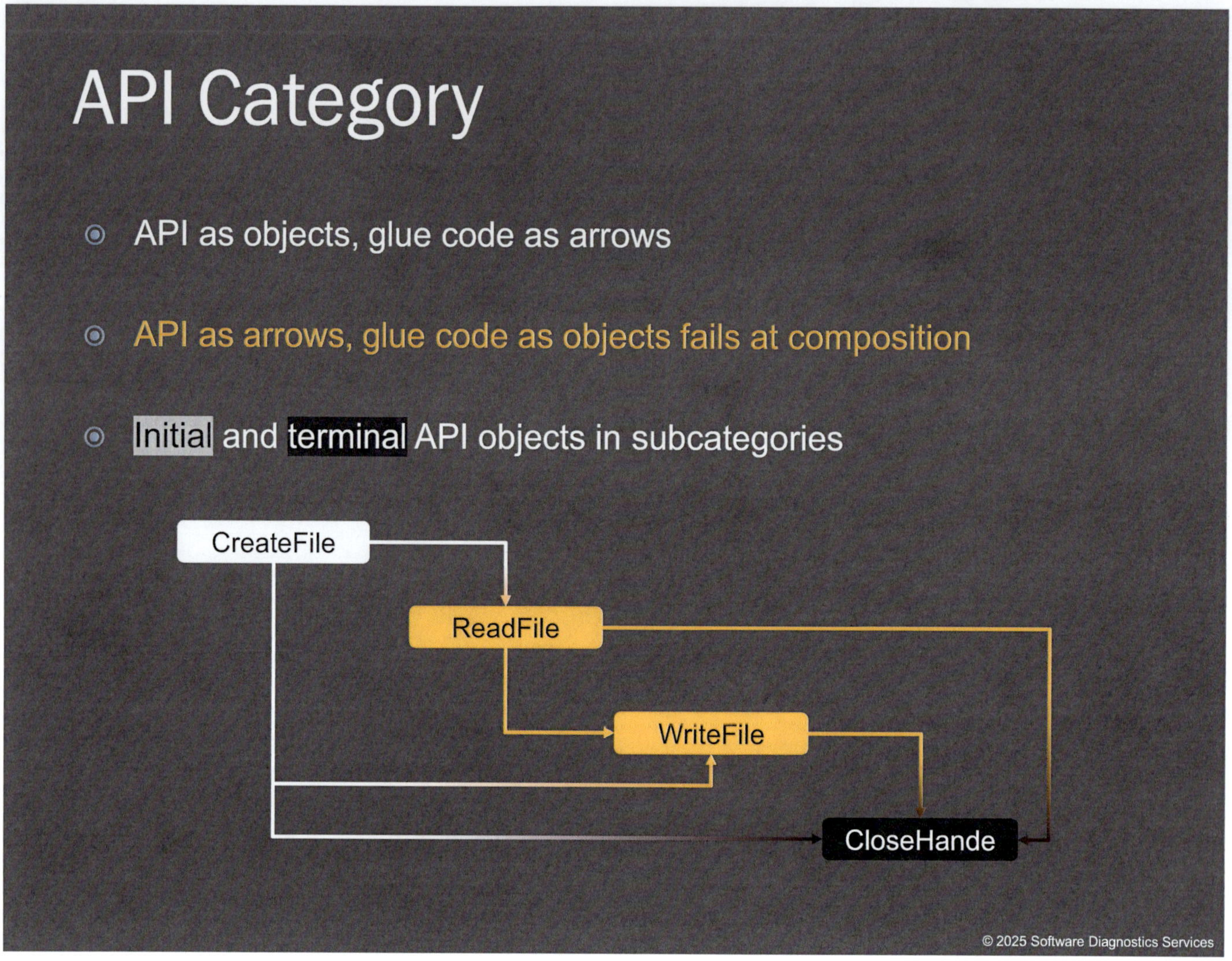

API Category

- API as objects, glue code as arrows
- API as arrows, glue code as objects fails at composition
- Initial and terminal API objects in subcategories

CreateFile

ReadFile

WriteFile

CloseHande

Let's translate all those informal abstract concepts into the API world. Again, in the API category, API functions are objects, and glue code is arrows. The other possible dual arrangement, API as arrows, is invalid due to impossible compositions: we cannot get API by composing two API calls. The result of the composition is an object, glue code. Some API functions can be initial or terminal in small subcategories, for example, *Create* and *Close*. In larger subcategories, there can be *Initialize* and *Uninitialize* functions.

API Functor

- Translates between API layers (different API)
- Stack trace as functor
- Translates between different API sequences
- Endofunctor – between the same API
- Translates between different code implementations

Different API layers could be different API categories or subcategories of the same category. All such translations are functors. It also suggests interpreting stack traces as functors. Finally, translation between different API sequences is also a functor that maps API calls to API calls and glue code to glue code.

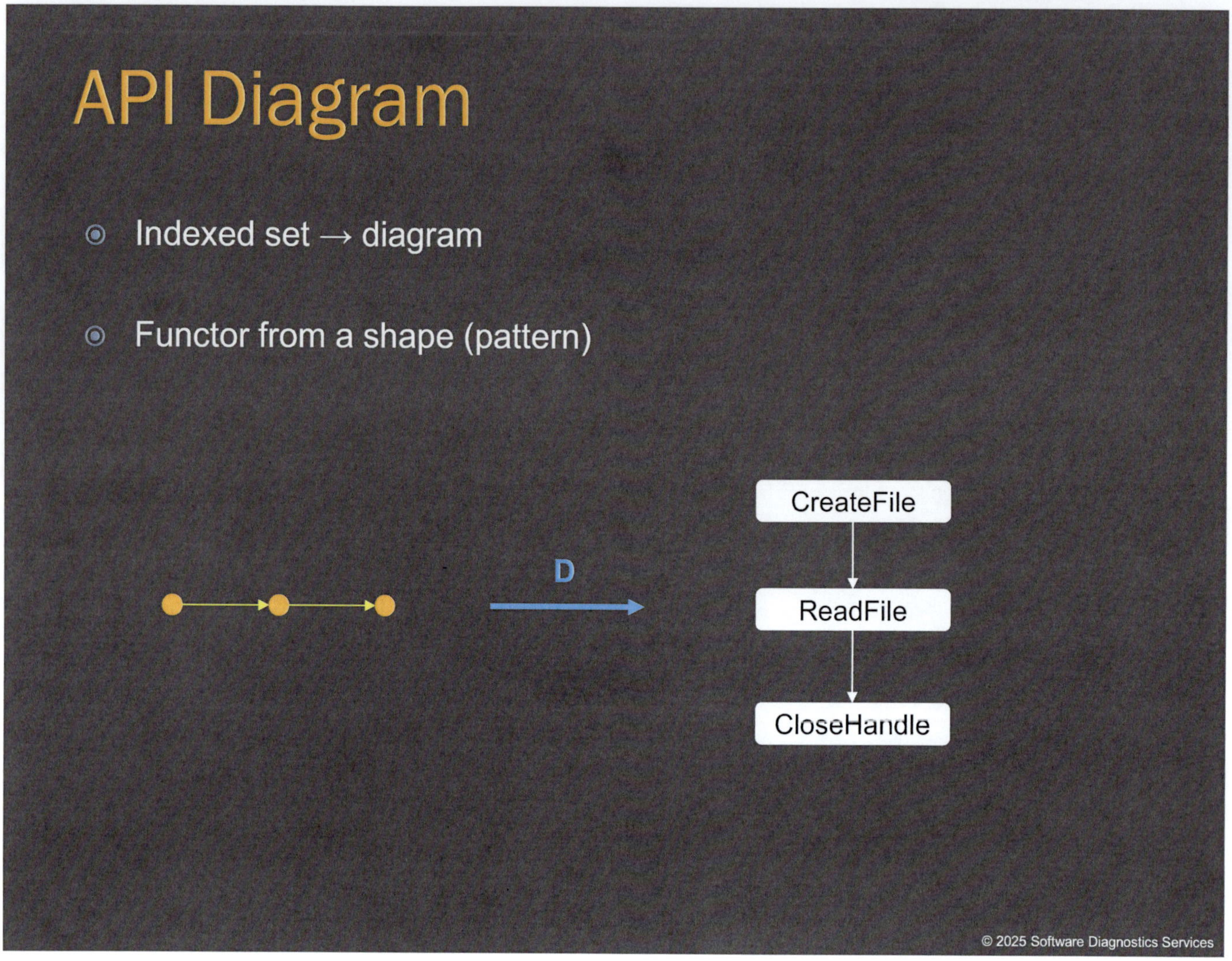

API Diagram

- Indexed set → diagram
- Functor from a shape (pattern)

D

CreateFile

ReadFile

CloseHandle

You may have noticed that category theory illustrations are similar to graphs and diagrams. In category theory, a diagram in some category **C** is a functor *D* from some shape category where we don't care about individual objects and arrows to that category **C**. They are categorical analogs to indexed sets where a shape is a set of indexes.

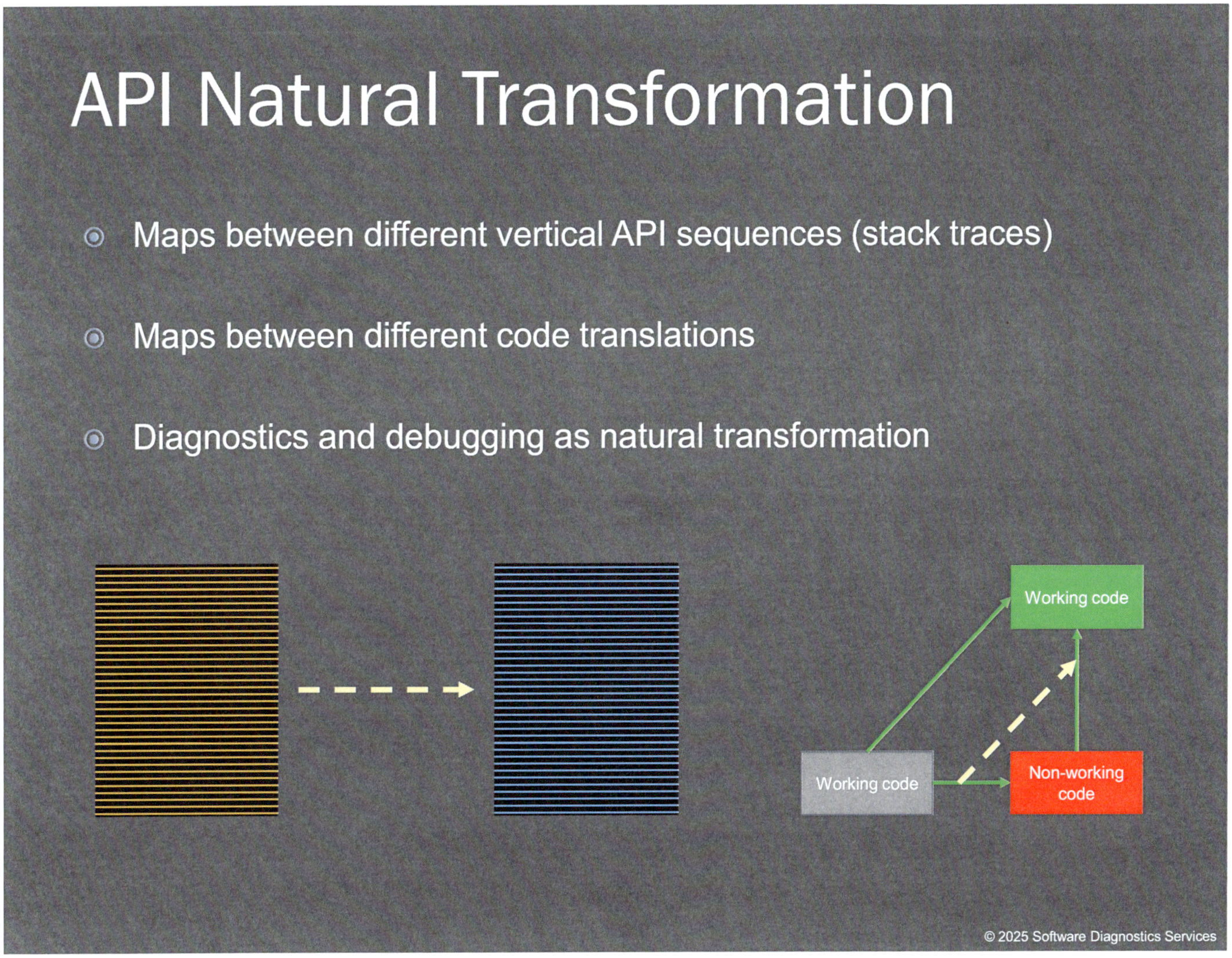

API Natural Transformation

- Maps between different vertical API sequences (stack traces)
- Maps between different code translations
- Diagnostics and debugging as natural transformation

Working code
Working code
Non-working code

If stack traces are functors, then maps between them are natural transformations. The same goes for maps between different code translations. Hence, it naturally suggests a categorical interpretation of diagnostics and debugging as natural transformation.

Cross-platform API

- Windows API / Linux API
- Similar diagrams
- Cross-platform development as a natural transformation

Diagrams are functors. Therefore, maps between diagrams are natural transformations. We can represent various API sequences in different operating systems as similar diagrams, enabling us to view cross-platform development as a natural transformation.

API Adjunction

- Navigation between different API sequences
- Call and return stack trace sequences, callbacks (when stack traces correspond to vertical API sequences)
- Back translation between traces/logs (when traces correspond to API horizontal sequences)

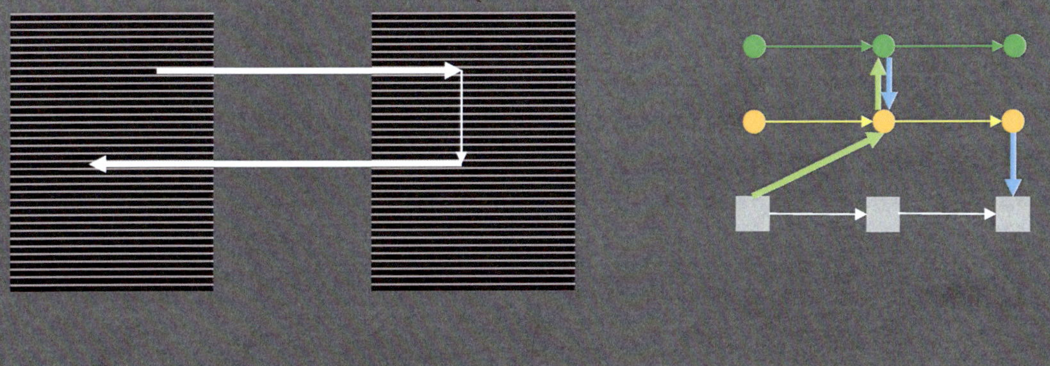

API adjunction formalizes back-and-forth navigation between different API sequences, be it traces and logs (horizontal API sequences) or call and return stack trace sequences (vertical API sequences). In some way, it corresponds to changes in perspective, especially when navigating around logs.

Informal n-API

- Arrows between arrows
- 1-API – normal API usage
- 2-API – diagnostics, debugging
- 3-API – higher diagnostics, debugging (debugging the debugging)
- ∞-API – for homework ☺

On this slide, we suggest the further informal application of n-category theory, where we consider arrows between arrows. 1-category here maps to normal API usage. 2-category corresponds to diagnostics and debugging. 3-category corresponds to higher level diagnostics and debugging, diagnosing diagnostics and debugging, debugging diagnostics and debugging.

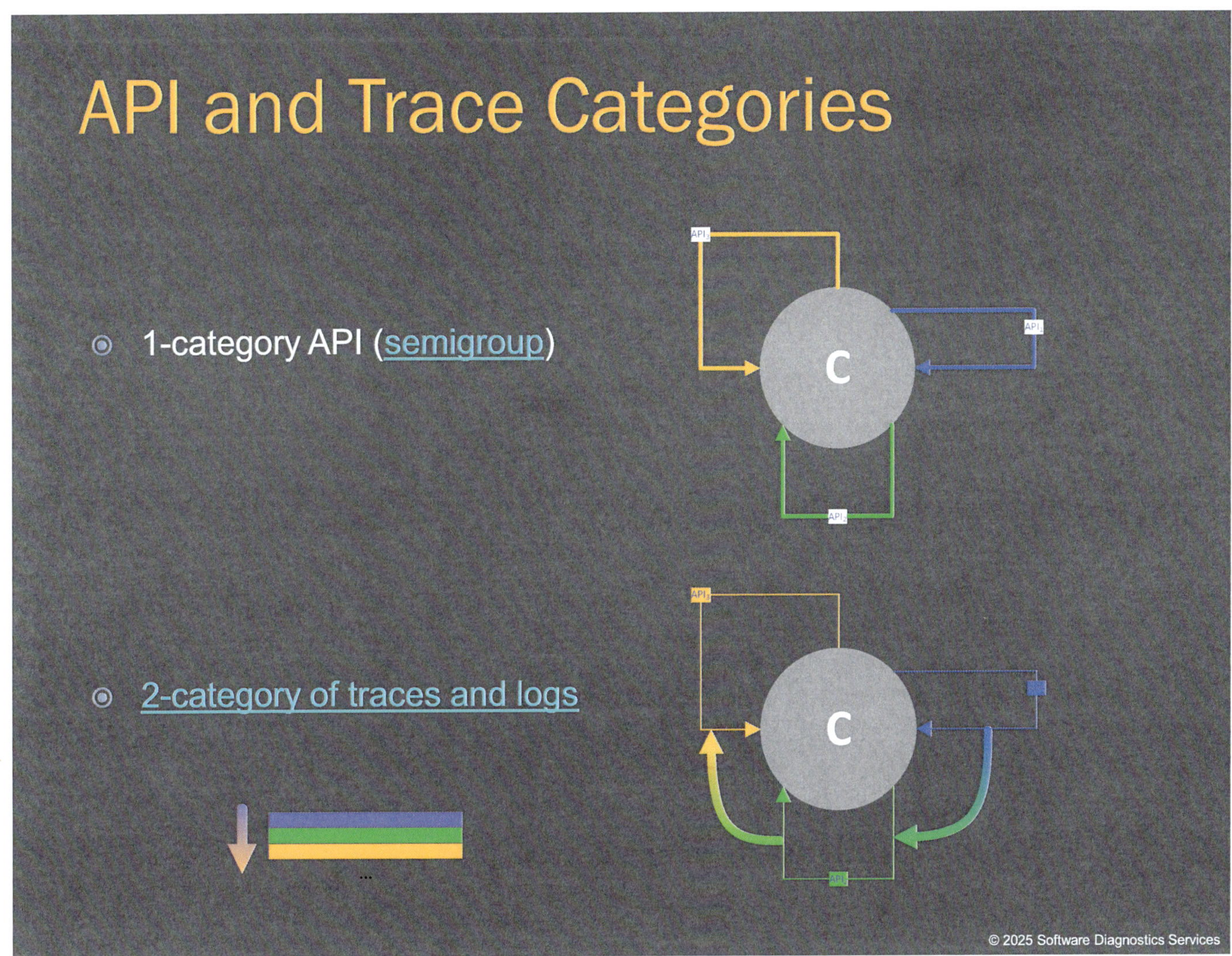

There can be various categories formed from API and code. We have already looked at one where the API calls are objects connected via code arrows. There can be a different view and a different category of one object, Code, and API calls as arrows. Arrows between arrows may correspond to traces and logs, forming the 2-category. The original motivation for this view comes from this link:

2-category of traces and logs
https://www.dumpanalysis.org/traces-logs-as-2-categories

semigroup
https://en.wikipedia.org/wiki/Semigroup

API I/O

- Categories – one input, one output
- Operads – many inputs, one output
- Properads – many inputs, many outputs

```
HANDLE CreateThread(
  [in, optional]  LPSECURITY_ATTRIBUTES   lpThreadAttributes,
  [in]            SIZE_T                  dwStackSize,
  [in]            LPTHREAD_START_ROUTINE  lpStartAddress,
  [in, optional]  __drv_aliasesMem LPVOID lpParameter,
  [in]            DWORD                   dwCreationFlags,
  [out, optional] LPDWORD                 lpThreadId
);
```

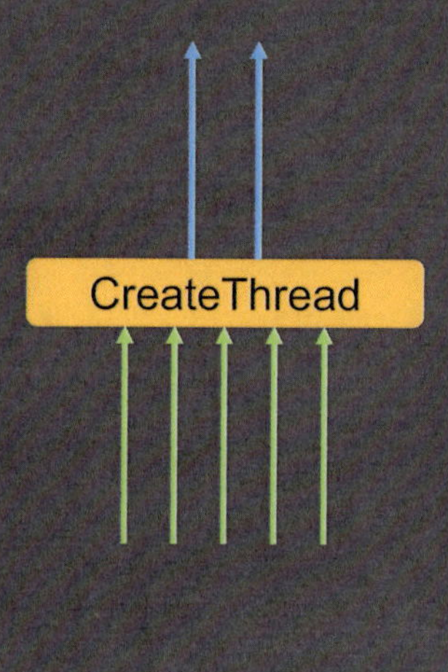

Now, we consider API functions in terms of their input/output. Since objects of categories have one input and one output, we abstract from API input and consider it as some glue code. In functional programming languages, there may be many inputs and one produced output, and these mathematical objects (operations) are called operads. In Windows API, outputs may be modified, producing new values, and therefore, we have several inputs and outputs, and these mathematical operations are called properads.

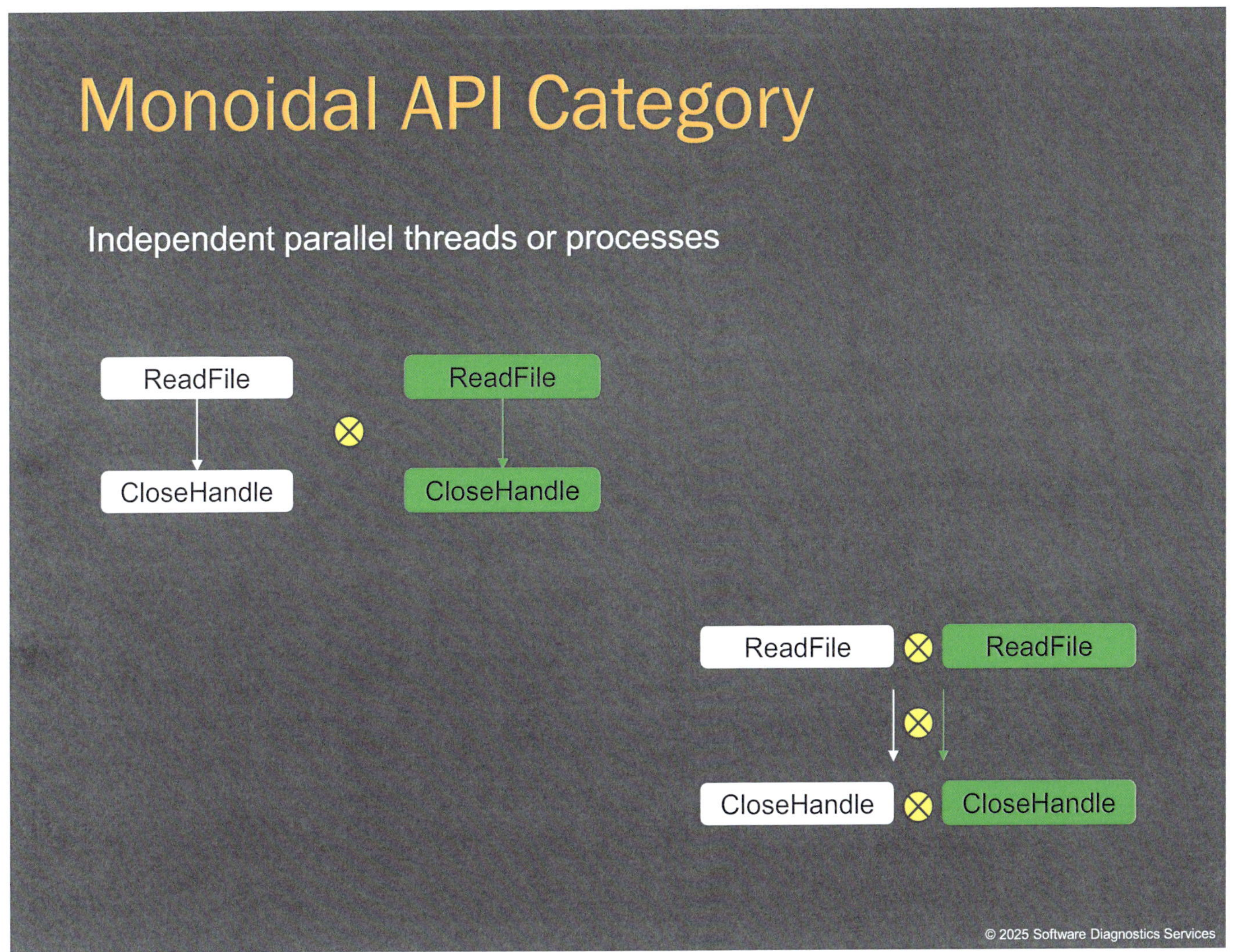

Finally, we consider the so-called monoidal categories for independent parallel threads or processes. In normal categories, the $api_2 \circ api_1$ combination corresponds to sequential API calls (api_1, then api_2), but in monoidal categories, the $api_2 \otimes api_1$ combination corresponds to independent API calls.

Regarding more information about categorical thinking and the applied side of categories, at the end of the training, I put a slide for further reading.

Windows API and Languages

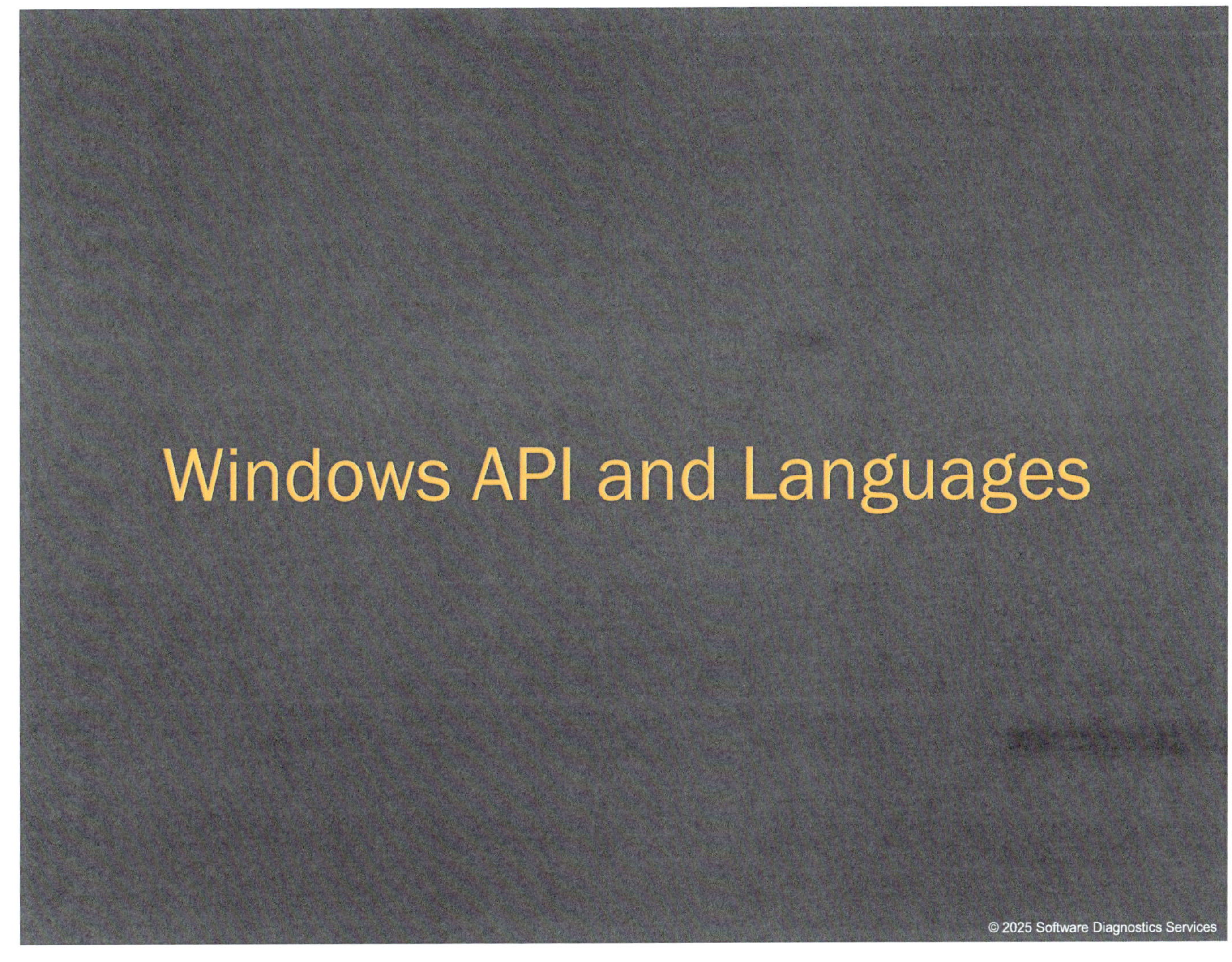
Windows API and Languages
© 2025 Software Diagnostics Services

Now, we look at how Windows API is used by various languages other than C and C++. The choice of languages was dictated by my own interests. Fortunately, all of them are mainstream. I've chosen the simple example of calling the *MessageBox* function. Slight modifications of the examples to crash programs are left for homework exercises.

API and C#

- P/Invoke example:

```
using System;
using System.Runtime.InteropServices;

class WapiTest
{
  [DllImport("user32.dll")]
  public static extern uint MessageBox(ulong hWnd, string message, string title, uint flags);

  public static void Main()
  {
    MessageBox(0, "Hello Windows API!", "From C#", 0);
  }
}
```


Our first language is C#. The old traditional way to call Windows API is to use the so-called **P/Invoke** mechanism. Here, we annotate an external function as a DLL import. To run this example, you need to install Visual Studio 2022 with .NET Desktop Development enabled. I created a simple .NET console app and replaced the contents of the .cs file with this example.

API Metadata

- Overview
- C#/Win32 P/Invoke Source Generator

However, manually creating Windows API function prototypes is very tedious and error-prone. Some structures and data types need to be accurately translated as well. Microsoft provides Windows API metadata for various languages and a P/Invoke source generator for C#.

Overview
https://github.com/microsoft/win32metadata

C#/Win32 P/Invoke Source Generator
https://github.com/microsoft/cswin32

API and Scala Native

- Scala Native
- Native code interoperability

My second language choice was Scala, the language I started learning more than 4 years ago. It was originally created for JVM, but there's also a Native implementation using LLVM. For it, in addition to POSIX library functions, on Windows platforms, it is possible to call Windows API. Since the choice of language is rather unusual, I created a special exercise for it that also details the required Scala Native installation steps. Over the last 4 years, Scala Native Windows support has really improved.

Scala Native
https://scala-native.org/en/stable/

Native code interoperability
https://scala-native.org/en/stable/user/interop.html

Exercise W9

- **Goal:** Install the Scala Native environment and write a simple program that uses Windows API
- \AWAPI-Dumps\Exercise-W9.pdf

Exercise W9

Goal: Install the Scala Native environment and write a simple program that uses Windows API.

1. Install JDK and set the JAVA_HOME variable. We used https://adoptium.net/en-GB and chose to set JAVA_HOME during installation.

```
C:\AWAPI-Dumps>java -version
openjdk version "17.0.5" 2022-10-18
OpenJDK Runtime Environment Temurin-17.0.5+8 (build 17.0.5+8)
OpenJDK 64-Bit Server VM Temurin-17.0.5+8 (build 17.0.5+8, mixed mode, sharing)
```

2. Install Scala with cs setup as recommended at https://www.scala-lang.org/download/. You need to run a command prompt as Administrator.

```
c:\AWMDA-Dumps>cs setup
Checking if a JVM is installed
Found a JVM installed under C:\Program Files\Eclipse Adoptium\jdk-17.0.5.8-hotspot.

Checking if ~\AppData\Local\Coursier\data\bin is in PATH
  Should we add ~\AppData\Local\Coursier\data\bin to your PATH? [Y/n] Y

Checking if the standard Scala applications are installed
  Installed ammonite
  Installed cs
  Installed coursier
  Installed scala
  Installed scalac
  Installed scala-cli
  Installed sbt
  Installed sbtn
  Installed scalafmt

c:\AWMDA-Dumps>scala -version
Scala code runner version: 1.5.4
Scala version (default): 3.6.2
```

3. Install Visual Studio 2022 Community with Desktop development with C++ if you have not done that already (or have the Professional edition).

4. Install LLVM-19.1.6-win64.exe from https://github.com/llvm/llvm-project/releases/tag/llvmorg-19.1.6 (we use the minimum required) and choose to set the path for all users:

```
C:\AWAPI-Dumps>clang --version
clang version 19.1.6
Target: x86_64-pc-windows-msvc
Thread model: posix
InstalledDir: C:\Program Files\LLVM\bin
```

5. Create SBT native project:

```
C:\AWAPI-Dumps>mkdir scala-wapi
```

```
C:\AWAPI-Dumps>cd scala-wapi

C:\AWAPI-Dumps\scala-wapi>sbt new scala-native/scala-native.g8
...
A minimal project that uses Scala Native.

name [Scala Native Seed Project]: wapi

Template applied in C:\AWAPI-Dumps\scala-wapi\.\wapi

C:\AWAPI-Dumps\scala-wapi>cd wapi
```

6. Open *\src\main\scala\Main.scala* and replace code with:

```
import scala.scalanative.unsafe._
import scala.scalanative.unsigned._

@link("user32")
@extern
object user32 {
  def MessageBoxA(hWnd: ULong, lpText: CString, lpCaption: CString, uType: UInt): Int = extern
}

object Main {
  def main(args: Array[String]): Unit =
    user32.MessageBoxA(0L.toULong, c"Hello Windows API!", c"From Scala", 0.toUInt)
}
```

7. Run the project:

```
C:\AWAPI-Dumps\scala-wapi\wapi>sbt run
[info] welcome to sbt 1.10.6 (Eclipse Adoptium Java 17.0.8)
...
[info] compiling 1 Scala source to C:\AWAPI-Dumps\scala-wapi\wapi\target\scala-3.3.3\classes
...
...
[info] Checking intermediate code (quick) (27 ms)
[info] Discovered 482 classes and 1928 methods after optimization
[info] Optimizing (debug mode) (2113 ms)
[info] Produced 41 LLVM IR files
[info] Generating intermediate code (4261 ms)
[info] Compiling to native code (16292 ms)
[info] Linking with [dbghelp, advapi32, user32]
[info] Linking native code (immix gc, none lto) (1199 ms)
[info] Postprocessing (0 ms)
[info] Total (26676 ms)
[success] Total time: 42 s, completed 25 Dec 2024, 10:08:28
```

You see the message box popup:

API and Golang

- windows package
- Example:

```
package main

import "unsafe"
import "golang.org/x/sys/windows"

func main() {
  var user32 = windows.NewLazyDLL("user32.dll")
  var procMessageBox = user32.NewProc("MessageBoxW")

  message, _ := windows.UTF16PtrFromString("Hello Windows API!")
  title, _ := windows.UTF16PtrFromString("From Golang")
  procMessageBox.Call(0, uintptr(unsafe.Pointer(message)), uintptr(unsafe.Pointer(title)), 0)
}
```


Another popular language choice is Golang, which I recently used in conjunction with C++ for some projects. There's a windows package, and its simple usage is illustrated on this slide.

windows package
https://pkg.go.dev/golang.org/x/sys/windows

API and Rust

- Rust for Windows
- Example:

```
use windows_sys::{
  core::*, Win32::UI::WindowsAndMessaging::*
};

fn main() {
  unsafe {
    MessageBoxW(0, w!("Hello Windows API!"), w!("From Rust"), 0);
  }
}
```

- Other bindings: winapi-rs

Another of my favorite languages is Rust. Since Microsoft loves it, too, it created bindings for Windows API based on API metadata. I created a simple cargo project using the UNICODE version of the *MessageBox* function.

Rust for Windows
https://github.com/microsoft/windows-rs

winapi-rs
https://crates.io/crates/winapi

API and Python

- ctypes library
- Example:

```
from ctypes import windll

windll.user32.MessageBoxW(0, "Hello Windows API!", "From Python", 0)
```


And finally, Python, where the ctypes library has Windows features illustrated on this slide. The code is the simplest among all examples. The provided library link documentation has additional examples, including Windows types.

ctypes library
https://docs.python.org/3/library/ctypes.html

Windows API Classes

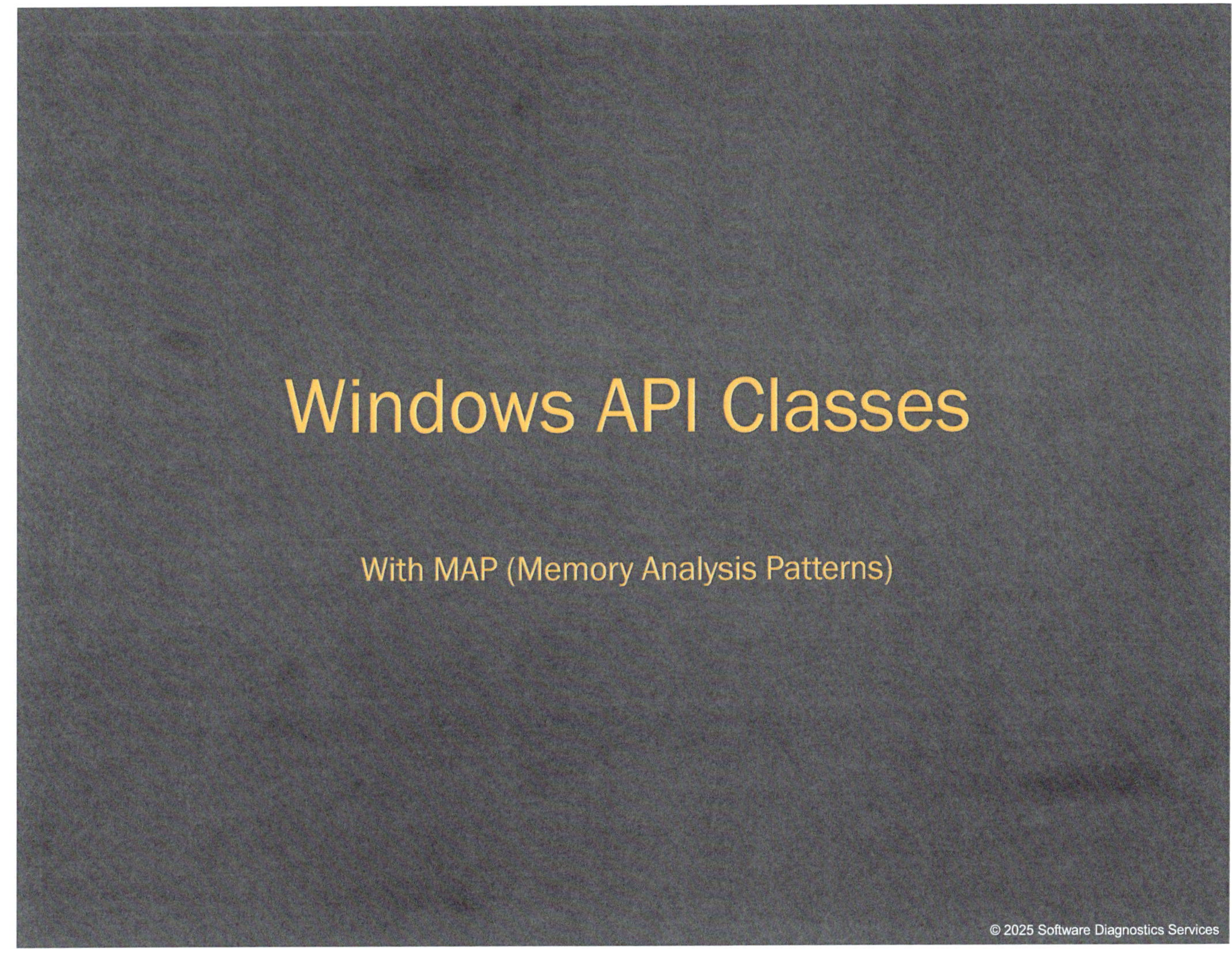
Windows API Classes
With MAP (Memory Analysis Patterns)
© 2025 Software Diagnostics Services

The final section of this training is a tour through selected Windows API classes. I don't use the term category to avoid confusion with category theory. Also, classes may contain API sets and subsets defined by a contract. We chose only API classes for which there's some correspondence with memory analysis patterns since our training is about diagnostics. To avoid repeating excellent Microsoft documentation, we also put links to the Microsoft learning site for overviews. The third edition of this training may expand the coverage.

General Resources

- API Index
- Windows SDK

These are links to general resources for study. Windows SDK contains many useful examples. Sometimes, it is also possible to copy/paste snippets of documentation examples into a default Visual C++ desktop project for a quick build, run, and inspection in WinDbg.

API Index
https://learn.microsoft.com/en-us/windows/win32/apiindex/windows-api-list

Windows SDK
https://developer.microsoft.com/en-US/windows/downloads/windows-sdk/

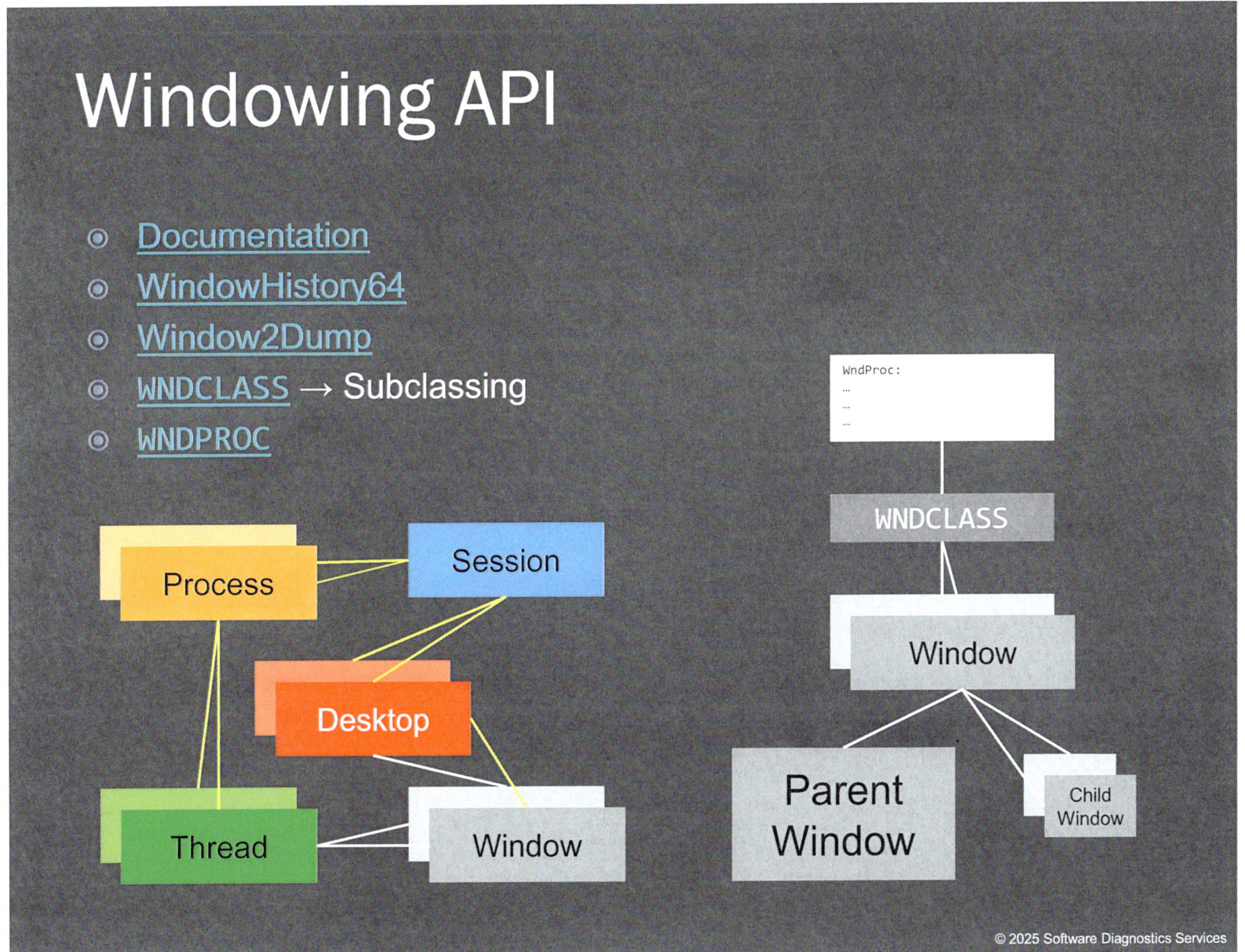

We start with Windowing API. We saw some examples, such as *CreateWindowExW*, in the previous exercise. From an object-oriented perspective, window objects are described by their window class, the so-called *WNDCLASS* structure. This structure contains a pointer to a window procedure that determines window behavior in response to window messages. We cover messaging in the next slide. This window procedure can be changed for exiting window classes to alter corresponding window behavior, for example, to draw it differently. It is the essence of Window subclassing. Windows are created by threads, and each thread may have several windows. Windows belong to a desktop, which belongs to an interactive user session. In the diagram on the left, solid yellow lines mean at least one instance. For example, a process has at least one thread. Windows may have parent windows and child windows. The *WindowHistory64* tool I wrote in the past is a great way to explore such relationships and how they change over time. The 64-bit version of this tool also monitors windows from 32-bit processes. It also has tooltips showing window information if you point to it. The *Window2Dump*, another tool with source code, can dump a process that belongs to a particular window, and it also has informational tooltips.

Documentation
https://learn.microsoft.com/en-us/windows/win32/winmsg/windows

WindowHistory64
https://support.citrix.com/article/CTX109235/windowhistory-tool

Window2Dump
https://bitbucket.org/softwarediagnostics/window2dump/

WNDCLASS
https://learn.microsoft.com/en-us/windows/win32/api/winuser/ns-winuser-wndclassw

WNDPROC
https://learn.microsoft.com/en-us/windows/win32/api/winuser/nc-winuser-wndproc

Messaging API

- Documentation
- `PostMessage` to thread message queue
- Thread message loop
- `SendMessage` directly to `WndProc` (Wait Chain)
- Message IPC (`WM_COPYDATA`)
- MessageHistory(64) → Hooking (Message Hooks, modeling example)
- Structure:
  - Message ID (`WM_xxx`, `EM_xxx`, `LB_xxx`, ...) – `WinUser.h`
  - 2 parameters (wParam and lParam)
  - `WM_LBUTTONDOWN`
- Blocking: Message Box, Dialog Box, Input Thread
- UI Problem Patterns: Error Message Box, Unresponsive Window

Windows receive commands and exchange information via messages. There's a message queue for each thread. The so-called **message loop** gets messages from the queue and dispatches them to the appropriate window procedure after an optional translation. You can place such loops anywhere in your thread code. If you post a message, it can be delayed in processing or even ignored. Another way is to call a window procedure directly. This is done by *SendMessage* API. The calling code is blocked, waiting for processing results; therefore, there's a possibility for freeze. It is also possible to send data between windows; the legacy interprocess communication mechanism is still used in many 3rd-party desktop applications, especially if windows belong to different processes. *MessageHistory* tool (or Microsoft graphical Spy++) is an example of message hooking where a hooking DLL is mapped to each process space to intercept messages. The latter also explains why 32-bit and 64-bit MessageHistory coexist. You cannot map a 32-bit DLL to a 64-bit process and vice versa. So, you need to run both in parallel. WindowHistory64, mentioned in the previous slide, achieves interoperability via a hidden 32-bit process that does 32-bit hooking and sends collected data to the 64-bit process to combine it with 64-bit window data. Windows messages have a very simple structure of 2 parameters. All messages have unique IDs you can find in *WinUser.h* file. There are some memory analysis patterns related to windowing and messaging. Such unchecked GUI activity can block non-interactive services that don't have access to user desktop sessions.

Documentation
https://learn.microsoft.com/en-us/windows/win32/winmsg/messages-and-message-queues

Wait Chain
https://www.dumpanalysis.org/blog/index.php/2010/12/16/crash-dump-analysis-patterns-part-42h/

WM_COPYDATA
https://learn.microsoft.com/en-us/windows/win32/dataxchg/wm-copydata

MessageHistory(64)
https://support.citrix.com/article/CTX111068/messagehistory

Hooking
https://learn.microsoft.com/en-us/windows/win32/winmsg/hooks

Message Hooks
https://www.dumpanalysis.org/blog/index.php/2010/07/06/crash-dump-analysis-patterns-part-100/

modeling example
https://www.dumpanalysis.org/blog/index.php/2010/07/13/models-for-memory-and-trace-analysis-patterns-part-3/

WM_LBUTTONDOWN
https://learn.microsoft.com/en-us/windows/win32/inputdev/wm-lbuttondown

Message Box
https://www.dumpanalysis.org/blog/index.php/2008/02/19/crash-dump-analysis-patterns-part-51/

Dialog Box
https://www.dumpanalysis.org/blog/index.php/2011/01/29/crash-dump-analysis-patterns-part-128/

Error Message Box
https://www.dumpanalysis.org/blog/index.php/2011/07/14/user-interface-problem-analysis-patterns-part-1/

Unresponsive Window
https://www.dumpanalysis.org/blog/index.php/2012/09/09/user-interface-problem-analysis-patterns-part-2/

GDI API

- Documentation
- Device contexts
- Handle Limit (User Space)
- Handle Limit (Kernel Space)
- Create → Delete
- Screen glitches

Graphics Device Interface API is a window and a device-independent way to draw graphics. It was the original graphics API in Windows. The main notion here is a device context on which you do graphical operations. You acquire it at the beginning and need to release it after usage; otherwise, there would be a leak. Various graphics primitives such as fonts, pens, and brushes have corresponding handles to refer to, and the number of them has a limit, so in the case of a leak, you may see screen glitches.

Documentation
https://learn.microsoft.com/en-us/windows/win32/gdi/windows-gdi

Device contexts
https://learn.microsoft.com/en-us/windows/win32/gdi/about-device-contexts

Handle Limit (User Space)
https://www.dumpanalysis.org/blog/index.php/2016/06/05/crash-dump-analysis-patterns-part-58b/

Handle Limit (Kernel Space)
https://www.dumpanalysis.org/blog/index.php/2008/04/09/crash-dump-analysis-patterns-part-58a/

GDI+ API

- Documentation
- C++ library
- gdiplus module
- Imports from gdi32

Later, Microsoft introduced a C++ library called GDI+ with exports from the *gdiplus* module. It is easier to use than GDI, especially for complex operations. The documentation says it communicated with drivers, but on inspection, it imports from the GDI32 module, so perhaps it fallbacks to the previous API if necessary.

Documentation
https://learn.microsoft.com/en-us/windows/win32/gdiplus/-gdiplus-gdi-start

Module/Library API

- libloaderapi.h
- `LoadLibraryW`, `GetProcAddess`, `FreeLibrary`
- Module analysis patterns
- Unloaded Module
- Dynamic linking (Missing Component)
- Static linking (Missing Component)

In addition to static linking, it is also possible to load modules dynamically and get access to exported APIs, including undocumented ones or APIs from lower layers. If modules are missing in the search path, you may be unable to launch a process in the case of static linking. Moreover, modules can be unloaded, and any dangling references to them may cause a crash. There are many module-related analysis patterns, so I grouped them into a subcatalog.

libloaderapi.h
https://learn.microsoft.com/en-us/windows/win32/api/libloaderapi/

LoadLibraryW
https://learn.microsoft.com/en-us/windows/win32/api/libloaderapi/nf-libloaderapi-loadlibraryw

GetProcAddess
https://learn.microsoft.com/en-us/windows/win32/api/libloaderapi/nf-libloaderapi-getprocaddress

FreeLibrary
https://learn.microsoft.com/en-us/windows/win32/api/libloaderapi/nf-libloaderapi-freelibrary

Module analysis patterns
https://www.dumpanalysis.org/blog/index.php/2012/07/15/module-patterns/

Unloaded Module
https://www.dumpanalysis.org/blog/index.php/2012/06/27/crash-dump-analysis-patterns-part-178/

Missing Component (Dynamic Linking)
https://www.dumpanalysis.org/blog/index.php/2008/04/22/crash-dump-analysis-patterns-part-59/

Missing Component (Static Linking)
https://www.dumpanalysis.org/blog/index.php/2008/06/12/crash-dump-analysis-patterns-part-59b/

Process/Thread API

- Overview
- Process analysis patterns
- Thread analysis patterns
- `CreateThread → CloseHandle`
- Zombie Processes
- Insufficient Memory (Handle Leak)

There are separate memory analysis patterns subcatalogs related to processes and threads. Zombie processes and thread handle leaks are signature problems.

Overview
https://learn.microsoft.com/en-us/windows/win32/procthread/processes-and-threads

Process analysis patterns
https://www.dumpanalysis.org/blog/index.php/2013/01/05/process-patterns/

Thread analysis patterns
https://www.dumpanalysis.org/blog/index.php/2013/01/05/thread-patterns/

Zombie Processes
https://www.dumpanalysis.org/blog/index.php/2008/02/28/crash-dump-analysis-patterns-part-54/

Insufficient Memory (Handle Leak)
https://www.dumpanalysis.org/blog/index.php/2007/07/15/crash-dump-analysis-patterns-part-13b/

Services API

- Documentation
- Sample service template (included in the book)
- Input Thread

Due to additional boilerplate code, it is a bit harder to write a service than just a normal process. In this book, I included a Visual Studio project I used in another *Accelerated Windows Memory Dump Analysis* course to model a blocked service thread due to expected user input.

Documentation
https://learn.microsoft.com/en-us/windows/win32/services/services

Security API

- Documentation
- Deviant Token

Security and identity API is one of the unexplored memory dump analysis pattern regions. I included it because there is an existing analysis pattern. Each process has a security token describing its security capabilities. Process threads inherit such a token but can have it changing, thus impersonating other users.

Documentation
https://learn.microsoft.com/en-us/windows/win32/security

Deviant Token
https://www.dumpanalysis.org/blog/index.php/2012/12/31/crash-dump-analysis-patterns-part-191/

IPC API

- Overview
- Wait Chain (Named Pipes)
- Coupled Processes
  - Strong
  - Weak
  - Semantics

Interprocess communication can take different forms, for example, via mapped files, shared memory regions, and pipes. In some sense, even synchronization notifications can be considered IPC. Various IPC mechanisms gave rise to different levels of process coupling expressed in corresponding memory analysis pattern variants.

Overview
https://learn.microsoft.com/en-us/windows/win32/ipc/interprocess-communications

Wait Chain (Named Pipes)
https://www.dumpanalysis.org/blog/index.php/2011/01/03/crash-dump-analysis-patterns-part-42i/

Strong
https://www.dumpanalysis.org/blog/index.php/2007/09/26/crash-dump-analysis-patterns-part-28/

Weak
https://www.dumpanalysis.org/blog/index.php/2010/04/07/crash-dump-analysis-patterns-part-28b/

Semantics
https://www.dumpanalysis.org/blog/index.php/2010/08/04/crash-dump-analysis-patterns-part-103/

RPC API

- Overview
- Wait Chain (RPC)
- Semantic Structure (PID.TID)
- LPC/ALPC/RPC patterns and case studies

Remote Procedure Calls are a way to call functions remotely. Such calls may form wait chains and block threads. If the RPC target is on the same machine, then RPC may be implemented internally using ALPC (**Advanced Local Procedure Call**) mechanisms. The analysis of them usually requires kernel memory dumps and complete memory dumps if user space stack traces are required. I've put a link to the corresponding case studies.

Overview
https://learn.microsoft.com/en-us/windows/win32/rpc/rpc-start-page

Wait Chain (RPC)
https://www.dumpanalysis.org/blog/index.php/2010/09/14/crash-dump-analysis-patterns-part-42g/

Semantic Structure (PID.TID)
https://www.dumpanalysis.org/blog/index.php/2011/02/19/crash-dump-analysis-patterns-part-130a/

LPC/ALPC/RPC patterns and case studies
https://www.dumpanalysis.org/blog/index.php/2011/11/14/rpc-lpc-and-alpc-patterns-and-case-studies/

Synchronization API

- syncapi.h
- Wait chain analysis patterns
- Deadlock and livelock analysis patterns
- Shared Structure
- `WaitOnAddress` - Deadlock (Futex)

There are many synchronization primitives and higher-level structures built on them, from events and mutexes to critical sections. For example, a thread may wait for another process or thread, and the latter, in turn, may wait for something else – these form the basis of **wait chains**. It is also possible to wait for each other – the essence of **deadlock**. The *WaitOnAddress* API is used for a fast user mutex (futex) implementation.

syncapi.h
https://learn.microsoft.com/en-us/windows/win32/api/synchapi/

Wait chain analysis patterns https://www.dumpanalysis.org/blog/index.php/2009/02/17/wait-chain-patterns/

Deadlock and livelock analysis patterns
https://www.dumpanalysis.org/blog/index.php/2009/02/17/deadlock-patterns/

Shared Structure
https://www.dumpanalysis.org/blog/index.php/2013/12/07/crash-dump-analysis-patterns-part-203/

WaitOnAddress
https://learn.microsoft.com/en-us/windows/win32/api/synchapi/nf-synchapi-waitonaddress

I/O API

- Device I/O ioapiset.h
- File I/O fileapi.h
- Blocking File
- Handle (Invalid Handle)
- `CreateFile` → `CloseHandle` (Handle Leak)

Hardware is visible to software as devices. Devices and files on them are referenced via handles. Especially for files, if handles are not closed, this may result in handle leaks. Threads may also be blocked on I/O.

ioapiset.h
https://learn.microsoft.com/en-us/windows/win32/api/ioapiset/

fileapi.h
https://learn.microsoft.com/en-us/windows/win32/api/fileapi/

Blocking File
https://www.dumpanalysis.org/blog/index.php/2011/06/25/crash-dump-analysis-patterns-part-145/

Invalid Handle
https://www.dumpanalysis.org/blog/index.php/2008/05/20/crash-dump-analysis-patterns-part-61/

Handle Leak
https://www.dumpanalysis.org/blog/index.php/2012/12/23/crash-dump-analysis-patterns-part-189/

Runtime API

- Reference (Windows Runtime)
- C/C++ runtime (ucrtbase, msvcrt, msvcp_win, msvcp110_win)
- Invalid Parameter
- C++ Exception
- Wait Chain (C++11, Condition Variable)

There are two different meanings to the word "runtime" here. One is a traditional C/C++ language runtime sense of using dynamically linked libraries. There are a few memory analysis patterns associated with it, including C++ exceptions. Windows also has its own **Runtime API**, including string manipulation functions via string handles.

Reference
https://learn.microsoft.com/en-us/windows/win32/winrt/reference

Invalid Parameter
https://www.dumpanalysis.org/blog/index.php/2017/05/14/crash-dump-analysis-patterns-part-117b/

C++ Exception
https://www.dumpanalysis.org/blog/index.php/2008/10/21/crash-dump-analysis-patterns-part-77/

Wait Chain (C++11, Condition Variable)
https://www.dumpanalysis.org/blog/index.php/2016/09/23/crash-dump-analysis-patterns-part-42n/

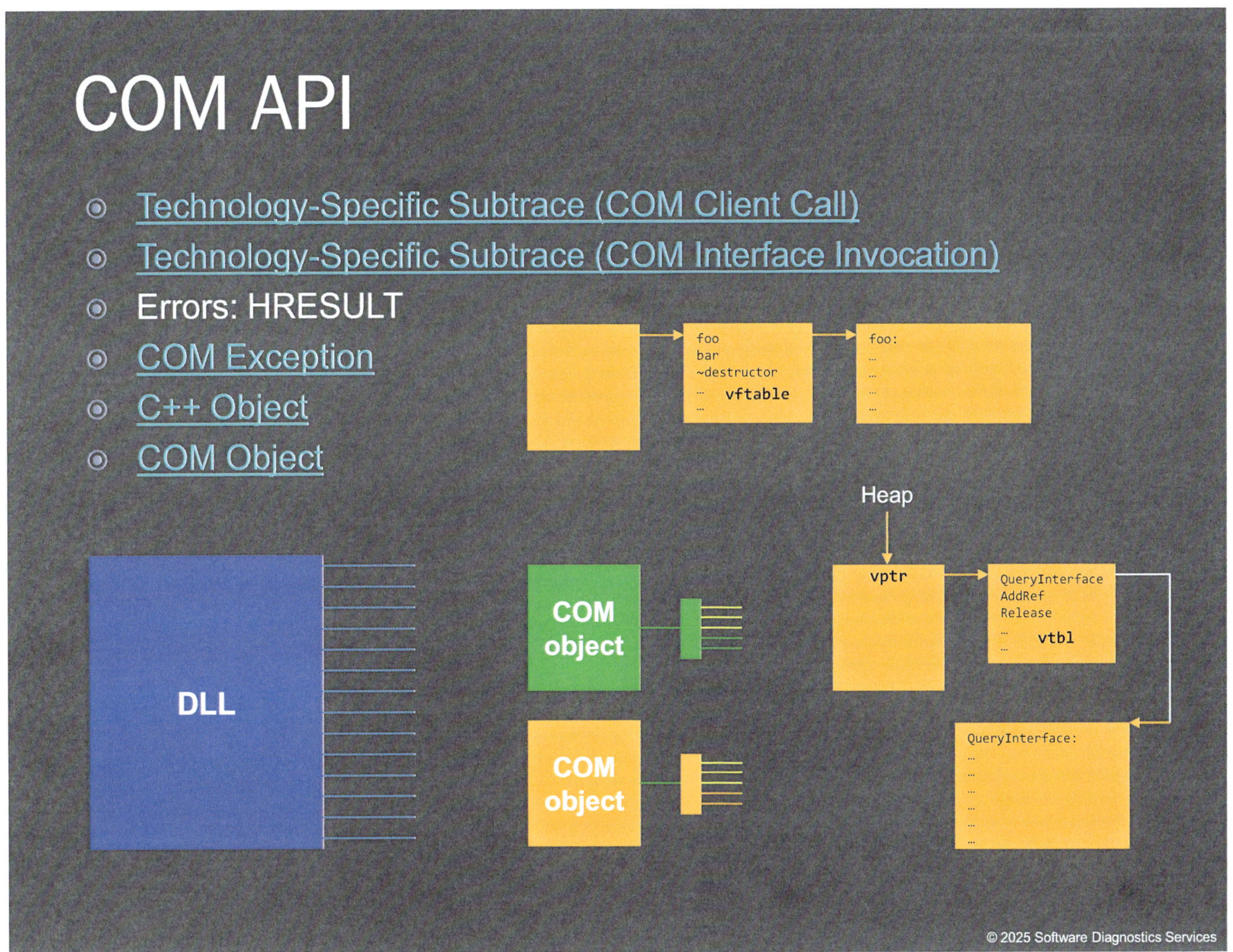

Now, we have come to a completely different way to represent API. It is called the **Component Object Model**. Traditional Modules (or Components) expose an undifferentiated collection of functions shown on the left. COM objects expose smaller sets of functions grouped into interfaces. Each interface must have 3 functions: *QueryInterface*, which allows interface inspection and discovery, and *AddRef* and *Release* for manual object lifetime management. The memory binary layout of such interfaces is compatible with C++ object virtual function table layout. In addition, COM interface functions return HRESULT errors, allowing for 3rd-party extensions. Although COM is not actively advertised by Microsoft now, it is a backbone for many other APIs, such as high-level multimedia and 3D graphics via DirectX, and we can see it a lot even in ordinary memory dumps, as in our final Exercise W10.

Technology-Specific Subtrace (COM Client Call)
https://www.dumpanalysis.org/blog/index.php/2015/05/10/crash-dump-analysis-patterns-part-127d/

Technology-Specific Subtrace (COM Interface Invocation)
https://www.dumpanalysis.org/blog/index.php/2011/01/15/crash-dump-analysis-patterns-part-127/

COM Exception
https://www.dumpanalysis.org/blog/index.php/2021/01/31/crash-dump-analysis-patterns-part-274/

C++ Object
https://www.dumpanalysis.org/blog/index.php/2021/01/25/crash-dump-analysis-patterns-part-273/

COM Object
https://www.dumpanalysis.org/blog/index.php/2022/11/27/crash-dump-analysis-patterns-part-282/

Exercise W10

- **Goal:** Find COM objects and their interfaces in raw stack regions
- **Memory Analysis Patterns:** Stack Trace Collection; Technology-Specific Subtrace (COM Client Call); Execution Residue (Unmanaged Space, User); COM Object; C++ Object
- \AWAPI-Dumps\Exercise-W10.pdf

Exercise W10

Goal: Find COM objects and their interfaces in raw stack regions.

Memory Analysis Patterns: Stack Trace Collection; Technology-Specific Subtrace (COM Client Call); Execution Residue (Unmanaged Space, User); COM Object; C++ Object.

1. Launch WinDbg.

2. Open \AWAPI-Dumps\Process\wordpad.DMP

3. We get the dump file loaded:

```
Microsoft (R) Windows Debugger Version 10.0.27725.1000 AMD64
Copyright (c) Microsoft Corporation. All rights reserved.


Loading Dump File [C:\AWAPI-Dumps\Process\wordpad.DMP]
User Mini Dump File with Full Memory: Only application data is available


************* Path validation summary **************
Response                         Time (ms)     Location
Deferred                                       srv*
Symbol search path is: srv*
Executable search path is:
Windows 10 Version 22000 MP (2 procs) Free x64
Product: WinNt, suite: SingleUserTS Personal
Edition build lab: 22000.1.amd64fre.co_release.210604-1628
Debug session time: Sat Oct 15 20:19:05.000 2022 (UTC + 0:00)
System Uptime: 0 days 0:27:57.202
Process Uptime: 0 days 0:00:26.000
................................................................
.....
Loading unloaded module list
.
For analysis of this file, run !analyze -v
win32u!NtUserGetMessage+0x14:
00007ffc`f3811414 ret
```

4. Open a log file using **.logopen**:

```
0:000> .logopen C:\AWAPI-Dumps\W10.log
Opened log file 'C:\AWAPI-Dumps\W10.log'
```

5. Find a COM thread in all thread stack traces:

```
0:000> ~*kL

.  0  Id: 12e8.928 Suspend: 0 Teb: 000000c0`033d2000 Unfrozen
 # Child-SP          RetAddr               Call Site
00 000000c0`0309f6c8 00007ffc`f586464e     win32u!NtUserGetMessage+0x14
01 000000c0`0309f6d0 00007ffc`acc20813     user32!GetMessageW+0x2e
02 000000c0`0309f730 00007ffc`acc20736     mfc42u!CWinThread::PumpMessage+0x23
```

```
03 000000c0`0309f760 00007ffc`acc1f2bc     mfc42u!CWinThread::Run+0x96
04 000000c0`0309f7a0 00007ff7`5764bcfd     mfc42u!AfxWinMain+0xbc
05 000000c0`0309f7e0 00007ffc`f4fd54e0     wordpad!__wmainCRTStartup+0x1dd
06 000000c0`0309f8a0 00007ffc`f62a485b     kernel32!BaseThreadInitThunk+0x10
07 000000c0`0309f8d0 00000000`00000000     ntdll!RtlUserThreadStart+0x2b

   1  Id: 12e8.1e2c Suspend: 0 Teb: 000000c0`033d4000 Unfrozen
 # Child-SP          RetAddr               Call Site
00 000000c0`0311f838 00007ffc`f62b6c2f     ntdll!NtWaitForWorkViaWorkerFactory+0x14
01 000000c0`0311f840 00007ffc`f4fd54e0     ntdll!TppWorkerThread+0x2df
02 000000c0`0311fb30 00007ffc`f62a485b     kernel32!BaseThreadInitThunk+0x10
03 000000c0`0311fb60 00000000`00000000     ntdll!RtlUserThreadStart+0x2b

   2  Id: 12e8.5f0 Suspend: 0 Teb: 000000c0`033d6000 Unfrozen
 # Child-SP          RetAddr               Call Site
00 000000c0`0319faa8 00007ffc`f62b6c2f     ntdll!NtWaitForWorkViaWorkerFactory+0x14
01 000000c0`0319fab0 00007ffc`f4fd54e0     ntdll!TppWorkerThread+0x2df
02 000000c0`0319fda0 00007ffc`f62a485b     kernel32!BaseThreadInitThunk+0x10
03 000000c0`0319fdd0 00000000`00000000     ntdll!RtlUserThreadStart+0x2b

   3  Id: 12e8.23c8 Suspend: 0 Teb: 000000c0`033d8000 Unfrozen
 # Child-SP          RetAddr               Call Site
00 000000c0`0347c888 00007ffc`f436ec7f     ntdll!NtAlpcSendWaitReceivePort+0x14
01 000000c0`0347c890 00007ffc`f43a1896     rpcrt4!LRPC_BASE_CCALL::SendReceive+0x12f
02 000000c0`0347c960 00007ffc`f435188c     rpcrt4!NdrpSendReceive+0xa6
03 000000c0`0347c990 00007ffc`f4351d3f     rpcrt4!NdrpClientCall2+0x43c
04 000000c0`0347cfa0 00007ffc`f5dac7b1     rpcrt4!NdrClientCall2+0x1f
05 (Inline Function) --------`--------     combase!BulkUpdateOIDs+0x7c
06 (Inline Function) --------`--------     combase!CRpcResolver::BulkUpdateOIDs+0x1ad
07 000000c0`0347cfd0 00007ffc`f5dc8a8d     combase!CROIDTable::ClientBulkUpdateOIDWithPingServer+0x821
08 (Inline Function) --------`--------
combase!CROIDTable::ForceImmediateBulkUpdateWithPingServerForAdds+0x16
09 000000c0`0347d220 00007ffc`f5dba353     combase!CStdMarshal::ConnectCliIPIDEntry+0x4ad
0a 000000c0`0347d2f0 00007ffc`f5dba4c6     combase!CStdMarshal::MakeCliIPIDEntry+0x133
0b 000000c0`0347d3f0 00007ffc`f5dbd005     combase!CStdMarshal::UnmarshalIPID+0x7a
0c 000000c0`0347d490 00007ffc`f5db13c1     combase!CStdMarshal::UnmarshalObjRef+0x1c5
0d (Inline Function) --------`--------     combase!UnmarshalSwitch+0xa1
0e (Inline Function) --------`--------     combase!UnmarshalObjRef+0x151
0f 000000c0`0347d570 00007ffc`f5de6145     combase!CoUnmarshalInterface+0x5e1
10 000000c0`0347d960 00007ffc`f437e637     combase!Ndr64ExtInterfacePointerUnmarshall+0x1b5
11 000000c0`0347d9d0 00007ffc`f437ec8c     rpcrt4!Ndr64TopLevelPointerUnmarshall+0x257
12 000000c0`0347da40 00007ffc`f4429fd8     rpcrt4!Ndr64TopLevelPointerUnmarshall+0x8ac
13 000000c0`0347dab0 00007ffc`f442baa4     rpcrt4!Ndr64pClientUnMarshal+0x278
14 000000c0`0347db30 00007ffc`f5de4a2c     rpcrt4!NdrpClientCall3+0x434
15 000000c0`0347deb0 00007ffc`f5e64662     combase!ObjectStublessClient+0x14c
16 000000c0`0347e240 00007ffc`ee0e67c1     combase!ObjectStubless+0x42
17 000000c0`0347e290 00007ffc`d85ea650     netprofm!CPubINetwork::GetNetworkConnections+0xc1
18 000000c0`0347e300 00007ffc`d85ea482     winspool!IsNetworkPPP+0x5c
19 000000c0`0347e3d0 00007ffc`d85eae71     winspool!GetCurrentNetworkIdInternal+0x21a
1a 000000c0`0347e480 00007ffc`d85ece9a     winspool!GetCurrentNetworkId+0x65
1b 000000c0`0347e530 00007ffc`d85d9e77     winspool!InternalGetDefaultPrinter+0xc6
1c 000000c0`0347e5c0 00007ffc`f5a1b29f     winspool!GetDefaultPrinterW+0xb7
1d 000000c0`0347e640 00007ffc`f5a1b86b     comdlg32!PrintBuildDevNames+0x7f
1e 000000c0`0347eaf0 00007ffc`f5a1b1f7     comdlg32!PrintDlgX+0x21b
1f 000000c0`0347efa0 00007ffc`acc2eb61     comdlg32!PrintDlgW+0x47
20 000000c0`0347f4b0 00007ffc`acc2eab1     mfc42u!CWinApp::UpdatePrinterSelection+0x51
21 000000c0`0347f640 00007ff7`57696522     mfc42u!CWinApp::GetPrinterDeviceDefaults+0x21
22 000000c0`0347f670 00007ff7`57696718     wordpad!CWordPadApp::CreateDevNames+0xb2
23 000000c0`0347f750 00007ffc`acc2f13d     wordpad!CWordPadApp::DoDeferredInitialization+0x18
24 000000c0`0347f780 00007ffc`f417dfb4     mfc42u!_AfxThreadEntry+0xdd
25 000000c0`0347f840 00007ffc`f417e08c     msvcrt!_callthreadstartex+0x28
26 000000c0`0347f870 00007ffc`f4fd54e0     msvcrt!_threadstartex+0x7c
27 000000c0`0347f8a0 00007ffc`f62a485b     kernel32!BaseThreadInitThunk+0x10
28 000000c0`0347f8d0 00000000`00000000     ntdll!RtlUserThreadStart+0x2b
```

```
   4  Id: 12e8.184c Suspend: 0 Teb: 000000c0`033da000 Unfrozen
 # Child-SP          RetAddr           Call Site
00 000000c0`034ff858 00007ffc`f3b7fb10 ntdll!NtWaitForMultipleObjects+0x14
01 000000c0`034ff860 00007ffc`f5d82748 KERNELBASE!WaitForMultipleObjectsEx+0xf0
02 000000c0`034ffb50 00007ffc`f5d825ba combase!WaitCoalesced+0xa4
03 000000c0`034ffde0 00007ffc`f5d823bc combase!CROIDTable::WorkerThreadLoop+0x5a
04 000000c0`034ffe30 00007ffc`f5d82339 combase!CRpcThread::WorkerLoop+0x58
05 000000c0`034ffea0 00007ffc`f4fd54e0 combase!CRpcThreadCache::RpcWorkerThreadEntry+0x29
06 000000c0`034ffed0 00007ffc`f62a485b kernel32!BaseThreadInitThunk+0x10
07 000000c0`034fff00 00000000`00000000 ntdll!RtlUserThreadStart+0x2b

   5  Id: 12e8.a78 Suspend: 0 Teb: 000000c0`033dc000 Unfrozen
 # Child-SP          RetAddr           Call Site
00 000000c0`0357f588 00007ffc`f62b6c2f ntdll!NtWaitForWorkViaWorkerFactory+0x14
01 000000c0`0357f590 00007ffc`f4fd54e0 ntdll!TppWorkerThread+0x2df
02 000000c0`0357f880 00007ffc`f62a485b kernel32!BaseThreadInitThunk+0x10
03 000000c0`0357f8b0 00000000`00000000 ntdll!RtlUserThreadStart+0x2b

   6  Id: 12e8.2558 Suspend: 0 Teb: 000000c0`033de000 Unfrozen
 # Child-SP          RetAddr           Call Site
00 000000c0`035ff9f8 00007ffc`f62b6c2f ntdll!NtWaitForWorkViaWorkerFactory+0x14
01 000000c0`035ffa00 00007ffc`f4fd54e0 ntdll!TppWorkerThread+0x2df
02 000000c0`035ffcf0 00007ffc`f62a485b kernel32!BaseThreadInitThunk+0x10
03 000000c0`035ffd20 00000000`00000000 ntdll!RtlUserThreadStart+0x2b

   7  Id: 12e8.534 Suspend: 0 Teb: 000000c0`033e0000 Unfrozen
 # Child-SP          RetAddr           Call Site
00 000000c0`0367e918 00007ffc`f3ebb92e win32u!NtGdiGetCharABCWidthsW+0x14
01 000000c0`0367e920 00007ffc`f3ebb896 gdi32full!LoadGlyphMetricsWithGetCharABCWidthsI+0x5e
02 000000c0`0367ecc0 00007ffc`f3ebb3e7 gdi32full!LoadGlyphMetrics+0x96
03 000000c0`0367ed00 00007ffc`e74c93dd gdi32full!CUspShapingFont::GetGlyphDefaultAdvanceWidths+0x147
04 000000c0`0367ed60 00007ffc`f3ec5e06 TextShaping!ShapingGetGlyphPositions+0x4ed
05 000000c0`0367ef60 00007ffc`f3ec9b2e gdi32full!ShlPlaceOT+0x256
06 000000c0`0367f180 00007ffc`f3ec91d8 gdi32full!RenderItemNoFallback+0x56e
07 000000c0`0367f2c0 00007ffc`f3ec90ab gdi32full!RenderItemWithFallback+0xe8
08 000000c0`0367f310 00007ffc`f3ec8e6f gdi32full!RenderItem+0x3b
09 000000c0`0367f360 00007ffc`f3ecb030 gdi32full!ScriptStringAnalyzeGlyphs+0x20f
0a 000000c0`0367f410 00007ffc`f3ec80e9 gdi32full!ScriptStringAnalyse+0x660
0b 000000c0`0367f5d0 00007ffc`f3ec7aee gdi32full!LpkCharsetDraw+0x5d9
0c 000000c0`0367f800 00007ffc`f585bf24 gdi32full!LpkDrawTextEx+0x5e
0d 000000c0`0367f870 00007ffc`f585bc10 user32!DrawTextExWorker+0x2f4
0e 000000c0`0367fb10 00007ffc`abda21cb user32!DrawTextW+0x40
0f 000000c0`0367fb80 00007ffc`abd9c3c7 UIRibbon!CreateFontBitmap+0x1ff
10 000000c0`0367fcd0 00007ffc`abd9c60d UIRibbon!CFontList::_GenerateBitmaps+0x93
11 000000c0`0367fd20 00007ffc`abd9c67f UIRibbon!CFontList::_LoadFonts+0xc1
12 000000c0`0367fd50 00007ffc`f4fd54e0 UIRibbon!CFontList::s_LoadFontsThreadProc+0x2f
13 000000c0`0367fd80 00007ffc`f62a485b kernel32!BaseThreadInitThunk+0x10
14 000000c0`0367fdb0 00000000`00000000 ntdll!RtlUserThreadStart+0x2b
```

6. Let's dump thread #0 execution residue but interpret memory values as pointers to memory with the latter having symbols if possible (the **dpp** WinDbg command). In such a way, we may be able to catch COM objects' vptr (the first member in binary object layout), which is the address of the vtbl or "vftable":

```
0:000> !teb
TEB at 000000c0033d2000
    ExceptionList:        0000000000000000
    StackBase:            000000c0030a0000
    StackLimit:           000000c003095000
    SubSystemTib:         0000000000000000
    FiberData:            0000000000001e00
    ArbitraryUserPointer: 0000000000000000
    Self:                 000000c0033d2000
```

```
    EnvironmentPointer:   0000000000000000
    ClientId:             00000000000012e8 . 0000000000000928
    RpcHandle:            0000000000000000
    Tls Storage:          000002a0b8be96d0
    PEB Address:          000000c0033d1000
    LastErrorValue:       0
    LastStatusValue:      0
    Count Owned Locks:    0
    HardErrorMode:        0

0:000> dpp 000000c003095000 000000c0030a0000
000000c0`03095000  00000000`00000000
000000c0`03095008  00000000`00000000
000000c0`03095010  00000000`00000000
000000c0`03095018  00000000`00000000
000000c0`03095020  00000000`00000000
000000c0`03095028  00000000`00000000
000000c0`03095030  00000000`00000000
[...]
000000c0`03097d80  000000c0`03099390 00000000`00000000
000000c0`03097d88  00000000`000000c0
000000c0`03097d90  000002a0`b8bda4e0 00007ffc`f4314838 msctf!CActiveLanguageProfileNotifySink::`vftable'
000000c0`03097d98  00007ffc`f62c7631 8b48c18b`41c88b44
000000c0`03097da0  000002a0`b8bda4d0 00000000`00000000
000000c0`03097da8  000002a0`b3d90000 00000000`00000000
000000c0`03097db0  000000c0`03097f61 67000000`00000000
000000c0`03097db8  00000000`00000000
000000c0`03097dc0  00000000`00000000
000000c0`03097dc8  00000000`00000030
[...]
000000c0`0309e8c0  00000000`00000000
000000c0`0309e8c8  00000000`00000000
000000c0`0309e8d0  000002a0`b3fee0e0 00007ff7`576ec9f0 wordpad!CWordPadDoc::`vftable'
000000c0`0309e8d8  00000000`00002722
000000c0`0309e8e0  00000000`ffffffff
[...]
000000c0`0309ee50  00000000`00000000
000000c0`0309ee58  00000000`00000363
000000c0`0309ee60  00000000`00000001
000000c0`0309ee68  00007ffc`f62defa7 000000e0`249c8d4c
000000c0`0309ee70  000002a0`b3dbfe90 00007ffc`d388fc70 msftedit!CTxtEdit::`vftable'
000000c0`0309ee78  000002a0`b3dbfe90 00007ffc`d388fc70 msftedit!CTxtEdit::`vftable'
000000c0`0309ee80  00000000`00000000
000000c0`0309ee88  00007ffc`d35f8081 00000080`24b48b48
[...]

0:000> !address 000002a0`b8bda4e0


Mapping file section regions...
Mapping module regions...
Mapping PEB regions...
Mapping TEB and stack regions...
Mapping heap regions...
Mapping page heap regions...
Mapping other regions...
Mapping stack trace database regions...
Mapping activation context regions...

Usage:                  Heap
Base Address:           000002a0`b8ba0000
End Address:            000002a0`b8c52000
```

```
Region Size:            00000000`000b2000 ( 712.000 kB)
State:                  00001000          MEM_COMMIT
Protect:                00000004          PAGE_READWRITE
Type:                   00020000          MEM_PRIVATE
Allocation Base:        000002a0`b8ba0000
Allocation Protect:     00000004          PAGE_READWRITE
More info:              heap owning the address: !heap -s -h 0x2a0b3d90000
More info:              heap segment
More info:              heap entry containing the address: !heap -x 0x2a0b8bda4e0

Content source: 1 (target), length: 77b20
```

```
0:000> dps 00007ffc`f4314838
00007ffc`f4314838  00007ffc`f426c870 msctf!CActiveLanguageProfileNotifySink::QueryInterface
00007ffc`f4314840  00007ffc`f4271770 msctf!CActiveLanguageProfileNotifySink::AddRef
00007ffc`f4314848  00007ffc`f426e710 msctf!CActiveLanguageProfileNotifySink::Release
00007ffc`f4314850  00007ffc`f4309500 msctf!CActiveLanguageProfileNotifySink::OnActivated
00007ffc`f4314858  00007ffc`f42665b0 msctf!CicBridge::CDummyIUnk::QueryInterface
00007ffc`f4314860  00007ffc`f42724a0 msctf!CBStoreHolderWin32::AddRef
00007ffc`f4314868  00007ffc`f426b350 msctf!CicBridge::CDummyIUnk::Release
00007ffc`f4314870  00007ffc`f42594c0 msctf!CStartReconversionNotifySink::QueryInterface
00007ffc`f4314878  00007ffc`f4271770 msctf!CActiveLanguageProfileNotifySink::AddRef
00007ffc`f4314880  00007ffc`f426e710 msctf!CActiveLanguageProfileNotifySink::Release
00007ffc`f4314888  00007ffc`f42fb6f0 msctf!CStartReconversionNotifySink::StartReconversion
00007ffc`f4314890  00007ffc`f42fb630 msctf!CStartReconversionNotifySink::EndReconversion
00007ffc`f4314898  00007ffc`f42deaa0 msctf!CClassFactory::QueryInterface
00007ffc`f43148a0  00007ffc`f42de9a0 msctf!CClassFactory::AddRef
00007ffc`f43148a8  00007ffc`f423b960 msctf!CClassFactory::Release
00007ffc`f43148b0  00007ffc`f426eb80 msctf!CClassFactory::CreateInstance
```

```
0:000> dps 00007ffc`d388fc70
00007ffc`d388fc70  00007ffc`d3641600 msftedit!CTxtEdit::QueryInterface
00007ffc`d388fc78  00007ffc`d36409d0 msftedit!CTxtEdit::AddRef
00007ffc`d388fc80  00007ffc`d3640a00 msftedit!CTxtEdit::Release
00007ffc`d388fc88  00007ffc`d35f60a0 msftedit!CTxtEdit::TxSendMessage
00007ffc`d388fc90  00007ffc`d3647110 msftedit!CTxtEdit::TxDraw
00007ffc`d388fc98  00007ffc`d3689280 msftedit!CTxtEdit::TxGetHScroll
00007ffc`d388fca0  00007ffc`d3689b80 msftedit!CTxtEdit::TxGetVScroll
00007ffc`d388fca8  00007ffc`d367c1b0 msftedit!CTxtEdit::OnTxSetCursor
00007ffc`d388fcb0  00007ffc`d37521e0 msftedit!CTxtEdit::TxQueryHitPoint
00007ffc`d388fcb8  00007ffc`d35dd230 msftedit!CTxtEdit::OnTxInPlaceActivate
00007ffc`d388fcc0  00007ffc`d3649c60 msftedit!CTxtEdit::OnTxInPlaceDeactivate
00007ffc`d388fcc8  00007ffc`d365d5b0 msftedit!CTxtEdit::OnTxUIActivate
00007ffc`d388fcd0  00007ffc`d365d5c0 msftedit!CTxtEdit::OnTxUIDeactivate
00007ffc`d388fcd8  00007ffc`d3690550 msftedit!CTxtEdit::TxGetText
00007ffc`d388fce0  00007ffc`d364a4e0 msftedit!CTxtEdit::TxSetText
00007ffc`d388fce8  00007ffc`d3752170 msftedit!CTxtEdit::TxGetCurTargetX
```

We see addresses of objects allocated from the heap (`000002a0`b8bda4e0` and `000002a0`b3dbfe90`), and the objects' first member is vptr, an address of vtbl (`00007ffc`f4314838` and `00007ffc`d388fc70`, respectively). The first three members of the interface function tables (vtbl) are *QueryInterface*, *AddRef*, and *Release*, as expected from any COM object. We see references to COM objects on a thread that doesn't have any COM processing on its current stack trace. Notice that not every `` `vftable' `` is COM object vtbl (it can be just a C++ object):

```
0:000> dps 00007ff7`576ec9f0
00007ff7`576ec9f0  00007ff7`576a4960 wordpad!CWordPadDoc::GetRuntimeClass
00007ff7`576ec9f8  00007ff7`576a3980 wordpad!CWordPadDoc::`vector deleting destructor'
```

```
00007ff7`576eca00  00007ff7`576a5d60 wordpad!CWordPadDoc::Serialize
00007ff7`576eca08  00007ff7`57689130 wordpad!CObject::AssertValid
00007ff7`576eca10  00007ff7`576891d0 wordpad!CObject::Dump
00007ff7`576eca18  00007ff7`576a49f0 wordpad!CWordPadDoc::OnCmdMsg
00007ff7`576eca20  00007ff7`5764aa00 wordpad!CDocument::OnFinalRelease
00007ff7`576eca28  00007ff7`5764a700 wordpad!CCmdTarget::IsInvokeAllowed
00007ff7`576eca30  00007ff7`5764a710 wordpad!CCmdTarget::GetDispatchIID
00007ff7`576eca38  00007ff7`5764a720 wordpad!CCmdTarget::GetTypeInfoCount
00007ff7`576eca40  00007ff7`5764a730 wordpad!CCmdTarget::GetTypeLibCache
00007ff7`576eca48  00007ff7`5764a740 wordpad!CCmdTarget::GetTypeLib
00007ff7`576eca50  00007ff7`576a4940 wordpad!CWordPadDoc::GetMessageMap
00007ff7`576eca58  00007ff7`5764a750 wordpad!CCmdTarget::GetCommandMap
00007ff7`576eca60  00007ff7`5764a590 wordpad!CCmdTarget::GetDispatchMap
00007ff7`576eca68  00007ff7`5764a760 wordpad!CCmdTarget::GetConnectionMap
```

7. Let's now dump thread #3 execution residue. We may expect COM object references there due to COM processing:

```
0:000> ~3s
ntdll!NtAlpcSendWaitReceivePort+0x14:
00007ffc`f63448c4 c3              ret

0:003> !teb
TEB at 000000c0033d8000
    ExceptionList:        0000000000000000
    StackBase:            000000c003480000
    StackLimit:           000000c00347a000
    SubSystemTib:         0000000000000000
    FiberData:            0000000000001e00
    ArbitraryUserPointer: 0000000000000000
    Self:                 000000c0033d8000
    EnvironmentPointer:   0000000000000000
    ClientId:             00000000000012e8 . 00000000000023c8
    RpcHandle:            0000000000000000
    Tls Storage:          000002a0b8be97f0
    PEB Address:          000000c0033d1000
    LastErrorValue:       14007
    LastStatusValue:      c0150008
    Count Owned Locks:    0
    HardErrorMode:        0

0:003> dpp 000000c00347a000 000000c003480000
000000c0`0347a000  00000000`00000000
000000c0`0347a008  00000000`00000000
000000c0`0347a010  00000000`00000000
000000c0`0347a018  00000000`00000000
000000c0`0347a020  00000000`00000000
000000c0`0347a028  00000000`00000000
000000c0`0347a030  00000000`00000000
[...]
000000c0`0347d690  000002a0`b8c1a040 00007ffc`f5f9a6b8 combase!CStdIdentity::`vftable'
000000c0`0347d698  000000c0`0347d630 00000001`574f454d
000000c0`0347d6a0  000002a0`b3dc36e0 00007ffc`f5f98430 combase!CObjectContext::`vftable'
000000c0`0347d6a8  000002a0`00000000
[...]
```

```
0:003> dps 00007ffc`f5f9a6b8
00007ffc`f5f9a6b8  00007ffc`f5d69540 combase!CStdIdentity::QueryInterface
[onecore\com\combase\dcomrem\stdid.cxx @ 1050]
00007ffc`f5f9a6c0  00007ffc`f5ded9b0 combase!CStdIdentity::AddRef
[onecore\com\combase\dcomrem\stdid.cxx @ 1072]
00007ffc`f5f9a6c8  00007ffc`f5dedb60 combase!CStdIdentity::Release
[onecore\com\combase\dcomrem\stdid.cxx @ 1078]
00007ffc`f5f9a6d0  00007ffc`f5f05b50 combase!CStdIdentity::CreateServer
[onecore\com\combase\dcomrem\stdid.cxx @ 1640]
00007ffc`f5f9a6d8  00007ffc`f5e77380 combase!CStdIdentity::IsConnected
[onecore\com\combase\dcomrem\stdid.cxx @ 1552]
00007ffc`f5f9a6e0  00007ffc`f5f05ee0 combase!CStdIdentity::LockConnection
[onecore\com\combase\dcomrem\stdid.cxx @ 1589]
00007ffc`f5f9a6e8  00007ffc`f5f05dd0 combase!CStdIdentity::Disconnect
[onecore\com\combase\dcomrem\stdid.cxx @ 1573]
00007ffc`f5f9a6f0  00007ffc`f5e24810 combase!CStdIdentity::GetConnectionStatus
[onecore\com\combase\dcomrem\stdid.hxx @ 104]
00007ffc`f5f9a6f8  00007ffc`f5f01550 combase!CStdIdentity::`scalar deleting destructor'
00007ffc`f5f9a700  00007ffc`f5e25420 combase!CStdIdentity::SetMapping
[onecore\com\combase\dcomrem\stdid.hxx @ 93]
00007ffc`f5f9a708  00007ffc`f5e253f0 combase!CStdIdentity::GetMapping
[onecore\com\combase\dcomrem\stdid.hxx @ 94]
00007ffc`f5f9a710  00007ffc`f5f05ec0 combase!CStdIdentity::GetServerObjectContext
[onecore\com\combase\dcomrem\stdid.hxx @ 95]
00007ffc`f5f9a718  00007ffc`f5d4b400 combase!CStdIdentity::GetWrapperForContext
[onecore\com\combase\dcomrem\stdid.cxx @ 2126]
[...]
```

```
0:003> dps 00007ffc`f5f98430
00007ffc`f5f98430  00007ffc`f5dacd30 combase!CObjectContext::QueryInterface
[onecore\com\combase\dcomrem\context.cxx @ 1536]
00007ffc`f5f98438  00007ffc`f5e25120 combase!CObjectContext::AddRef
[onecore\com\combase\dcomrem\context.cxx @ 1551]
00007ffc`f5f98440  00007ffc`f5d8e990 combase!CObjectContext::Release
[onecore\com\combase\dcomrem\context.cxx @ 1577]
00007ffc`f5f98448  00007ffc`f5e769e0 combase!CObjectContext::SetProperty
[onecore\com\combase\dcomrem\context.cxx @ 3760]
00007ffc`f5f98450  00007ffc`f5ef7000 combase!CObjectContext::RemoveProperty
[onecore\com\combase\dcomrem\context.cxx @ 3913]
00007ffc`f5f98458  00007ffc`f5dffb00 combase!CObjectContext::GetProperty
[onecore\com\combase\dcomrem\context.cxx @ 3841]
00007ffc`f5f98460  00007ffc`f5ef57e0 combase!CObjectContext::EnumContextProps
[onecore\com\combase\dcomrem\context.cxx @ 3961]
00007ffc`f5f98468  00007ffc`f5e20e90 combase!CObjectContext::Freeze
[onecore\com\combase\dcomrem\context.cxx @ 4131]
00007ffc`f5f98470  00007ffc`f5ef57a0 combase!CObjectContext::DoCallback
[onecore\com\combase\dcomrem\context.cxx @ 4170]
00007ffc`f5f98478  00007ffc`f5ef7140 combase!CObjectContext::SetContextMarshaler
[onecore\com\combase\dcomrem\context.cxx @ 3376]
00007ffc`f5f98480  00007ffc`f5df7d50 combase!CObjectContext::GetContextMarshaler
[onecore\com\combase\dcomrem\context.cxx @ 3403]
...
```

8. We close logging before exiting WinDbg:

```
0:000> .logclose
Closing open log file C:\AWAPI-Dumps\W10.log
```

Networking API

- winsock2.h
  - send and WSASend
- Winsock SPI
- High Contention (Sockets)

Windows sockets API consists of traditional Unix-like names such as connect, send, receive, and extended API having a **WSA** prefix. I recently added the **High Contention** memory analysis pattern variant related to sockets.

winsock2.h
https://learn.microsoft.com/en-us/windows/win32/api/winsock2/

Winsock SPI
https://learn.microsoft.com/en-us/windows/win32/winsock/winsock-spi

High Contention (Sockets)
https://www.dumpanalysis.org/blog/index.php/2022/11/27/crash-dump-analysis-patterns-part-29f/

Console API

- Documentation
- Input Thread
- Main Thread

You also find the *ReadConsole* function in various input and main threads of console applications.

Documentation
https://learn.microsoft.com/en-us/windows/console/

Main Thread
https://www.dumpanalysis.org/blog/index.php/2007/10/23/crash-dump-analysis-patterns-part-32/

Process Heap API

- heapapi.h
- Used by C/C++/Rust runtime like malloc
- `HeapAlloc` is redirected to `ntdll!RtlAllocateHeap`
- There can be several process heaps
- Large allocations use Virtual Memory API
- Leaks
  - Memory Leak (Process Heap)
  - Relative Memory Leak
  - Memory Fluctuation (Process Heap)
- Corruption / Double Free / Invalid Parameter

WinDbg Commands

```
0:000> !heap
```


Finally, we come to dynamic memory allocations. Each process may have several heaps, and they can be used by higher-level C/C++ runtime libraries, for example, to implement *malloc* or *operator new* calls. Large heap allocations use the lower granularity API we look at in the next slide. And, of course, there are corruption and memory leaks associated with improper heap usage.

heapapi.h
https://learn.microsoft.com/en-us/windows/win32/api/heapapi/

Memory Leak (Process Heap)
https://www.dumpanalysis.org/blog/index.php/2007/08/06/crash-dump-analysis-patterns-part-20a/

Relative Memory Leak
https://www.dumpanalysis.org/blog/index.php/2016/09/15/crash-dump-analysis-patterns-part-243/

Memory Fluctuation (Process Heap)
https://www.dumpanalysis.org/blog/index.php/2014/10/04/crash-dump-analysis-patterns-part-211/

Corruption
https://www.dumpanalysis.org/blog/index.php/2006/10/31/crash-dump-analysis-patterns-part-2/

Double Free
https://www.dumpanalysis.org/blog/index.php/2007/08/19/crash-dump-analysis-patterns-part-23a/

Invalid Parameter
https://www.dumpanalysis.org/blog/index.php/2010/11/29/crash-dump-analysis-patterns-part-117/

Virtual Memory API

- memoryapi.h
- Used by large heap allocations and .NET
- Different memory protections
- Insufficient Memory (Reserved Virtual Memory)
- Insufficient Memory (Region)
- Memory Leak (Regions)
- Insufficient Memory (Module Fragmentation) – 32-bit

WinDbg Commands

```
0:000> !address
```


Virtual memory allocations have at least 4KB (one page) granularity, and they also allow different levels of protection, such as readonly. Of course, memory leaks are also possible. There was an interesting module fragmentation pattern in 32-bit times when a hooking module was loaded in the middle of virtual address space, preventing large contiguous virtual memory allocation.

memoryapi.h
https://learn.microsoft.com/en-us/windows/win32/api/memoryapi/

Insufficient Memory (Reserved Virtual Memory)
https://www.dumpanalysis.org/blog/index.php/2012/04/09/crash-dump-analysis-patterns-part-13h/

Insufficient Memory (Region)
https://www.dumpanalysis.org/blog/index.php/2014/10/27/crash-dump-analysis-patterns-part-13k/

Memory Leak (Regions)
http://www.dumpanalysis.org/blog/index.php/2014/10/30/crash-dump-analysis-patterns-part-20e/

Insufficient Memory (Module Fragmentation)
https://www.dumpanalysis.org/blog/index.php/2008/03/18/crash-dump-analysis-patterns-part-13e/

Strings API

- Documentation
- Local Buffer Overflow
- Shared Buffer Overwrite
- strsafe.h

Windows API includes functions for safe working with strings.

Documentation
https://learn.microsoft.com/en-us/windows/win32/menurc/strings

Local Buffer Overflow
https://www.dumpanalysis.org/blog/index.php/2007/11/14/crash-dump-analysis-patterns-part-36/

Shared Buffer Overwrite
https://www.dumpanalysis.org/blog/index.php/2010/10/18/crash-dump-analysis-patterns-part-110/

strsafe.h
https://learn.microsoft.com/en-us/windows/win32/menurc/strsafe-ovw

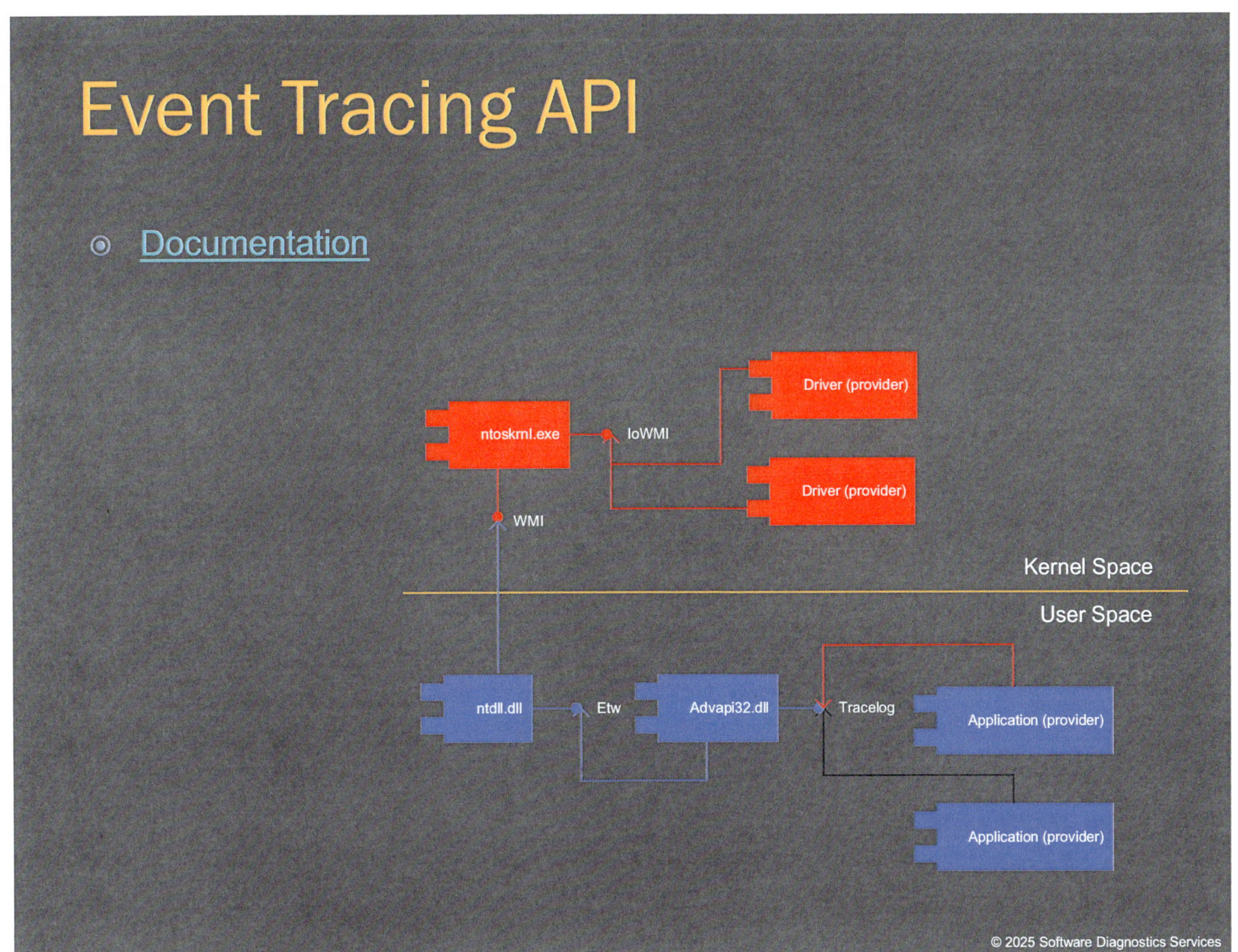

Event Tracing API is used for tracing and logging (producers) and trace and log collection and analysis (consumers).

Documentation
https://learn.microsoft.com/en-us/windows/win32/etw/event-tracing-portal

Error Handling & Debugging API

- Documentation
- Handled Exception
- Last Error Collection

Finally, we provide a reference to error handling and basic debugging API.

Documentation
https://learn.microsoft.com/en-us/windows/win32/Debug

Last Error Collection
https://www.dumpanalysis.org/blog/index.php/2008/08/05/crash-dump-analysis-patterns-part-74/

Handled Exception
https://www.dumpanalysis.org/blog/index.php/2011/10/17/crash-dump-analysis-patterns-part-152a/

Structures, Types, and Variables

- Injected Symbols
- Windows Data Types

At the beginning of this training, we mentioned links and WinDbg commands for the basic type system. Later, we mentioned Windows protocol references. This slide contains a reference for Windows types, including security structures and types used in various protocol implementations. The slide also contains a link for the Injected Symbols memory analysis pattern useful when certain Windows data structure definitions are not available in symbol files associated with modules present in memory dumps. In such a case, we load an unrelated module that has a corresponding symbol file containing required structure definitions. The *Accelerated Windows Memory Dump Analysis* training course mentioned in the later resources slide also has a corresponding exercise for 64-bit process memory dump. Another training course, *Accelerated Disassembly, Reconstruction, and Reversing*, has some exercises showing how to map header information to binary data.

Injected Symbols
https://www.dumpanalysis.org/blog/index.php/2013/02/27/crash-dump-analysis-patterns-part-197/

Windows Data Types
https://learn.microsoft.com/en-us/openspecs/windows_protocols/ms-dtyp

References and Resources

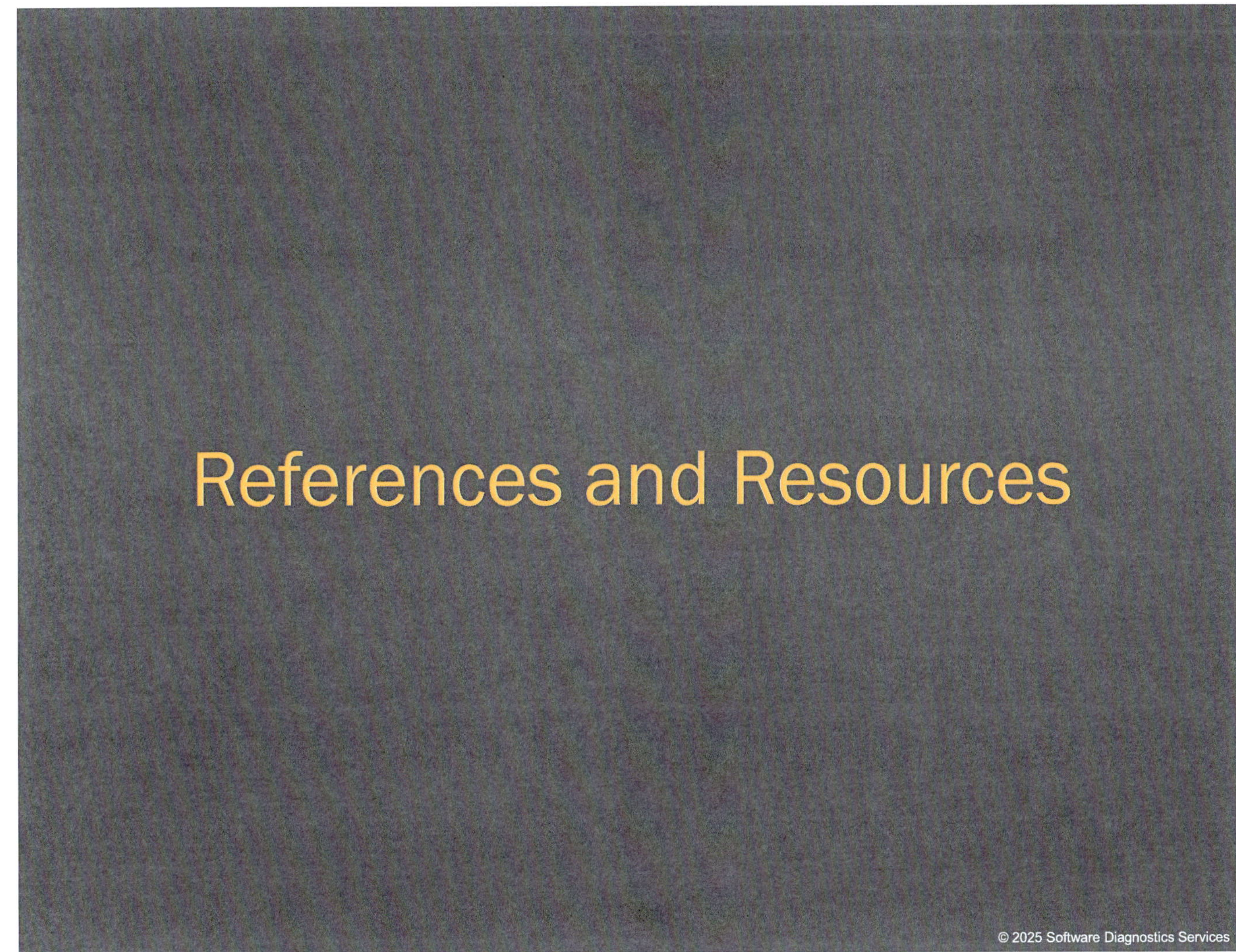
References and Resources
© 2025 Software Diagnostics Services

Now a few slides about references and resources for further reading.

Reading Windows-based Code

- Legacy Windows code and C language
  - Part 1
  - Part 2
  - Part 3
  - Part 4
  - Part 5
  - Part 6

In 2004, I developed a few training sessions for reading Windows-based code. It combines Windows-specific types with required C language knowledge to follow basic SDK samples.

Part 1
https://patterndiagnostics.com/Training/Reading-Windows-based-Code-Part1.pdf

Part 2
https://patterndiagnostics.com/Training/Reading-Windows-based-Code-Part2.pdf

Part 3
https://patterndiagnostics.com/Training/Reading-Windows-based-Code-Part3.pdf

Part 4
https://patterndiagnostics.com/Training/Reading-Windows-based-Code-Part4.pdf

Part 5
https://patterndiagnostics.com/Training/Reading-Windows-based-Code-Part5.pdf

Part 6
https://patterndiagnostics.com/Training/Reading-Windows-based-Code-Part6.pdf

Resources (Construction)

- Learning DCOM
- Programming Windows, 5th Edition
- Subclassing and Hooking with Visual Basic: Harnessing the Full Power of VB/VB.NET
- Windows Graphics Programming: Win32 GDI and DirectDraw
- Introduction to 3D Game Programming with DirectX 12
- Windows via C/C++
- Windows System Programming, 4th Edition
- Windows 10 System Programming
- Concurrent Programming on Windows
- The Old New Thing (also for postconstruction)
- Introducing Windows 7 for Developers
- Fundamentals of Audio & Video Programming for Games
- Software Application Development: A Visual C++, MFC, and STL Tutorial
- Windows Native API Programming

I also compiled a list of books useful for studying various Windows API classes from a software construction perspective. Some of them I read from cover to cover, including the previous editions. I haven't included various DirectX APIs in this training due to the lack of the corresponding memory analysis patterns. Direct X uses COM, and I included a DirectX 12 book I partially read a few years ago when I was doing some DirectX-related maintenance. I added a high-level multimedia API book (which also uses COM), and other relatively modern stuff can be found in the Windows 7 book. If you are interested in maintaining MFC apps, then the last monumental book is recommended, as it uses traditional legacy MFC-style development. Think MFC – think Windows API since MFC is a high-level wrapper around Desktop Windows API.

The Old New Thing
https://devblogs.microsoft.com/oldnewthing/

Resources (Postconstruction)

- WinDbg Help / WinDbg.org (quick links)
- DumpAnalysis.org / SoftwareDiagnostics.Institute / PatternDiagnostics.com
- Debugging.TV / YouTube.com/DebuggingTV / YouTube.com/PatternDiagnostics
- Practical Foundations of Windows Debugging, Disassembling, Reversing, Second Edition
- Software Diagnostics Library
- Encyclopedia of Crash Dump Analysis Patterns, Third Edition
- Trace, Log, Text, Narrative, Data
- Memory Dump Analysis Anthology (Diagnomicon)

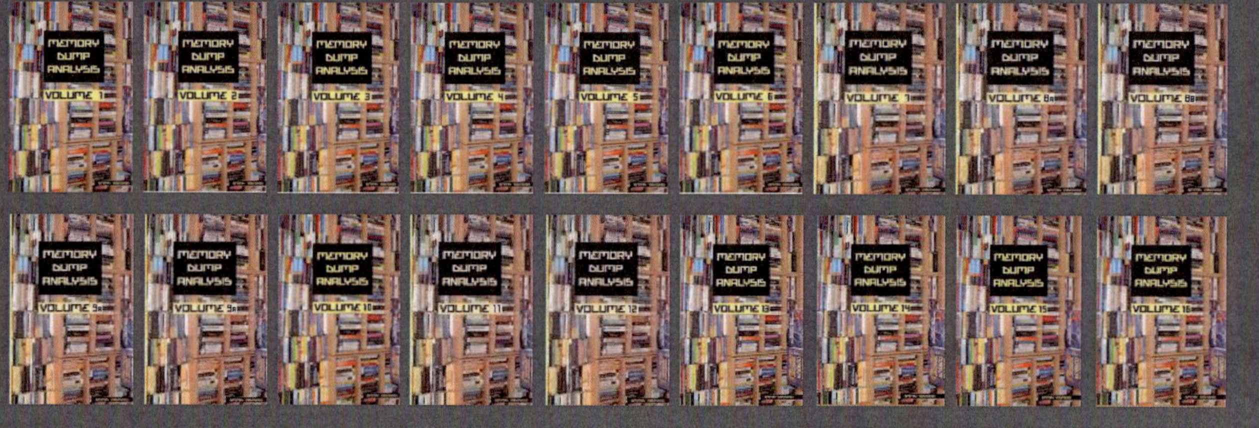

If you don't have experience with assembly language, then the Practical Foundation book teaches you assembly language from scratch in the context of WinDbg.

WinDbg quick links
http://WinDbg.org

Software Diagnostics Institute
https://www.dumpanalysis.org

Software Diagnostics Services
https://www.patterndiagnostics.com

Software Diagnostics Library
https://www.dumpanalysis.org/blog

Memory Dump Analysis Anthology (Diagnomicon)
https://www.patterndiagnostics.com/mdaa-volumes

Debugging.TV
http://debugging.tv/
https://www.youtube.com/DebuggingTV
https://www.youtube.com/PatternDiagnostics

Practical Foundations of Windows Debugging, Disassembling, Reversing, Second Edition
https://www.patterndiagnostics.com/practical-foundations-windows-debugging-disassembling-reversing

Encyclopedia of Crash Dump Analysis Patterns, Third Edition
https://www.patterndiagnostics.com/encyclopedia-crash-dump-analysis-patterns

Trace, Log, Text, Narrative, Data
https://www.patterndiagnostics.com/trace-log-analysis-pattern-reference

Resources (Training)

- Accelerated Windows Memory Dump Analysis, Sixth Edition
- Accelerated Windows Malware Analysis with Memory Dumps, Third Edition
- Accelerated Windows Debugging[4], Fourth Edition
- Accelerated Disassembly, Reconstruction and Reversing, Third Edition
- Accelerated Windows Trace and Log Analysis
- Accelerated Rust Windows Memory Dump Analysis

This slide shows various courses I developed over time. We mentioned some ADDR patterns from the *Accelerated Disassembly, Reconstruction, and Reversing* course.

Accelerated Windows Memory Dump Analysis, Sixth Edition
https://www.patterndiagnostics.com/accelerated-windows-memory-dump-analysis-book

Accelerated Windows Malware Analysis with Memory Dumps, Third Edition
https://www.patterndiagnostics.com/accelerated-windows-malware-analysis-book

Accelerated Windows Debugging[4], Fourth Edition
https://www.patterndiagnostics.com/accelerated-windows-debugging-book

Accelerated Disassembly, Reconstruction and Reversing, Second Revised Edition
https://www.patterndiagnostics.com/accelerated-disassembly-reconstruction-reversing-book

Accelerated Windows Trace and Log Analysis
https://www.patterndiagnostics.com/accelerated-windows-software-trace-analysis-book

Accelerated Rust Windows Memory Dump Analysis
https://www.patterndiagnostics.com/accelerated-rust-windows-memory-dump-analysis-book

Resources (Category Theory)

Applied category theory books that have chapters explaining category theory:

- Conceptual Mathematics: A First Introduction to Categories
- The Joy of Abstraction: An Exploration of Math, Category Theory, and Life
- Category Theory for Programmers
- Categories for Software Engineering
- An Invitation to Applied Category Theory: Seven Sketches in Compositionality
- Life Itself: A Comprehensive Inquiry Into the Nature, Origin, and Fabrication of Life
- Category Theory for the Sciences
- Conceptual Mathematics and Literature: Toward a Deep Reading of Texts and Minds
- Diagrammatic Immanence: Category Theory and Philosophy
- Mathematical Mechanics: From Particle To Muscle
- Memory Evolutive Systems; Hierarchy, Emergence, Cognition
- Mathematical Structures of Natural Intelligence
- Sheaf Theory Through Examples
- Visual Category Theory
- Monoidal Category Theory

As I promised when introducing category theory, this slide references applied category theory books. I also highlighted three books that influenced me; for example, they provided a deeper understanding of applied aspects, resulting in a better understanding of mathematical aspects.

Category Theory for Programmers
https://github.com/hmemcpy/milewski-ctfp-pdf

Visual Category Theory
https://dumpanalysis.org/visual-category-theory

www.ingramcontent.com/pod-product-compliance
Lightning Source LLC
Chambersburg PA
CBRC091939210326
41598CB00012B/864

* 9 7 8 1 9 1 2 6 3 6 8 8 4 *